To my wife, Pauline

ELEMENTARY MODERN PHYSICS

ATAM P. ARYA
West Virginia University

ADDISON-WESLEY PUBLISHING COMPANY
Reading, Massachusetts · Menlo Park, California
London · Amsterdam · Don Mills, Ontario · Sydney

This book is in the
ADDISON-WESLEY SERIES IN PHYSICS

Third printing, October 1976

Printed in the United States of America. Published simultaneously in Canada. Library of Congress Catalog Card No. 73-1466.

ISBN 0-201-00304-X
CDEFGHIJ-MA-7987

PREFACE

This book was written for students of science and engineering at the level of the sophomore or junior year in college. It may be used as a text for a course that is taken after the student has had two semesters of university physics or as a text for a separate course in modern physics. The only prerequisites are a knowledge of general classical physics and of introductory calculus.

Most of the physics developed in the present century is called modern physics. We cannot, of course, deal with all the details of modern physics in this elementary text. However, students should be exposed not only to the basic essentials of modern physics, but also to many diverse topics which will give them some insight into (1) the present status of physics and (2) what might be expected in a decade or so. For this reason, the book introduces a wide range of interesting topics. For example, under relativity, we include a brief discussion of *tachyons*, the *twin paradox*, and *general relativity*. When we come to atomic physics, we touch upon *spin–orbit coupling* and give an introductory treatment of the *quantum theory of the hydrogen atom* as well as of *penetrating and nonpenetrating orbits*. Because of the increasing interest in molecular and solid-state physics, we have put in three chapters with special emphasis on *lasers*, *crystal structure*, *band theory*, *p-n junctions* and their applications, and *superconductivity*. The material on nuclear physics includes a discussion of *radioactive dating*, the *Mössbauer effect*, *exchange forces*, *non-conservation of parity*, and so forth. In addition, we give a full exposition, at an elementary level, of *elementary particle physics*. After discussing detectors and accelerators in Chapter 13, we devote all of Chapter 15 to particle physics, which leads the way to *quarks*, with which we end the chapter.

Whenever it was possible to do so in a natural and unobtrusive manner, I have incorporated historical background into the text. One of the main aims has been to explain the physical significance of mathematical equations, and to show derivations and results of experiments. For a student who wants to read an original work, a list of references is provided at the end of each chapter. For those interested in further study and detail, a list of suggestions for further reading is included at the end of each chapter. Short summaries are also given at the end of each chapter.

If one is to fully appreciate physics, one must learn to solve problems involving physics. The student need not solve the most difficult problems; it is enough that he solve simple problems. In this way he will increase his understanding of basic concepts. I have therefore included a generous sampling of problems, of varying degrees of difficulty. There are problems for every student. Even an average student, if he is exposed to solved problems, with solutions which explain the basic principles and use of mathematical expressions, can find problem-solving both easy and interesting. For this reason, I have included about sixty worked-out examples, which are sprinkled throughout the text. There are also many illustrations, both line drawings and original photographs.

ACKNOWLEDGEMENTS

From the time the rough draft was prepared to the final publication, the author and the editor (Mr. Allan M. Wylde), continuously sought advice and suggestions from many sources so that the book would meet the requirements of elementary modern physics courses as taught at different colleges and universities. For this the author is indebted to Professor Mason R. Yerian of Stanford University, Professor E. Coleman of the University of Minnesota, Professor Richard Madey of Kent State University, Professor Paul Stoler of Rensselaer Polytechnic Institute, Professor Joseph Priest of Miami University, Professor Ronald A. Brown of State University of New York at Oswego, and Professor Robert J. Schwensfeir, Jr., of Bucknell University.

My thanks to the Editorial and Production departments of Addison-Wesley, and especially to Mr. Allan M. Wylde, who initiated this project and saw it to completion with keen interest throughout.

Finally, my appreciation goes to my wife, Pauline, for her assistance and patience in discussing and proofreading the manuscript several times.

Morgantown, West Virginia
January 1974

A. P. A.

DESCRIPTION AND SUGGESTED USE OF THIS TEXT

DESCRIPTION. Most developments in physics that have taken place in the twentieth century are classified as modern physics: relativity, quantum mechanics, atomic physics, molecular physics, solid-state physics, nuclear physics, and elementary-particle (or high-energy) physics. These topics are distributed over 15 chapters, as follows.

Chapter 1. *Relativity.* Objects moving at very high speeds, speeds approaching that of light.

Chapters 2 and 3. *Waves and particles and quantum mechanics.* On the microscopic scale, there is essentially no difference in the behavior of waves and particles. Describing these two mathematically leads to the development of quantum mechanics.

Chapter 4. *Atomic structure.* The circumstances that led to the development in the structure of the atom.

Chapters 5, 6, and 7. *Atomic spectra.* The arrangement of electrons in the atoms, energy-level diagrams, possible transitions, and effects of magnetic fields on these atoms; also x-ray spectra.

Chapter 8. *Molecular structure and spectra.* The structure and spectra of molecules; *lasers* and the three statistics.

Chapters 9 and 10. *Solid-state physics.* Formation of solids; crystal structure; theories of solids and superconductivity.

Chapters 11, 12, and 14. *Nuclear physics.* These chapters deal with basic properties of nuclei, nuclear decays, and nuclear reactions.

Chapter 13. *Detectors and accelerators.* Serve both nuclear physics and elementary-particle physics.

Chapter 15. *Elementary-particle physics.* The discoveries of the particles, theories for classifications of particles; quarks.

USE. The amount of material to be covered in any course naturally depends on the time available. The following are purely suggestions and guidelines which may be changed (without loss of continuity) to suit the tastes and needs of different students and instructors.

A. *For a 15-week (3 hr) or 10-week (4 hr) course:* Fourteen of the fifteen chapters may be covered, provided that all the sections marked with stars are omitted.

B. *For 15-week (4 hr) course:* All fifteen chapters may be covered.

CONTENTS

CHAPTER 1
THE SPECIAL THEORY OF RELATIVITY

The burning sun and the stars generate tremendous amounts of energy, according to Einstein's mass–energy relation $E = mc^2$. Our sun converts 4.5×10^9 *kg (or* $4\frac{1}{2}$ *million tons) per second to produce* 4×10^{26} *joules of energy per second. At this rate, the mass of the sun will not decrease appreciably, even in several billion years. Photograph shows spiral nebula in Virgo, seen edge on.* (*Photograph from Mount Wilson and Palomar Observatories*)

1.1 INTRODUCTION

Before the beginning of the twentieth century, two branches of physics—mechanics and electromagnetism—had developed quite independently of each other, and there seemed to be no strong relation between the two. The laws of mechanics and of electromagnetism had been verified to such an extent that physicists were sure that there would be no further modifications or improvements of them. But to everyone's surprise, early in this century, physicists were faced with many new and basic problems. First, they found that Newton's second law of motion, which had been so well established for objects moving with *low* speeds, did not give the correct results when applied to objects moving with *high* speeds—speeds comparable to the speed of light, c (3×10^8 m/sec). Second, they found that, for two observers in relative motion, one could not use the same set of transformation equations to transform the laws of mechanics and electromagnetism from the frame of reference of one observer to the frame of reference of the other observer.

These and other difficulties were overcome by the formulation of the *special theory of relativity* by Einstein[1] in 1905. This theory deals with objects or observers (or frames of reference) that are moving with uniform velocities relative to each other. Later on, in 1915, Einstein[2] developed the *general theory of relativity*, which deals with accelerated frames of reference. Most of this chapter will be devoted to the special theory of relativity, although a brief mention of the general theory of relativity will be made at the end.

1.2 GALILEAN TRANSFORMATIONS AND THEIR LIMITATIONS

According to *Newton's first law* (or the *law of inertia*), a system at rest will remain at rest or a system in uniform motion will remain in uniform motion *if no net external force acts on the system.* Such systems in which the law of inertia holds are called *inertial systems.* For all practical purposes, a set of coordinate axes attached to the earth may be regarded as an inertial system, provided we neglect the small acceleration resulting from the rotational and orbital motion of the earth. (An ideal inertial system is a coordinate frame of reference fixed in space with respect to fixed stars.) We shall treat any system which is moving with a uniform

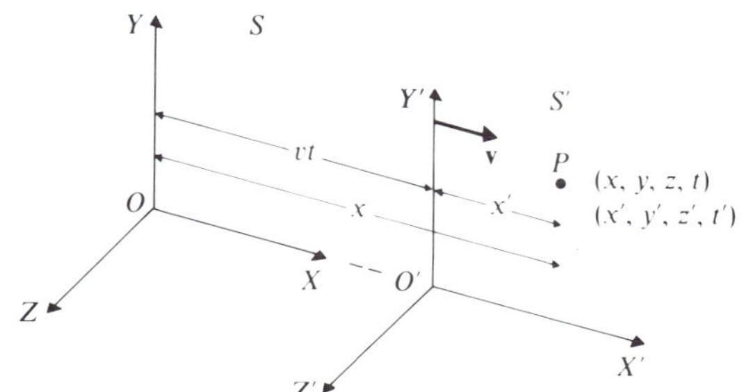

Fig. 1.1 (x, y, z, t) and (x', y', z', t') are the space–time coordinates of an event occurring at P, as measured by observers at rest in the inertial systems S and S', respectively. The system S' moves with a velocity $\mathbf{v}$ relative to the system S along the XX' axes.

velocity (relative to any other system) as an inertial system. Let us see how we transfer the coordinates of an *event* (a physical phenomenon) from one inertial system to another inertial system which is moving with a uniform velocity relative to the first. In Newtonian mechanics such transformations are made by the Galilean transformation equations discussed below.

Let one set of coordinate axes XYZ be located in an inertial system S and another set $X'Y'Z'$ in an inertial system S' which is moving with respect to system S with a velocity v along the XX' axes, as shown in Fig. 1.1. The origins of the two inertial systems coincide at $t = t' = 0$. Let the coordinates of an event taking place at some point P be (x, y, z, t) and (x', y', z', t') in the inertial systems S and S', respectively. According to Fig. 1.1, these coordinates are related by the *Galilean coordinate transformation equations:*

$$\boxed{\begin{aligned} x' &= x - vt \\ y' &= y \\ z' &= z \\ t' &= t \end{aligned}} \quad \text{or} \quad \boxed{\begin{aligned} x &= x' + vt \\ y &= y' \\ z &= z' \\ t &= t' \end{aligned}} \tag{1.1}$$

Note that in Newtonian relativity, we always assume that $t' = t$. If we differentiate the above equations, assuming that d/dt and d/dt' are identical, we get the following *velocity transformation equations*:

$$\begin{aligned} \frac{dx'}{dt'} &= \frac{dx}{dt} - v \\ \frac{dy'}{dt'} &= \frac{dy}{dt} \\ \frac{dz'}{dt'} &= \frac{dz}{dt} \end{aligned} \quad \text{that is,} \quad \boxed{\begin{aligned} u'_x &= u_x - v \\ u'_y &= u_y \\ u'_z &= u_z \end{aligned}} \tag{1.2}$$

Differentiating once again, we get the *acceleration transformation equations*:

$$\begin{aligned} \frac{d^2x'}{dt'^2} &= \frac{d^2x}{dt^2} \\ \frac{d^2y'}{dt'^2} &= \frac{d^2y}{dt^2} \\ \frac{d^2z'}{dt'^2} &= \frac{d^2z}{dt^2} \end{aligned} \quad \text{that is,} \quad \begin{aligned} a'_x &= a_x \\ a'_y &= a_y \\ a'_z &= a_z \end{aligned} \tag{1.3}$$

or in vector form

$$\mathbf{a}' = \mathbf{a}. \tag{1.4}$$

That is, the accelerations are the same as viewed from either inertial system. We can state this fact by saying that the acceleration is *invariant* with respect to Galilean transformations. Actually, we can go a step further and show that an equation which describes an event in one reference frame does not change its form when transferred to another reference frame by using Galilean transformation. For example, the components of a force **F** acting on a particle of mass m at a point P in reference system S may be written as

$$F_x = m\frac{d^2x}{dt^2}, \qquad F_y = m\frac{d^2y}{dt^2}, \qquad F_z = m\frac{d^2z}{dt^2}. \tag{1.5}$$

Using the values of x, y, z, and t from Eq. (1.1) or directly from Eq. (1.3), we get (assuming mass is invariant)

$$F'_x = m\frac{d^2x'}{dt'^2}, \qquad F'_y = m\frac{d^2y'}{dt'^2}, \qquad F'_z = m\frac{d^2z'}{dt'^2} \tag{1.6}$$

and this implies that the form of the equations (Newton's second law in this case) has not changed under Galilean transformation. We say that *Newton's law is invariant under Galilean transformation.*

The *principle of Newtonian relativity* states that, for all inertial systems which are in uniform relative motion, the equations of motion remain invariant. Since the forms of the laws remain the same, we cannot, by performing mechanical experiments, determine which of the two reference frames is at rest and which is moving. As a matter of fact, all the laws of classical mechanics are invariant under Galilean transformation. Example 1.1 illustrates the invariance of the form of the conservation of linear momentum and kinetic energy under Galilean transformation.

Example 1.1. Consider a collision between two masses m_1 and m_2 moving with velocities u_1 and u_2 along the X-axis in an inertial system S. After the collision, the velocities of the two masses are U_1 and U_2 still along the X-axis. The conservation of linear momentum and kinetic energy (Fig. 1.2) may be written as

$$m_1u_1 + m_2u_2 = m_1U_1 + m_2U_2 \tag{1.7}$$

$$\tfrac{1}{2}m_1u_1^2 + \tfrac{1}{2}m_2u_2^2 = \tfrac{1}{2}m_1U_1^2 + \tfrac{1}{2}m_2U_2^2 \tag{1.8}$$

Let us observe this collision from another inertial system S' moving with a velocity **v** along the XX' axes, as shown in Fig. 1.2. Let us assume that the velocities of the two masses m_1 and m_2 before and after the collision in S' are u'_1, u'_2, and U'_1, U'_2, respectively. Using the Galilean velocity transformations given by Eq. (1.2), we substitute $u_1 = u'_1 + v$, $u_2 = u'_2 + v$, $U_1 = U'_1 + v$, and $U_2 = U'_2 + v$ in Eq. (1.7), resulting in

$$m_1(u'_1 + v) + m_2(u'_2 + v) = m_1(U'_1 + v) + m_2(U'_2 + v),$$

which, on simplification, yields

$$m_1u'_1 + m_2u'_2 = m_1U'_1 + m_2U'_2. \tag{1.9}$$

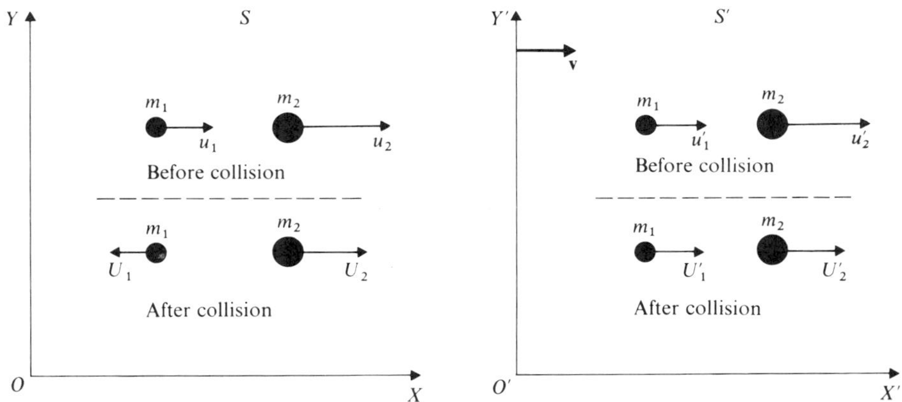

Fig. 1.2 Collision between two masses m_1 and m_2 as viewed from inertial systems S and S' moving with a relative velocity $\mathbf{v}$ along the XX' axes.

Similarly, Eq. (1.8) becomes

$$\tfrac{1}{2}m_1(u_1' + v)^2 + \tfrac{1}{2}m_2(u_2' + v)^2 = \tfrac{1}{2}m_1(U_1' + v)^2 + \tfrac{1}{2}m_2(U_2' + v)^2,$$

which, on simplifying and using Eq. (1.9), yields

$$\tfrac{1}{2}m_1 u_1'^2 + \tfrac{1}{2}m_2 u_2'^2 = \tfrac{1}{2}m_1 U_1'^2 + \tfrac{1}{2}m_2 U_2'^2. \tag{1.10}$$

Note that Eqs. (1.9) and (1.10) in system S' have exactly the same form as Eqs. (1.7) and (1.8) in system S, thereby showing that the form of conservation of linear momentum and kinetic energy are invariant under Galilean transformations.

A key question to ask at this stage is: Do these Galilean transformations hold good in the case of electromagnetism as well? Unfortunately, the laws of electromagnetism are not invariant under Galilean transformations. For example, a spherical electromagnetic wave propagating with a constant speed c in the reference system S may be given by

$$x^2 + y^2 + z^2 - c^2t^2 = 0. \tag{1.11}$$

For this equation to be invariant, its form in the system S' should be

$$x'^2 + y'^2 + z'^2 - c^2t'^2 = 0. \tag{1.12}$$

Substituting x, y, z, and t from Eq. (1.1) into Eq. (1.11) yields*

$$(x' + vt)^2 + y'^2 + z'^2 - c^2t'^2 = 0,$$

---

* Note that we are making use of the experimental fact[3] that c is invariant, i.e., it is the same in all inertial systems.

which implies that

$$x^2 + y^2 + z^2 - c^2t^2 \neq x'^2 + y'^2 + z'^2 - c^2t'^2.$$

Thus Galilean transformations hold good for classical mechanics, but not for electromagnetism. We encounter further difficulties when we apply the Galilean transformation equations to the following experimental situations, which are designed to find a reference system which is an absolute or rest reference system.

1.3 SEARCH FOR AN ABSOLUTE FRAME OF REFERENCE

In 1860, James C. Maxwell presented his theory of electromagnetism, which summarized all the laws of electricity and magnetism. They became known as *Maxwell's equations.* His theory also predicted the existence of electromagnetic waves, which propagated through space with a speed of 3×10^8 m/sec. Since this is the same speed as the measured speed of light, people thought that light might be an electromagnetic wave. When Heinrich Hertz produced electromagnetic waves in 1888 in the laboratory, this achievement, coupled with additional experimental and theoretical calculations, seemed to establish beyond doubt that light consists of transverse electromagnetic waves.

Maxwell's theory did not require that there be a medium for the propagation of electromagnetic waves. However, because of other mechanical wave phenomena which do require a medium (for example, air is needed to support sound vibrations, while water is needed to support water waves), physicists thought it necessary to assign a medium to support light and other electromagnetic waves. Nineteenth century physicists named this medium *luminiferous ether*, or just *ether*. They assumed that ether fills all space as well as vacuum. And they had to assign some strange properties to ether. For example, it had to be transparent and massless, so that electromagnetic waves could travel through a vacuum. Yet, on the contrary, it had to have elastic properties to sustain the vibrations of the wave motion. The fact that ether transmitted only transverse vibrations implied that it had to have a very high rigidity (the same as that of steel); and yet planets, as well as other objects, of necessity moved through the ether freely.

This ether hypothesis leads us to the following two alternatives.

i) *The stationary-ether hypothesis:* We assume that the ether is at rest with respect to the bodies moving through it. Thus we call the "ether frame" the *rest frame* or *absolute frame.* The velocity of light is always c in this frame.

ii) *The ether-drag hypothesis:* We assume that the ether is being dragged along with the bodies which move through it.

Let us now discuss two experiments: (A) the stellar aberration, which definitely supports the stationary-ether hypothesis, and (B) the Michelson–Morley experiment, which cannot be explained by the stationary-ether hypothesis. This really puts us in doubt as to the existence of ether as an absolute frame of reference or medium.

A. Stellar aberration

A variation of the apparent direction of the stars at different times of the year is called the aberration of light. It was first observed in 1727 by the astronomer Bradley.[4]

Let us look through a telescope at a distant star located overhead (at the zenith). If we assume that there is no relative motion between the earth and the ether (which is the same as saying that the ether is being dragged along with the earth), then the light coming from the star will enter the telescope only if the telescope is held in a vertical position. But this is not true, because it is found experimentally that, if one wishes to observe a star, the telescope must be inclined at an angle α, as shown in Fig. 1.3.

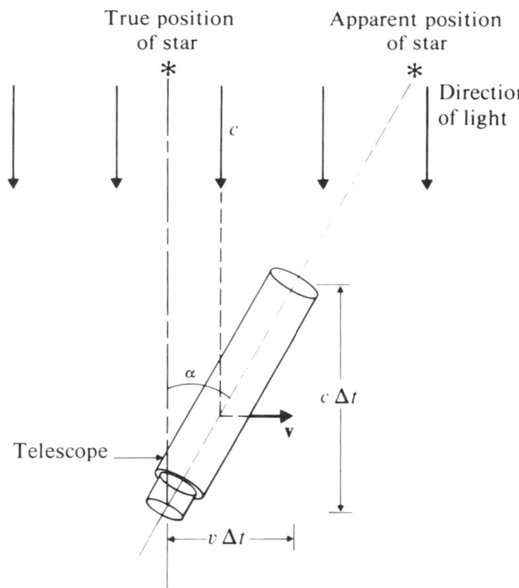

Fig. 1.3 To view a star located overhead, one must tilt the telescope at an angle α, as shown.

We can calculate the value of the angle α if we assume that the ether is stationary, while the earth is moving through it with a relative velocity $\mathbf{v}$. Thus, if it takes a time Δt for the light to travel from the star to the telescope, the light will travel a distance $c\,\Delta t$, while the telescope (or earth) will move a distance $v\,\Delta t$. Thus the angle at which the telescope must be inclined is, as shown in Fig. 1.3,

$$\tan\alpha = \frac{v\,\Delta t}{c\,\Delta t} = \frac{v}{c} = \frac{3 \times 10^4\ \text{m/sec}}{3 \times 10^8\ \text{m/sec}} = 10^{-4}.$$

That is,

$$\alpha \simeq 20 \cdot 5'' \text{ of arc.} \tag{1.13}$$

In order to measure α experimentally, a scientist fixes a telescope on earth at one position so as to view a particular star. After six months, the earth is moving in the opposite direction, as shown in Fig. 1.4(a), and hence, to view the same star, he has to readjust the telescope. The difference between the two angular positions of the telescope is 2α, and this angle should be equal to the apparent change in the position of the star in six months.

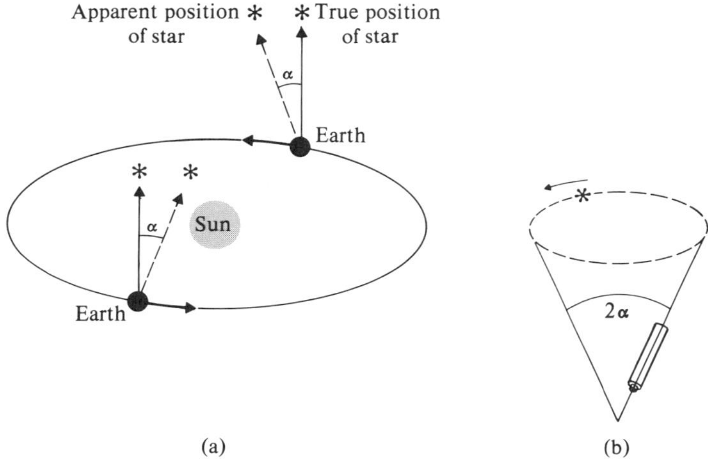

Fig. 1.4 (a) Variation in the apparent direction of a star as viewed by a telescope over a period of six months. (b) During one year, the moving telescope traces a cone of angle $2\alpha = 41''$, while the apparent position of the star traces out a circular path.

Observations made on different stars confirm the magnitude of the aberration given by Eq. (1.13), and in a period of one year the moving telescope on earth traces a cone of $2\alpha = 41''$, as shown in Fig. 1.4(b). Thus we can draw the following conclusion.

The earth is moving through a stationary ether with a velocity of 3×10^4 m/sec *relative to the rest ether frame (which may be considered to be the absolute reference frame).*

B. The Michelson–Morley experiment

We can confirm the existence of an absolute frame of reference (stationary ether) if we can measure the absolute velocity of earth with respect to the stationary ether. In 1887, A. A. Michelson and E. W. Morley[5] performed an experiment with just shahibai

this aim in mind. They assumed that the ether was at rest, and that light signals traveled with a velocity $c = 3 \times 10^8$ m/sec with respect to the stationary ether. They also assumed that the relative velocity of light with respect to the moving earth was different from c, and should depend on the velocity of the earth. Hence, by using light signals, they felt that they should be able to derive the absolute velocity of the earth with respect to ether. The experimental setup they used had to be very sensitive, because the orbital velocity of the earth, $v = 3 \times 10^4$ m/sec, is only 10^{-4} of the speed of light.

i) *The Michelson interferometer.* The experimental setup used by Michelson and Morley[6] consisted of a very precise optical instrument which has come to be known as the *Michelson interferometer.* Figure 1.5 shows the basic outline of the equipment. A beam of monochromatic light of wavelength λ, from source S, after passing through a lens L, falls on a half-silvered plate P, which splits the one beam into two beams. The part being transmitted is called beam 1, and the part being reflected is called beam 2. These beams, after being reflected from mirrors M_1 and M_2, respectively, arrive at plate P, into the telescope T, and to the eye at E. Beam 2 goes through the plate P three times, while beam 1 goes through only once. So, to make the optical paths of the two beams equal, a compensating plate CP is placed in the path of beam 1. When beams 1 and 2 arrive at the telescope T, they produce

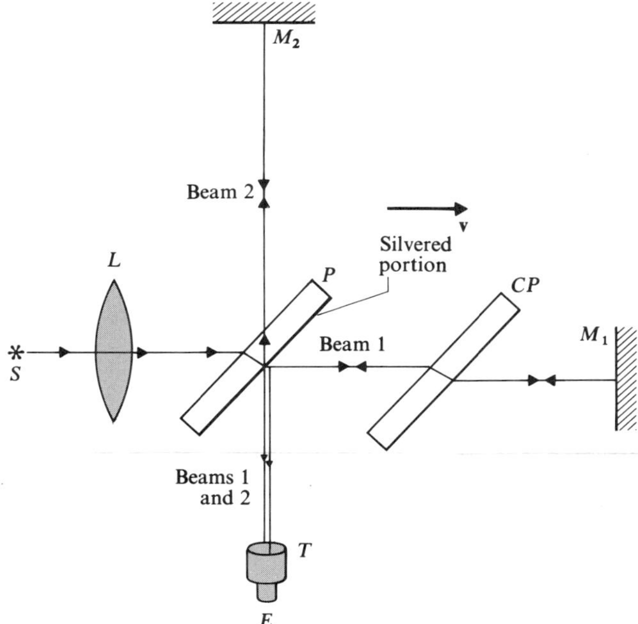

Fig. 1.5 Schematic diagram of the Michelson–Morley experiment.

interference fringes of the type shown in Fig. 1.6. Michelson and Morley, in order to reduce mechanical vibrations, mounted their interferometer on a large stone that floated in a tank of mercury.

If the optical paths of the two beams arriving at E are exactly equal (that is, if the two beams are *in phase*), a bright fringe will be produced; this is called *constructive interference.* If one mirror, say M_1, is moved a distance $\lambda/4$, this will produce a path difference of $\lambda/2$ between the two beams (the two beams are then out of phase). This will result in a dark fringe at E, and we have *destructive interference.* Thus, by moving mirror M_1, we can make the fringes move past a reference mark, say across crosshairs cc and $c'c'$, as shown in Fig. 1.6, and hence we can count the fringes.

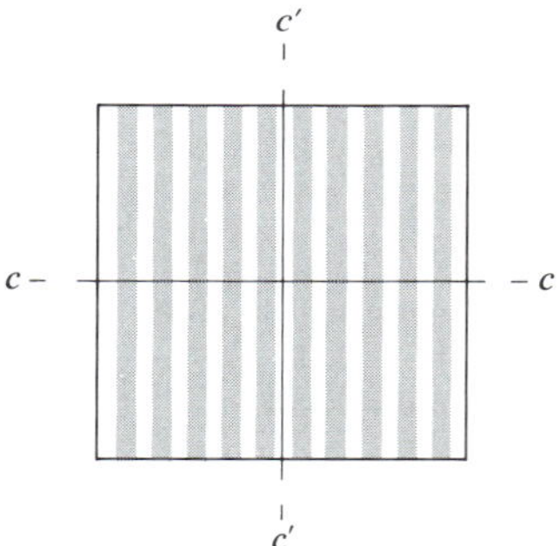

Fig. 1.6 A typical pattern of fringes observed in the Michelson–Morley experiment.

ii) *The experiment itself.* Let us assume that the earth is moving to the right, as shown in Fig. 1.5, with a velocity **v** with respect to the stationary ether. The interferometer is arranged in such a way that $PM_1 = PM_2 = L$, and the arm PM_1 is parallel to **v**. If we use Galilean transformations, the velocity of a light signal traveling from P to M_1 is $c - v$, while from M_1 to P (after reflection from M_1), its velocity is $c + v$. Thus the time t_1 for this round trip is

$$t_1 = \frac{L}{c - v} + \frac{L}{c + v} = \frac{2Lc}{c^2 - v^2} = \frac{2L}{c}\bigg/\left(1 - \frac{v^2}{c^2}\right). \tag{1.14}$$

Beam 2 follows the path shown in Fig. 1.7(a). The component of velocity perpendicular to the direction of motion of the interferometer is obtained by vector addition, and is $\sqrt{c^2 - v^2}$ (Fig. 1.7b). Thus the time t_2 which beam 2 takes to make the round trip PM_2P is

$$t_2 = \frac{2L}{\sqrt{c^2 - v^2}} = \frac{2L}{c}\bigg/\sqrt{1 - \frac{v^2}{c^2}}. \tag{1.15}$$

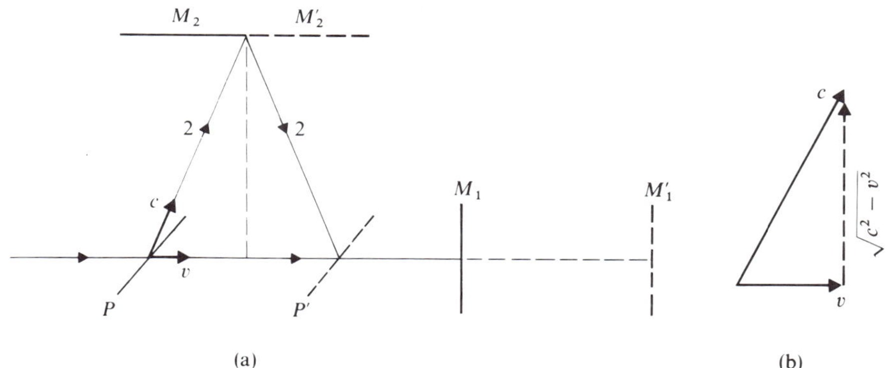

Fig. 1.7 (a) The path of beam 2 while the interferometer is moving with velocity **v** parallel to PM_1. (b) Diagram showing the resultant of velocities **c** and **v**.

The difference between the time taken by beam 1 and the time taken by beam 2 is

$$\Delta t = t_1 - t_2 = \frac{2L}{c}\left[\left(1 - \frac{v^2}{c^2}\right)^{-1} - \left(1 - \frac{v^2}{c^2}\right)^{-1/2}\right]. \tag{1.16}$$

Since v/c is very small, we can use binomial expansion to expand the two expressions in parentheses, keeping only the first two terms in each expansion. That is,

$$\begin{aligned}\Delta t &= \frac{2L}{c}\left[\left(1 + \frac{v^2}{c^2}\cdots\right) - \left(1 + \frac{1}{2}\frac{v^2}{c^2}\cdots\right)\right] \\ &\simeq \frac{2L}{c}\left(\frac{1}{2}\frac{v^2}{c^2}\right) = \frac{Lv^2}{c^3}.\end{aligned} \tag{1.17}$$

If we rotate the interferometer through an angle of 90°, the two beams interchange their roles, and hence the total retardation in time is $2\,\Delta t$. Thus, if we use Eq. (1.17), the number ΔN of interference fringes that shift across the crosshairs in the eyepiece of the telescope while the interferometer is being rotated is

$$\boxed{\Delta N = \frac{\text{Path difference}}{\text{Wavelength}} = \frac{2\,\Delta tc}{\lambda} = \frac{2Lv^2}{\lambda c^2}.} \tag{1.18}$$

In their experiment, Michelson and Morley reflected beams 1 and 2 back and forth eight times, to achieve a path length of about 10 m. Using a light source of $\lambda = 5000$ Å, they expected a fringe shift of

$$\Delta N = \frac{2 \times 10\text{ m} \times (3 \times 10^4\text{ m/sec})^2}{5.0 \times 10^{-7}\text{ m}\,(3 \times 10^8\text{ m/sec})^2} = 0.4. \tag{1.19}$$

The Michelson interferometer is capable of detecting fringe shifts as small as 1/100 of a fringe. Hence Michelson and Morley should have had no difficulty in measuring 0.4 of a fringe shift while they were using a continuous rotation (from 0° to 90°) of the interferometer. But they could detect no fringe shift! They repeated the experiment at different places, at different seasons, and at different times of the day. Still no fringe shift. Finally they concluded that there was no fringe shift at all. That is

$$\Delta t = t_1 - t_2 = 0 \qquad \text{or} \qquad t_1 = t_2. \tag{1.20}$$

iii) *Interpretation of the experiment.* The negative result of the Michelson–Morley experiment was a disappointment to all the believers of the stationary-ether hypothesis, because this meant that it was not possible to measure the absolute velocity of the earth with respect to the (assumed stationary) ether.

However, we can explain the negative result if we give up the Galilean transformation and assume that the speed of light is invariant, i.e., that the velocity of light is the same in *all* inertial frames. This is a fact that has been experimentally verified.[3] Thus, for beam 1, $t_1 = 2L/c$, and for beam 2, $t_2 = 2L/c$. Therefore

$$\Delta t = t_1 - t_2 = 0.$$

Note, from Eq. (1.18), that ΔN approaches zero as v approaches zero, and hence no fringe shift is to be expected. Thus we may say that the Galilean transformations are good when v is small, but must be modified for large v.

C. Conclusions

From these two experiments and our discussion, we may draw the following conclusions.

1) The stationary-ether hypothesis can explain stellar aberration, but not the Michelson–Morley experiment, which implies that we are unable to locate the absolute (or preferred) frame of reference (ether frame), if there is one.

2) The Michelson–Morley experiment indicates that the speed of light is invariant. Hence the laws of electromagnetism are correct, and need no modifications.

3) The Galilean transformations are good for mechanics, but not for electromagnetism.

Thus we should look for a different set of transformation equations, which will be good for both mechanics and electromagnetism. Of course, these new transformations will have to be such that the speed of light remains invariant.

1.4 EINSTEIN'S POSTULATES AND LORENTZ TRANSFORMATIONS

A. Einstein's postulates

In 1905 the difficulties we have been talking about were overcome by Albert Einstein,[1] when he set forth the Special Theory of Relativity in his paper entitled "On the Electrodynamics of Moving Bodies." This paper might be described as a great leap forward for physics. Einstein's assumptions can be stated in the form of two basic postulates.

Postulate I. *The Principle of Equivalence* (*or Relativity*)
The laws of physics are the same in all inertial frames (frames moving with uniform translational motion with respect to each other). That is, all inertial frames are equivalent.

Postulate II. *The Principle of the Constancy of the Speed of Light*
The speed of light in free space (vacuum) is always a constant equal to c, and is independent of the relative motion of the inertial systems, the source, and the observer.

Postulate I implies that laws of mechanics as well as electromagnetism are invariant under all transformations. This brings about a degree of unity between the two. It also states that the absolute (or preferred) inertial frame (such as the ether frame) does not exist. Postulate II is merely a statement of an experimental fact.

Although these two postulates look very simple, they have far-reaching consequences, as we shall see.

First let us derive the transformation equations for inertial frames of reference moving with uniform relative velocities. And let us at the same time require that these transformations be applicable to both Newtonian mechanics and electromagnetism. Einstein himself derived just such transformations. They are called *Lorentz transformations*, after a physicist named H. A. Lorentz, who, in 1890, originated them as part of his theory of electromagnetism.

B. Lorentz coordinate transformations

Consider an inertial system S at rest. Consider another inertial system S', moving with a uniform translational velocity $\mathbf{v}$ along the XX' axes, as shown in Fig. 1.8. There are two observers. One observer is at rest with respect to the S system and

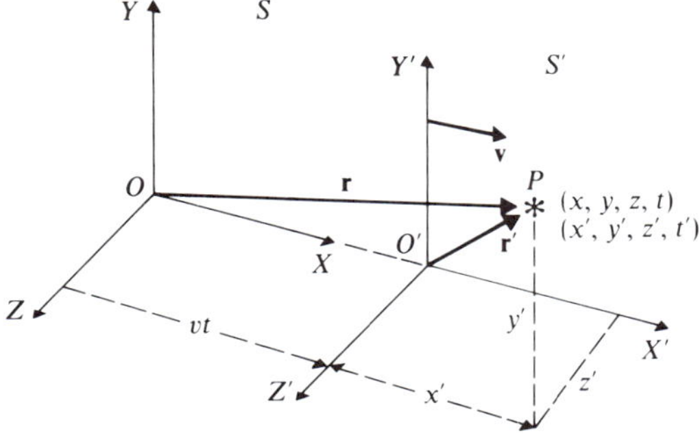

Fig. 1.8 The coordinates of a point P as measured from two inertial systems S and S'. System S' is moving with respect to S with a velocity $\mathbf{v}$ along the XX' axes.

the other observer is at rest with respect to the S' system. The coordinates of the two inertial systems coincide at $t = t' = 0$. At this very same time, a flash of light is emitted from the common origin of S and S'. After a certain time, the light signal reaches point P, as shown in Fig. 1.8. This point is at a distance $\mathbf{r}$ from the origin of the S system, and a distance $\mathbf{r}'$ from the origin of the S' system. According to Newtonian relativity, the speed of the signal would be different for the two observers, although the time taken to reach point P would be the same. But, according to postulate II of the special theory of relativity, the speed of light c must be the same in any inertial system. Hence the times which the two systems take to travel the distances r and r' should be different, so that

$$r = ct, \tag{1.21a}$$

$$r' = ct'. \tag{1.21b}$$

It is because of the relaxation of the condition $t = t'$ and the requirement that c be constant that we obtain altogether new transformations. Equations (1.21) may be rewritten as

$$x^2 + y^2 + z^2 = c^2t^2, \tag{1.22a}$$

$$x'^2 + y'^2 + z'^2 = c^2t'^2. \tag{1.22b}$$

Since the relative motion of the systems S and S' is along only the XX' axes, the directions perpendicular to the motion should be left unaffected by the new transformations. Hence

$$y = y' \qquad \text{and} \qquad z = z'. \tag{1.23}$$

Assuming that the points are located along the XX' axes (so that $y = y' = 0$, $z = z' = 0$) reduces Eq. (1.22) to

$$x^2 - c^2t^2 = 0, \tag{1.24a}$$

$$x'^2 - c^2t'^2 = 0. \tag{1.24b}$$

That is,

$$x^2 - c^2t^2 = x'^2 - c^2t'^2. \tag{1.24c}$$

We are interested in finding x' and t' in terms of x and t. That is,

$$\begin{aligned} x' &= x'(x, t), \\ t' &= t'(x, t). \end{aligned} \tag{1.25}$$

If we assume that the transformations are linear, we may write

$$x' = a_{11}x + a_{12}t, \tag{1.26a}$$

$$t' = a_{21}x + a_{22}t, \tag{1.26b}$$

where a_{11}, a_{12}, a_{21}, a_{22} are constants to be evaluated. [It is necessary to have linear transformations in order for one event in one system to be interpreted as one event in the other system; quadratic transformations imply more than one event in the other system.]

Let us consider the motion of the origin of the S' system for which $x' = 0$ at $t' = 0$. According to S, the system appears to be moving with a velocity v, so that $x = vt$. We can obtain this from Eq. (1.26a) by writing it in the form $x' = a_{11}(x - vt)$ so that, when $x' = 0$, $x = vt$. Thus we must conclude that $a_{12} = -va_{11}$. Therefore

$$x' = a_{11}(x - vt), \tag{1.27a}$$

$$t' = a_{21}x + a_{22}t. \tag{1.27b}$$

Substituting for x' and t' into Eq. (1.24c) and rearranging, we get

$$(a_{11}^2 - c^2a_{21}^2 - 1)x^2 - 2(va_{11}^2 + c^2a_{21}a_{22})xt - (c^2a_{22}^2 - v^2a_{11}^2 - c^2)t^2 = 0. \tag{1.28}$$

Since this is identically zero, the coefficients must vanish. That is,

$$\begin{aligned} a_{11}^2 - c^2a_{21}^2 &= 1 \\ va_{11}^2 + c^2a_{21}a_{22} &= 0 \\ c^2a_{22}^2 - v^2a_{11}^2 &= c^2 \end{aligned} \tag{1.29}$$

Solving these equations and, for simplicity, defining $\beta = v/c$ and $\gamma = 1/\sqrt{1 - \beta^2}$, we obtain

$$a_{11} = a_{22} = \boxed{\frac{1}{\sqrt{1 - (v^2/c^2)}} = \frac{1}{\sqrt{1 - \beta^2}} = \gamma}$$

and

$$a_{21} = \frac{-v/c^2}{\sqrt{1 - (v^2/c^2)}} = -\frac{\beta/c}{\sqrt{1 - \beta^2}} = -\frac{\beta\gamma}{c}. \tag{1.30}$$

Thus, using these values in Eqs. (1.27) together with Eqs. (1.23), we obtain the *Lorentz coordinate transformation equations* (from system S to system S'):

$$\begin{aligned} x' &= \frac{x - vt}{\sqrt{1 - (v^2/c^2)}} \\ y' &= y \\ z' &= z \\ t' &= \frac{t - (v/c^2)x}{\sqrt{1 - (v^2/c^2)}} \end{aligned} \quad \text{or} \quad \boxed{\begin{aligned} x' &= \gamma(x - vt) \\ y' &= y \\ z' &= z \\ t' &= \gamma\left(t - \frac{\beta}{c}x\right) \end{aligned}} \tag{1.31}$$

We obtain the inverse transformations (from system S' to system S) by replacing v by $-v$ and interchanging primed and unprimed coordinates. That is,

$$
\begin{aligned}
x &= \frac{x' + vt'}{\sqrt{1-(v^2/c^2)}} \\
y &= y' \\
z &= z' \\
t &= \frac{t' + (v/c^2)x'}{\sqrt{1-(v^2/c^2)}}
\end{aligned}
\quad \text{or} \quad
\boxed{
\begin{aligned}
x &= \gamma(x' + vt') \\
y &= y' \\
z &= z' \\
t &= \gamma\left(t' + \frac{\beta}{c}x'\right)
\end{aligned}
}
\tag{1.32}
$$

An important point to be noted in these equations is that t and t' are no longer independent of space coordinates. That is, t is a function of t' and x', while t' is a function of t and x.

When the situation involves very low velocities (i.e., in Newtonian or classical mechanics), any new transformation equations must reduce to Galilean transformations. We can easily see that this is the case if we assume that $v \to 0$, and hence that $v/c \ll 1$, $v^2/c^2 \lll 1$, and also that $v/c^2 \lll 1$. Then Eqs. (1.31) reduce to

$$x' = x - vt, \qquad y' = y, \qquad z' = z, \qquad t' = t,$$

which are the Galilean transformations. Thus we conclude:

$$\text{Limit}_{v \to 0} \text{ (Lorentz transformations)} = \text{(Galilean transformations)}. \tag{1.33}$$

Lorentz transformations also put a limit on the maximum value of v. Thus v must be less than c, or the quantity $\sqrt{1-(v^2/c^2)}$ would be an imaginary quantity. That would make space and time coordinates imaginary, which is physically not possible.

Hence, in a vacuum, nothing can move with a velocity greater than the velocity of light.

C. Lorentz velocity transformations

Using Lorentz coordinate transformations, we can derive Lorentz velocity transformations, which replace the Galilean velocity transformations given by Eq. (1.2). Let us consider two inertial systems S and S' moving with a relative velocity $\mathbf{v}$ along the XX' axes. Consider a particle at P, which is moving in space and has velocity $\mathbf{u}$ (u_x, u_y, u_z), as measured by an observer in system S, and velocity $\mathbf{u}'$ (u'_x, u'_y, u'_z) as measured by an observer in system S'. The aim is to find relations between the components

$$u_x = \frac{dx}{dt}, \qquad u_y = \frac{dy}{dt}, \qquad u_z = \frac{dz}{dt},$$

and the components

$$u'_x = \frac{dx'}{dt'}, \qquad u'_y = \frac{dy'}{dt'}, \qquad u'_z = \frac{dz'}{dt'}.$$

Differentiating Eqs. (1.31), we obtain

$$\begin{aligned} dx' &= \gamma(dx - v\,dt), \\ dy' &= dy, \\ dz' &= dz, \\ dt' &= \gamma\left(dt - \frac{v}{c^2}\,dx\right). \end{aligned} \tag{1.34}$$

Thus the components of $\mathbf{u}'$ are given by

$$u'_x = \frac{dx'}{dt'} = \frac{\gamma(dx - v\,dt)}{\gamma[dt - (v\,dx/c^2)]} = \frac{(dx/dt) - v}{1 - (v/c^2)(dx/dt)}$$

or

$$u'_x = \frac{u_x - v}{1 - (vu_x/c^2)}$$

and similarly

$$u'_y = \frac{u_y}{\gamma[1 - (vu_x/c^2)]} \tag{1.35}$$

and

$$u'_z = \frac{u_z}{\gamma[1 - (vu_x/c^2)]}$$

These are called the *Lorentz velocity transformations.* As before, we can obtain the inverse transformation—that is, u_x, u_y, u_z in terms of u'_x, u'_y, and u'_z by replacing v by $-v$, and interchanging primed and unprimed coordinates. That is,

$$\begin{aligned} u_x &= \frac{u'_x + v}{1 + (vu'_x/c^2)} \\ u_y &= \frac{u'_y}{\gamma[1 + (vu'_x/c^2)]} \\ u_z &= \frac{u'_z}{\gamma[1 + (vu'_x/c^2)]} \end{aligned} \tag{1.36}$$

Note that—unlike the space-coordinate transformations for y' and z', which did not involve x—the velocity components u'_y and u'_z do depend on the velocity u_x.

Two important consequences may be drawn from the Lorentz velocity transformations. (i) Nothing can move in a vacuum with a velocity greater than the velocity of light. (ii) The velocity of light is independent of the relative motion of the source or of the observer (Postulate II).

These points are demonstrated in the following example.

Example 1.2. Consider two rockets, A and B, each moving with velocity $0.9c$ relative to the earth, and approaching each other according to an observer standing on the earth. Calculate the relative velocity of B with respect to A.

Let the earth be a stationary system S, the rocket A be the system S' moving with velocity $v = 0.9c$ along the XX' axes, and the rocket B be the point P moving in space with $u_x = -0.9c$, as shown in Fig. 1.9. Then, according to the first of Eqs. (1.35),

$$u'_x = \frac{u_x - v}{1 - (vu_x/c^2)} = \frac{-0.9c - 0.9c}{1 - [(0.9c)(-0.9c)/c^2]} = -\frac{1.8}{1.81}c,$$

$$u'_x = -0.9945c.$$

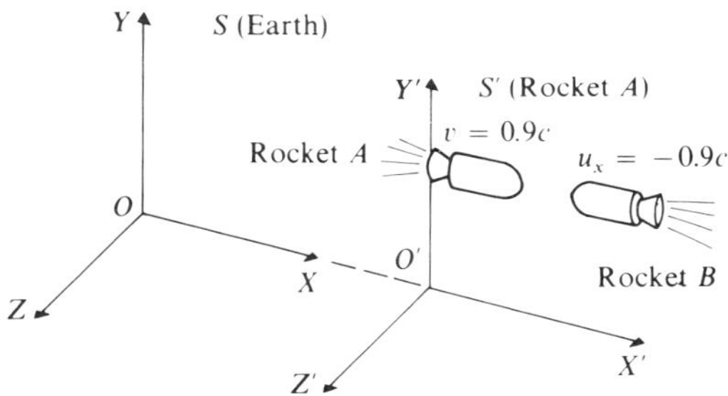

Fig. 1.9 Rockets A and B, each moving with a velocity $0.9c$, are observed approaching each other. The observer on earth is in system S, while the rocket A is in system S', moving with velocity $v = +0.9c$ along the XX' axes. The rocket B is at point P, moving in space with velocity $u_x = -0.9c$ with respect to the earth.

Thus the rocket B relative to an observer in rocket A is moving with a velocity $0.9945c$ along the $-X'$ axis. Thus $u'_x < c$, and will always be less than c. According to Newtonian relativity, the velocity of B with respect to A would be $-0.9c - (0.9c) = -1.8c$. That is, $|-1.8c|$, which is greater than c.

Note that, if $u_x = c$, then $|u'_x| = c$. That is, the speed of light is independent of the motion of the source and of the observer.

1.5 LENGTH, TIME, AND SIMULTANEITY IN RELATIVITY

When we think about the Lorentz transformations, we suddenly realize that length, time interval, and simultaneity have meanings altogether different from their meanings in classical physics. In relativity, not only is there no absolute reference frame, but there is no absolute simultaneity. These quantities (length, time, and simultaneity) depend on the relative motion of observers. According to Einstein, there is no such thing as absolute motion, except that the speed of light in vacuum is the same in all inertial systems.

Let us pause a moment and demonstrate these points.

a) Length contraction

Consider two observers at rest in inertial systems S and S'. The observer in S' has a rod of length L_0 lying parallel to the X' axis. That is, $L_0 = x'_2 - x'_1$ is the same for both observers when they are at rest relative to each other. Now let us assume that the S' system starts moving with a velocity $\mathbf{v}$ along the XX' axes. To the observer in S' the length of the rod is still L_0, but to the observer in S the length of the rod is $L = x_2 - x_1$, where we must find x_2 and x_1 by making use of the Lorentz transformations Eq. (1.31). That is,

$$x'_2 = \gamma(x_2 - vt_2), \qquad x'_1 = \gamma(x_1 - vt_1). \tag{1.37}$$

On subtraction, this gives

$$x'_2 - x'_1 = \gamma[(x_2 - x_1) - v(t_2 - t_1)]. \tag{1.38}$$

Since the observer in the S system must measure the two ends of the rod simultaneously, it means that $t_1 = t_2$, and Eq. (1.38) yields

$$x_2 - x_1 = \frac{1}{\gamma}(x'_2 - x'_1),$$

where $x'_2 - x'_1 = L_0$, $1/\gamma = \sqrt{1 - \beta^2}$, and $x_2 - x_1 = L$ is the length as measured by an observer in S. Thus

$$\boxed{L = \frac{L_0}{\gamma} = L_0\sqrt{1 - \beta^2}.} \tag{1.39}$$

Since the quantity $\sqrt{1 - \beta^2}$ is always less than unity (and $\gamma > 1$), the length L is less than L_0. That is, to the observer in S, the rod looks as though it is contracted. (Note that, since $y' = y$ and $z' = z$, there is no observed change in length in these directions.)

The effect is reciprocal. If S has a rod of length L_0, while S' is moving and looks at the rod, it will appear contracted to him. That is, it will appear to be of length $L_0\sqrt{1 - \beta^2}$.

Thus the measured length of an object is maximum when the object is at rest relative to the observer, and appears contracted by a factor $\sqrt{1 - \beta^2}$ to an observer who is in motion relative to the object.

The length L_0, measured in the frame in which it is at rest, is called the *proper length*, and the frame is the *proper reference frame*, while L is the *improper length* and the corresponding frame is the *improper reference frame*.

Example 1.3. According to an observer at rest in the system S', a rod 1 m long is at rest and makes an angle of 45° with the X'-axis. Let us calculate the length of this rod and the angle which it makes with the X-axis according to an observer in the system S. The system S' is moving with a velocity $v = (\sqrt{3}/2)c$ with respect to S along the XX' axes.

Let us say that the length of the rod at rest with respect to S' is L_0, and that it is resolved into two components, L_{0x} and L_{0y}, as shown in Fig. 1.10. Thus

$$L_{0x} = L_0 \cos\theta_0, \qquad L_{0y} = L_0 \sin\theta_0.$$

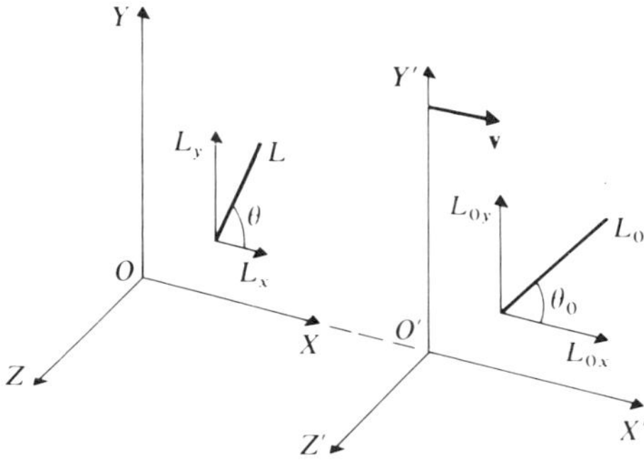

Fig. 1.10 A rod of length L_0 is at rest in inertial system S'. As observed from the inertial system S, its length is L, as shown, and to the observer in S, it appears contracted.

As viewed from system S, the vertical component L_{0y} which is perpendicular to **v** remains unchanged. That is,

$$L_y = L_{0y} = L_0 \sin\theta_0.$$

The horizontal component which is parallel to **v** will be contracted as observed from system S, and is given by Eq. (1.39) to be

$$L_x = \sqrt{1-\beta^2}\, L_{0x} = \sqrt{1-\beta^2}\, L_0 \cos\theta_0.$$

Thus, according to the observer in S, the length L and the angle θ which it makes with the X-axis is

$$L = \sqrt{L_x^2 + L_y^2} = \sqrt{[\sqrt{1-\beta^2}\, L_0\cos\theta_0]^2 + (L_0\sin\theta_0)^2}$$

$$= L_0\sqrt{1-\beta^2\cos^2\theta_0}$$

and

$$\tan\theta = \frac{L_y}{L_x} = \frac{L_0\sin\theta_0}{\sqrt{1-\beta^2}\, L_0\cos\theta_0} = \frac{\tan\theta_0}{\sqrt{1-\beta^2}}.$$

In the present situation $L_0 = 1$ meter, $\theta_0 = 45°$, and $v = (\sqrt{3}/2)c$. Therefore

$$\sqrt{1 - \beta^2} = \sqrt{1 - (\sqrt{3}/2)^2} = \sqrt{1 - \tfrac{3}{4}} = \sqrt{0.25} = 0.5.$$

Also $\cos\theta_0 = \cos 45° = 1/\sqrt{2}$; $\tan\theta_0 = \tan 45° = 1$. Hence

$$L = L_0\sqrt{1 - \beta^2\cos^2\theta_0} = 1\ \text{m}\sqrt{1 - \tfrac{3}{4} \times \tfrac{1}{2}} = 0.625\ \text{m},$$

$$\tan\theta = \frac{\tan\theta_0}{\sqrt{1 - \beta^2}} = \frac{1}{0.5} = 2 \quad \text{or} \quad \theta = 63°\ 27'.$$

Note that contraction has taken place only along the X-axis. That is,

$$L_x = \sqrt{1 - \beta^2}\, L_{0x} = \sqrt{1 - \beta^2}\, L_0 \cos\theta_0 = 0.5 \times 1 \times \frac{1}{\sqrt{2}} = 0.353\ \text{m}.$$

---

b) Dilation of time (slowing down of clocks)

A time interval, like a length interval, is not absolute. Consider, for example, two different events. The time intervals between these two events are registered on two different clocks, in inertial systems S and S', but *both* clocks are observed from system S. From the fourth equation in (1.32), we have

$$t_1 = \frac{t_1' + (v/c^2)x_1'}{\sqrt{1 - \beta^2}} \quad \text{and} \quad t_2 = \frac{t_2' + (v/c^2)x_2'}{\sqrt{1 - \beta^2}}$$

and hence

$$t_2 - t_1 = \frac{(t_2' - t_1') + (v/c^2)(x_2' - x_1')}{\sqrt{1 - \beta^2}}.$$

According to this equation, it is the observer in system S who looks at the clock of S' [which must be at rest during the observation (that is, $x_1' = x_2'$)], reads a time interval of $t_2' - t_1'$, and then compares it with his own time interval $(t_2 - t_1)$. Thus

$$t_2 - t_1 = \frac{t_2' - t_1'}{\sqrt{1 - \beta^2}}. \tag{1.40}$$

Or, if we define $t_2' - t_1' = T_0$ as the interval between the two events as recorded by an observer in S' who is at rest with respect to its clock, and $t_2 - t_1 = T$ as the interval as recorded by an observer in S who is in motion (with velocity $-v$ with respect to S' and its clock), we obtain the following relation:

$$T = \frac{T_0}{\sqrt{1 - \beta^2}} = \gamma T_0. \tag{1.41}$$

Since $\sqrt{1 - \beta^2}$ is always less than unity and hence $\gamma > 1$, the interval T is longer than T_0. Thus, to the observer in S (who is moving with respect to S' with a velocity $-v$ along the XX' axes), the clock of S' seems to be running slow as compared to his own clock. The effect is reciprocal. That is, when the observer in S' looks at the clock in S, it appears to be running slow as compared to his own.

Thus a clock appears to run at its fastest when it is at rest relative to the observer, and its rate seems to be slowed down by a factor of $1/\sqrt{1 - \beta^2}$ *when the clock is moving with a velocity* **v** *relative to the observer.*

The time interval measured in the reference frame in which the clock is at rest—that is, T_0—is called the *proper time*, while T is the *improper time*.

Example 1.4. According to an observer on earth, a car covers 1 mile in 50 sec on a straight stretch of highway. How long will it take to cover this distance according to an observer with his own clock on a space ship which is receding from earth with a speed of $0.95c$: (a) perpendicular to the line of motion of the car, and (b) along the same line of motion as the car?

The proper time T_0 is 50 sec. Thus, according to the observer in the space ship, $T = \gamma T_0$, where $\gamma = 1/\sqrt{1 - \beta^2} = 1/\sqrt{1 - (0.95)^2} = 3.2$. Therefore

$$T = \gamma T_0 = 3.2 \times 50 = 160 \text{ sec.}$$

a) There will be no contraction in the length of the distance the car travels if it is assumed that the space ship is receding from earth in a direction perpendicular to the line of motion of the car. However, according to the observer in the space ship, it takes the car 160 sec to go 1 mile.

b) If the space ship is receding along the same line of motion as the car, there will be a contraction in the length of the distance the car travels, given by

$$L = L_0/\gamma = 1 \text{ mile}/3.2 = 0.31 \text{ mile.}$$

Thus, according to the observer in the space ship, the car will travel a distance of only 0.31 mile in 160 sec. That is, it will travel 1 mile in 517 sec or 8.6 min.

---

c) Simultaneity

According to Newtonian mechanics, time is assumed to be absolute; that is, $t' = t$. Hence the same time scale applies to all inertial reference frames. This concept of absolute time leads to the concept of *absolute simultaneity*. Thus, in classical mechanics, when we say "Events A and B are simultaneous, i.e., they occur at the same time," we do not mention any particular frame of reference, because the events are considered to be simultaneous in all inertial systems. The reason this is true in classical mechanics is that there is an implied assumption that information is conveyed with infinite speed.

However, in relativity, nothing moves with a speed greater than the speed of light. Hence, as in the case of length and time, there is no such thing as absolute simultaneity. Events which are simultaneous to an observer in one inertial system may not necessarily be simultaneous to an observer in another inertial system which is moving with a uniform velocity with respect to the first. According to Einstein, the definition of simultaneity is as follows.

Two events taking place in a particular frame of reference at points $A(x_1, t_1)$ and $B(x_2, t_2)$ are simultaneous if light signals simultaneously emitted from the geometrical midpoint between x_1 and x_2 arrive at x_1 at t_1 and at x_2 at t_2.

Thus, in Einstein's theory of relativity, simultaneity does not have an absolute meaning. The reason for this is that the coordinates t and t' not only are unequal, but also depend on the space coordinates x and x', as given by Lorentz transformations. For example: (i) Two events A and B happening at x_1 and x_2 in the system S are simultaneous in S because $t_1 = t_2$, but are not simultaneous in S' because, according to Eq. (1.31), $t_1' \neq t_2'$. (ii) Two events A and C happening at x in system S at times t_1 and t_2 appear from S' not only to be happening at different times but also at different locations. That is, $t_1' \neq t_2'$ and $x_1' \neq x_2'$. These concepts have been confirmed in many physical situations.[7,8]

1.6 MASS AND MOMENTUM IN RELATIVITY

According to classical theory the linear momentum $\mathbf{p}$ of a body is given by

$$\mathbf{p} = m_0\mathbf{v}, \tag{1.42}$$

where m_0 is the inertial mass of the body and $\mathbf{v}$ is its velocity. Also, for an isolated system of particles with masses $m_{01}, m_{02}, \ldots, m_{0i}$ moving with velocities $\mathbf{v}_1, \mathbf{v}_2, \ldots, \mathbf{v}_i$, respectively, and on which there are no external forces acting, the conservation of linear momentum requires that

$$m_{01}\mathbf{v}_1 + m_{02}\mathbf{v}_2 + \cdots + m_{0i}\mathbf{v}_i = \sum_i m_{0i}\mathbf{v}_i = \text{constant}. \tag{1.43}$$

There is a definite reason for using the subscript zero to describe the masses. In Newtonian mechanics the mass is considered to be a constant quantity, independent of the velocity with which it is moving, as assumed in Eqs. (1.42) and (1.43). But in relativity theory, the mass, like length and time intervals, is not a constant. It changes with the velocity with which it is moving. Of course, at low velocities, the mass of an object does not change appreciably, and the above equations from classical mechanics are approximately correct. But when an object is moving with high velocity, these equations no longer hold. One may use the conservation of linear momentum—which is an inviolate principle in physics—in conjunction with the Lorentz transformations to arrive at an expression for the variation of mass with velocity. Let us now put these ideas to work for us, and see what happens.

Consider a perfectly inelastic collision between two identical balls in the inertial system S' which is moving relative to system S with a velocity $\mathbf{v}$ along the XX' axes, as shown in Fig. 1.11. The balls in S' each have a mass m' and are moving with velocities u' and $-u'$ along the X'-axis. The masses of these two balls as viewed from the system S are m_1 and m_2 and their velocities are u_1 and u_2, respectively. After the collision the two balls stick together, and are at rest in the system S', but appear to be moving with velocity $\mathbf{v}$ with respect to the system S, as shown in Fig. 1.11. Thus the conservation of linear momentum as applied to the inertial systems S and S' separately yields

$$m'u' + m'(-u') = 0 \qquad \text{in } S', \tag{1.44}$$

$$m_1u_1 + m_2u_2 = (m_1 + m_2)v \qquad \text{in } S, \tag{1.45}$$

where, according to the Lorentz velocity transformations given by Eqs. (1.36), u_1 and u_2 are

$$u_1 = \frac{u' + v}{1 + (u'v/c^2)} \tag{1.46}$$

and

$$u_2 = \frac{-u' + v}{1 - (u'v/c^2)}. \tag{1.47}$$

If we let $M = m_1 + m_2$, we may rewrite Eq. (1.45) as

$$m_1(u_1 - u_2) = M(v - u_2) \tag{1.48a}$$

or

$$m_2(u_1 - u_2) = M(u_1 - v). \tag{1.48b}$$

Dividing one by the other and substituting for u_1 and u_2 from Eqs. (1.46) and (1.47), we have

$$\frac{m_1}{m_2} = \frac{v - u_2}{u_1 - v} = \frac{1 + (u'v/c^2)}{1 - (u'v/c^2)}. \tag{1.49}$$

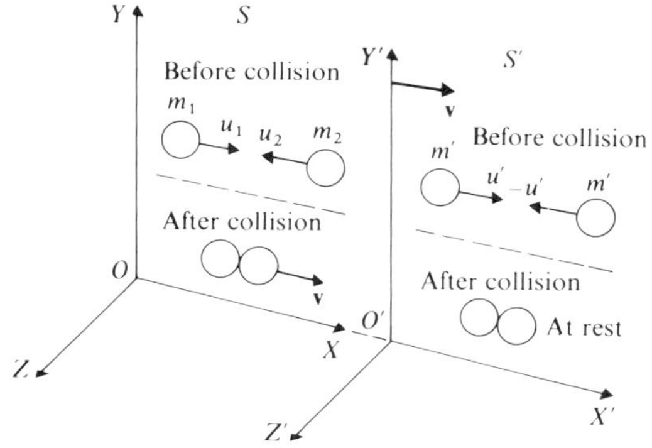

Fig. 1.11 A completely inelastic collision between two masses, each of mass m', taking place in the S' system, as viewed from the S system before and after the collision.

Using the value of u_1 from Eq. (1.46) in the following expression,

$$\sqrt{1-(u_1^2/c^2)} = \sqrt{1-\frac{1}{c^2}\left[\frac{u'+v}{1+(u'v/c^2)}\right]^2} = \frac{\sqrt{[1-(v^2/c^2)][1-(u'^2/c^2)]}}{1+(u'v/c^2)}$$

or

$$1+\frac{u'v}{c^2} = \frac{\sqrt{[1-(v^2/c^2)][1-(u'^2/c^2)]}}{\sqrt{1-(u_1^2/c^2)}}. \tag{1.50}$$

Similarly

$$1-\frac{u'v}{c^2} = \frac{\sqrt{[1-(v^2/c^2)][1-(u'^2/c^2)]}}{\sqrt{1-(u_2^2/c^2)}}. \tag{1.51}$$

Substituting Eqs. (1.50) and (1.51) into Eq. (1.49), we get

$$\frac{m_1}{m_2} = \frac{\sqrt{1-(u_2^2/c^2)}}{\sqrt{1-(u_1^2/c^2)}}$$

or

$$m_1\sqrt{1-\frac{u_1^2}{c^2}} = m_2\sqrt{1-\frac{u_2^2}{c^2}} = \text{constant} = m_0, \tag{1.52}$$

where m_0 is the rest mass of the identical balls. Thus, in general, if a mass is moving with a velocity v relative to an observer,

$$m\sqrt{1-\frac{v^2}{c^2}} = m_0$$

or

$$\boxed{m = \frac{m_0}{\sqrt{1-(v^2/c^2)}} = \frac{m_0}{\sqrt{1-\beta^2}} = \gamma m_0.} \tag{1.53}$$

This equation states that the mass of an object is not constant in general. Only when the object is at rest is its mass constant and equal to m_0. When the object starts to move with a velocity v with respect to the observer, its mass appears to be increased to m. This is called the *relativistic mass*, which is given by Eq. (1.53). *Thus m_0, and not m, is an invariant quantity in relativity theory.* In classical physics, however, there is not much difference between m and m_0 because, as $v \to 0$, from Eq. (1.53), $m \to m_0$. From Table 1.1, it is obvious that m does not differ by more than 0.5% from m_0 for v/c less than $\frac{1}{10}$. If $v \to c$, $m \to \infty$, which means that, if

Table 1.1

Variation of Mass with Velocity

v/c	0.01	0.02	0.05	0.10	0.50	0.80	0.90	0.98	0.99	1.00
m/m_0	1.00005	1.0001	1.0013	1.005	1.15	1.666	2.30	5.00	7.10	∞

we wanted a material particle to travel at the speed of light, we would have to increase its mass to infinity!

Of course that would be impossible, and hence this is another reason for saying that c is a limiting velocity impossible for a material particle to achieve (except tachyons, which we'll talk about in Section 1.12). Incidentally, the expression in Eq. (1.53) for the variation of mass with velocity has been confirmed experimentally. We'll discuss one of these experiments in Section 1.9.

In view of the above discussion, let us redefine linear momentum and conservation of linear momentum according to relativity theory, by the following equations, respectively:

$$\mathbf{p} = m\mathbf{v} = \gamma m_0 \mathbf{v}, \tag{1.54}$$

$$\sum_i m_i \mathbf{v}_i = \sum_i \gamma_i m_{0i} \mathbf{v}_i = \text{constant.} \tag{1.55}$$

Example 1.5. At what speed will the mass of a proton become double its rest mass?

From Eq. (1.53), we know that the rest mass and the moving mass are related by

$$m = \gamma m_0 = \frac{m_0}{\sqrt{1 - (v^2/c^2)}}.$$

Then, for $m = 2m_0$, we have

$$\gamma = \frac{1}{\sqrt{1 - (v^2/c^2)}} = \frac{m}{m_0} = 2.$$

That is, $v = \dfrac{\sqrt{3}}{2} c = 0.866c.$

Note that this is true for a particle of any rest mass: electron, proton, neutron, and so forth.

1.7 RELATIVISTIC MECHANICS

The change in the definition of mass and momentum, as discussed in Section 1.6, requires modification of our classical definition of Newton's second law, and also our definitions of kinetic and total energy. According to Eq. (1.54), that is,

$$\mathbf{p} = m\mathbf{v} = \frac{m_0 \mathbf{v}}{\sqrt{1 - (v^2/c^2)}} = \gamma m_0 \mathbf{v}, \tag{1.56}$$

where $\mathbf{p}$ has the following three components:

$$p_x = \gamma m_0 v_x, \qquad p_y = \gamma m_0 v_y, \qquad p_z = \gamma m_0 v_z, \tag{1.57}$$

with

$$\gamma = \frac{1}{\sqrt{1 - (v^2/c^2)}} \qquad \text{and} \qquad v^2 = v_x^2 + v_y^2 + v_z^2. \tag{1.58}$$

In classical physics, according to Newton's second law, the force **F** acting on a body is defined as the time rate of change of momentum. However, in relativistic mechanics, it is the time rate of change of relativistic momentum. That is,

$$\mathbf{F} = \frac{d\mathbf{p}}{dt} = \frac{d}{dt}(m\mathbf{v}) = \frac{d}{dt}(\gamma m_0 \mathbf{v}), \tag{1.59}$$

or, since m is no longer a constant,

$$\mathbf{F} = \frac{d(m\mathbf{v})}{dt} = m\frac{d\mathbf{v}}{dt} + \mathbf{v}\frac{dm}{dt}, \tag{1.60}$$

which, in the classical limit, $v \ll c$, makes $dm/dt \to 0$. Hence Eq. (1.60) takes the familiar form

$$\mathbf{F} = m\frac{d\mathbf{v}}{dt} = m\mathbf{a}.$$

Suppose that a particle of rest mass m_0 is acted on by a force **F** through a distance x in time t along the X-axis, and that it attains a final velocity v. As in classical mechanics, the kinetic energy K may be defined as the work done by the unbalanced force **F**. That is,

$$\begin{aligned} K &= \int_0^x F\,dx = \int_0^x \frac{d}{dt}(mv)\,dx = \int_0^v \frac{dx}{dt}\,d(\gamma m_0 v) \\ &= m_0 \int_0^v v d(\gamma v). \end{aligned} \tag{1.61}$$

Note that

$$d\gamma = d\left(1 - \frac{v^2}{c^2}\right)^{-1/2} = \frac{v}{c^2}\left(1 - \frac{v^2}{c^2}\right)^{-3/2} dv$$

and that

$$\begin{aligned} d(\gamma v) &= \gamma\,dv + v\,d\gamma \\ &= \left[\left(1 - \frac{v^2}{c^2}\right)^{-1/2} dv + \frac{v^2}{c^2}\left(1 - \frac{v^2}{c^2}\right)^{-3/2} dv\right] = \frac{dv}{[1 - (v^2/c^2)]^{3/2}}. \end{aligned}$$

Substituting for $d(\gamma v)$ in Eq. (1.61) and integrating, we obtain

$$\begin{aligned} K &= m_0 \int_0^v v d(\gamma v) = m_0 \int_0^v \frac{v dv}{[1 - (v^2/c^2)]^{3/2}} \\ &= m_0 c^2 \left[\frac{1}{\sqrt{1 - (v^2/c^2)}} - 1\right] = \frac{m_0 c^2}{\sqrt{1 - (v^2/c^2)}} - m_0 c^2. \end{aligned}$$

That is,

$$\boxed{K = mc^2 - m_0 c^2 = \gamma m_0 c^2 - m_0 c^2 = (\gamma - 1) m_0 c^2,} \tag{1.62}$$

which is quite different from the classical expression $K = \frac{1}{2}m_0v^2$. But, of course, Eq. (1.62) does reduce to the classical value for $v/c \ll 1$, that is, for low speeds, as can be shown by expanding γ with the help of the binomial theorem and keeping only the first two terms. Thus

$$\lim_{v/c \ll 1} K = m_0c^2(\gamma - 1) = m_0c^2\left[\frac{1}{\sqrt{1-(v^2/c^2)}} - 1\right]$$
$$= m_0c^2\left[\left(1 - \frac{v^2}{c^2}\right)^{-1/2} - 1\right] \simeq m_0c^2\left[\left(1 + \frac{1}{2}\frac{v^2}{c^2} + \cdots\right) - 1\right]$$
$$= \tfrac{1}{2}m_0v^2. \tag{1.63}$$

Let us return to Eq. (1.62), which states that the increase in kinetic energy of a particle is due to an increase in mass of the particle. If we define $\Delta m = m - m_0$, then

$$K = \Delta mc^2 = (m - m_0)c^2. \tag{1.64}$$

This relation implies that energies may be assigned to masses in relativistic mechanics. The quantity m_0c^2, which does not depend on the speed of the particle, is called the *rest-mass energy*, and is denoted by E_0. By analogy, mc^2 $(= K + m_0c^2)$, which is the sum of the rest-mass energy and the kinetic energy, is denoted by E, and is called the *total energy*. Thus Eq. (1.62) takes the form

$$E = mc^2 = K + E_0 = \gamma m_0c^2 = \gamma E_0, \tag{1.65}$$

where

$$E_0 = m_0c^2 \tag{1.66}$$

and

$$E = mc^2. \tag{1.67}$$

Equation (1.67), which states the relationship between mass and energy, is of course the famous *Einstein's mass–energy relation.*

If the particle is moving in a conservative field, and hence has a potential energy V as well, then the total energy is given by

$$E = K + V + E_0. \tag{1.68}$$

Now let us think about this equivalence between mass and energy. Obviously the separate laws for the conservation of mass and the conservation of energy must be replaced in relativity theory by a single law, called the *conservation of mass–energy* or *the law of conservation of total relativistic energy*, according to which *the total relativistic energy is invariant under a Lorentz transformation.* That is, for an isolated system, the total relativistic energy is the same as the energy observed from any inertial system. Hence

Total energy = rest energy + kinetic energy + potential energy = constant.

There is a simple and very useful relation between relativistic momentum p, rest-mass energy E_0, and total energy E which we can obtain as follows. Let us start with

$$m = \frac{m_0}{\sqrt{1 - (v^2/c^2)}}.$$

Squaring and rearranging, we get

$$m^2\left(1 - \frac{v^2}{c^2}\right) = m_0^2.$$

Multiplying both sides by c^4, we get

$$m^2c^4 - m^2v^2c^2 = m_0^2c^4.$$

Substituting for $p = mv$, $E_0 = m_0c^2$, and $E = mc^2$, we obtain the required relation

$$\boxed{E^2 = p^2c^2 + m_0^2c^4} \tag{1.69}$$

or

$$E^2 = p^2c^2 + E_0^2.$$

At this stage let us put together the results derived in relativistic mechanics and discuss some useful approximations. We obtained

$$m = \frac{m_0}{\sqrt{1 - (v^2/c^2)}} = \gamma m_0 \tag{1.53}$$

$$p = mv = \frac{m_0 v}{\sqrt{1 - (v^2/c^2)}} = \gamma m_0 v \tag{1.56}$$

$$K = mc^2 - m_0c^2 = m_0c^2(\gamma - 1) \tag{1.62}$$

$$E = mc^2 = K + E_0 = \gamma m_0c^2 = \gamma E_0 \tag{1.65}$$

$$E_0 = m_0c^2 \tag{1.66}$$

and

$$E^2 = p^2c^2 + E_0^2. \tag{1.69}$$

Example 1.6 will presently illustrate the use of these equations.

Now let us talk about some rather interesting approximations that would result from the following limitations: (i) $v \ll c$, (ii) $v \simeq c$, and (iii) $v = c$.

i) *Classical limit:* $v \ll c$ (or $\gamma \to 1$). That is, if the velocity of an object is very small compared with the velocity of light, the particle may be treated classically. The relativistic equations then reduce to the equations of classical mechanics:

$$m \simeq m_0, \qquad p \simeq m_0 v, \qquad K \simeq \tfrac{1}{2} m_0 v^2. \tag{1.70}$$

In order to decide whether to use the classical expressions or not, we consider

$$\gamma = \frac{E}{m_0 c^2} = \frac{E_0 + K}{m_0 c^2} = 1 + \frac{K}{m_0 c^2}. \tag{1.71}$$

We can use classical mechanics only if $\gamma \to 1$, which is possible if $K \ll m_0 c^2$. For example, if $v/c < \frac{1}{10}$ or $K/m_0 c^2 < \frac{1}{200}$, using the expressions in Eq. (1.70) results in an error of less than 1%.

ii) *Relativistic limit:* $v \simeq c$ (or $\gamma \gg 1$). This is the extreme relativistic region. From Eq. (1.71), E (or mc^2) and K are very large compared with $m_0 c^2$. The relativistic equations reduce to the following simplified form:

$$p \simeq \frac{E}{c}, \qquad K \simeq E. \tag{1.72}$$

Thus, for example, $v/c > 0.99$ means that $E/m_0 c^2 > 7$, and the use of $E = pc$ does not introduce an error greater than 1%.

iii) *Extreme relativistic limit:* $v = c$ (or $\gamma = \infty$). If a particle is moving with a velocity equal to the velocity of light, $m = m_0/\sqrt{1 - (v^2/c^2)}$ yields $m = \infty$, which is not possible. Thus we may say that no material particle can be made to move with a velocity equal to that of light. But there is another interesting alternative. The difficulty that $m = \infty$ for $v = c$ is removed if we assume that $m_0 = 0$. Then we get $m = 0/0$, which is indeterminate. Thus only those particles which have zero rest mass can travel with $v = c$, and hence in this case

$$m_0 = 0, \qquad E = pc, \qquad K = E. \tag{1.73}$$

That is, this particle has momentum and energy, but no rest mass. According to classical mechanics, there can be no such particle. But, according to Einstein, a particle which has characteristics described by Eq. (1.73) is indeed a reality; it is called a *photon* (see Chapter 2).

1.8 MASS AND BINDING ENERGY

Before we discuss some of the situations in which the applicability of Einstein's mass–energy relationship $E = mc^2$ becomes obvious, we have to set forth the *units* of mass and energy commonly used in atomic and nuclear physics.

A convenient unit of energy is the *electron volt* (eV). This is defined as the energy acquired by a particle carrying a charge e (which is the charge carried by one electron) when accelerated through a potential difference of one volt. Since the charge carried by one electron is 1.602×10^{-19} coulomb, then

$$1 \text{ eV} = 1.602 \times 10^{-19} \text{ coul} \times 1 \text{ volt} = 1.602 \times 10^{-19} \text{ joule}. \tag{1.74}$$

The bigger units of energy commonly used are as follows.

One kilo electron volt = 1 keV = 10^3 eV
One million electron volts = 1 MeV = 10^6 eV
One giga electron volt = 1 GeV = 10^9 eV

One giga electron volt is also called one billion electron volts, and is denoted by 1 BeV. Such high energies are easily produced. For example, the AG synchrotron at Brookhaven has a capability of producing a 30-BeV proton beam.

The masses of the particles in atomic and nuclear physics are given in terms of the recently introduced *unified mass unit* u. *One atomic mass unit* u *is defined as one-twelfth of the mass of the natural carbon-12 atom* (isotope ^{12}C). [Note that this unit replaces the older unit amu (still in use, however) which is defined as one-sixteenth of the mass of the natural oxygen-16 isotope.]

Avogadro's number, $N_A = 6.02252 \times 10^{23}$, gives the number of atoms in one mole, that is, in 12 g of atomic carbon there are N_A atoms. Therefore

$$1 \text{ u} = \frac{1}{12} \frac{12 \text{ gram}}{6.02252 \times 10^{23}} = 1.66043 \times 10^{-24} \text{ g}$$

or

$$1 \text{ u} \simeq 1.66 \times 10^{-27} \text{ kg}. \tag{1.75}$$

Now let us use $E = mc^2$ to find a relation between the atomic mass unit u and the energy unit MeV:

$$\begin{aligned} E = mc^2 &= 1 \text{ u} \times c^2 \\ &= \frac{(1.66043 \times 10^{-27} \text{ kg}) \times (3.0 \times 10^8 \text{ m/sec})^2}{(1.60 \times 10^{-19} \text{ joules/eV})(10^6 \text{ eV/MeV})} \\ &= 931.478 \text{ MeV}. \end{aligned}$$

Therefore

$$1 \text{ u} = 931.478 \text{ MeV}/c^2 \tag{1.76}$$

Usually we say that 1 u is equal to 931.5 MeV, without mentioning c^2.

Here are the rest masses, in units of energy, of a few of the particles we'll be dealing with.

Rest mass of a proton = m_p = 1.00759 u = 938.256 MeV.
Rest mass of a neutron = m_n = 1.00898 u = 939.550 MeV.
Rest mass of an electron = m_e = 0.00055 u = 0.511006 MeV.
Rest mass of a deuteron = m_d = 2.01410 u = 1875.580 MeV.

Even a small amount of mass, if completely converted into energy, produces a tremendous amount of energy. For example, 1 g of any material produces

$$\begin{aligned} E = mc^2 &= 0.001 \text{ kg } (3 \times 10^8 \text{ m/sec})^2 \\ &= 9 \times 10^{13} \text{ joules} \\ &= 25{,}000 \text{ megawatt hours.} \end{aligned}$$

Example 1.6. An electron is accelerated from rest through a potential difference $\Delta V = 10^7$ volts. Find (a) its kinetic energy, (b) its total relativistic energy, (c) its mass, (d) its speed, and (e) its momentum.

a) Kinetic energy $=$ work done $= K_f - K_i$

$$= K_f - 0 = K_f = e\,\Delta V = e \times 10^7 \text{ V} = 10 \text{ MeV}.$$

b) $E = K + m_0c^2 = 10 \text{ MeV} + \frac{1}{2} \text{ MeV} = 10.5 \text{ MeV}.$

c) From Eq. (1.62),

$$K = mc^2 - m_0c^2 = (\gamma - 1)m_0c^2.$$

Therefore

$$\gamma = \frac{K}{m_0c^2} + 1 = \frac{10}{\frac{1}{2}} + 1 = 20 + 1 = 21.$$

Since the rest mass of the electron is $m_0 = 9.1 \times 10^{-31}$ kg,

$$m = \gamma m_0 = 21 \times 9.1 \times 10^{-31} \text{ kg} = 1.91 \times 10^{-29} \text{ kg}.$$

d) $$\gamma = \frac{1}{\sqrt{1 - \beta^2}} = \frac{1}{\sqrt{1 - (v^2/c^2)}} = 21$$

or

$$\sqrt{1 - \frac{v^2}{c^2}} = \frac{1}{21}, \qquad \frac{v^2}{c^2} = 1 - \left(\frac{1}{21}\right)^2 = 1 - \frac{1}{441}.$$

$$v = 0.9983c.$$

e) $p = mv = \gamma m_0 v = 21 \times 9.1 \times 10^{-31} \text{ kg} \times 0.9983 \times 3 \times 10^8$ m/sec, or we could say

$$p = \sqrt{\frac{E^2 - E_0^2}{c^2}} = \frac{\sqrt{E^2 - E_0^2}}{c} = \frac{\sqrt{(10.5)^2 - (0.5)^2}}{c} \text{ MeV} = 10.485 \frac{\text{MeV}}{c}.$$

Or, assuming an extreme relativistic case, we would have

$$p \simeq \frac{E}{c} = 10.5 \frac{\text{MeV}}{c}.$$

The equivalence of mass and energy, and the law of conservation of mass–energy, can be well illustrated by their application to a so-called bound system.

Bound system. *A bound system is one in which the rest mass of a combined system is less than the sum of the separate masses that comprise it.*

In a bound system, the potential energy is negative, and hence corresponds to some sort of attractive force which holds the masses in the system together. For example, consider two masses m_1 and m_2 which are brought together to form a mass M, which will hold together only if M is less than $(m_1 + m_2)$. That is,

$$m_1 + m_2 = M + \frac{E_b}{c^2}\text{(released)}, \tag{1.77}$$

where E_b is the amount of energy which the system releases due to the decrease in the rest-mass energies. In other words, the amount of energy

$$\boxed{E_b = (m_1 + m_2 - M)c^2 = \Delta mc^2} \tag{1.78}$$

is released during the formation of the bound system, as shown in Fig. 1.12(a). This energy E_b is called the *binding energy* because it is responsible for holding the parts of the system together. Thus, if we supply energy E_b equal to the binding energy, we can break apart system M into m_1 and m_2:

$$M + \frac{E_b\text{ (added)}}{c^2} = m_1 + m_2, \tag{1.79}$$

as shown in Fig. 1.12(b).

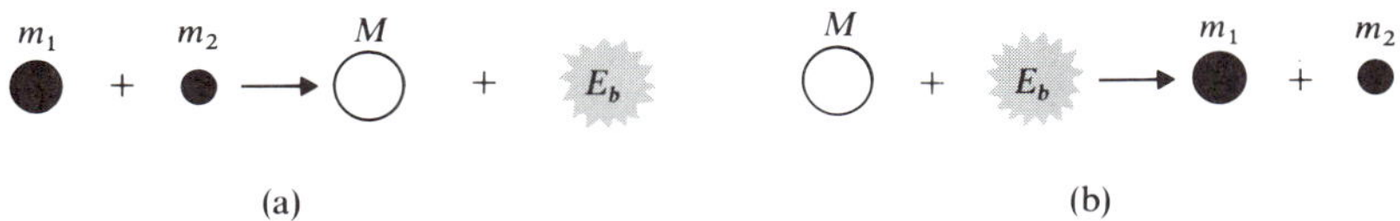

Fig. 1.12 (a) Two particles of masses m_1 and m_2 combine to form a stable bound system of mass M, provided there is simultaneous release of energy E_b. (b) If energy E_b, equal to the binding energy, is absorbed by M, it leads to the separation of M into its constituents m_1 and m_2.

The relation expressed by Eq. (1.78) or (1.79) is one of the most important consequences of Einstein's mass–energy relation. It is used constantly in atomic and nuclear physics, where it has been tested to an accuracy of better than 0.1%. The use of this relation is demonstrated in the following example.

Example 1.7. A deuteron, ${}^2_1H$, consisting of a proton and a neutron, is an isotope of hydrogen. It is the simplest of the stable bound nuclear systems. The rest masses of a deuteron, a proton, and a neutron are 2.01410 u (= 1875.58 MeV), 1.007825 u (= 938.26 MeV), and 1.008665 u (= 939.55 MeV), respectively. Thus the sum of the masses of the proton and the neutron is 938.26 + 939.55 = 1877.81 MeV. From Eq. (1.78), we may write

$$\begin{aligned} E_b &= (m_p + m_n - M_d)c^2 \\ &= (1877.81 - 1875.58) = 2.23 \text{ MeV}, \end{aligned}$$

which means that when a deuteron is formed by combining a proton and a neutron, an amount of energy equal to 2.23 MeV is released. We say that 2.23 MeV is the binding energy of a deuteron.

Now suppose we wanted to break a deuteron into a proton and a neutron. We would have to add 2.23 MeV to the deuteron:

$$M_d + \frac{2.23 \text{ MeV}}{c^2} = m_p + m_n.$$

We can supply the amount of energy 2.23 MeV by hitting the deuteron with 2.23-MeV gamma rays. [Actually slightly more than this amount of energy is needed, to account for the recoil energy.] This process is called the *photodisintegration* of the deuteron, and has been accomplished experimentally.

1.9 EXPERIMENTAL VERIFICATION OF THE RELATIVITY THEORY

Scientists have tested the special theory of relativity over and over again by applying it to many different situations. Because of the limitations of this book, we shall discuss only the following: (a) variation of mass with velocity, (b) length contraction and time dilation in meson decay.

a) Variation of mass with velocity

Consider a particle of rest mass m_0 and charge q, moving with velocity $\mathbf{v}$ in electric and magnetic fields $\mathbf{E}$ and $\mathbf{B}$. The force acting on the particle (the Lorentz force) is given by

$$\mathbf{F} = q\mathbf{E} + q\mathbf{v} \times \mathbf{B}. \tag{1.80}$$

If the assumption that the force is equal to the time rate of change of momentum is correct, the motion of this particle in relativity should be described by the following equation:

$$q\mathbf{E} + q\mathbf{v} \times \mathbf{B} = \frac{d}{dt}(m\mathbf{v}) = m_0 \frac{d}{dt}\left(\frac{\mathbf{v}}{\sqrt{1 - (v^2/c^2)}}\right). \tag{1.81}$$

As a special case, suppose that $\mathbf{E} = 0$ and that the particle with velocity $\mathbf{v}$ moves in a plane at right angles to $\mathbf{B}$. The magnetic force acting on the charge q is

$$F_{\text{mag}} = |q\mathbf{v} \times \mathbf{B}| = qvB. \tag{1.82}$$

This provides the centripetal force (Fig. 1.13) which keeps the particle moving in a circle:

$$F_{\text{cent}} = \frac{mv^2}{r}, \tag{1.83}$$

where r is the radius of the circle and m is the moving mass. We can equate the two forces

$$qvB = \frac{mv^2}{r}, \tag{1.84}$$

which may be written as

$$qBr = mv = m_0\gamma v = p \tag{1.85}$$

or

$$\frac{q}{m_0} = \frac{\gamma v}{rB} = \frac{v}{rB\sqrt{1 - (v^2/c^2)}}. \tag{1.86}$$

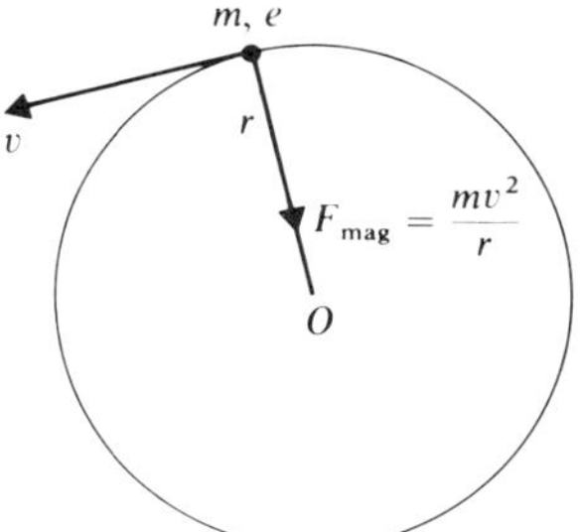

Fig. 1.13 Path of a charged particle moving perpendicular to a magnetic field B is a circle of radius r such that $F_{\text{mag}} = qvB = mv^2/r$.

In 1909, in order to experimentally confirm the variation of mass with velocity (which automatically verifies the form of relativistic momentum given by Eq. 1.85), a German physicist named Bucherer[9] performed an experiment in which he used the expression in Eq. (1.86). He obtained a beam of fast-moving electrons from a radioactive source, and passed it through a combination of electric and magnetic fields, which acted as a velocity selector. The emerging beam entered a magnetic field in which it bent into an arc of a circle of radius r, and was received on a photographic film.

For different values of v, r is different. When we know e, v, and B, and measure r, we can calculate the value of e/m_0 (since $q = e$) from Eq. (1.86). The results of Bucherer's experiment are shown in Table 1.2, in which we can see plainly that it is e/m_0 that is constant and not e/m; this verifies the relation $m = m_0/\sqrt{1 - (v^2/c^2)}$.

Table 1.2

Results of Bucherer's Experiments in Verifying $m = \gamma m_0$

$\dfrac{v}{c}$	Measured value of $e/m = v/rB$ (coul/kg)	Computed value of e/m_0 $(= e\gamma/m = e/m\sqrt{1-(v^2/c^2)})$ (coul/kg)
0.3173	1.661×10^{11}	1.752×10^{11}
0.3787	1.630×10^{11}	1.761×10^{11}
0.4281	1.590×10^{11}	1.760×10^{11}
0.5154	1.511×10^{11}	1.763×10^{11}
0.6870	1.283×10^{11}	1.767×10^{11}

Many more experiments have been performed since 1909, using different particles; some of these results are shown in Fig. 1.14. The formula $m = \gamma m_0$ has been tested with an accuracy better than 0.1%.

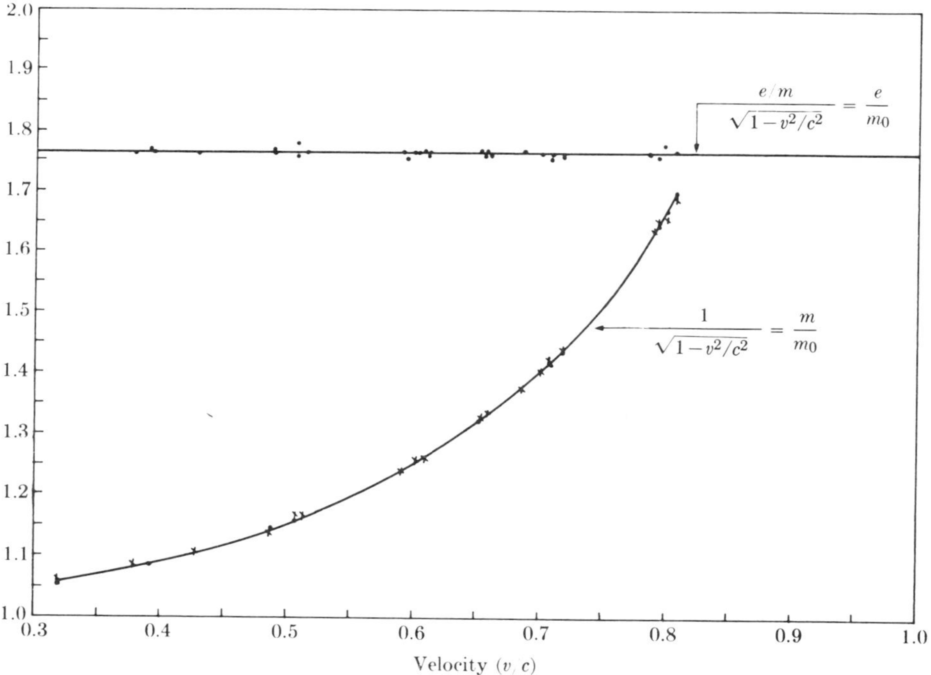

Fig. 1.14 The experimental points (the dots) obtained by Bucherer and others verify the variation of mass with velocity. The solid curves are theoretical plots of m/m_0 and e/m_0 versus (v/c).

b) Length contraction and time dilation in meson decay

The rest mass of a positive or negative pi meson (π^+ or π^-) is 279 times the rest mass of an electron. Mesons are, however, unstable, and half of them disintegrate (or decay) in 1.77×10^{-8} sec, which is called the *half-life*, $T_{1/2}$, of the decaying mesons. Thus if we have N mesons, after 1.77×10^{-8} sec, there will be only $N/2$ left; after $2 \times 1.77 \times 10^{-8}$ sec, there will be only $\frac{1}{2}(N/2) = N/4$ left; and after $3 \times 1.77 \times 10^{-8}$ sec, only $\frac{1}{2}(\frac{1}{2})(N/2) = N/8$ left.

These mesons, when produced in the laboratory by bombarding a beryllium target by a beam of very fast-moving protons, have a velocity of

$$v = 0.99c = 2.97 \times 10^8 \text{ m/sec}.$$

In a time equal to one half-life, the beam of mesons should travel a distance $x = vt = 2.97 \times 10^8 \text{ m/sec} \times 1.77 \times 10^{-8} \text{ sec} = 5.25$ m before the intensity of the beam is reduced to one-half. Values that have been experimentally measured in laboratories show that a beam of mesons travels a distance of 39 m before the intensity of the beam is reduced to half the value it originally had at the beryllium target. The discrepancy is easily explained if we make use of the proper frames of reference in our calculations.

i) *Pi meson frame of reference* (*length contraction*). The half-life of the pi meson, $T_{1/2} = 1.77 \times 10^{-8}$ sec, is the amount of life which the meson has left to it when it is at rest, i.e., as measured in the frame of reference of the pi meson. When the meson moves with a velocity $v = 0.99c$, it is the same thing as if the laboratory were moving with a velocity of $0.99c$ with respect to the meson at rest. The distance of 39 meters traveled by the meson in the laboratory frame of reference, when measured by the meson in its own frame of reference, is

$$L_{\text{mes}} = L_{\text{lab}}/\gamma = \sqrt{1 - (v^2/c^2)}L_{\text{lab}} = \sqrt{1 - (0.99)^2} \times 39 \text{ m} = 5.3 \text{ m},$$

which agrees with the measured value in the frame of reference of the pi meson.

ii) *Laboratory frame of reference* (*time dilation*). An alternative viewpoint is to assume that the meson does travel 39 m in the laboratory frame of reference, but that the half-life 1.77×10^{-8} sec measured by an observer in the meson's frame of reference looks dilated to an observer in the laboratory frame of reference. That is,

$$\Delta t_{\text{lab}} = \gamma\, \Delta t_{\text{mes}} = \frac{T_{1/2}}{\sqrt{1 - (v^2/c^2)}}$$

$$= \frac{1.77 \times 10^{-8} \text{ sec}}{\sqrt{1 - (0.99)^2}} = 1.3 \times 10^{-7} \text{ sec}.$$

Thus a meson traveling with a velocity $v = 0.99c$ in time Δt_{lab} will travel a distance of

$$L_{\text{lab}} = 0.99 \times 3 \times 10^8 \text{ m/sec} \times 1.3 \times 10^{-7} \text{ sec} = 39 \text{ m},$$

which agrees with the experimentally measured value in the frame of reference of the laboratory.

Other experimental proofs relating to the special theory of relativity include verification of the mass–energy relation, the relativistic Doppler effect, the addition of velocities, and many others.

*1.10 INVARIANT QUANTITIES; ENERGY–MOMENTUM TRANSFORMATIONS

In classical mechanics the distance r of a point (x, y, z) from the origin in the inertial system S is given by

$$r^2 = x^2 + y^2 + z^2.$$

The distance r' of a point from the origin in any other inertial system S' moving with a uniform velocity with respect to S is

$$r'^2 = x'^2 + y'^2 + z'^2.$$

The assumption of classical mechanics is that this distance is invariant. That is,

$$r'^2 = r^2 \qquad \text{if} \qquad v \ll c,$$

and the origins of S and S' coincide at $t = t' = 0$. Or we may generalize, and state that the distance between any two points in three-dimensional space is independent of the relative velocity of the observer, provided that $v \ll c$. Thus

$$\text{Space invariant means } \Delta r^2 = \Delta r'^2. \tag{1.87}$$

Another invariant quantity as a result of Newtonian mechanics is the time interval between any two events. That is,

$$\text{Time invariant means } \Delta t = \Delta t'. \tag{1.88}$$

Equation (1.88) means that the time interval between any two events is the same, regardless of the frame of reference of the observer.

From Eqs. (1.87) and (1.88), we may conclude that, in Newtonian mechanics, space and time are two independent quantities and are separately invariant.

The theory of relativity is based on the experimentally proved fact: *The speed of light is constant, regardless of the relative motion of the source or the observer.* We may state this differently by saying that *the speed of light is a scalar invariant quantity.* As we saw, in relativity, unlike the case in classical mechanics, space and time are interdependent, and are related by Lorentz transformation equations which use four-dimensional space with three space coordinates x, y, z, and the fourth imaginary coordinate (the time coordinate) ict (where $i = \sqrt{-1}$). [Note that, for consistency, the time coordinate ict is used instead of t.] Though it is not a real

* For explanation of starred sections, see paragraphs on Use of This Text at the beginning of the book.

space, all the formal properties of three-dimensional space may be extended to four-dimensional space. Thus the position of a point in this four-dimensional space is expressed by a *space-time four-vector* **s** whose components are x, y, z, and ict, and its length is given by

$$s^2 = x^2 + y^2 + z^2 + (ict)^2. \tag{1.89}$$

This length is an invariant quantity in relativity. That is,

$$\begin{aligned} s^2 &= x^2 + y^2 + z^2 + (ict)^2 \\ &= x'^2 + y'^2 + z'^2 + (ict')^2 = s'^2, \end{aligned} \tag{1.90}$$

where $\mathbf{s}'$ is the same space-time four-vector measured in the inertial system S' which is moving relative to the inertial system S.

In general we may state that, in relativity, *an interval between any two events in four-dimensional space is invariant.*

That is:

$$\text{Space time invariance means } \Delta s^2 = \Delta s'^2 \tag{1.91}$$

or

$$\begin{aligned} &(x_2 - x_1)^2 + (y_2 - y_1)^2 + (z_2 - z_1)^2 + (ict_2 - ict_1)^2 \\ &\qquad = (x_2' - x_1')^2 + (y_2' - y_1')^2 + (z_2' - z_1')^2 + (ict_2' - ict_1')^2. \end{aligned}$$

Thus the two separate space and time invariants of Newtonian mechanics are replaced, in relativity theory, by one: Eq. (1.91).

In addition to the space–time four-vector, there is a momentum–energy four-vector in relativity theory that is also invariant. Starting with Eq. (1.69), that is,

$$E^2 = p^2c^2 + E_0^2,$$

where $p^2 = p_x^2 + p_y^2 + p_z^2$, we may write it as

$$-\frac{E_0^2}{c^2} = p^2 - \frac{E^2}{c^2}$$

or

$$\left(\frac{iE_0}{c}\right)^2 = p_x^2 + p_y^2 + p_z^2 + \left(\frac{iE}{c}\right)^2. \tag{1.92}$$

Since m_0 is constant, $E_0 = m_0c^2$ is also constant, and so is $E^2 - p^2c^2$. From Eq. (1.92), it is obvious that the three components of momentum, p_x, p_y, p_z, together with a fourth (energy) component (iE/c) form an invariant quantity. We may say that the dynamical state of a particle is described by a single four-vector

$(p_x, p_y, p_z, iE/c)$. The assumption of relativistic dynamics is that the momentum–energy four-vector in any inertial system is the same invariant quantity. That is:

Momentum–energy invariance means

$$(iE_0/c)^2 = p_x^2 + p_y^2 + p_z^2 + (iE/c)^2 = p_x'^2 + p_y'^2 + p_z'^2 + (iE'/c)^2. \tag{1.93}$$

Actually the analogy between the space–time four-vector and the momentum–energy four-vector is so close that we can write the momentum–energy transformation equations by simply replacing x, y, z and ict by p_x, p_y, p_z, and iE/c, respectively, in Eq. (1.31) for Lorentz coordinate transformations. Thus the *Lorentz momentum–energy transformation equations are*:

$$\begin{aligned} p_x' &= \gamma\left(p_x - \frac{vE}{c^2}\right) & \quad & p_x = \gamma\left(p_x' + \frac{vE'}{c^2}\right) \\ p_y' &= p_y & \text{and} \quad & p_y = p_y' \\ p_z' &= p_z & & p_z = p_z' \\ E' &= \gamma(E - p_x v) & & E = \gamma(E' + p_x' v) \end{aligned} \tag{1.94}$$

where $\gamma = 1/\sqrt{1 - (v^2/c^2)}$ and v is the velocity of S' with respect to S along the XX' axes.

The momentum–energy transformation equations are very useful for studying problems involving the collisions of particles in nuclear and high-energy physics.

1.11 THE TWIN PARADOX[10]

The so-called *twin paradox* or *clock paradox* is one of the topics in relativity that has been discussed widely in the last decade. The paradox may be stated as follows.

Two clocks, R and M, are synchronized to start with. Both clocks are at rest. All of a sudden, clock M starts moving with uniform velocity, makes a round trip to a nearby planet in outer space, and then returns to R. According to the theory of time dilation, the moving clock M loses time compared to the stationary clock R, which is the same thing as saying that the moving clock has been running slow.

Now let's use everyday language to state the problem another way: Suppose that a very dedicated astronaut and his wife have a pair of twins, Robert and Mike. After much discussion, they decide that Robert will stay at home, while Mike will be placed in a space ship (moving with uniform velocity v comparable to c), make a trip to a planet P, and then return to earth. Exit Mike. We lower the curtain to denote the passage of x years. Finally—hurrah!—Mike returns home. And there is poor old Robert, white-haired and bent and wrinkled. Mike, who still looks young and vigorous, can hardly believe that this old man is his twin brother!

Let us say that during these x years Robert has aged T_R and Mike has aged T_M. Then according to the theory of time dilation

$$T_M = T_R\sqrt{1 - v^2/c^2}.$$

That is,

$$T_R = T_M/\sqrt{1 - v^2/c^2} \qquad \text{or} \qquad T_M < T_R. \tag{1.95}$$

But, according to the special theory of relativity, the motion of Mike in his space ship is relative, and hence the effect ought to be reciprocal. Thus *the paradox would be that Mike would expect Robert to be younger, while Robert would expect Mike to be younger*, when they met again. We can overcome the difficulty if we realize that Mike does not move with uniform relative velocity throughout his trip. He must accelerate or decelerate at different stages: when he takes off, when he turns around at the planet P for the homeward journey, and finally when he slows down while he is approaching earth again. Thus the fallacy is that Mike does not remain in the same inertial frame throughout his trip; hence the aging effect is not reciprocal.

Is it really true that the space-traveling twin Mike ages less than the homebody twin Robert? Well, look at it this way: There is no difference between physical clocks and biological clocks. Suppose we take the heartbeat to be a clock. Then Mike's heartbeat will be slower than Robert's heartbeat. So the answer is yes!

1.12 TACHYONS[11,12,13]

As we have said, Einstein's Special Theory of Relativity, since it was first proposed in 1905, has been experimentally tested many times and is well established. One of the implications of the theory has been that nothing, not even a tiny particle, can travel faster than the speed of light, because it would take an infinite amount of energy to get the particle accelerated up to the speed of light. But this has not stopped scientists from making serious suggestions to the effect that particles can indeed travel faster than light. They say that there's no reason why a particle cannot exist which is already traveling faster than light (so that we needn't supply infinite energy to speed it up). Such particles are called *tachyons*, from the Greek word *tachys*, meaning "swift."

Scientists assume that the rest mass of a tachyon is imaginary, while the momentum and energy are real. They have to assume this, because rest mass is not directly observable, while momentum and energy are. So if we wish to grant that tachyons exist, we have to go along with this.

Another interesting property of a tachyon is that, as it loses energy, it speeds up, until it is traveling infinitely fast. When it gets to this point, it has no energy at all. Figure 1.15 shows the change in momentum and energy with velocity of a

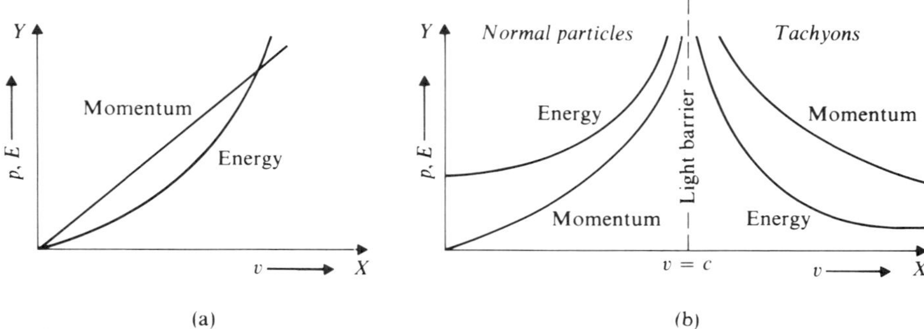

Fig. 1.15 (a) The plots of p and E versus v for $v \ll c$, for which Newtonian mechanics is applicable. (b) The plots of p and E versus v for normal particles for which $v < c$ and special relativity is applicable, and for tachyons, for which $v > c$. Note that, for tachyons, as energy decreases, v increases.

freely moving particle. In (a), $v \ll c$, and the curves conform to Newtonian mechanics. In (b), due to the special theory of relativity, as v increases, momentum and energy increase and approach—but never reach—the asymptote, where $v = c$. Normal particles (for which $v < c$) lie to the left of the dashed line $v = c$, while tachyons (for which $v > c$) lie to the right.

Physicists have tried to produce tachyons in the laboratory, but without success. At Princeton University, for example, T. Alvager and M. N. Kreisler[12] surrounded with some material a source of gamma rays (gamma rays are electromagnetic waves of very high frequencies) such as cesium-134. They expected to create a pair of equally and oppositely charged tachyons, T_+ and T_-. However, since a tachyon loses all its energy in a distance of one thousandth of a centimeter, they were not successful in detecting tachyons.

There are some indications that Einstein's Special Theory of Relativity may break down in certain regions of outer space, such as inside a quasar, in which matter is in a highly compressed state. This leads to evidence of the existence of tachyons in the interior of quasars, but this has not been established as yet.

If in the future tachyons *are* detected and their existence established, this will lead to modifications of present-day theories of physics.

*1.13 GENERAL RELATIVITY AND THE EQUIVALENCE PRINCIPLE

Not all the difficulties encountered in classical mechanics are satisfactorily answered by the special theory of relativity. Let us consider two problems that are not.

1) Newton's law of gravitation states that

$$F_g = m_g \frac{GM}{r_e^2},$$

where m_g is the mass of an object, M is the mass of the earth, r_e is the distance between the center of the earth and the mass m_g, and G is the universal gravitational constant. This law implies action-at-a-distance, which means that the force between the two masses is conveyed with an infinite velocity. This contradicts the special theory of relativity, in which nothing moves with a velocity greater than the velocity of light. Also, according to Newton, the inertial mass m_i given by $F_i = m_i a$ is completely identical to the gravitational mass m_g.

2) It is a basic assumption of the special theory of relativity that all inertial frames are equivalent if they are moving with uniform velocities relative to one another. This assumption certainly excludes all accelerated (or noninertial) frames of reference; and in a sense it makes inertial frames unique. Hence it contradicts the basic assumption—that there is no absolute reference system—on which the special theory of relativity is based.

In order to overcome these difficulties, Einstein,[2] in 1907, proposed the *principle of equivalence*, and later, in 1916, developed the *general theory of relativity*. Let us now discuss these briefly.

Newton showed that if there is any difference between the inertial mass and the gravitational mass of an object, it is less than one part in 1000. Bessels calculated the time period of a pendulum by the expression

$$T = 2\pi \sqrt{\frac{l}{g}\frac{m_i}{m_g}} \tag{1.96}$$

instead of assuming that $m_i = m_g$ and using the expression

$$T = 2\pi \sqrt{\frac{l}{g}}, \tag{1.97}$$

where l is the length of the pendulum. Thus he showed the equivalence between the inertial and gravitational masses. If there was any difference between the two, it was less than one part in 6×10^4. R. V. Eötvös,[14] in 1922, performed even more precise experiments which involved measuring the direction of the freely hanging pendulum and hence the resultant between the forces F_i and F_g. These experiments showed the equivalence between the two kinds of masses, and proved that any difference is less than one part in 2×10^8. Using modern techniques, R. H. Dicke[15] performed the same types of experiments, and showed that the difference between m_i and m_g is less than one part in 10^{10}.

The equivalence of the inertial and the gravitational mass follows from the equivalence principle, which may be stated as follows.

Equivalence principle: *The effects produced by the gravitational field are identical to those produced by acceleration and are completely undistinguishable, i.e., one is equivalent to the other.*

For example, an astronaut sitting in his space ship on earth (i.e., in an inertial frame) feels the force $F = mg$, where M is the mass of the astronaut. Then suppose he leaves the earth (gets out of earth's gravitational field), and moves in outer space with an acceleration (noninertial frame) $a = -g$. He still feels the same force, and obtains the same results for any experiment, whether he performs it while at rest on earth or in an accelerated motion in outer space.

As another example, take a satellite going around the earth in an orbit. Inside the satellite there is weightlessness, because the gravitational force has been transformed by the acceleration $a = \omega^2 r = GM/r^2$, where r is the distance from the center of the earth to the satellite.

These considerations led Einstein to state that *there is no absolute accelerated frame of reference*, just as in special relativity there is no absolute velocity frame. The fact that all accelerated motion is relative led Einstein to form the *general theory of relativity*, which not only takes into consideration the equivalence principle, but also extends it to nonuniform (inhomogeneous) gravitational fields.

The purpose of the mathematical formulation of the general theory of relativity is twofold. (1) It keeps the laws of physics invariant. (2) It leads to the formulation of a theory of gravitation in which gravitational effects propagate with the speed of light.

The following are some of the most interesting and intricate experiments which have been used to test the above principles.

a) The gravitational red shift

b) Deflection of light in a gravitational field

c) Precession of the perihelion of mercury

The first two may be explained by making use of only the equivalence principle, while the third uses the general theory of relativity. We cannot do justice to these experiments in this text, but the interested reader should look into advanced books under the heading Suggested Readings at the end of this chapter.

SUMMARY

Galilean transformations are good for mechanics (only if $v \ll c$), but not for electromagnetism.

The stationary-ether hypothesis explains the stellar aberration, but fails to explain the negative results of the Michelson–Morley experiment.

These difficulties were overcome with the development of the Special Theory of Relativity by Albert Einstein, in 1905. He introduced two postulates: the principle of equivalence and the principle of the constancy of the speed of light. This resulted in the following transformation equations:

$$\text{Lorentz coordinate transformations}\begin{cases} x' = \gamma(x - vt) \\ y' = y \\ z' = z \\ t' = \gamma\left(t - \dfrac{\beta}{c}x\right) \end{cases}$$

where $\beta = v/c$ and $\gamma = 1/\sqrt{1 - \beta^2}$.

$$\text{Lorentz velocity transformations}\begin{cases} u'_x = \dfrac{u_x - v}{1 - (vu_x/c^2)} \\ u'_y = \dfrac{u_y}{\gamma[1 - (vu_x/c^2)]} \\ u'_z = \dfrac{u_z}{\gamma[1 - (vu_x/c^2)]} \end{cases}$$

The relationships between classical and relativistic forms are as follows.

Quantity	*Classical form*	*Relativistic form*
Length	$L = L_0$	$L = L_0/\gamma = L_0\sqrt{1 - \beta^2}$
Time	$T = T_0$	$T = \gamma T_0 = T_0/\sqrt{1 - \beta^2}$
Mass	$m = m_0$	$m = \gamma m_0 = \dfrac{m_0}{\sqrt{1 - \beta^2}}$
Momentum	$p = m_0 v$	$p = \gamma m_0 v = mv$
Newton's law	$F = m_0 \dfrac{dv}{dt}$	$F = \dfrac{d}{dt}(mv)$
Kinetic energy	$K = \frac{1}{2}m_0 v^2$	$K = mc^2 - m_0c^2 = (\gamma - 1)m_0c^2$
Rest energy	$E_0 = 0$	$E_0 = m_0c^2$
Total energy	$E = K$	$E = mc^2 = E_0 + K = \gamma E_0$

The relation between E_0, E, and p is

$$E^2 = p^2c^2 + E_0^2 = p^2c^2 + m_0^2c^4.$$

The binding energy of a system is

$$E_b = (m_1 + m_2 - M)c^2 = \Delta mc^2.$$

★An interval between any two events in four-dimensional space is invariant. The momentum–energy four-vector is also invariant.

★Momentum–energy transformation equations are as follows.

$$p'_x = \gamma\left(p_x - \frac{vE}{c^2}\right)$$

$$p'_y = p_y$$

$$p'_z = p_z$$

$$E' = \gamma(E - p_x v)$$

PROBLEMS

1. An inertial system S' is moving with respect to another inertial system S along the XX' axes with a velocity of 50 ft/sec. The origins of the two coincide at $t = t' = 0$. A particle in the system S is described by the equation

$$x = a + bt + ct^2,$$

where a, b, and c are constants, x is in feet, and t is in seconds. Using Galilean transformations, describe the position, velocity, and acceleration of this particle as viewed by an observer in the system S'.

2. Show the invariance of the conservation of angular momentum under Galilean transformations.

3. A river of width L flows with a velocity $\mathbf{v}$ as shown in the figure. Two swimmers A and B can swim with a velocity c relative to the river bank. Swimmer A makes the round trip PQP, where $PQ = L$, in time t_1. Swimmer B makes the round trip PRP, where $PR = L$, in time t_2. Calculate t_1 and t_2 in terms of L, c, and v. After we have measured t_1 and t_2, how do we calculate v from the known values of c and L?

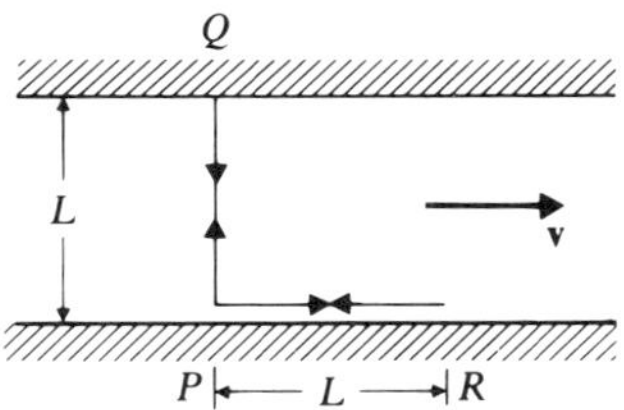

4. Given that the coordinates (x, y, z, t) of an event in system S are $(10^5, 10^5, 0, 10^{-3})$. What are the coordinates of this event in a system S' which is moving with a velocity $\frac{3}{4}c$ along the XX' axes?

5. A rod 1 m in length is lying along the X-axis in the inertial system S. This length is being measured by an observer in a system S' which is moving along the XX' axes with a velocity v. What should the value of v be so that to an observer in S' the length of the rod looks to be (a) 0.99 m, (b) 0.50 m?

6. A space ship is traveling at a speed of $\frac{3}{5}c$ toward the moon. What is the ratio of the diameter of the moon as measured by an observer in the space ship to the diameter of the moon measured by another observer on earth?

7. A rocket is moving radially away from the earth with a velocity of $v = (\sqrt{3}/2)c$. An astronaut in the rocket measures the length of an external antenna on the rocket. He finds that it is 10 ft long and makes an angle of 45° with the rocket. What is the length and inclination of the antenna as calculated by an observer on earth?

8. A rod 1 m long in system S makes an angle of 30° with the X-axis. What should be the velocity of an observer in system S' moving along the XX' axes so that to him the rod looks to be making an angle of 45°?

9. A cube is moving with a velocity v in the direction parallel to one of its edges. Show that the volume V of the cube and its density ρ are given by

$$V = V_0\sqrt{1 - \beta^2} \qquad \text{and} \qquad \rho = \gamma^2 m_0/V_0,$$

where m_0 and V_0 are the rest mass and rest volume of the cube, respectively.

10. An observer in a superjet moving with a speed of $0.80c$ sees a man on earth run a race which lasts 10 seconds. For how long does the runner himself think that he ran in the race? [Note that $\Delta t'$ is given, and that we want to find Δt.]

11. A boy runs a 100-m dash in 10 sec as measured by an observer on earth. How long does it take according to an observer in a space ship moving with a speed of $0.98c$? What is the length of the track according to the moving observer?

12. Two observers S and S' synchronize their clocks; that is, $t = t' = 0$ at $x = x' = 0$. If an event in system S' lasted for 30 sec according to the observer in the S system, how long did it last according to the observer in the system S'?

13. The time period T_0 of a pendulum is 2 sec; i.e., it takes 2 sec to complete one cycle. What is the time period of this pendulum when measured by an observer whose inertial system is moving with a speed of $0.9c$ with respect to the inertial system of the pendulum?

14. According to an observer on earth, a laser pulse starting from earth reaches the moon in 1.21 sec. How much time will this pulse take to reach the moon according to an observer moving in the same direction as the pulse with a speed (a) $0.1c$, (b) $0.5c$, (c) $0.99c$?

15. A space ship is traveling at a speed of 3×10^7 m/sec with respect to the ground. An astronaut in this space ship measures his heartbeat as one per second. With each heartbeat, a light signal is emitted from the space ship, and is observed by the observers in the ground control station. What is the interval between heartbeats according to the persons watching in the ground control station?

16. A particle called a mu meson (μ meson) carries a charge of one electron, although it has a rest mass of about 207 times the rest mass of an electron. Fast-moving mu mesons form a part of cosmic radiation, and can be produced in the laboratory as well. The mean life of μ mesons almost at rest in the laboratory is found to be 2.2×10^{-6} sec, while the average life of fast-moving μ mesons observed in cosmic rays is 1.1×10^{-5} sec. Calculate the following.

a) The speed of the cosmic-ray μ meson

b) The distance traveled by the cosmic-ray μ meson through the atmosphere in its average lifetime according to an observer in (i) the laboratory frame of reference, and (ii) the μ-meson frame of reference.

17. Using relativistic formulas for the addition of velocities, show that (a) $u'_x = c$ if either $u_x = c$ or $v = c$, (b) $u'_x < c$ only if both $u_x < c$ and $v < c$.

18. An observer is approaching a flashing light signal with a velocity of $c/2$. What is the velocity of the light signal as measured by the approaching observer?

19. Two electrons, each moving with a velocity $0.9c$, are moving toward each other. What is the velocity of one electron with respect to the other according to (a) Galilean transformations, (b) Lorentz transformations?

20. An inertial system S' is moving with a velocity $0.9c$ relative to the inertial system along the XX' axes. A particle moving in the system S has velocity components of $u_x = 0.5c$ and $u_y = 0.5c$. Find the magnitude and direction of the velocity of the particle as measured by observers in the systems S and S', using (a) Galilean velocity transformations, (b) Lorentz velocity transformations.

21. A particle in an inertial system S is moving with a velocity of $0.8c$ making an angle of $45°$ with the X-axis. Find the magnitude and direction of the velocity of the particle as observed by an observer in an inertial system S' which is moving relative to the system with a velocity of $0.8c$ along the XX' axes.

22. Calculate the momentum and the total energy of (a) an electron, (b) a proton, each moving with a speed of $0.9c$.

23. Calculate the mass and the momentum of a 1-GeV (a) proton, (b) electron.

24. A particle has a kinetic energy equal to half its rest-mass energy. Calculate the speed and the momentum of the particle.

25. Show that even if the relativistic mass m is used in the expression for kinetic energy $K = \frac{1}{2}mv^2$, it is still not correct.

26. Calculate (a) the speed, (b) the relativistic momentum, (c) the kinetic energy, and (d) the total relativistic energy of an electron that has been accelerated from rest through a potential difference of 5×10^6 volts.

27. Calculate the amount of energy used in accelerating an electron and a proton from (a) $0.5c$ to $0.9c$, (b) $0.9c$ to $0.99c$.

28. What should be the value of v so that the use of Galilean transformation equations does not introduce an error of more than (a) 1%, (b) 5%?

29. A particle of rest mass m_0 approaches head-on another particle of mass m_0, each moving with velocity $\frac{3}{4}c$. Calculate (a) the relativistic velocity, (b) the relativistic mass of each particle with respect to the other.

30. The rate at which the earth receives energy from the sun is $\sim 1.33 \times 10^3$ watts/m^2. Calculate the yearly fractional loss of the mass of the sun in providing this energy. The mass of the sun is 2×10^{30} kg; its distance from the earth is 1.55×10^{11} m. How long do you think the sun will keep on providing this energy? [Assume a constant rate of radiation.]

31. Show that the relativistic momentum of a particle may be written as

$$p = \frac{(2E_0K + K^2)^{1/2}}{c}$$

What is the value of p for (a) $\beta \to 0$, (b) $\beta \to 1$?

32. Show that the radius of the path of a particle of charge q moving in a circle at right angles to a uniform magnetic field of flux density B is given by

$$r = \frac{(2E_0K + K^2)^{1/2}}{qcB}$$

33. Calculate the kinetic energy and the radius of the circle of the path of a proton having a momentum of 2.50×10^{-22} kg-m/sec in a magnetic field of 1 weber/m^2.

34. An electron starting from rest is accelerated through a potential difference of 2×10^6 volts and then enters a magnetic field of 2.0 webers/m^2. What is the radius of curvature (a) classically, (b) relativistically?

35. A helium atom consists of two protons, two neutrons, and two electrons. Its mass is 4.0026 u. [*Note:* The symbol u was defined in Section 1.8.] Calculate the binding energy of the atom, neglecting the small binding energy of the electrons.

★36. Show that the speed of light is an invariant quantity. To do this, use the Lorentz velocity transformations in the relation

$$u_x^2 + u_y^2 + u_z^2 = u_x'^2 + u_y'^2 + u_z'^2 = c^2.$$

37. Make plots of E versus v for an electron and a proton. How do these compare with the plots from classical mechanics?

38. Make plots of E versus p for an electron and a proton. How do these compare with the plots from classical mechanics?

REFERENCES

1. Albert Einstein, *Ann. Physik* **17,** 549, 891 (1905)
2. Albert Einstein, *Ann. Physik* **49,** 749 (1916)
3. E. R. Cohen and J. W. M. DuMond, *Rev. Mod. Phys.* **37,** 537 (1965)
4. A. Stewart, "The Discovery of Stellar Aberration," *Sci. Am.*, March 1964
5. A. A. Michelson and E. W. Morley, *Am. J. Sci.* **134,** 333 (1887)
6. R. S. Shankland, "The Michelson–Morley Experiment," *Sci. Am.*, November 1964
7. R. S. Shankland, *Am J. Phys.* **31,** 47 (1963)
8. H. Erlichson, *Am. J. Phys.* **35,** 89 (1967)
9. A. H. Bucherer, *Ann. Physik* **28,** 513 (1909)
10. J. Bronowski, "The Clock Paradox," *Sci. Am.*, February 1963
11. J. G. Taylor, "Particles Faster than Light," *Sci. Am.* **43,** September 1969
12. T. Alvager and M. N. Kriesler, *Phys. Rev.*, **171,** 1357 (1968)

13. M. N. Kriesler, *Amer. Sci.* **61,** page 201, 1973
14. R. V. Eötvös, D. Pekar, and E. Fekete, *Ann. Physik*, **LXVIII,** 11 (1922)
15. R. H. Dicke, *Am. J. Phys.* **35,** 559 (1967)

Suggested Readings

1. E. P. Ney, *Electromagnetism and Relativity*, New York: Harper and Row, 1962
2. Robert Resnick, *Introduction to Special Relativity*, New York: John Wiley, 1968
3. Atam P. Arya, *Fundamentals of Atomic Physics*, Chapters II and III; Boston, Mass: Allyn and Bacon, 1971
4. A. P. French, *Special Relativity*, New York: W. W. Norton, 1966
5. Albert Einstein, *The Meaning of Relativity*, Princeton, N.J.: Princeton University Press, 1956
6. Albert Einstein, *et al.*, *The Principle of Relativity*, New York: Dover Publications, 1923 (a collection of papers on special and general relativity)
7. Resource Letter SRT-1 on Special Theory of Relativity, *Am. J. Phys.* **30,** 462 (1962) (gives a complete reference list)

CHAPTER 2
THE CONCEPTS OF WAVES AND PARTICLES

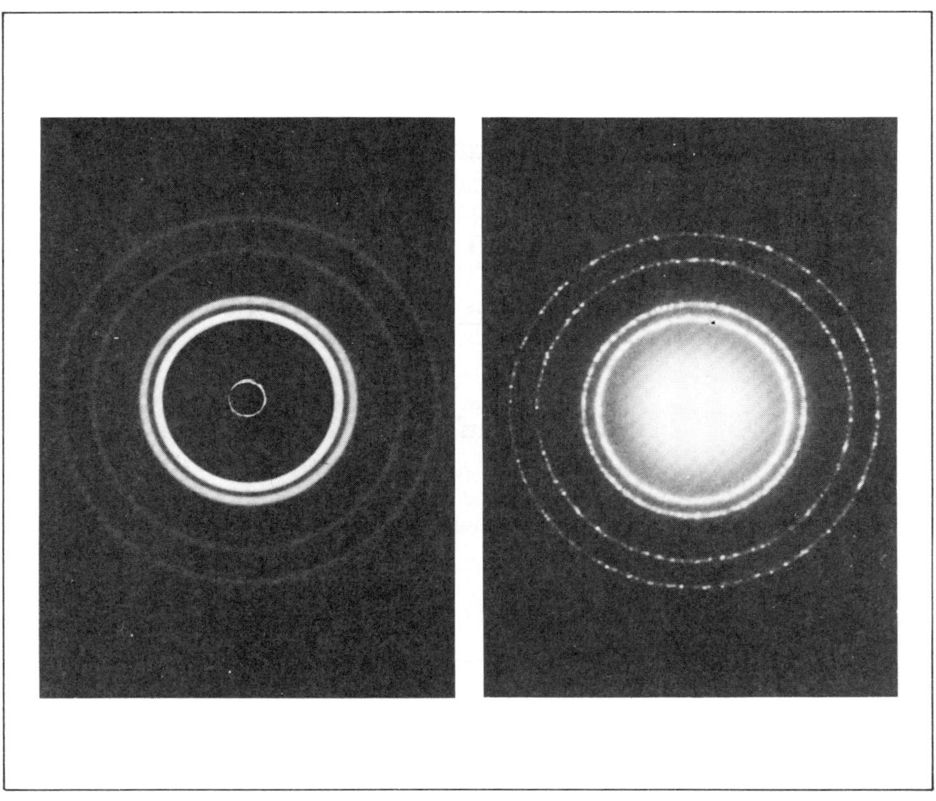

Similarity between diffraction patterns produced by 600-eV *electrons from an aluminum-foil target and x-rays of wavelength* 0.71 Å *also from an aluminum-foil target demonstrates the wave nature of electrons.* (*Courtesy of Film Studio, Education Development Center*)

2.1 QUANTIZATION OF CHARGE AND MASS

Dalton's atomic hypothesis, which he introduced in 1802, indicated that all elements were made up of a combination of different numbers of atoms; in those days people thought the atom was the smallest indivisible unit of matter. Even though we now know that the atom is not the smallest unit of mass, we also know that all matter is composed of discrete individual particles. Thus we may say that *matter is quantized.* Can the same property be attributed to electricity? In other words, does electricity come in multiples of some smallest unit of charge, or is it something continuous? The answer is that charge, like matter, is *quantized.* Sometimes the quantization of matter and of charge is called the *atomicity of matter and of charge.*

Early interest in electricity was shown by Thales of Miletus (*c.* 600 B.C.), but the first serious work was undertaken around 1600 by William Gilbert and in 1733 by Charles Dufay. Benjamin Franklin, in 1747, investigated somewhat detailed properties of electrostatics. The work that led to the concept of quantization of charge was by the English physicist Michael Faraday in his famous work on electrolysis (1831–1837), in which he established the Faraday laws of electrolysis and the quantization of charge. In 1897 J. J. Thomson[1] discovered the electron and measured its charge-to-mass ratio. Then, in 1909, R. A. Millikan[2] measured the charge on the electron by means of the Millikan oil-drop method. This confirmed the quantization of charge. The conclusion from such experiments was that the charge q on any particle or body was an integral multiple of charge e,

$$q = ne, \qquad n = 1, 2, 3, \ldots, \tag{2.1}$$

where e is the charge on the electron. Its currently accepted value is

$$e = (1.60210 \pm 0.00002) \times 10^{-19} \text{ coul.} \tag{2.2}$$

Thus it was established that both charge and mass are quantized.

2.2 QUANTIZATION OF RADIATION

In Chapter 1, we saw that the Lorentz transformation equations apply equally to mechanics and to electromagnetism. This indicates that there is some sort of close unity between the two. And just now we said that not only the charge, but the mass of any mechanical system—atoms, molecules, etc.—is also quantized, and comes in multiples of the masses of electrons, protons, and neutrons. Now: Is there any characteristic of electromagnetic radiation that also exhibits quantization properties? As a matter of fact, with respect to energy, there is a much closer relation between mechanical and electromagnetic systems. According to classical electromagnetic theory, electromagnetic energy is carried by waves. This leads us to the conclusion that energy flows continuously. But we shall demonstrate in this chapter that the flow of electromagnetic energy is not continuous, but is in the form of discrete bundles of energy. That is, we say that electromagnetic energy is quantized, and also that an electromagnetic system can exist only in certain discrete energy states. Furthermore, we shall show that even mechanical systems can exist only in discrete energy states. Thus once again we find a sort of unity between electromagnetic and mechanical systems as far as their energies are concerned. Let us briefly summarize, from the historical point of view, the circumstances that led to the quantization-of-energy hypothesis.

According to Newton's corpuscular theory (1675), light consisted of tiny particles emitted by a source. Even though Huygens had suggested the wave theory of light (1678), it was not accepted till the beginning of the nineteenth century, when Young showed (1801) that diffraction experiments could be explained only by the wave theory, not by the corpuscular theory. The wave nature of light was further confirmed by Maxwell's development of equations for the electromagnetic field and their subsequent verification by Hertz in 1887, proving that light consists of electromagnetic waves. The physicists at that time thought that the wave theory of light (or of electromagnetic radiation) was here to stay. But, to everyone's surprise, it started showing signs of weakness after only ten years.

Early in the twentieth century, many scientists made experimental observations connected with electromagnetic radiation which could not be explained by the wave theory of light. These experiments concerned the following phenomena.

a) Blackbody radiation spectrum
b) Photoelectric effect
c) X-ray spectra
d) Compton scattering
e) Optical line spectra

Thus the wave theory of light could explain interference, diffraction, and polarization, but not experiments (a) to (e). In order to explain the blackbody radiation spectrum, Max Planck[3] in 1901 introduced the quantum hypothesis, which was eventually used in explaining all the experimental facts from (a) to (e).

Planck's hypothesis (*original form*). *If a physical system executes a simple harmonic motion in one dimension with frequency* ν, *it can take only those energy values* E *which are given by the relation*

number of quantums

$$E = nh\nu, \qquad \text{where } n = 1, 2, 3, \ldots \tag{2.3}$$

and h is a constant called *Planck's constant*.

Thus Planck implied that the system can exist only in certain discrete energy states given by Eq. (2.3). Such states are called *quantum states*, while n is called the *quantum number*. Furthermore, *the oscillators emit or absorb energy in bundles of size* $h\nu$, *i.e., the radiation emitted or absorbed is also quantized.* The value of h is found to be 6.625×10^{-34} joule-sec.

Figure 2.1 shows the distinction between the classical theory and the Planck quantization hypothesis as applied to the harmonic oscillator. As shown in

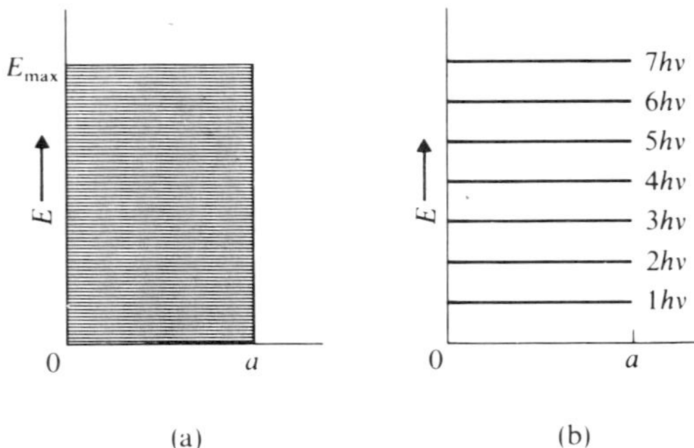

Fig. 2.1 Possible energy values E of a one-dimensional simple-harmonic oscillator: (a) According to classical theory, E can have any value between 0 and E_{max}. (b) According to Planck's quantum hypothesis, E can take only discrete energy values.

Fig. 2.1(a), according to the classical theory the oscillator may take any energy value (continuous range) between 0 and E_{max}. But Fig. 2.1(b) shows that, according to the quantum hypothesis, only discrete energy values are allowed.

Example 2.1. A radio station operates on a frequency of 98 megacycles and radiates a power of 200 kW. How many quanta of energy are emitted per second?

According to the quantum hypothesis, the quantum energy E is given by

$$\begin{aligned} E &= h\nu \\ &= (6.625 \times 10^{-34} \text{ joule-sec})(98 \times 10^{6} \text{ cycles/sec}) \\ &= 6.49 \times 10^{-26} \text{ joule/quantum.} \end{aligned}$$

$$\begin{aligned} \text{Number of quanta emitted per sec} &= \frac{\text{power}}{\text{quantum energy}} \\ &= \frac{200 \text{ kW}}{(6.49 \times 10^{-26} \text{ joule/quantum})} \\ &= \frac{200 \times 10^{3} \text{ joule/sec}}{(6.49 \times 10^{-26} \text{ joule/quantum})} \\ &= 3.09 \times 10^{30} \text{ quanta/sec.} \end{aligned}$$

★2.3 BLACKBODY RADIATION

Any object in thermal equilibrium at any temperature is constantly emitting and absorbing radiation. However, not all materials are equally capable of absorbing and emitting radiation in different parts of the spectrum (i.e., in the frequency range from zero to infinity). According to *Kirchhoff's law*, any object which is a good absorber of radiation of a particular wavelength is also a good emitter of radiation of the same wavelength. A body which absorbs radiation of all wavelengths is called a *blackbody*. A close approximation to a perfect blackbody is a cavity made out of a hollow container of any material, say iron or copper, with a narrow opening (as shown in Fig. 2.2) and with its inside painted with lampblack. Any radiation incident on this hole will enter the cavity, be reflected from the walls of the cavity as shown in Fig. 2.2, and eventually be absorbed. There is slight chance that much of the incident radiation energy will escape out of the hole. And if this container is heated to any temperature T, and we analyze the radiation coming out of the hole, it will contain radiation of all frequencies (or wavelengths).

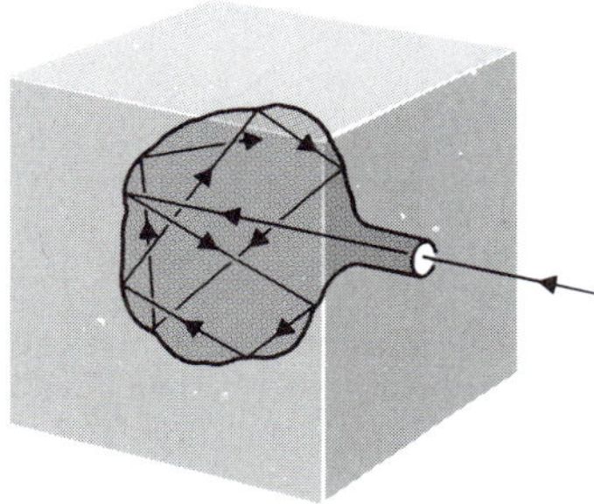

Fig. 2.2 A cavity with a narrow opening is almost equivalent to a perfect blackbody.

So, too, the radiation emitted from a blackbody has a *continuous spectrum*, i.e., it contains radiation of all frequencies (within a certain range). Let $I(\nu)\,d\nu$ be the intensity of the radiation emitted between the frequencies ν and $\nu + d\nu$. Experimental measurements result in a plot of $I(\nu)$ versus ν, as shown in Fig. 2.3, for different temperatures.

Let us note some simple characteristics from these plots: (a) The distribution of frequencies is a function of the temperature of the blackbody. (b) The total amount of radiation emitted $\int_0^\infty I(\nu)\,d\nu$ increases with increasing temperature. (c) The position of the peak maximum shifts toward higher frequencies with increasing equilibrium temperature.

Without going into any mathematical details, let us say that classical electromagnetic theory (or wave theory) together with classical thermodynamics does not explain the characteristics of the blackbody spectrum, but Planck's quantum hypothesis, together with classical thermodynamics, can.

According to classical wave theory, at an equilibrium temperature T, the electromagnetic radiation inside the cavity of the blackbody forms standing waves. The number of standing waves that can fit into a cavity of given size (i.e., the number of possible modes in the cavity) depends on the wavelength. If the wavelength is small, the number of possible modes is large; for large wavelengths the number of

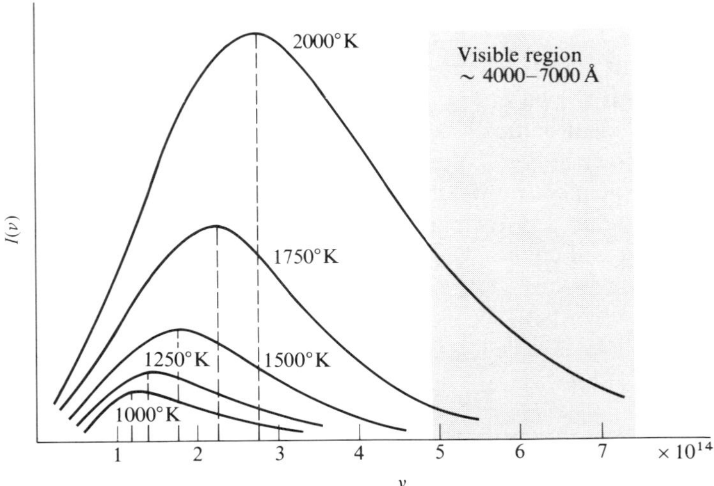

Fig. 2.3 The spectral distribution of radiation from a blackbody at different temperatures. Note that the maximum in each curve shifts to higher ν with increasing T.

possible modes is small. Detailed geometrical consideration carried out by Rayleigh and Jeans [4,5] indicate that this increase in number of modes is proportional to $1/\lambda^2$ or ν^2. Also, when the cavity is in thermal equilibrium, each of the standing waves must be assigned an average kinetic energy kT. This leads to the *Rayleigh–Jeans* law,[4,5] given by

$$I(\nu)\, d\nu = \frac{8\pi\nu^2}{c^3} kT\, d\nu. \tag{2.4}$$

That is, $I(\nu) \propto \nu^2$, and the plot of $I(\nu)$ versus ν is as shown in Fig. 2.4. When we compare this theory with the experimental data, we see that the agreement between theory and experiment is good at low frequencies, but not at high frequencies (or short wavelengths). This disagreement at high frequencies (in the ultraviolet region) is called the *ultraviolet catastrophe*, because according to classical theory, at high frequencies $\int_0^\infty I(\nu)\, d\nu$ will be infinitely large, in contradiction with the experimentally measured value.

In 1900, the German theoretical physicist Max Planck (1858–1947) announced[3] that, by assuming electromagnetic radiation to be emitted or absorbed in bundles of size $h\nu$, where h is Planck's constant, one could correctly predict the blackbody spectrum. Such a bundle of energy is called a *quantum*. For a given frequency ν, all the quanta have the same energy, but quanta of high frequencies have high energies and those of low frequencies have low energies. Thus, when a blackbody is in thermal equilibrium, the atoms and the molecules in the cavity will emit radiation only if they have energy in excess of $h\nu$. For low

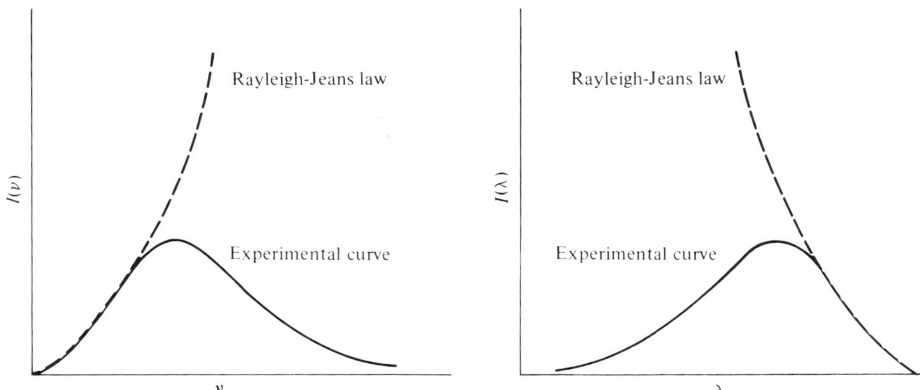

Fig. 2.4 Comparison between the experimental spectral distribution of radiation from a blackbody and the Rayleigh–Jeans law (the classical prediction). Note that the theory completely disagrees at higher ν (or lower λ), leading to "ultraviolet catastrophe."

ν, many atoms and molecules might have this excess energy, but as ν increases the number of atoms or molecules having energies in excess of $h\nu$ decreases because the bundles (or quanta) become progressively bigger. Thus $I(\nu)$ does not increase with ν for large ν, but decreases, and eventually goes to zero asymptotically, thus avoiding the ultraviolet catastrophe predicted by classical theory. (In classical wave theory, there is no restriction on the amount of energy that could build up.)

Planck made use of the Maxwell–Boltzmann distribution, according to which, if there are N_0 molecules in a system in equilibrium at absolute temperature T, the number of molecules N_n with energy E_n is given by

$$N_n = N_0 \exp(-E_n/kT), \tag{2.5}$$

where k is the Boltzmann constant. Combining Eq. (2.5) with his quantum hypothesis $E_n = nh\nu$, and calculating the mean energy, Planck arrived at the following expression for the distribution of the maximum intensity of radiation in the spectrum of the blackbody:

$$I(\nu) = \frac{8\pi h\nu^3}{c^3} \frac{1}{e^{(h\nu/kT)} - 1}, \tag{2.6}$$

which is the famous *Planck radiation law*. Figure 2.5 compares this theoretical formula with the experimental points, and we can see that the agreement is perfect. The value of the Planck constant h obtained by making a fit to Planck's experimental data is

$$h = 6.63 \times 10^{-34} \text{ joule-sec},$$

while the most accurate modern accepted value[6] is

$$h = (6.62517 \pm 0.0005) \times 10^{-34} \text{ joule-sec}. \tag{2.7}$$

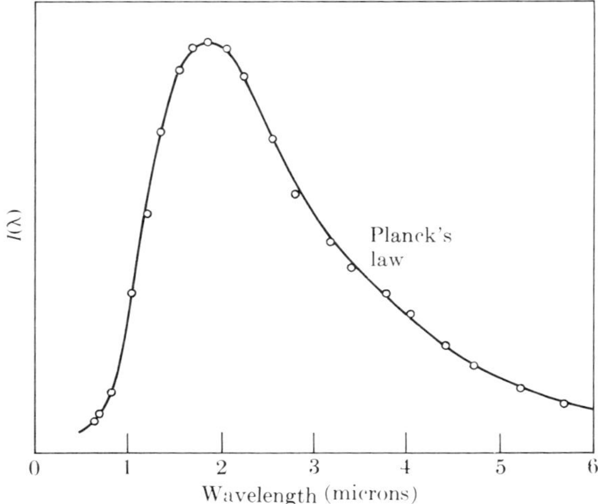

Fig. 2.5 Comparison between the experimental spectral distribution of radiation from a blackbody (open dots) and the theoretical Planck's radiation law (continuous curve). The agreement between the two confirms the quantum hypothesis. [From: Richtmyer, Kennard, and Lauritsen, *Introduction to Modern Physics*, 6th edition, McGraw-Hill, New York, 1969).

The success of Planck's quantum hypothesis in explaining the spectrum of the blackbody had far-reaching consequences. For one thing, it makes us doubt the wave nature of radiation supported by electromagnetic theory by implying that energy is discrete and not continuous. It was hard to accept this, because the theory of the wave nature of electromagnetic radiation was so clearly able to explain interference, diffraction, and polarization of light waves. But when (in 1905) Einstein was able to explain the photoelectric effect by using the quantum hypothesis, physicists were compelled to accept the truth of the quantum hypothesis. From 1905 to 1925, physicists had to contend with both the wave nature (classical theory) and the quantum nature (old quantum theory) of electromagnetic radiation. Two conflicting theories of electromagnetic radiation to explain different things!

2.4 THE PHOTOELECTRIC EFFECT

The ejection of electrons from a surface by the action of light (or electromagnetic radiation) is called the *photoelectric effect*. It was discovered by Heinrich Hertz[7] in 1887, when he was trying to confirm the existence of electromagnetic waves as predicted by the Maxwell theory in 1884. Ironically this same photoelectric effect was later used by Einstein[8] in 1905 to contradict many aspects of the classical electromagnetic theory as discussed here. A complete investigation of the characteristics of the photoelectric effect was carried out by R. A. Millikan.[9]

a) Experiment Figure 2.6 shows an experimental arrangement used for studying the photoelectric effect. It consists of a glass vacuum tube T in which are placed a metallic plate M and a charge-collecting plate C. When a monochromatic beam of light shines through the quartz window onto the plate M, electrons are ejected from the metallic surface. The collector C, if it is at a positive potential V with respect to M (which is at zero potential), collects these ejected electrons, which produces a current i_p, which is measured by the galvanometer G. Increasing V increases i_p until i_p reaches a constant value, i.e., it approaches saturation.

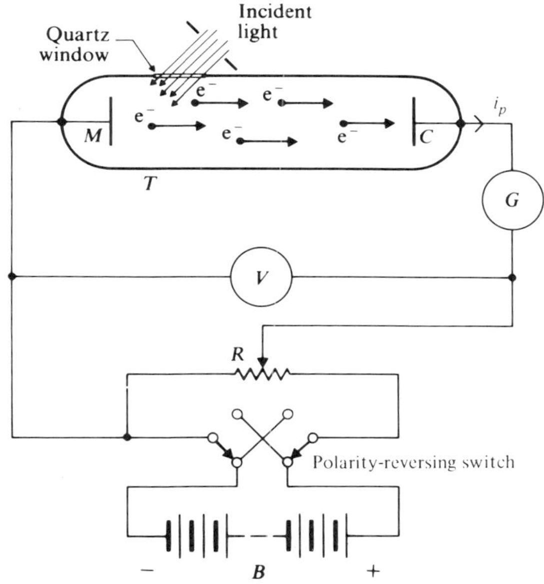

Fig. 2.6 An experimental arrangement used for investigating various characteristics of the photoelectric effect. The voltage V can be varied by means of sliding contact R.

If we apply a negative potential to C by using the reversing switch, the electrons are repelled by C. Only those electrons whose kinetic energy is greater than $e|V|$ will be able to reach the collector C and produce a current measured by G. The potential for which no electrons reach the plate C—and hence $i_p = 0$—is called the *stopping potential*, V_0. The relation between the maximum kinetic energy K_{max} of the electrons and the stopping potential V_0 is

$$K_{max} = \tfrac{1}{2}mv_{max}^2 = e|V_0|. \tag{2.8}$$

By varying the experimental conditions, one can obtain the following results.

1) If ν is kept constant, the photoelectric current i_p increases with increasing intensity I of the incident radiation.
2) Photoelectrons are emitted within less than 10^{-9} sec after the surface is illuminated by light; i.e., there is no time lag between illumination and emission.
3) For a given surface, the emission of photoelectrons takes place *only* if the frequency of the incident radiation is equal to or greater than a certain minimum frequency ν_0, called the *threshold frequency*; ν_0 is different for different surfaces.
4) The maximum kinetic energy, K_{max}, of photoelectrons is independent of the intensity I of the incident light. This is obvious from Fig. 2.7, which shows that the stopping potential is the same for light of three different intensities but of the same frequency. Also $K_{max} = |eV_0|$.

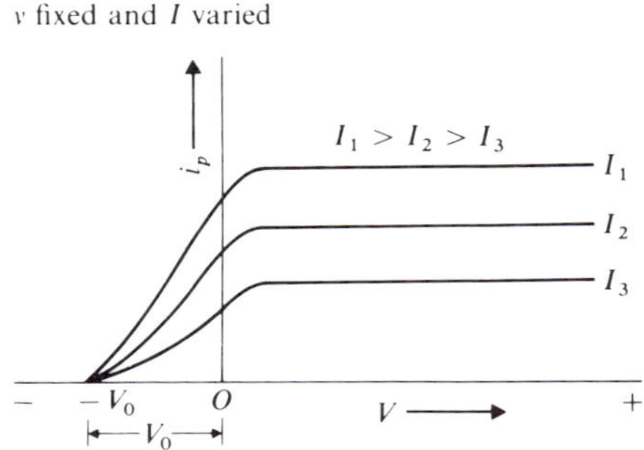

Fig. 2.7 The stopping potential V_0 is the same for beams of different intensities so long as the frequency ν of the incident beam remains the same.

5) The maximum kinetic energy, K_{max}, of photoelectrons depends on the frequency of the incident radiation. This is obvious from Fig. 2.8, which shows that the stopping potential is different for different ν, even though I is the same.
6) There is a linear relation between K_{max} and ν, as shown in Fig. 2.9 for the surfaces of three different metals: cesium, potassium, and tungsten. This relation is mathematically expressed as

$$K_{max} = a\nu + b, \tag{2.9}$$

where a is the slope of the straight line, and is the same for all surfaces, while b is the intercept, and is different for different metals.

b) Explanation by classical theory The wave theory (or the classical electromagnetic theory) predicts (1) (that i_p increases with I), and contradicts (2). According to the wave theory, radiation of very weak intensity should take much longer to accumulate enough energy to pull the electron from the surface. As far as

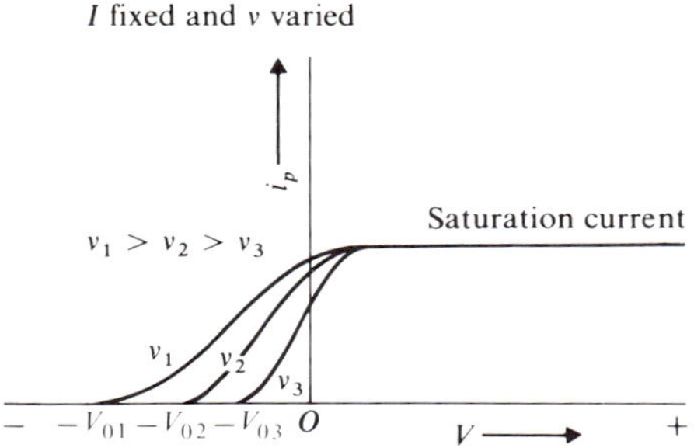

Fig. 2.8 The plots show that the stopping potentials and hence the kinetic energies of the emitted electrons are functions of the frequency of the incident light even though the intensities are the same.

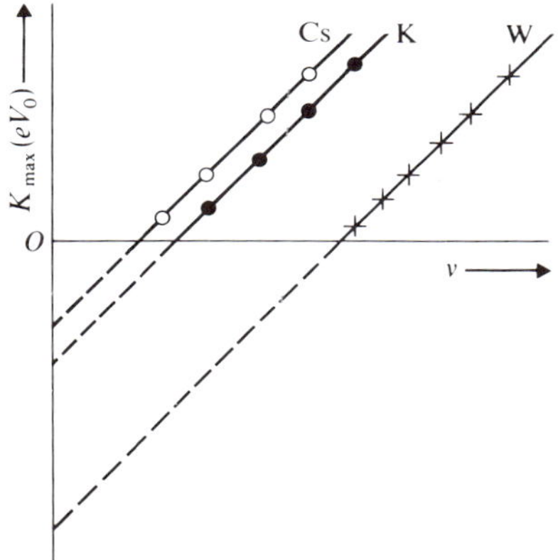

Fig. 2.9 Plots of the measured values of K_{max} $(= eV_0)$ versus ν for different metals show the relation between K_{max} and ν; that is, $K_{max} = a\nu + b$.

(3) is concerned, according to the classical theory there should be no minimum (threshold) frequency for emission of photoelectrons. If the incident light shines for a long enough time, it should be able to pull the electrons from the surface. The classical theory contradicts points (4) and (5), but does not predict (6) at all.

c) Explanation by quantum theory In 1905, Albert Einstein gave the right explanation of the photoelectric effect by adopting Planck's quantum hypothesis and applying it to electromagnetic radiation (for this, he received a Nobel prize in 1921). The quantum hypothesis says that electromagnetic radiation of frequency ν incident on a metallic surface consists of bundles of energy. Each bundle of energy is called a *quantum* or a *photon*. (The quantum hypothesis is also called the photon hypothesis.) If an electron absorbs a photon of energy $h\nu$, in order to escape from the metallic surface it uses up an amount of energy, w, called the *work function* of the metal, while the rest of the energy (equal to $h\nu - w$) appears as the kinetic energy K_{max} of the electron. Thus

$$\boxed{K_{max} = h\nu - w,} \tag{2.10}$$

which is like the form given by Eq. (2.9). Comparison shows that h is the constant slope a, while $-w$ is the intercept b, which is different for different metals. Figure 2.10 shows the relation between these different quantities. (The value of Planck's constant h obtained by making a best fit to the experimental data of the photoelectric effect agrees with the value of h obtained by fitting Planck radiation-law to the spectrum of the blackbody radiation.) Thus the quantum hypothesis has successfully explained (6), while (1) to (5) in the above listing follow very simply: (1) If the intensity of the beam increases, the number of photons in the beam striking the surface also increases, leading to an increase in photoelectron emission and hence in the current i_p. (2) Either the electron absorbs the photon and is pulled out of the surface at once, or not at all. Thus there should be no time lag between absorption and emission. (3) According to Eq. (2.10), for an electron to be just released from the surface—that is, $K_{max} = 0$—the photon must have a threshold energy $h\nu_0$ given by $h\nu_0 = w$; that is, an energy equal to the work function (or binding energy of the electron) of the surface of the metal (Fig. 2.10). From (4) and (5) and from Eq. (2.10), it is obvious that K_{max} is independent of I, but depends on ν.

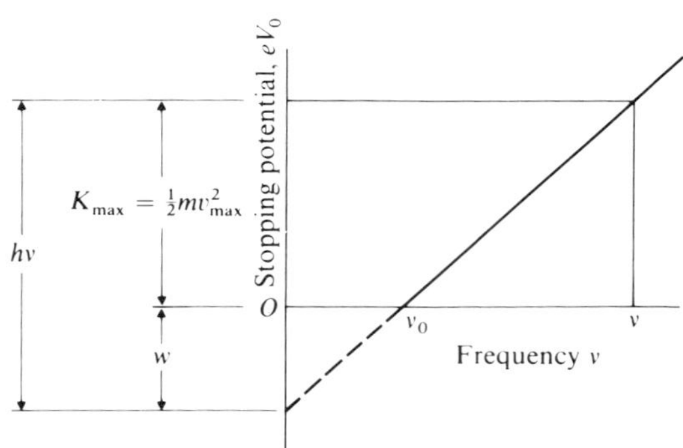

Fig. 2.10 A plot of K_{max} versus ν, showing the relation between the incident energy $h\nu$, the maximum kinetic energy K_{max}, the work function w, the threshold frequency ν_0, and the stopping potential eV_0.

d) Conclusion Now where do we stand? Well, the explanation of the photoelectric effect which uses the quantum theory points to the success of the photon hypothesis. Also, according to Einstein, energy is quantized and each photon carries an amount of energy $h\nu$ (and not $nh\nu$, where n is an integer).

[*Note:* A free electron does not absorb a photon because it does not conserve both momentum and energy. In the photoelectric effect, even though the bound electron absorbs the photon, a tiny fraction of the energy does go to the recoiling atom (in order to conserve both energy and momentum), but it is so small that it may be neglected.]

Example 2.2. Light of wavelength $\lambda = 5893$ Å is incident on a potassium surface. The stopping potential for the emitted electrons is 0.36 volt. Calculate the maximum energy of the photoelectron, the work function, and the threshold frequency.

From Eq. (2.8), the maximum energy K_{max} of the photoelectron is

$$K_{max} = eV_0 = 0.36 \text{ eV},$$

which, when we substitute in Eq. (2.10), $K_{max} = h\nu - w$, gives us

$$eV_0 = h\nu - w,$$

from which we obtain

$$\begin{aligned} w &= h\nu - eV_0 \\ &= \frac{hc}{\lambda} - eV_0 \\ &= \frac{(6.625 \times 10^{-34} \text{ joule-sec})(3 \times 10^8 \text{ m/sec})}{(5893 \times 10^{-10} \text{ m})(1.6 \times 10^{-19} \text{ joule/eV})} - 0.36 \text{ eV} \\ &= 2.15 \text{ eV} - 0.36 \text{ eV} = 1.79 \text{ eV}. \end{aligned}$$

Thus the work function is 1.79 eV.

The threshold frequency is given by the relation $h\nu_0 = w$:

$$\begin{aligned} \nu_0 &= \frac{w}{h} \\ &= \frac{(1.79 \text{ eV})(1.6 \times 10^{-19} \text{ joule/eV})}{6.625 \times 10^{-34} \text{ joule-sec}} \\ &= 4.33 \times 10^{14} \text{ cycles/sec.} \end{aligned}$$

2.5 THE CONTINUOUS X-RAY SPECTRUM

Wilhelm Roentgen,[10] in 1895, discovered that when fast-moving electrons strike a metallic target, a highly penetrating radiation is emitted. Because nobody knew what this radiation was, he named it x-rays. These days we know that x-rays are electromagnetic waves, with wavelengths between 0.1 Å and 100 Å. We can readily understand how x-rays are produced if we remember that, according to electromagnetic theory, an accelerated charged particle emits electromagnetic radiation. When a fast-moving electron, with kinetic energy K, strikes a target, there is an attractive force between the negative electron and the positive charge of the nucleus of the atom. As shown in Fig. 2.11, this force changes the path of the electron, which is equivalent to saying that the electron has been decelerated (or accelerated). Since the electron has undergone acceleration, it emits electromagnetic radiation. The loss in kinetic energy $(K_1 - K_2)$ of the electron appears as electromagnetic radiation of energy $h\nu$, which we call x-rays.

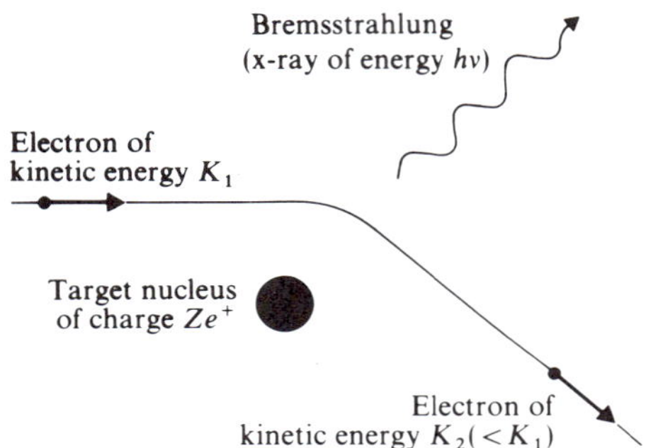

Fig. 2.11 The production of x-rays by the Bremsstrahlung process.

This process of radiation being produced by an accelerated charged particle is called *Bremsstrahlung*, a German word meaning *slowing-down radiation*. Now we're going to show that, according to the quantum hypothesis, these x-rays consist of photons. First let's discuss the experimental arrangement, and then the characteristics of these x-rays.

Figure 2.12 shows a sketch of a high-vacuum x-ray tube. Electrons are emitted from a cathode C, which is heated indirectly by filament F connected to battery B. These electrons are accelerated in a vacuum by a high potential difference V of several thousand volts applied between the cathode C and the anode T (the target). The kinetic energy K of the electrons when they strike the target is given by

$$K = eV. \tag{2.11}$$

These electrons hit the target, where they lose 98% of their kinetic energy because the collisions cause them to produce heat. The rest of their kinetic energy makes up x-rays produced by the Bremsstrahlung process. These x-rays are emitted in all directions. Since they are very penetrating and dangerous to health, the tube is surrounded by lead shielding. A well-collimated beam of x-rays is obtained through a small hole in the shielding (Fig. 2.12).

Figure 2.13 shows a typical x-ray spectrum; here we see the intensity $I(\nu)$

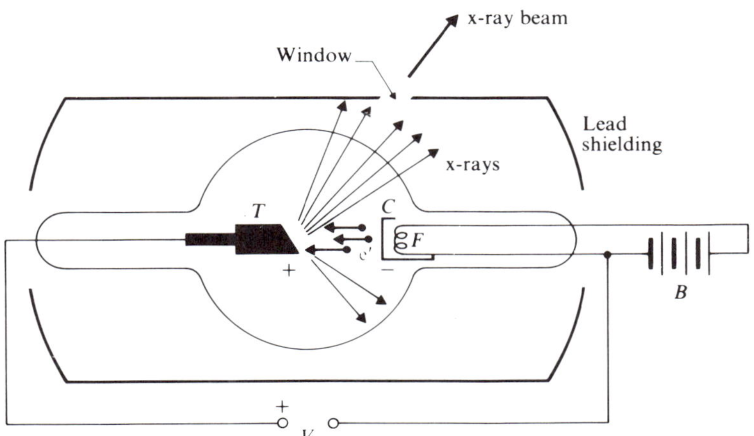

Fig. 2.12 An outline of an x-ray tube for production of x-rays by slowing down fast-moving electrons in the target T.

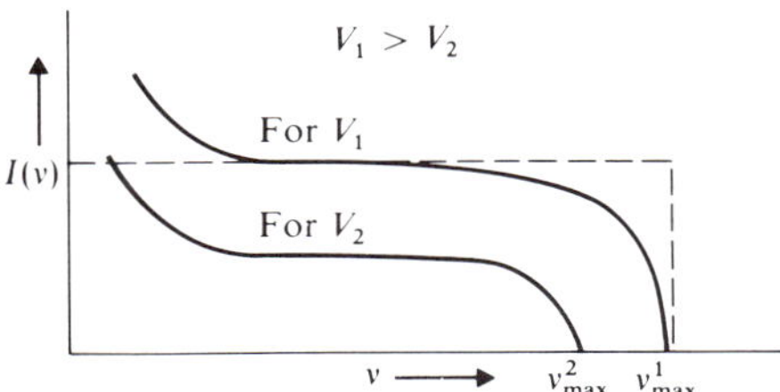

Fig. 2.13 The plot of x-ray intensity $I(\nu)$ versus frequency ν observed in a continuous x-ray spectrum at two different operating voltages.

versus ν plotted for two different operating voltages. Figure 2.14 shows plots of $I(\nu)$ versus ν for two different targets, tungsten (W) and molybdenum (Mo); the x-ray tube operates at the same voltage in both cases. Outstanding features of the x-ray spectra are as follows.

a) The x-ray spectrum from any target has a continuous distribution of radiation of all frequencies up to a certain maximum frequency ν_{max}. The value of ν_{max} does not depend on the target material, as is obvious from Fig. 2.14, but on the potential difference V, as shown in Fig. 2.13. Also ν_{max} is directly proportional to V. That is,

$$\frac{\nu_{max}}{V} = \text{constant}. \tag{2.12}$$

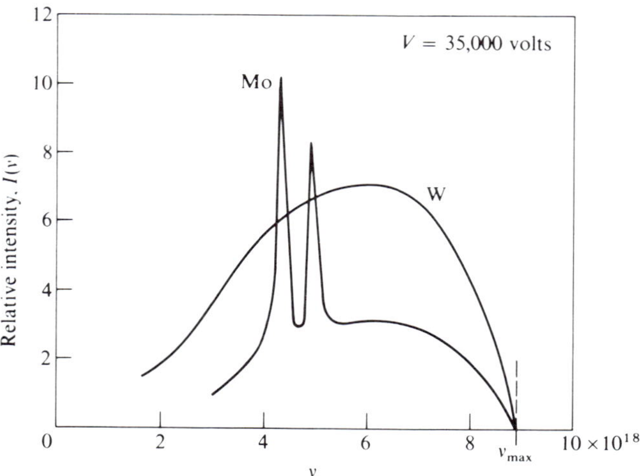

Fig. 2.14 The plots of relative intensity $I(\nu)$ versus ν observed in the x-ray spectra of tungsten (W) and molybdenum (Mo) operating at the same voltage in both cases. Note that the maximum frequency ν_{max} is the same for both targets.

b) As Fig. 2.14 shows, in the case of Mo there are discrete lines superimposed on the continuous x-ray spectrum. The position of those lines is not affected by changing V, but the lines do appear at different positions in different materials. These are the *characteristic x-ray lines*. They depend on the nature of the target material, and we shall talk about them in detail in Chapter 7.

The existence of the high frequency limit ν_{max} and its dependence on V given by Eq. (2.12) cannot be explained by the classical electromagnetic theory, but the photon hypothesis does offer an explanation: The incident electron can produce any number of photons, but if it loses all its energy in producing a single photon of energy $h\nu_{max}$, the following relation should hold:

$$h\nu_{max} = eV, \tag{2.13}$$

where e and h are universal constants. Hence we may write

$$\frac{\nu_{max}}{V} = \frac{e}{h} = \text{constant}, \tag{2.14}$$

which is the experimentally observed relation stated in Eq. (2.12). Using the relation $\lambda\nu = c$, we may write Eq. (2.14) as

$$\frac{c}{\lambda_{min} V} = \frac{e}{h}$$

or

$$\lambda_{\min} = \frac{hc}{eV}. \tag{2.15}$$

Substituting for h, c, and e, we obtain a convenient and useful relation

$$\lambda_{\min} = \frac{1.24 \times 10^{-4}}{V}\ \text{cm}, \tag{2.16}$$

where V is in volts.

This method has been used for accurate determination of e/h, and agrees with the values obtained by other methods. This points once again to the success of the photon hypothesis.

Example 2.3. To produce x-rays of 1.377 Å wavelength from a copper target in an x-ray tube, one must operate the tube at a voltage of 9000 volts. Calculate the ratio h/e.

From Eq. (2.13), we obtain

$$h\nu_{\max} = eV$$

or

$$\frac{h}{e} = \frac{V}{\nu_{\max}} = \frac{V\lambda_{\min}}{c}$$

$$= \frac{(9000\ \text{volts})(1.377 \times 10^{-10}\ \text{m})}{(3 \times 10^{8}\ \text{m/sec})}$$

$$= 4.13 \times 10^{-15}\ \text{volt-sec.}$$

Since e is found to be 1.602×10^{-19} coul, Planck's constant is given by

$$\begin{aligned} h &= 4.13 \times 10^{-15}\ \text{volt-sec} \times 1.602 \times 10^{-19}\ \text{coul} \\ &= 6.616 \times 10^{-34}\ \text{volt-sec coul} \\ &= 6.616 \times 10^{-34}\ \text{joule-sec.} \end{aligned}$$

This agrees very well with a more accurate value of 6.625×10^{-34} joule-sec.

2.6 THE PHOTON

Before we investigate any more phenomena which lead to the success of the photon hypothesis, let's pause and look into the characteristics of the photon. Actually the radiation from a blackbody (the thermal radiation), the visible light used in photoelectric effects, and x-rays are only three different regions of the large range of electromagnetic radiation shown in Fig. 2.15. Note that the frequency range (or wavelength range) of the visible light is very narrow. If the photon hypothesis applies to any range of electromagnetic radiation, it should apply to the whole

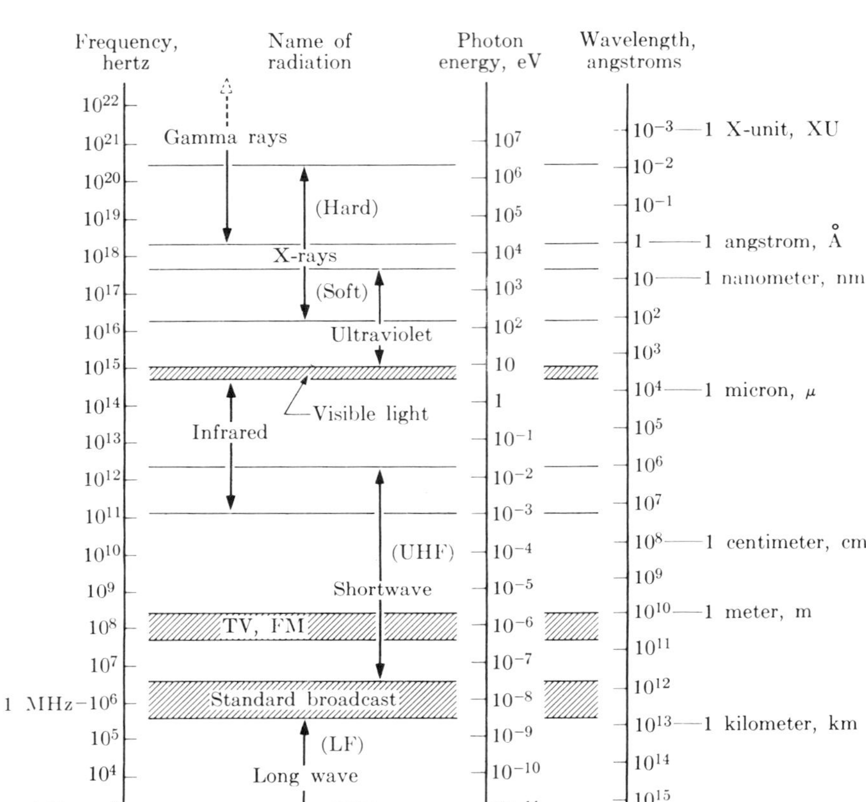

Fig. 2.15 Visible light forms a very small fraction of the whole range of electromagnetic radiation shown.

range. That is, electromagnetic radiation of any frequency should consist of photons, not waves. Let us look into the mass, energy, and momentum of these photons.

The photon (or quantum) is a packet or bundle of energy moving with the speed of light c. Hence, according to the formula $m = m_0/\sqrt{1 - (v/c)^2}$ of the special theory of relativity, the moving mass of the photon is $m = m_0/0 = \infty$, which is impossible. But, if we assume that m_0 is zero, then $m = 0/0$, which is an indeterminate quantity. The fact that the rest mass of the photon is zero should not particularly disturb us, because photons are never found at rest. For a photon of energy $E = h\nu$, if m is the moving mass of the photon, according to the special theory of relativity, $E = mc^2$. Hence

$$E = h\nu = mc^2 \tag{2.17}$$

or

$$m = \frac{h\nu}{c^2} = \frac{E}{c^2}. \tag{2.18}$$

From the relation

$$E^2 = p^2c^2 + m_0^2c^4,$$

since $m_0 = 0$, the energy of the photon is also given by

$$E = pc \tag{2.19}$$

and its momentum is

$$p = \frac{E}{c} = \frac{mc^2}{c} = mc. \tag{2.20}$$

Also

$$p = \frac{E}{c} = \frac{h\nu}{c}. \tag{2.21}$$

Thus, if a photon of frequency ν is to be treated as a particle, then

$$\boxed{m_0 = 0, \qquad E = h\nu, \qquad m = h\nu/c^2, \qquad p = h\nu/c.} \tag{2.22}$$

We shall find these characteristics of photons useful when we discuss the Compton effect, which, in a striking manner, establishes the photon hypothesis.

2.7 THE COMPTON EFFECT

The Compton effect, or Compton scattering, concerns the scattering of x-rays (or of electromagnetic waves in general) by free electrons. According to the classical electromagnetic theory, when an electromagnetic wave of frequency ν is incident on free charged particles such as electrons, the charges absorb electromagnetic radiation and oscillate with frequency ν. These oscillating charges, in turn, reradiate electromagnetic waves of the same frequency ν. This type of scattering, in which there is no change in wavelength, is called *coherent scattering*. It has been observed in the visible region of radiation, and also at longer wavelengths.

When we observe the scattering of radiation of very short wavelengths, such as x-rays, the predictions of classical theory fail. The scattered x-rays are found to consist of two frequencies: the original frequency ν (or wavelength λ) and another frequency, $\nu' < \nu$ (or wavelength $\lambda' > \lambda$). The wavelength λ is called the *unmodified wavelength*, while the wavelength λ' is called the *modified wavelength*. This type of scattering is called *incoherent scattering*.

The correct explanation of the presence of the modified wavelength in x-ray scattering was given by the American physicist A. H. Compton,[11] in 1922 (for which he received a Nobel prize in 1927). By adopting the quantum hypothesis and applying the laws of conservation of momentum and energy, Compton was able to calculate the change in frequency or wavelength of radiation of very short wavelengths. This led to the confirmation of the photon hypothesis, i.e., establishing the particle-like nature of electromagnetic radiation.

Figure 2.16 shows an experimental arrangement for observing Compton scattering: A beam of monochromatic x-rays is obtained from the source S, which is an x-ray tube. After it has been well collimated by means of a slit system C, it is made incident on a target T, made of something such as carbon, in which the loosely bound electrons are assumed to be almost free. (The energy of an x-ray of 1 Å wavelength is 12.4 keV, as compared to the binding energy of one of the outer electrons of carbon, which is 11 eV, and negligible.) The wavelengths of the scattered x-ray radiation are measured at different angles. The results are shown in Fig. 2.17: λ indicate the unmodified wavelengths, λ' indicate the modified ones. The points are the experimentally measured values, while the continuous curves are the values Compton predicted (as calculated below) by using the quantum hypothesis.

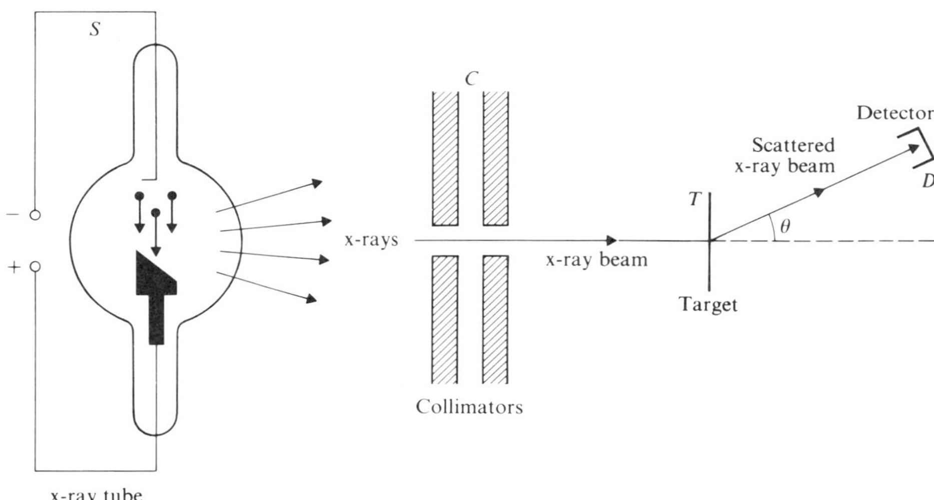

Fig. 2.16 Schematic sketch of an experimental arrangement for observing Compton scattering.

Figure 2.18 shows schematically the collision between the incident x-ray and a free electron at rest. According to the quantum hypothesis, the incident x-ray is a photon of energy $h\nu$ and momentum $h\nu/c$ incident on an electron of rest-mass energy m_0c^2, as shown in Fig. 2.18(a). After the elastic collision (or Compton

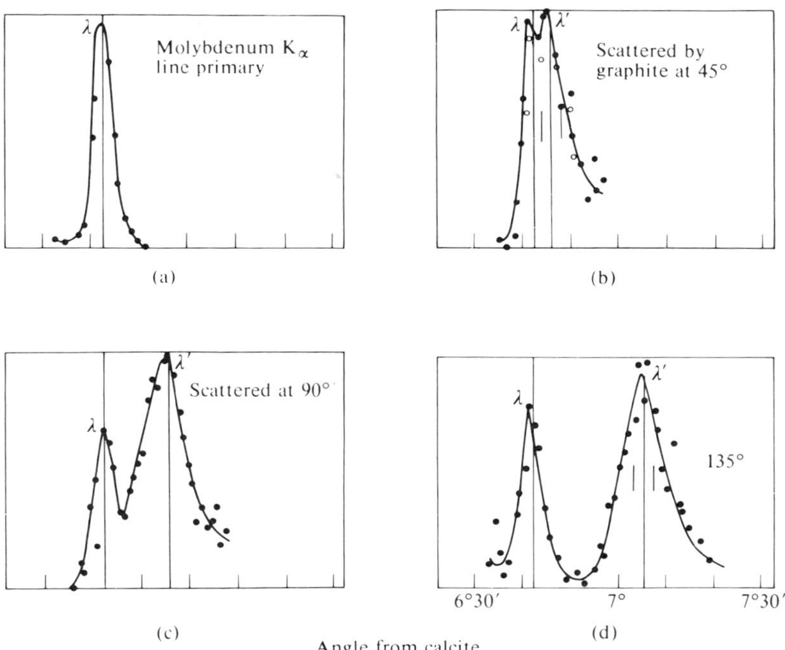

Fig. 2.17 Intensity versus wavelength of scattered x-rays (Compton scattering) at different angles. The modified wavelength λ' and unmodified wavelength λ are both shown. (From A. H. Compton, *Phys. Rev.* **22**, 411 (1923)).

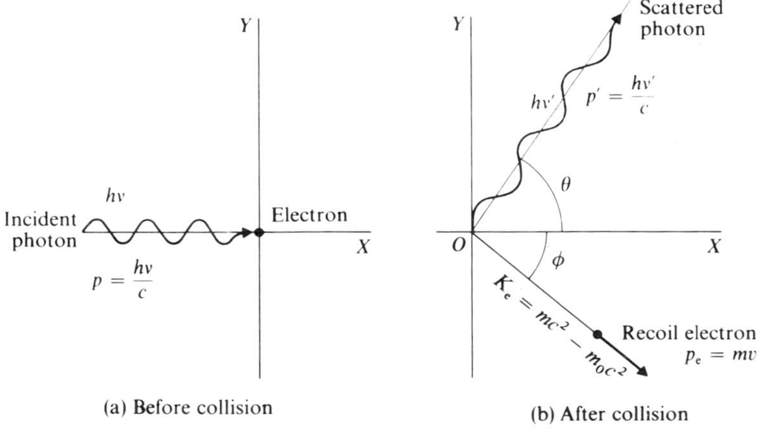

Fig. 2.18 The scattering of x-rays may be treated as a collision between an incident photon and a free electron: (a) before the collision, (b) after the collision.

scattering) between the photon and the electron (see Fig. 2.18b), the scattered photon (i.e., the scattered x-ray) has energy $h\nu'$ ($< h\nu$) and momentum $h\nu'/c$ ($< h\nu/c$). Thus the wavelength λ' of the scattered photon is larger than λ. The decrease in the energy of the photon, $h\nu - h\nu'$, appears as the kinetic energy K_e of the recoil electron. That is, if m is the moving mass of the electron, then, according to relativity,

$$K_e = mc^2 - m_0c^2 = m_0c^2\left(\frac{1}{\sqrt{1-\beta^2}} - 1\right), \tag{2.23}$$

where $\beta = v/c$, v being the velocity of the recoil electron. Its relativistic momentum p_e is

$$p_e = mv = m\beta c = \frac{m_0\beta c}{\sqrt{1-\beta^2}}. \tag{2.24}$$

Let the scattered photon and the recoiling electron make angles θ and ϕ, respectively, with the direction of the incident photon, as shown in Fig. 2.18(b). Let $p = h\nu/c$, $p' = h\nu'/c$, and $p_e = mv$ be the momenta of the incident photon, scattered photon, and the recoiling electron, respectively. Resolving $\mathbf{p}$, $\mathbf{p}'$, and $\mathbf{p}_e$ into X and Y components and applying the conservation of linear momentum yields

$$\frac{h\nu}{c} = \frac{h\nu'}{c}\cos\theta + p_e\cos\phi, \tag{2.25}$$

$$0 = \frac{h\nu'}{c}\sin\theta - p_e\sin\phi, \tag{2.26}$$

while the conservation of energy gives

$$h\nu = h\nu' + K_e. \tag{2.27}$$

The above three equations contain four variables, θ, ϕ, $h\nu'$, and K_e. If we eliminate one of these, say ϕ, then we can determine the other three. Thus we rewrite Eqs. (2.25) and (2.26),

$$p_ec\cos\phi = h\nu - h\nu'\cos\theta, \tag{2.28}$$

$$p_ec\sin\phi = h\nu'\sin\theta. \tag{2.29}$$

Squaring and adding these equations, we obtain

$$p_e^2c^2 = (h\nu)^2 + (h\nu')^2 - 2(h\nu)(h\nu')\cos\theta. \tag{2.30}$$

The total energy E_e of the recoiling electron is

$$E_e = K_e + m_0c^2 = (h\nu - h\nu') + m_0c^2, \tag{2.31}$$

which from the theory of relativity is also equal to

$$E_e = \sqrt{p_e^2c^2 + m_0^2c^4}. \tag{2.32}$$

Equating Eqs. (2.31) and (2.32) and squaring, we get

$$p_e^2c^2 = (h\nu - h\nu')^2 + 2(h\nu - h\nu')m_0c^2. \tag{2.33}$$

Again equating the values of $p_e^2c^2$ from Eqs. (2.30) and (2.33) and solving for $h\nu'$, we obtain

$$\boxed{h\nu' = \frac{h\nu}{1 + (h\nu/m_0c^2)(1 - \cos\theta)}.} \tag{2.34}$$

By using the relation $\lambda\nu = \lambda'\nu' = c$ in Eq. (2.34), we may write the change in wavelength $\Delta\lambda = \lambda' - \lambda$, due to the Compton effect, as

$$\boxed{\Delta\lambda = \lambda' - \lambda = \frac{h}{m_0c}(1 - \cos\theta).} \tag{2.35}$$

Thus the modified wavelength λ' is longer than the incident (unmodified) wavelength λ. The change in wavelength $\lambda' - \lambda$ depends on the rest mass of the free electron m_0 and the scattering angle θ, but not on the wavelength of the incident photon. The quantity h/m_0c has the dimensions of length and is called the *Compton wavelength*. Substituting for h, m_0, and c in Eq. (2.35), we find that $h/m_0c = 0.02426$ Å and $\lambda' = \lambda + 0.02426(1 - \cos\theta)$ Å. As shown in Fig. 2.17, these results do agree with measured values of the modified wavelengths at different angles.

Thus we have seen that the quantum hypothesis successfully explains the presence of the modified wavelengths at different angles, but it must also explain the presence of *un*modified ones (i.e., the coherent scattering predicted by classical theory) at different angles. This is accounted for by the photon hypothesis, as follows. So far we have assumed that the scattering takes place because of the collision of the incident photon and a free electron. But it may also take place when the incident photon collides with one of the tightly bound inner electrons of the atom. In that case, the whole atom of mass M recoils, instead of just the electron of mass m_0. Thus, in Eq. (2.35), the m_0 is replaced by M. But M for the lightest atom (hydrogen) is 1836 times m_0; hence the Compton wavelength h/Mc is 0.0000133 Å or smaller. Thus $\Delta\lambda$ in the case of $\theta = 180°$ (where it is a maximum) is negligible. That is why the scattering of photons from the inner bound electrons does not result in a change of wavelength. This explains the presence of unmodified wavelengths, together with modified wavelengths at all the angles, as shown in Fig. 2.17.

Note that quantum effects come into existence because of the presence of h. Thus, if $h \to 0$ or $m_0 \to \infty$, according to Eq. (2.35), $\lambda' \to \lambda$. This implies that in the macroscopic region in which $h \to 0$ or the particles are large, quantum theory reduces to classical theory. That is,

$$\lim_{h \to 0 \text{ or } m \to \infty} \text{quantum theory} \to \text{classical theory}. \tag{2.36}$$

But so far as the microscopic region is concerned, the classical theory must be replaced by the quantum theory.

Example 2.4. Suppose that x-rays of 100 keV energy are incident on a target, and undergo Compton scattering. Calculate (a) the energy of the x-rays scattered at an angle of 30° to the direction of the incident, (b) the energy of the recoiling electron, (c) the angle of the recoiling electron.

a) In Eq. (2.35), we have

$$\lambda' - \lambda = \frac{h}{m_0 c}(1 - \cos\theta).$$

When we substitute $\lambda = hc/E$ and $\lambda' = hc/E'$, where E and E' are the energies of the incident and scattered x-rays, that is, $E = h\nu$ and $E' = h\nu'$, we get

$$\frac{1}{E'} - \frac{1}{E} = \frac{1}{m_0 c^2}(1 - \cos\theta), \tag{2.37}$$

where $E = 100$ keV, $m_0 c^2 = 510$ keV, and $\cos\theta = \cos 30° = 0.866$. Therefore

$$\frac{1}{E'} - \frac{1}{100 \text{ keV}} = \frac{1}{510 \text{ keV}}(1 - 0.866)$$

$$= \frac{0.134}{510 \text{ keV}} = \frac{1}{3810 \text{ keV}}$$

or

$$\frac{1}{E'} = \frac{1}{100 \text{ keV}} + \frac{1}{3810 \text{ keV}}.$$

Hence $E' = 97.5$ keV.

b) The kinetic energy K of the recoiling electron is

$$K = E - E' = (100 - 97.5) \text{ keV} = 2.5 \text{ keV}.$$

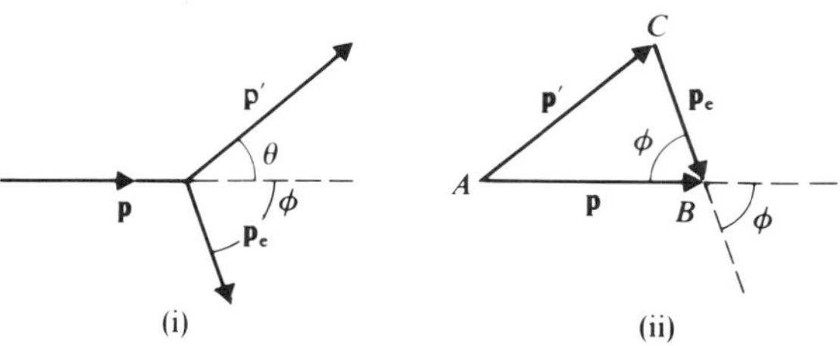

c) Figure (i) shows that $\mathbf{p}$, $\mathbf{p}'$, and $\mathbf{p}_e$ are the momenta of the incident x-ray, the scattered x-ray, and the recoiling electron, while Fig. (ii) shows the triangular relations between these quantities. In order to find ϕ, we apply the law of cosines to triangle ABC,

$$p'^2 = p^2 + p_e^2 - 2pp_e \cos\phi,$$

or, rearranging and using $E = pc$ and $E' = p'c$ for photons, we get

$$\cos\phi = \frac{E^2 - E'^2 + p_e^2c^2}{2Ep_ec}. \tag{2.38}$$

But, for an electron,

$$E^2 = p_e^2c^2 + E_0^2,$$

where $E_0 = m_0c^2$ and $E = K + E_0$. Therefore $(K + E_0)^2 = p_e^2c^2 + E_0^2$ gives

$$p_ec = K\left(1 + \frac{2E_0}{K}\right)^{1/2},$$

which, when we substitute in Eq. (2.38), yields

$$\cos\phi = \frac{E^2 - E'^2 + K^2[1 + (2E_0/K)]}{2EK[1 + (2E_0/K)]^{1/2}}. \tag{2.39}$$

Substituting for $E = 100$ keV, $E' = 97.5$ keV, $K = 2.5$ keV, $E_0 = 510$ keV, we get

$$\cos\phi = 0.29, \qquad \phi = 73°.$$

2.8 THE deBROGLIE HYPOTHESIS

Let us recapitulate. Initially we said that the interference and diffraction experiments—followed by the Maxwell electromagnetic theory and its experimental verification by Hertz in 1889—had firmly established the wave nature of electromagnetic radiation. Then came new experimental facts—such as the spectrum of blackbody radiation, the photoelectric effect, the continuous x-ray spectrum, and the Compton effect—in which the wave theory failed. Physicists were compelled to adopt the photon or quantum hypothesis, which implies assigning some sort of particle nature to electromagnetic radiation. This dual nature—wave and quantum—of electromagnetic radiation continued to be generally accepted from 1905 to 1924.

In 1924, the French physicist Louis deBroglie,[12] after looking deeply into the special theory of relativity and the photon hypothesis, suggested in his doctoral dissertation that there was a more fundamental relation between waves and particles. Thus, according to the special theory of relativity for a photon,

$$E = h\nu \tag{2.40}$$

and its momentum p is

$$p = \frac{h}{\lambda}. \tag{2.41}$$

Note that the left sides of these two equations contain E and p, which are characteristics of the particles, and the right sides contain ν and λ, which are the characteristics of the waves. These two sets of quantities are connected through Planck's constant h. Louis deBroglie suggested that the problem is symmetrical and that the dual (wave and particle) nature of electromagnetic radiation may be extended to material particles such as electrons, protons, neutrons, etc., the difference being that photons have $m_0 = 0$ and $v = c$, while the material particles have $m_0 \neq 0$ and $v < c$.

The above type of argument led deBroglie[12] to suggest the following hypothesis applicable to all matter, radiation as well as particles.

a) *If there is a particle of momentum p, its motion is associated with (or guided by) a wave of wavelength*

$$\lambda = \frac{h}{p}. \tag{2.42}$$

b) *If there is a wave of wavelength λ, the square of the amplitude of the wave at any point in space is proportional to the probability of observing, at that point in space, a particle of momentum*

$$p = \frac{h}{\lambda}. \tag{2.43}$$

Though the above hypotheses look very unfamiliar, they had far-reaching consequences. In fact, they revolutionized the whole of physics. The implication is that, depending on the circumstances, a wave may be observed as a particle, while a particle may be observed as a wave! We have already seen how electromagnetic waves may be observed as photons in different situations. Thus in order to validate the deBroglie hypothesis, it remains to be shown that in some experimental situations particles may behave like waves. The outstanding characteristics of electromagnetic waves are the interference and diffraction phenomena. If we could show that a beam of particles also exhibits a diffraction pattern, we would have proved the dual nature of matter. Let us first discuss the diffraction of waves.

Example 2.5. Calculate the deBroglie wavelength associated with the following.

a) A golf ball of mass 50 g moving with a velocity of 20 m/sec.

b) A proton moving with a velocity of 2200 m/sec.

c) An electron moving with a kinetic energy of 10 eV.

a) According to deBroglie's hypothesis

$$\lambda = \frac{h}{p} = \frac{h}{mv}$$

$$= \frac{6.625 \times 10^{-34} \text{ joule-sec}}{0.05 \text{ kg} \times 20 \text{ m/sec}}$$

$$= 6.625 \times 10^{-34} \frac{\text{joule-sec}^2}{\text{kg-m}}$$

$$= 6.625 \times 10^{-34} \text{ m},$$

which is too small to be detected.

b) The mass of a proton is 1.67×10^{-27} kg. Therefore the deBroglie wavelength of a proton is

$$\lambda = \frac{h}{p} = \frac{h}{mv}$$

$$= \frac{6.625 \times 10^{-34} \text{ joule-sec}}{1.67 \times 10^{-27} \text{ kg} \times 2200 \text{ m/sec}}$$

$$= 1.91 \times 10^{-10} \text{ m} = 1.91 \text{ Å}.$$

This wavelength is of the order of x-ray wavelengths. Note that we have used the non-relativistic expression for momentum because, for this velocity, the kinetic energy is much smaller than the rest-mass energy of the proton.

c) For an electron of kinetic energy 10 eV, we can still use classical mechanics because 10 eV is much less than 510 keV, the rest mass of the electron. Thus $K = \frac{1}{2}mv^2 = p^2/2m$, or $p = \sqrt{2mK}$. From deBroglie's hypothesis,

$$\lambda = \frac{h}{p} = \frac{h}{\sqrt{2mK}}.$$

But $K = 10$ eV. Therefore

$$\lambda = \frac{6.625 \times 10^{-34} \text{ joule-sec}}{\sqrt{2(9.11 \times 10^{-31} \text{ kg})(10 \text{ eV})(1.602 \times 10^{-19} \text{ joule/eV})}}$$

$$= 3.9 \times 10^{-10} \text{ m} = 3.9 \text{ Å}.$$

If the kinetic energy of the electron becomes comparable to its rest mass, we must use the relativistic expression for momentum.

2.9 THE DIFFRACTION OF WAVES

The spatial variation in intensity (with alternating maxima and minima) observed after a beam of light has passed through a slit is called *diffraction*. From our study of general physics, we have seen that mechanical diffraction gratings having as many as 12,000 lines per centimeter are used to produce diffraction of visible light. If the beam is incident normally to the diffraction grating, the condition for diffraction maxima is given by

$$d \sin \theta = n\lambda, \tag{2.44}$$

where $n = 1, 2, 3, \ldots$ is the order of the diffraction maximum, λ is the wavelength of the incident electromagnetic wave, θ is the diffraction angle, and d is the width of a single transparent or opaque line on the grating. Thus, for example, for 12,000 lines per centimeter, $d = 1/12{,}000 \sim 10^{-4}$ cm. From Eq. (2.44), if $n = 1$, $\sin \theta = \lambda/d$ requires that, if we are to see diffraction successfully, d should be of the order of λ (if $d \gg \lambda$, $\sin \theta \simeq 0$ and $\theta \simeq 0°$, and hence no diffraction pattern appears). In the case of visible light, $\lambda \sim (4000\text{–}7000)$ Å and for the mechanical grating $d \sim 10^{-4}$ cm $= 1000$ Å. Hence a diffraction pattern will be observed.

Now suppose that we want to look at the diffraction of x-rays which are electromagnetic radiation of wavelengths between 0.1 Å and 100 Å. It is quite obvious that the mechanical grating could not be used for this, because λ as compared to d is smaller by a factor of 10 to 10^4.

In 1912, Max von Laue suggested that a crystalline solid, in which the atoms are arranged in a regular array and are a few angstroms apart, might be used as a three-dimensional grating for the diffraction of x-rays. Figure 2.19 shows a portion of a single crystal of rock salt (NaCl) in which the Na^+ and Cl^- ions are arranged in a simple cubic lattice, placed alternatively a distance d apart, d being the edge length of each small cube. Combinations of atoms may be used to define families of parallel planes, called the *Bragg planes*. Three such families of Bragg planes are shown in Fig. 2.20.

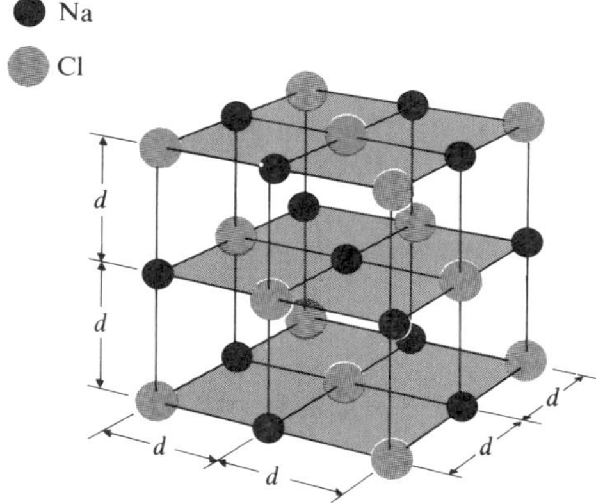

Fig. 2.19 A portion of a single crystal of rock salt (NaCl) in which Na^+ and Cl^- ions are arranged in a simple cubic lattice.

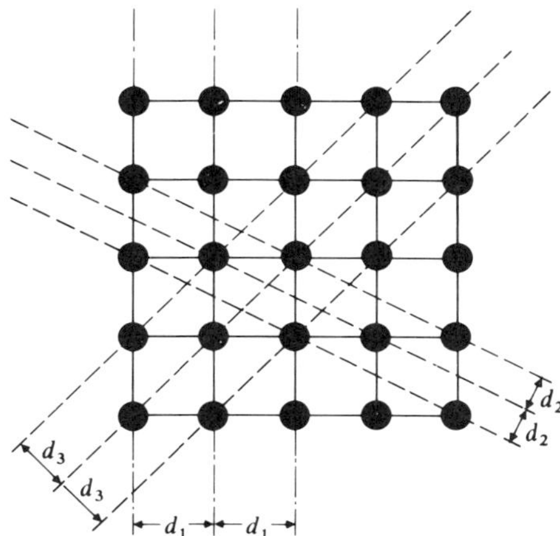

Fig. 2.20 Parallel lines passing through atoms define Bragg planes. Three sets of Bragg planes are shown.

To understand how diffraction takes place, let us consider a layer of atoms in a certain plane of a crystal, as shown in Fig. 2.21. When a beam of monochromatic x-rays of wavelength λ is incident on such a crystal, each atom will scatter a small fraction of the incident beam as a spherical wave, the phase of which is determined by the position of the scattering atom. In certain directions there is a constructive interference between the scattered radiation, while in other directions there is a destructive interference, thus leading to the diffraction pattern. The English

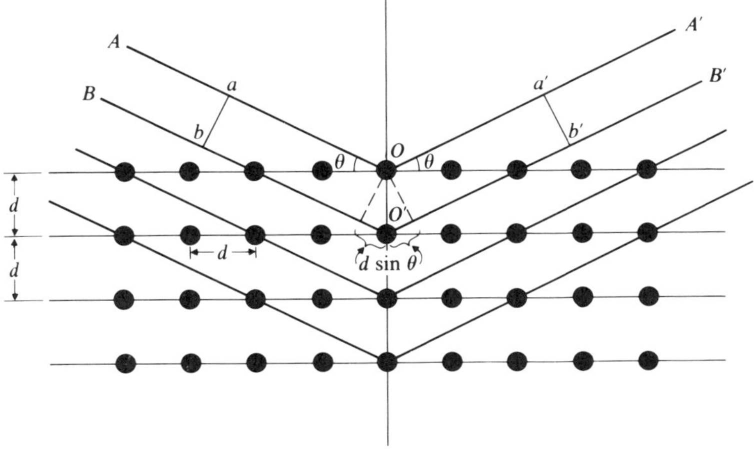

Fig. 2.21 Diffraction of x-rays from a layer of atoms in a certain plane of a crystal.

physicist W. L. Bragg derived the condition for constructive interference, as follows.

Look again at Fig. 2.21, and consider two incident rays scattered from two different atoms in a plane. The scattered rays, if they are in phase, produce constructive interference. Thus the paths AOA' and $BO'B'$ of two beams must differ by integral multiplies of λ'. The path difference is

$$BO'B' - AOA' = 2d \sin \theta. \tag{2.45}$$

The Bragg condition for being in phase for constructive interference leads to the *Bragg law*,

$$\boxed{2d \sin \theta = n\lambda,} \tag{2.46}$$

where $n = 1, 2, 3, 4, \ldots$.

Figure 2.22 shows the experimental arrangement of an x-ray crystal spectrometer. A beam of x-rays from the tube, after passing through the lead collimator W and slit S, is incident on a crystal C. The scattered beam is detected by a detector D, which may be an ionization chamber. Investigators use diffraction patterns from different Bragg planes to study the structure of crystals.

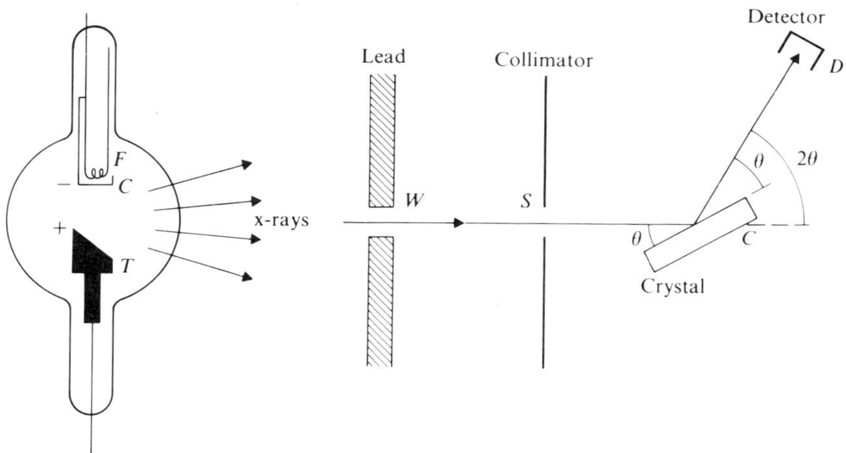

Fig. 2.22 Experimental arrangement of an x-ray crystal spectrometer for investigations of x-ray diffraction.

Example 2.6. One of the most accurate methods of determining Avogadro's number N_A is diffraction of x-rays from a crystal grating. X-rays of wavelength 1.3922 Å are reflected from the face of a NaCl crystal. The first-order reflection is observed at an angle of

14° 17′ 26″. From these data, calculate the lattice spacing and Avogadro's number. The molecular weight of NaCl is 58.454 and its density is 2.163 g/cm^3.

From Eq. (2.46), for $n = 1$, the lattice spacing is given by

$$d = \lambda/2 \sin \theta = \frac{1.3922 \text{ Å}}{2[\sin (14° 17' 26'')]}$$

$$= \frac{1.3922 \times 10^{-10} \text{ m}}{2(0.2468)} = 2.820 \text{ Å}.$$

The number of molecules per unit volume, n, is given by

$$n = \frac{\rho N_A}{M},$$

where ρ is the density and M is the molecular weight.

If the spacing in a cubic crystal such as NaCl is d, then the volume occupied by each ion is d^3; the number of ions per unit volume is $1/d^3$. Since there are two ions (Na^+ and Cl^-) in each molecule,

$$n = \frac{2\rho N_A}{M} = \frac{1}{d^3}.$$

Hence

$$N_A = \frac{M}{2\rho d^3} = \frac{58.454 \text{ g/mole}}{2(2.163 \text{ g/cm}^3)(2.830 \times 10^{-8} \text{ cm})^3}$$

$$= 6.02 \times 10^{23} \text{ atoms/mole} = 6.02 \times 10^{26} \text{ atoms/kmole}.$$

Now let's consider a diffraction pattern obtained from a thin foil instead of a single crystal. Figure 2.23 shows the experimental arrangement. The foil consists of microcrystals which are randomly oriented. The incident beam is strongly

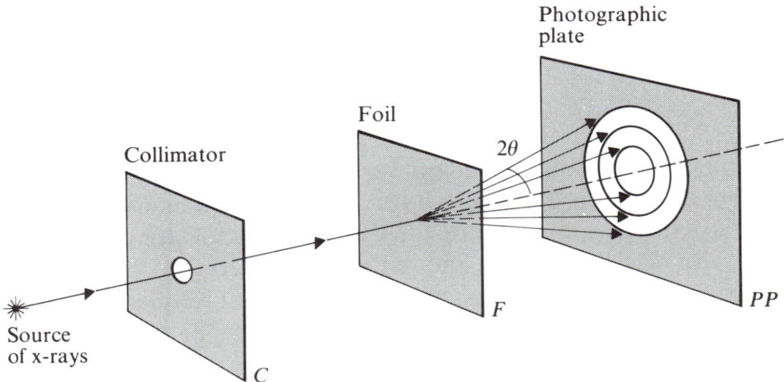

Fig. 2.23 Experimental arrangement for observing diffraction of x-rays from a thin foil which consists of microcrystals.

diffracted by only those microcrystals for which the Bragg condition is satisfied. The diffracted beam coming out of the other side is concentrated in a conical shell making an angle 2θ with the incident beam, as shown in Fig. 2.23. Because of the fact that there are several Bragg planes, many concentric circles of varying intensities, each corresponding to a definite set of Bragg planes, are obtained on the photographic plate placed a distance r from the crystal. Figure 2.24 is an example of an x-ray diffraction pattern obtained from a sample of polycrystalline aluminum. These patterns are called *powder patterns* or *Debye–Scherrer rings.*

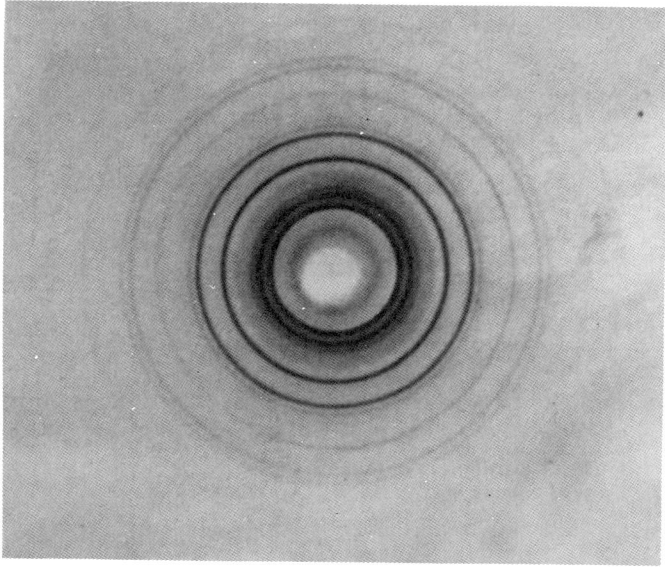

Fig. 2.24 X-ray diffraction pattern from a sample of polycrystalline aluminum. [Courtesy of Mrs. M. H. Read, Bell Telephone Laboratory, Murray Hill, N.J.]

2.10 DIFFRACTION OF PARTICLES

In order to confirm the deBroglie hypothesis, we must show that particles such as electrons, protons, neutrons, atoms, and others exhibit diffraction patterns similar to those of x-rays (or any other electromagnetic radiation) discussed in Section 2.9. To get some idea, let us calculate the deBroglie wavelength associated with an electron which has been accelerated through a potential difference of V. Let m be the mass and v the velocity of such an electron. Then, nonrelativistically,

$$\tfrac{1}{2}mv^2 = eV$$

or

$$v = \sqrt{\frac{2\,eV}{m}}\,. \tag{2.47}$$

If we use this value, the deBroglie wavelength is given by

$$\lambda = \frac{h}{p} = \frac{h}{mv} = \frac{h}{\sqrt{2m\,eV}}.$$

That is,

$$\lambda = \frac{6.62 \times 10^{-34}\text{ joule-sec}}{\sqrt{2 \times 9.108 \times 10^{-34}\text{ kg} \times 1.6 \times 10^{-19}\text{ coul} \times V}} = \sqrt{\frac{150}{V}}\,10^{-8}\text{ cm}.$$

Or, if V is expressed in volts,

$$\lambda = \sqrt{\frac{150}{V}}\text{ Å}. \tag{2.48}$$

Thus, if $V = 150$ volts, $\lambda = 1$ Å. That is, electrons which have been accelerated through a potential drop of 150 volts have such a momentum that the deBroglie wavelength associated with them is 1 Å. Since the wavelength of such an electron is comparable to that of x-rays, it should be possible to cause diffraction of these electrons by using a crystal as a diffraction grating.

The first confirmation that there is a wave associated with the electron came through the experiment of C. J. Davisson and L. H. Germer,[13] in 1927, working at the Bell Telephone Laboratories in the United States. At the time they were involved in measuring the emission of secondary electrons. (Secondary electrons are those electrons which are emitted when a beam of fast electrons strikes a solid.) Figure 2.25 shows their experimental arrangement. The electrons accelerated through a potential difference V by the electron gun G fell on a piece of nickel metal. Davisson and Germer measured the intensity of the scattered electrons as a function of the angle ϕ and plotted it in the form of a polar diagram. That is, at any angle ϕ the intensity is proportional to the magnitude of r, as shown in Fig. 2.25. The distribution of electrons observed was what one would expect from classical theory.

At one time, experimenters who were using nickel had to clean the oxidation from the surface by baking it in a high-temperature oven. After Davisson and Germer did this and repeated the experiment, they got quite different results. There was a pronounced peak (as shown in Fig. 2.26) at an angle of 50° when the voltage through which the electron beam was accelerated was 54 volts. What had happened was that the heating of the crystal resulted in orienting the microcrystals in the nickel sample in such a way that it became one large crystal (see Fig. 2.26). The waves associated with the electrons in the beam were satisfying Bragg's law, which caused a diffraction peak. Let us now prove this.

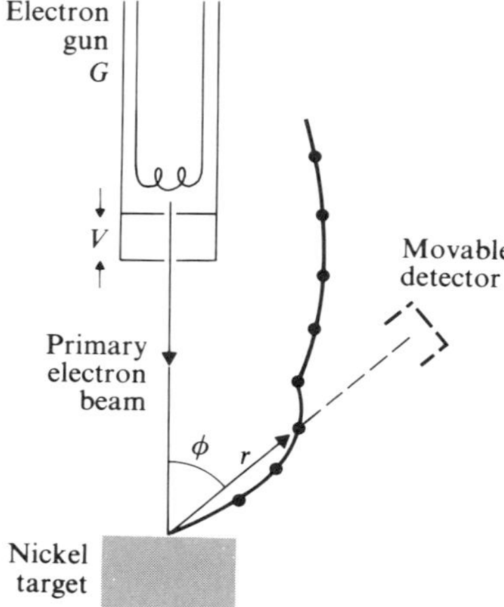

Fig. 2.25 Experimental arrangement of C. Davisson and L. H. Germer for investigation of the distribution of secondary electrons.

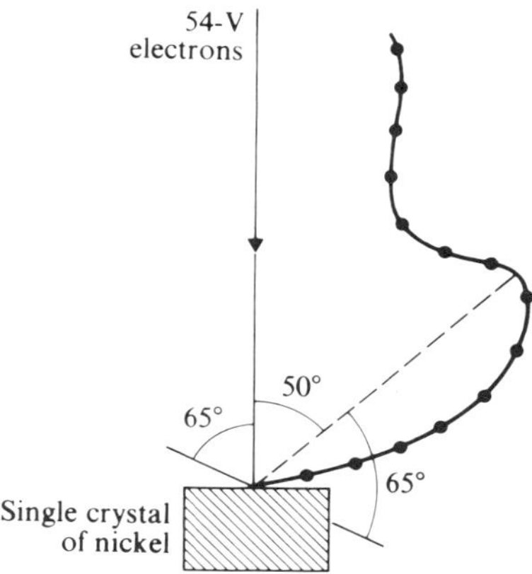

Fig. 2.26 Distribution of electrons that have been scattered from nickel target, after the nickel had been heated, clearly demonstrates the presence of a diffraction peak.

For $\phi = 50°$, $\theta = (180 - 50)/2 = 65°$, and d for nickel is 0.91 Å; hence, for $n = 1$,

$$n\lambda = d \sin \theta$$

gives

$$\lambda = 2d \sin \theta = 2 \times 0.91 \times \sin 65° = 1.65 \text{ Å}.$$

Thus, according to Bragg's law, the wavelength associated with the electrons should be 1.65 Å.

According to deBroglie's hypothesis, for 54 volts, the wavelength of the electron from Eq. (2.48) should be

$$\lambda = \sqrt{\frac{150}{V}}\ \text{Å} = \sqrt{\frac{150}{54}}\ \text{Å} = 1.67\ \text{Å},$$

which agrees with the above value. Hence the deBroglie hypothesis is confirmed. That is, Davisson and Germer had observed waves associated with the electrons!

In the same year—1927—G. P. Thomson,[14] son of J. J. Thomson, by using an experimental arrangement similar to that used for diffraction of x-rays by foils, observed the diffraction pattern of very fast electrons (~ a few keV) incident on a thin foil. The diffraction pattern he observed was similar to that of x-rays. Since then, there have been many experiments to confirm the wave aspects of particles, such as neutrons and atoms, etc. For example, Fig. 2.27 shows diffraction patterns of electrons incident on different metallic foils, and also the diffraction pattern of x-rays.

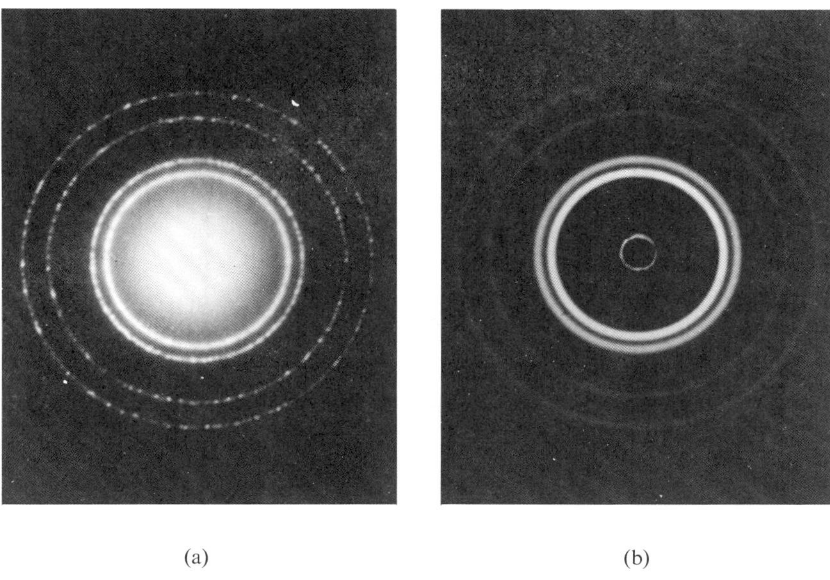

(a) (b)

Fig. 2.27 Similarity between diffraction patterns produced by 600-eV electrons from an aluminum-foil target and x-rays of wavelength 0.71 Å also from an aluminum-foil target demonstrates the wave nature of electrons. [Courtesy of Film Studio, Education Development Center]

SUMMARY

According to Planck's hypothesis, a physical system executing a simple harmonic motion can have only the following discrete energy values:

$$E = nh\nu, \qquad \text{where } n = 1, 2, 3, \ldots,$$

and energy is emitted or absorbed in bundles, called *photons*, of energy $h\nu$.

★The spectral distribution of a blackbody is given by the Planck radiation law:

$$I(\nu) = \frac{8\pi h\nu^3}{c^3} \frac{1}{e^{(h\nu/kT)} - 1}.$$

In the photoelectric effect, the relationship between the incident photon energy $h\nu$, the work function w, and the kinetic energy K_{max} of the emitted electron is

$$K_{\text{max}} = h\nu - w,$$

while the stopping potential V_0 is

$$e|V_0| = K_{\text{max}} = \tfrac{1}{2}mv_{\text{max}}^2.$$

The maximum energy of x-rays produced by bombarding a target with fast-moving electrons is

$$h\nu_{\text{max}} = eV \qquad \text{or} \qquad \lambda_{\text{min}} = \frac{1.24 \times 10^{-4}\ \text{cm-volt}}{V}.$$

A photon of frequency ν is treated like a particle with the following characteristics:

$$m_0 = 0, \qquad E = h\nu, \qquad m = \frac{h\nu}{c^2}, \qquad p = \frac{h\nu}{c}.$$

In Compton scattering, the incident photon of energy $h\nu$ loses a part of its energy to the almost-free electron, and the energy $h\nu'$ of the photon scattered at an angle θ is

$$h\nu' = \frac{h\nu}{1 + (h\nu/m_0c^2)(1 - \cos\theta)}$$

or the change in wavelength is

$$\Delta\lambda = \lambda' - \lambda = \frac{h}{m_0c}(1 - \cos\theta) = 0.02426(1 - \cos\theta)\ \text{Å}.$$

According to deBroglie's hypothesis, the motion of a particle of momentum p is guided by a wavelength λ, where

$$\lambda = \frac{h}{p}.$$

In a wave of wavelength λ, the square of the amplitude of the wave is proportional to the probability of finding a particle of momentum p, where

$$p = \frac{h}{\lambda}.$$

If a light beam is incident normally to the mechanical diffraction grating, the condition for maximum position is

$$d \sin\theta = n\lambda, \qquad n = 1, 2, 3, 4, \ldots$$

The Bragg condition for constructive interference leads to the Bragg law

$$2d \sin\theta = n\lambda, \qquad \text{where } n = 1, 2, 3, \ldots$$

The deBroglie wavelength of an electron accelerated through a potential difference of V is

$$\lambda = \frac{h}{mv} = \frac{h}{\sqrt{2\,\text{me}V}} = \sqrt{\frac{150}{V}}\ \text{Å}.$$

PROBLEMS

1. What is the energy, in electron volts, of a quantum of wavelength $\lambda = 5500$ Å?
2. In order for a normal eye to be sensitive to a visible light, it must receive 100 photons per second. Assuming that the wavelength of the visible light is 5550 Å, how many watts of power at the threshold are received by the eye?
3. ★Show that the Planck radiation formula, Eq. (2.6), reduces to the Rayleigh–Jeans law, Eq. (2.4), for low frequencies or long wavelengths.
4. Light of wavelength 4350 Å is incident on a sodium surface for which the threshold wavelength of the photoelectrons is 5420 Å. Calculate the work function in electron volts, the stopping potential, and the maximum velocity of the photoelectrons emitted.
5. Show that it is impossible for a photon to collide with a free electron and lose all its energy and momentum to the electron.
6. When light of wavelength 3132 Å falls on a cesium surface, a photoelectron is emitted for which the stopping potential is 1.98 volts. Calculate the maximum energy of the photoelectron, the work function, and the threshold frequency.
7. A beam of light of wavelength 5550 Å strikes a metal surface for which the threshold is 7320 Å. Calculate the maximum kinetic energy of the emitted electron and the stopping potential.
8. The work function of tungsten is 4.53 eV. If ultraviolet light of wavelength 1500 Å is incident on this surface, does it cause photoelectron emission? If so, what is the kinetic energy of the emitted electron?

9. The work function of potassium is 2.20 eV. What should be the wavelength of the incident electromagnetic radiation so that the photoelectrons emitted from potassium will have a maximum kinetic energy of 4 eV? Also calculate the threshold.

10. Calculate the value of Planck's constant from the following data, assuming that the electronic charge e has a value of 1.602×10^{-19} coulomb: A surface when irradiated with light of wavelength 5896 Å emits electrons for which the stopping potential is 0.12 volts. When the same surface is irradiated with light of wavelength 2830 Å, it emits electrons for which the stopping potential is 2.20 volts. Also calculate the work function and threshold frequency for the surface.

11. What voltage should be applied to an x-ray tube in order to produce x-rays of wavelength 0.1 Å?

12. Electrons in an x-ray tube are accelerated through a potential difference of 3000 volts. Suppose that these electrons are slowed down in a target. What is the minimum wavelength of the x-rays produced?

13. Calculate the wavelength associated with photons of energies 1×10^{-19} joule and 5×10^{-19} joule. What are the energies, in eV, of these photons?

14. Calculate the energy, wavelength, and frequency of a photon which has the same momentum as that of an electron with a kinetic energy of 1 MeV.

15. Show that, in Compton scattering, the kinetic energy of the recoiling electron is given by

$$K_e = h\nu\left[\frac{\alpha(1 - \cos\theta)}{1 + \alpha(1 - \cos\theta)}\right],$$

where $\alpha = h\nu/m_0c^2$. Also show that the maximum value of K_e is given by

$$K_{e\,\max} = h\nu\left[\frac{2\alpha}{1 + 2\alpha}\right].$$

16. Prove the following relation between θ and ϕ in Compton scattering:

$$\cos\theta = \left[1 - \frac{2}{(1 + \alpha)^2\tan^2\phi + 1}\right].$$

17. X-rays with energies of 200 keV are incident on a target and undergo Compton scattering. Calculate (a) the energy of the x-rays scattered at an angle of 60° to the incident direction. (b) The energy of the recoiling electrons. (c) The angle of the recoiling electrons.

18. X-rays of wavelength 1 Å are scattered by a carbon target. (a) Calculate the wavelength of the scattered x-ray at 90°. (b) What is the maximum energy of the recoiling electron for this situation?

19. What is the energy of the Compton-scattered electrons detected at an angle of 45° to the direction of the incident photon in terms of the incident energy? (See Problems 2.15 and 2.16.)

20. X-rays of wavelength 0.5 Å undergo Compton scattering when they strike a target. Calculate the wavelength and the energy of the scattered photons at an angle of 45° with the direction of incidence. Also calculate the energy of the Compton electron for this case.

21. What must the relation be, in Compton scattering, between $h\nu$ and θ_0 so that both the scattered photon and the recoil electron satisfy the relation $\theta = \phi = \theta_0$? If $h\nu$ = 10 MeV, what is θ_0?

22. Photons of 500 keV are incident on a target of almost-free electrons. If the energy of the scattered photon at 45° is 250 keV, what is the angle and the energy of the recoiling electron?

23. What is the energy of an incident photon, given that the Compton scattering of this photon produces a recoiling electron of maximum energy which, when it enters a magnetic field of 0.02 weber/m^2, moves in a circle of radius 1 cm?

24. Calculate the deBroglie wavelength associated with the following.

    a) An electron with a kinetic energy of 1 eV
    b) An electron with a kinetic energy of 510 keV
    c) A neutron moving with a velocity of 2200 m/sec
    d) An automobile of 2 tons moving with a speed of 60 miles/hr

25. Calculate the deBroglie wavelength associated with (a) a proton, and (b) a neutron of kinetic energy 15 MeV.

26. What should be the kinetic energy of (a) a proton and (b) an electron so that the deBroglie wavelength associated with either is 5000 Å?

27. Through what potential difference should an electron be accelerated so that the deBroglie wavelength associated with it is 0.1 Å?

28. Calculate the deBroglie wavelength of (a) a hydrogen molecule, and (b) an oxygen atom at room temperature, i.e., moving with thermal velocities.

29. What should be the kinetic energy of an electron so that its wavelength is the same as that of x-rays produced in an x-ray tube operating at 60,000 volts?

30. X-rays of wavelength 0.3 Å are incident on a crystal for which the lattice spacing is 0.5 Å. Calculate the angles at which the first two Bragg diffractions are observed.

31. Calculate Avogadro's number from the following data. The distance between the adjacent planes of a KCl crystal is 3.14 Å. The density and molecular weight of the KCl crystal are 1.98×10^3 kg/m^3 and 74.55, respectively.

32. Electrons, after being accelerated through a potential difference of 100 volts, strike a NaCl cubic crystal for which the lattice spacing is 2.82 Å. Calculate the angle of the first-order diffraction for which the Bragg condition is satisfied.

33. A beam of electrons with kinetic energy of 150 eV is incident on a nickel crystal whose spacing is 2.15×10^{-10} m. Calculate the angle of the first- and second-order diffraction of electrons.

34. A beam of 1-keV electrons is incident on a metallic foil in which the lattice spacing of the randomly oriented crystal is 3.8 Å. Calculate the angular diffraction of the first-order diffracted beam.

35. Electrons of 100 eV are incident on a gold foil whose lattice spacing in the microcrystals is 4.07 Å. Calculate the angle of the first-order diffraction.

36. Thermal neutrons (kinetic energy = 0.025 eV) undergo diffraction from a crystal of lattice spacing 2 Å. Calculate the angle of the maximum first-order diffraction.

37. What is the maximum energy of neutrons which may be filtered through a crystal grating with spacing of 3.5 Å between the reflecting planes? Assume the angle of diffraction to be 2°. [This method of filtering is sometimes used to obtain a beam of neutrons that is almost monoenergetic.]

REFERENCES

1. J. J. Thomson, *Phil. Mag.* **5,** 44 (1897)
2. R. A. Millikan, *Electrons (+ and −), Protons, Photons, Mesotrons, and Cosmic Rays*; Chicago: University of Chicago Press, 1947
3. Max Planck, *Ann. Phys.* **4,** 553 (1901); *Verhandl. dent. Physik. Ges.*, **2,** 202, 237 (1900)
4. Lord Rayleigh, *Phil. Mag.* **49,** 539 (1900)
5. J. H. Jeans, *Phil. Mag.* **10,** 91 (1905)
6. E. R. Cohen and J. W. M. Dumond, *Revs. Mod. Phys.* **37,** 537 (1965)
7. Heinrich Hertz, *Ann. Physik* **31,** 983 (1887)
8. Albert Einstein, *Ann. Physik* **17,** 132 (1905)
9. R. A. Millikan, *Phys. Rev.* **7,** 355 (1916)
10. Wilhelm Roentgen, *Sitz-ber. Phys.-Med. Ges.*, Wurzburg (1895); *Electrician*, Jan. 24, April 24 (1896)
11. A. H. Compton, *Phys. Rev.* **21,** 715 (1923); and **22,** 409 (1923)
12. Louis deBroglie, *Ann. Physik* **3,** 22 (1925)
13. C. J. Davisson and L. J. Germer, *Phys. Rev.* **30,** 705 (1927)
14. G. P. Thomson, *Nature* **120,** 802 (1927)

SUGGESTED READING

1. William F. Magie, *A Source Book in Physics*, Cambridge, Mass.: Harvard University Press, 1963; "Temperature Radiation" by Josef Stefan
2. Atam P. Arya, *Fundamentals of Atomic Physics*, Chapters IV and V; Boston, Mass.: Allyn and Bacon, 1971
3. M. Shamos, editor, *Great Experiments in Physics*, New York: Holt, Rinehart, and Winston, 1938
4. A. B. Arons and M. B. Peppard, "Einstein's Proposal of the Photon Concept," *Am. J. Phys.* **33,** 367 (1965)
5. A. Compton, "The Scattering of X-Rays as Particles," *Am. J. Phys.* **29,** 817 (1961)
6. L. deBroglie, *Physics and Microphysics*, New York: Torchbooks, Harper and Row, 1960
7. Charles F. Meyer, *The Diffraction of Light, X-Rays, and Material Particles*, Chicago: University of Chicago Press, 1934
8. G. Thomson, "Early Work in Electron Diffraction," *Am. J. Phys.* **29,** 821 (1961)

CHAPTER 3
INTRODUCTORY QUANTUM MECHANICS

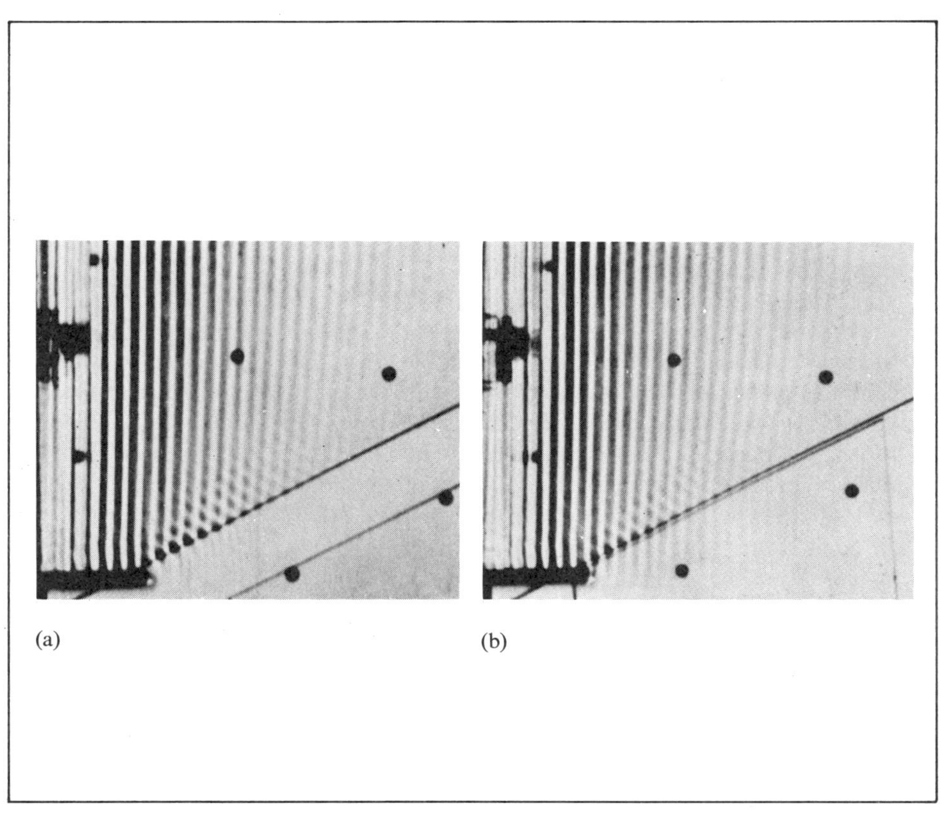

Penetration of a barrier by water in a ripple tank. (a) Incident waves are totally reflected from a very wide gap of deep water. (b) Incident waves are partly transmitted when deep water gap is very narrow, as shown. (Courtesy Film Studio, Education Development Center, Newton, Mass.)

3.1 BOHR'S PRINCIPLE OF COMPLEMENTARITY

Classical physics (or macroscopic physics) tells us that energy may be transported from one point to another either by means of waves, as in the case of water waves, or particles as in the case of a bullet transferring energy from the gun to the target. The propagation of water waves in a pond may be explained by the *wave model*, while that of the bullet may be explained by the *particle model*. These two models also apply to situations that are much less obvious. For example, we can use the wave model to explain the propagation of sound waves in an elastic medium (though this is not visible to the eye as are water waves). Similarly, according to the kinetic theory of gases, if we treat the molecules as hard spheres hitting the walls (although we cannot see the molecules), we can use the particle model to explain the pressure exerted on the walls of the container. As a matter of fact, any energy-transport phenomenon, whether visible to the naked eye or not, can be explained either by the wave model or by the particle model, but not by both. This is because *the particle model and the wave model are mutually contradictory*. We cannot simultaneously apply both particle and wave descriptions to the same phenomenon. Let us see now how these two models apply to the situations discussed in the previous chapter on the dual nature of electromagnetic radiation and material particles.

Let's take electromagnetic radiation first. If we are dealing with interference or diffraction, or if we are describing electromagnetic radiation by Maxwell's theory, we must make use of the wave model. On the other hand, in order to explain the photoelectric effect and the Compton effect, we must treat electromagnetic radiation as particles and make use of the particle model, i.e., the photon hypothesis. Once again we may point out that, even though it is essential to assign both wave and particle aspects to electromagnetic radiation, in any given experimental situation only one model can be applied. Never both! Accordingly, in 1928, Niels Bohr[1] stated the *principle of complementarity: The wave and particle aspects of electromagnetic radiation are complementary*. That is, any given experimental measurement involving electromagnetic radiation can be completely explained by one model or the other.

We can see how Bohr's principle of complementarity applies to the dual nature —wave and particle—of material particles such as electrons, protons, and others. The measurement of e/m and other characteristics of cathode rays clearly established the particle aspects of the electron. On the other hand, we said in Chapter 2 that electrons or any other material particles must be assigned waves of deBroglie wavelength ($\lambda = h/p$) in order to explain the diffraction of material particles. Thus once again the principle of complementarity seems to be at work. In some experiments the wave nature is suppressed, while in others the particle nature is suppressed, but the two models complement each other in explaining the properties of material particles.

When we speak of photons, we know that it is the electromagnetic wave that is associated with them. In the case of material particles, we may know their wavelengths from the deBroglie hypothesis, but we do not know the nature of these waves. The waves which guide the motion of (or are associated with) particles are called *matter waves*. Let us now deal with these in some detail.

3.2 WAVE-PACKET DESCRIPTION OF MATERIAL PARTICLES

A typical wave motion along the x-axis is described by an equation of the form

$$y(x, t) = A \sin (kx - \omega t), \tag{3.1}$$

where $y(x, t)$ may represent the displacement of a string, or the electric field $\mathscr{E}(x, t)$ of an electromagnetic wave, or it may stand for the pressure $P(x, t)$ in a sound wave; k $(= 2\pi/\lambda)$ is the propagation constant and ω $(= 2\pi\nu)$ is the angular frequency. In analogy to these quantities, we may replace $y(x, t)$ by $\Psi(x, t)$, which is called the *wave function* of the wave associated with a material particle, and which is given by

$$\Psi(x, t) = A \sin (kx - \omega t). \tag{3.2}$$

But we encounter two difficulties with this equation. One of the features that distinguishes a wave from a particle is that a wave has a continuity in space and hence is unlocalized, i.e., it is spread out; while a material particle is always localized in space. Therefore Eq. (3.2) could not represent a particle which is localized in space. Also the propagation velocity, the so-called *phase velocity*, v_{ph}, is given by

$$v_{ph} = \frac{\omega}{k} = \frac{2\pi\nu}{2\pi/\lambda} = \nu\lambda. \tag{3.3}$$

For a particle, using the relations $E = h\nu$ and $p = h/\lambda$, we get

$$v_{ph} = \frac{E}{h}\frac{h}{p} = \frac{E}{p}.$$

Also using $E = mc^2$ and $p = mv$, where m is the mass and v is the velocity of the particle, gives

$$v_{ph} = \frac{E}{p} = \frac{mc^2}{mv} = \frac{c^2}{v}.$$

That is,

$$\boxed{v_{ph} = \frac{c^2}{v}.} \tag{3.4}$$

For a photon, $v = c$ and hence $v_{ph} = c$, as it should be. But for a material particle v is always less than c, and hence v_{ph} is greater than c. For Eq. (3.2) to represent a particle, the velocity of the wave should be equal to the velocity of the particle.

Thus we have shown the inadequacy of Eq. (3.2) to represent a wave associated with a material particle. But the localization can be achieved and the difficulty connected with Eq. (3.4) overcome if we assume that the wave function $\Psi(x, t)$

actually represents a group of waves of different frequencies and wavelengths. Thus, for example,

$$\Psi(x, t) = \sum_{k_i=k}^{k_i=k+\Delta k} A(k_i) \sin(k_i x - \omega_i t). \tag{3.5}$$

The range of Δk needed in the summation depends on the degree of localization of the particle in space. The aim is to combine many waves of different frequencies and amplitudes so that the resultant has a high value of amplitude near the vicinity of the particle and is zero everywhere else.

As an example, Fig. 3.1(a) shows the result of summing of seven different waves, resulting in what is called a *wave packet*. It is shown at the bottom of this figure. All we can say is that the particle is somewhere in the wave packet and that the wave packet moves with the velocity of the particle, carrying the particle with it. In time t_1 the wave packet has moved a distance $v_0 t_1 = (p/m)t_1 = (\hbar k_0/m)t_1$, as shown in Fig. 3.1(b). If we want to localize the particle better than this, we will have to increase the range of Δk. In order to analyze this situation, we need to use Fourier series and the Fourier integral, which are outside the scope of this book. But when we use only a simple case, we shall show that the *group velocity*, v_g—or the velocity of the wave packet—is the same as that of the particle.

For simplicity, let us consider two waves of equal amplitudes, but slightly different frequencies:

$$\Psi_1(x, t) = A \sin(kx - \omega t) \tag{3.6}$$

and

$$\Psi_2(x, t) = A \sin[(k + \Delta k)x - (\omega + \Delta\omega)t]. \tag{3.7}$$

Adding these two waves and using trigonometric relationships, we get

$$\begin{aligned}\Psi(x, t) &= \Psi_1(x, t) + \Psi_2(x, t)\\ &= A \sin(kx - \omega t) + A \sin[(k + \Delta k)x - (\omega + \Delta\omega)t]\\ &= 2A \cos\left[\frac{\Delta k}{2} x - \frac{\Delta\omega}{2} t\right] \sin\left[\left(k + \frac{\Delta k}{2}\right) x - \left(\omega + \frac{\Delta\omega}{2}\right) t\right].\end{aligned}$$

Letting $k + (\Delta k/2) \simeq k$ and $\omega + (\Delta\omega/2) \simeq \omega$, we get

$$\begin{aligned}\Psi(x, t) &= 2A \cos\left(\frac{\Delta k}{2} x - \frac{\Delta\omega}{2} t\right) \sin(kx - \omega t)\\ &= A_m \sin(kx - \omega t),\end{aligned} \tag{3.8}$$

which represents a wave of original frequency ω, but with amplitude modulated as shown in Fig. 3.2. Even though individual waves still travel with phase velocity v_{ph}, the amplitude modulation

$$A_m = 2 \cos\left(\frac{\Delta k}{2} x - \frac{\Delta\omega}{2} t\right) \tag{3.9}$$

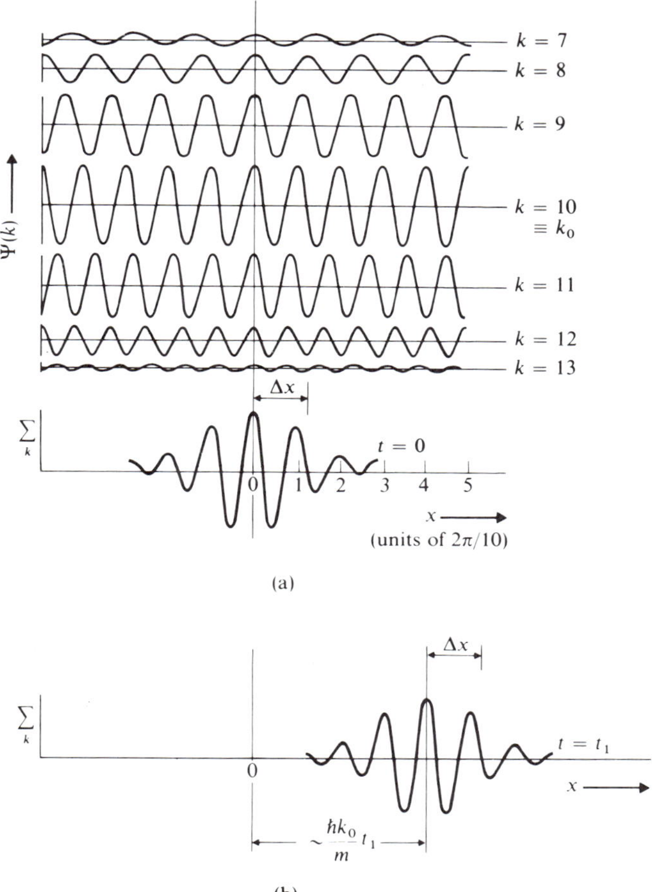

Fig. 3.1 a) Formation of a wave packet by combination of seven waves, with $k = 7$ to $k = 13$. The average wave number is $k_0 = 10$. b) In time t_1, the wave packet moves a distance $(\hbar k_0/m)t_1$. (Reprinted by permission from: Charles W. Sherwin, *Introduction to Quantum Mechanics*, Holt, Rinehart and Winston, 1959.)

travels with a group velocity v_g given by

$$v_g = \frac{\Delta\omega/2}{\Delta k/2} = \frac{\Delta\omega}{\Delta k},$$

or in the limit as $\Delta k \to 0$,

$$v_g = \frac{d\omega}{dk}. \tag{3.10}$$

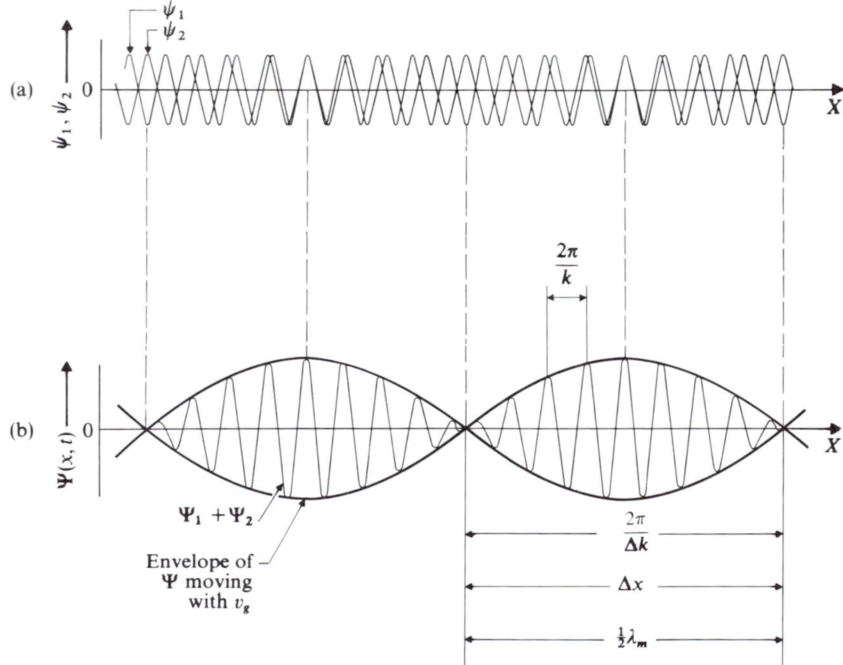

Fig. 3.2 Two waves of slightly different frequencies and wavelengths traveling in the x-direction as shown in (a) produce maxima and minima when added together as shown in (b).

Using the relations $E = h\nu = \hbar\omega$, where $\hbar = h/2\pi$ and $p = h/\lambda = \hbar k$, we get

$$\boxed{v_g = \frac{d\omega}{dk} = \frac{dE}{dp}.} \tag{3.11}$$

But from $E = \sqrt{m_0^2c^4 + p^2c^2}$, we obtain

$$\frac{dE}{dp} = \frac{pc^2}{\sqrt{m_0^2c^4 + p^2c^2}} = \frac{pc^2}{E} = \frac{mvc^2}{mc^2} = v$$

or

$$v_g = v. \tag{3.12}$$

That is, the velocity v_g of the group (or the velocity of the wave packet) is the same as that of the particle, and hence the packet can guide the motion of the particle and also localize it.

3.3 STATISTICAL INTERPRETATION OF THE WAVE FUNCTION

Of course, we must investigate further the relation between the wave function $\Psi(x, t)$ of a particle and its location. Once again, we can obtain a meaningful interpretation of the wave function $\Psi(x, t)$ by drawing an analogy with the amplitude of the electric field of the electromagnetic radiation, i.e., by replacing $y(x, t)$ by $\mathscr{E}(x, t)$ in Eq. (3.1).

Consider a beam of monochromatic electromagnetic radiation incident at a right angle to a screen. The intensity I of illumination, defined as the energy per unit area per unit time, is given by

$$I = \epsilon_0 \mathscr{E}^2 c, \tag{3.13}$$

where ϵ_0 is the permittivity of free space, $\mathscr{E}$ is the magnitude of the instantaneous electric field on the screen, and c is the velocity of light. Another viewpoint is to treat electromagnetic radiation as consisting of photons, i.e., the screen receives discrete bundles of energy $h\nu$. Let us say that the photon flux, or the number of photons falling on a unit area in a unit time, is N. Then we may write the intensity as

$$I = Nh\nu. \tag{3.14}$$

Equating the above two equations, we have

$$I = Nh\nu = \epsilon_0 \mathscr{E}^2 c, \qquad N = (\epsilon_0 c/h\nu)\mathscr{E}^2.$$

That is,

$$N \propto \mathscr{E}^2. \tag{3.15}$$

This is an interesting result. The left side of the equation represents the particle model of electromagnetic radiation, while the right side represents the wave model. For a large flux there is no difficulty in explaining the illumination of the screen with either model. It is true that each photon produces a discrete flash of light. But when a large number of photons fall on a screen, the discrete and randomly distributed flashes merge into a uniform illumination of the screen. The same result is obtained by considering $\mathscr{E}^2$.

Now suppose that the intensity of the electromagnetic radiation falling on the screen is decreased to such an extent that only a few photons are received on a unit area in a unit time. If we keep a record of the arrival of these photons on the screen, we may conclude the following.

It is not possible to predict the exact position and time of arrival of each single photon on the screen. The distribution of photons on the screen is completely random, but the average number of photons arriving per unit area per unit time is constant and predictable.

Thus if the screen is replaced by a photographic plate and is exposed to this weak radiation for a long time, the result will be a uniform illumination of the plate.

From all this it is clear that the situation is similar to one in statistical mechanics concerning the kinetic theory of gases, in which one cannot predict the exact behavior of individual particles in a system, but *can* calculate the overall average effects. In situations like these, we have to use the concepts of probability. Thus we may say that the photon flux gives only the probability of observing a photon and not the exact location and time of arrival of each photon on the screen. But according to Eq. (3.15), $N \propto \mathscr{E}^2$. Therefore

$$\mathscr{E}^2 \propto \text{probability of observing a photon.} \tag{3.16}$$

Thus: The square of the magnitude of the electric field strength at any point in space is proportional to the probability of observing a photon at that point.

Now we can extend these concepts to the wave packets associated with material particles. These considerations led Born,[2] in 1926, to state the statistical interpretation of the wave function $\Psi(x, y, z, t)$ and to give a precise meaning to the square of the absolute value of the wave function. According to Born:[2]

The wave function $\Psi(x, y, z, t)$ is a quantity such that the product

$$|\Psi(x, y, z, t)|^2 \, dv \tag{3.17}$$

is the probability of observing a particle in a volume element $dv = dx\,dy\,dz$ at the time t.

If the system is stationary—i.e., if it is independent of time—we may write the probability of observing a particle as

$$|\psi(x, y, z)|^2 \, dv = \psi^*(x, y, z)\psi(x, y, z) \, dv, \tag{3.18}$$

where $\psi(x, y, z)$ is a time-independent wave function and $\psi^*(x, y, z)$ is its complex conjugate. In the case of one dimension, we may state

$$\left.\begin{array}{l}\text{The probability of observing a particle}\\ \text{between } x \text{ and } x + dx \text{ is}\end{array}\right\} \propto |\psi(x)|^2 \, dx. \tag{3.19}$$

Thus, wherever ψ is large, the probability of finding the particle is large, and vice versa.

There is one fundamental difference between Ψ describing a matter wave (or wave packet) and $\mathscr{E}$ describing the electric field. $\mathscr{E}$ as well as $\mathscr{E}^2$ have physical meaning, while Ψ has no physical meaning; $|\Psi|^2$ has. The branch of physics which deals with the problems of finding Ψ and thus describing the behavior of wave packets (and hence the motion of particles) is called *quantum mechanics or wave mechanics.* This exploration was done independently by Erwin Schrödinger[3] (1926) and by Max Born, Werner Heisenberg, and Pascual Jordan[4] (1925). We'll talk about it later.

When we say that each photon or material particle is assigned a wave function, we imply that a statistical property is assigned to each individual particle. To

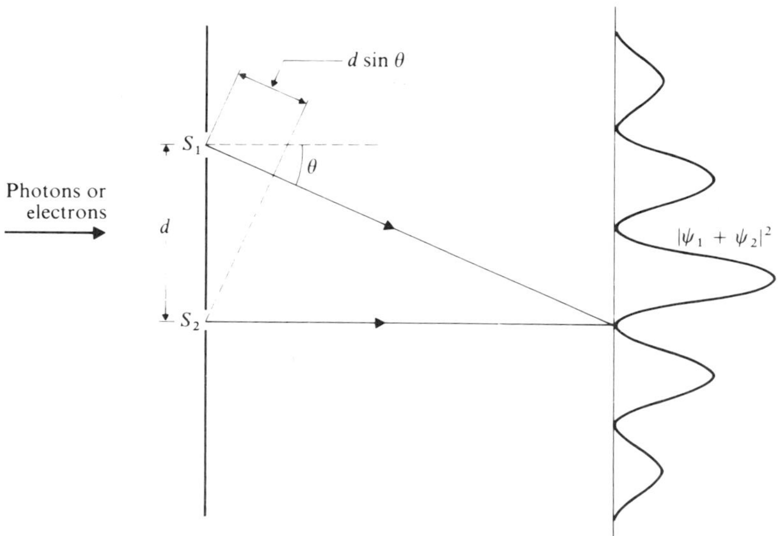

Fig. 3.3 Double-slit diffraction pattern of particles or photons. The first minimum occurs at an angle such that $d \sin\theta = \lambda/2$.

demonstrate this and other points made above, let's look at the familiar double-slit experiment, shown in Fig. 3.3. The points of zero intensity are those at which the waves arriving from the two slits are 180° out of phase.

Let us see what happens now if we replace the strong source by a very weak source of photons or electrons, and the screen by a photographic plate. If the plate is exposed for a very short time, the pattern obtained is as shown in Fig. 3.4(a); note the scattered discrete points. With increasing time of exposure, we get patterns of the type shown in Fig. 3.4(b) and (c). Eventually an extremely long exposure yields the typical pattern shown in Fig. 3.4(d) (similar to the one shown in Fig. 3.3). Note that for a short exposure the photon nature of the electromagnetic radiation is quite obvious, but for long exposure, the fluctuations cancel out and the quantum nature disappears. Thus it is the total number and not the rate of photons or electrons falling on the plate which determines the pattern. That is, $\mathscr{E}^2$ (and hence $|\Psi|^2$) is proportional to the probability of observing a photon in a volume dv. Thus $\mathscr{E}^2$ is large where photons are likely to be found on the screen and small where they are less likely to be found.

We may go a step further and reduce the intensity of photons to one photon per unit area per unit time. According to our classical thinking, the photon can pass through one slit or the other, but *not* both. Question: Is the passage of a photon from one slit effected whether the other slit is opened or closed? Answer: Yes. When one uses a beam of extremely weak intensity (say one photon per unit

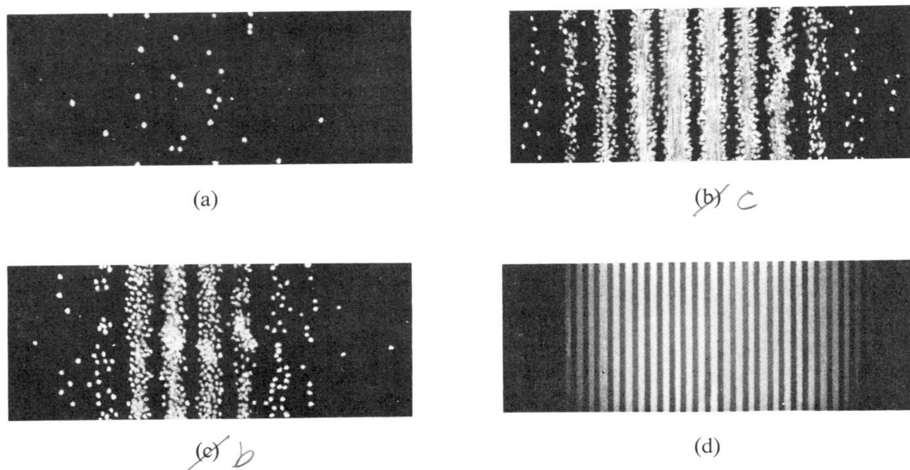

Fig. 3.4 Growth of a double-slit pattern on a photographic film when a beam of electrons or photons is incident on a double-slit system. (a) The pattern obtained when the film is struck by 28 electrons, (b) 1000 electrons, (c) about 10,000 electrons. (The dots are much bigger than actual size.) Note that there are no dots in the region of interference minima. (From Elisha R. Huggins, *Physics 1*, New York: W. A. Benjamin, 1968, page 510) (d) Double-slit pattern obtained when the film is exposed to millions of electrons or photons.

area per unit time) and a very long exposure, the pattern is a typical double-slit pattern *if* both slits are open; otherwise it is a typical single-slit pattern. This implies that each photon must be ascribed some statistical properties that are definitely sensitive to whether both slits are open, or just one.

3.4 THE HEISENBERG UNCERTAINTY PRINCIPLE

In the preceding sections we succeeded in assigning some sort of wave function to material particles. Often in this text we'll be able to see the gains made by doing this. But this gain is not without a sacrifice, which is that there's less accuracy in the simultaneous determination of certain pairs of quantities. We shall see this presently.

The fact that the quantity $|\Psi(x, t)|^2 \Delta x$ must be interpreted as a probability that the particle is within the region between x and $x + \Delta x$ implies that there is an uncertainty in the location of the position of the particle. Of course, if the wave packet is very narrow (see Fig. 3.2), Δx will become smaller, and hence the uncertainty in the position will be less. However, to construct a narrow wave packet requires that the range of wave numbers Δk between k and $k + \Delta k$ be very wide. Thus

$$\Delta x \propto \frac{1}{\Delta k}. \tag{3.20}$$

Without an exact knowledge of the wave packet, we may approximate this as

$$\Delta x\, \Delta k \approx 1. \tag{3.21}$$

Since $p = h/\lambda = (2\pi/\lambda)(h/2\pi) = k\hbar$, we can substitute $\Delta k = \Delta p/\hbar$ in the above equation to obtain

$$\Delta x\, \Delta p \approx \hbar.$$

Since this is the lowest limit of accuracy, we may write more generally

$$\boxed{\Delta x\, \Delta p \geq \hbar.} \tag{3.22}$$

More sophisticated calculations show that the lower limit of accuracy is $\frac{1}{2}\hbar$, but we shall use $\hbar$. [An alternative derivation using a specific example of a single slit is given at the end of this section.] Equation (3.22) implies that if the particle is completely localized, that is, if $\Delta x = 0$, then $\Delta p = \infty$, and hence the momentum of the particle is completely unknown. Similarly, if the momentum of the particle is known precisely, that is, if $\Delta p = 0$, then $\Delta x = \infty$, and hence the particle is completely unlocalized and spread out over all space. Thus when we try to simultaneously measure the position and momentum of a particle, the product of the uncertainties Δx and Δp cannot be less than $\hbar$. The relation in Eq. (3.22) was given by Werner Heisenberg[5] in 1927, and is known as the *Heisenberg uncertainty principle* (or indeterminacy principle). It can be stated as follows.

It is impossible to measure simultaneously and precisely both the position and the momentum of a particle.

This limit of accuracy is not due to any technical difficulties in the experiment; but it is inherent in the dual nature—wave and particle—of both electromagnetic radiation and material particles.

In the case of three dimensions, if p_x, p_y, and p_z are components of the momentum of a particle, then Eq. (3.22) takes the form

$$\begin{aligned} \Delta x\, \Delta p_x &\geq \hbar, \\ \Delta y\, \Delta p_y &\geq \hbar, \\ \Delta z\, \Delta p_z &\geq \hbar. \end{aligned} \tag{3.23}$$

Example 3.1. Using the Heisenberg uncertainty principle, calculate the minimum energy of an electron when it is trapped in a crystal lattice vacancy. Assume that the electron can move in a spherical volume of radius r_0.

According to the uncertainty relation,

$$\Delta x\, \Delta p \geq \hbar, \qquad \Delta p \geq \hbar/\Delta x.$$

The minimum value of p cannot be less than Δp. That is,

$$p_{\min} = \Delta p$$

and also

$$\Delta x = \text{diameter of sphere} = 2r_0, \qquad p_{\min} = \frac{\hbar}{2r_0}.$$

Therefore

$$E_{\min} = \frac{(p_{\min})^2}{2m_0} = \frac{\hbar^2}{8m_0 r_0^2},$$

where m_0 is the mass of the electron.

Let us assume that $r_0 = 4 \times 10^{-8}$ cm, that is, that the neighboring atoms are a distance r_0 apart. Therefore

$$E_0 = \frac{(1.054 \times 10^{-34}\ \text{J-sec})^2}{8 \times 9.1 \times 10^{-31}\ \text{kg} \times (4 \times 10^{-10}\ \text{m})^2}$$

$$= 0.097 \times 10^{-19}\ \text{J}$$

$$\simeq 0.061\ \text{eV}.$$

Two other pairs of quantities for which accuracy is limited in the simultaneous determination of both quantities are: (1) angular momentum and angle, and (2) energy and time. Suppose we have a particle at a particular angular position θ with an angular momentum L_θ; then the limits in the uncertainties $\Delta\theta$ and ΔL_θ are given by the relation

$$\boxed{\Delta\theta\ \Delta L_\theta \geq \hbar.} \tag{3.24}$$

Similarly, if a particle is in a given energy state E at time t, then the product of the uncertainties ΔE and Δt is given by

$$\boxed{\Delta t\ \Delta E \geq \hbar.} \tag{3.25}$$

Thus, according to Heisenberg, the measurement of the energy of a particle within a time Δt must be uncertain by an amount ΔE, the product being given by Eq. (3.25). Equation (3.25) may be derived from Eq. (3.22) by using the relation $E = p^2/2m$, which gives $\Delta E = p\ \Delta p/m = v\ \Delta p$, and $t = x/v$, which gives $\Delta t = \Delta x/v$.

Before we leave the subject of the uncertainty principle, let us discuss an experimental situation from which Eq. (3.22) follows automatically. Consider the diffraction of a beam of electrons by a single slit, as shown in Fig. 3.5. Suppose a single electron traveling along the Y-axis passes through the slit of width d. One can't say exactly where the electron crosses the slit. Thus the position of the

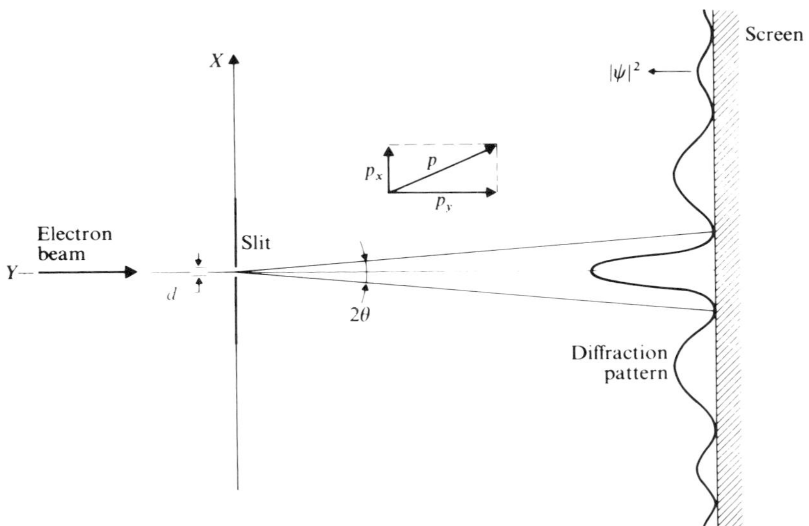

Fig. 3.5 Diffraction pattern of a beam of monoenergetic electrons incident on a single slit of width d.

electron is uncertain by the width of the slit, and we may write

$$\Delta x = d. \tag{3.26}$$

While the wave packet of the electron is passing through the slit, it is disturbed. Hence we obtain a diffraction pattern on the screen instead of a spot. This means that we are uncertain as to the momentum of the electron in the X-direction. The chances are that the electron will land within an angle 2θ formed by the central maximum of the diffraction pattern and the first minimum on either side. Thus if the initial momentum is p, its x-component after diffraction from the slit is $p \sin \theta$. Thus

$$\Delta p_x \simeq p \sin \theta. \tag{3.27}$$

But according to the diffraction condition for the first maximum, $d \sin \theta = \lambda$ or $\sin \theta = \lambda/d$, which gives

$$\Delta p_x \simeq p \frac{\lambda}{d} = \frac{h}{\lambda} \frac{\lambda}{d} = \frac{h}{d},$$

which, when we substitute for d from Eq. (3.26), yields

$$\Delta x \, \Delta p_x \simeq h \geq \hbar$$

which agrees with Eq. (3.22).

By now we ought to realize that uncertainties are inherent properties of the system under observation, and not merely technical limitations of a given experiment.

3.5 THE SCHRÖDINGER WAVE EQUATION

In classical mechanics, Newton's laws describe the motion of particles. In classical electromagnetism, the Maxwell field equations describe electromagnetic fields. At this point, we need a mathematical structure based on the premise of the dual nature of matter. This mathematical structure must make use of the wave function associated with matter waves, as discussed in the previous sections. Such a mathematical formulation—known as *wave mechanics* or *quantum mechanics*—was developed in 1926 by Erwin Schrödinger[3]. At the same time, Werner Heisenberg, Max Born, and Pascual Jordan[4] proposed another formulation, called *matrix mechanics*. At first glance the two appeared to be quite different, but eventually scientists realized that the results obtained by the two different treatments were essentially the same.

In this section we shall develop the Schrödinger wave equation, and in subsequent sections apply it to simple situations. Although the predictions of this theory may look quite strange, they agree with the experimental results. Let us be aware, at the outset, that Schrödinger's wave equation, like Newton's law, cannot be derived. There are two distinct approaches to obtaining it: the plausibility argument and the postulative approach. We shall take the latter approach.

Schrödinger described the amplitude of a matter wave by a complex quantity $\Psi(x, y, z, t)$, called the *wave function* of the state of the system. That is, it describes the particular dynamical system under observation. If the system does not change with time, we call it a *stationary state*, and the wave function, which is now independent of time, is denoted by $\psi(x, y, z)$. The basic postulates of quantum theory given below apply to the motion of a time-independent single particle, but are equally applicable to more general cases.

The postulates of quantum mechanics may be stated as follows.

Postulate 1 *Wave Functions to Describe Physical Systems*

For the wave function ψ to describe any physical system, it must be well-behaved, i.e., it must be finite, single-valued, and continuous everywhere. This is true only if the following so-called boundary conditions are satisfied.

1. ψ must be continuous, and
2. its derivative ψ' (with respect to space coordinates) must also be continuous everywhere, except where the potential $V(x, y, z)$ is infinite.
3. ψ must vanish at infinity, that is, $\psi \to 0$ as $x, y, z \to \pm\infty$.

In addition to this, the wave function ψ must satisfy the Born condition that $\psi^*\psi\, dx\, dy\, dz$ is the probability of observing a particle in a volume element $dx\, dy\, dz$. Since the particle must be somewhere in space, the integrated probability must be equal to unity. That is,

$$\int_{\text{All space}} \psi^*\psi\, dx\, dy\, dz = 1. \tag{3.28}$$

This is called the *normalization condition*, and it adds another boundary condition to the three already stated above.

If there are n particles in space, Eq. (3.28) takes the form

$$\int_{\text{All space}} \psi^*\psi\, dx\, dy\, dz = n, \tag{3.29}$$

where ψ is now the unnormalized wave function, and the normalized wave function is $\psi/\sqrt{n}$. The quantity $(1/\sqrt{n}) = N$ is called the *normalization constant.*

Postulate II *Operators for Observable Quantities*

There is an operator corresponding to every observable quantity. (In general, an operator is anything that is capable of doing something to a function, such as the multiplication operator $\times$, the differentiation operator d/dr, the integration operator $\int dr$, etc.) In quantum mechanics, the choice of the operator is arbitrary, but it must satisfy the condition that when it operates on a wave function it gives the observable quantity times the wave function. Thus, if O is an operator corresponding to the observable o, then

$$O\psi = o\psi. \tag{3.30}$$

Table 3.1 gives a few quantum-mechanical operators corresponding to different classical observable quantities. Starting with these, we can derive many other operators. (Note that i stands for the imaginary quantity $\sqrt{-1}$.)

The wave functions which result in the observable quantities according to Eq. (3.30) are called *eigenfunctions* (*eigen* means "proper"), the quantities o are called the *eigenvalues*, and the equation itself is called the *eigenvalue equation.* For example, if O is the momentum operator, when it operates on the eigenfunction of a particle in a given state, it yields the momentum of the particle in that state times the wave function.

To solve a problem in quantum mechanics means to find the eigenvalues and eigenfunctions by solving the Schrödinger wave equation.

★***Postulate III*** *The Expectation Value*

The average value, or the expectation value, of an observable quantity o corresponding to the operator O of a physical system in the state $\psi(x, y, z)$ is

Table 3.1

The Quantum-Mechanical Operators

Classical quantity		Quantum-mechanical operator
Position	x, y, z	x, y, z
Momentum	p	$p = -i\hbar\nabla = -i\hbar$ grad
x-component	p_x	$p_x = -i\hbar \dfrac{\partial}{\partial x} = \dfrac{\hbar}{i}\dfrac{\partial}{\partial x}$
y-component	p_y	$p_y = -i\hbar \dfrac{\partial}{\partial y} = \dfrac{\hbar}{i}\dfrac{\partial}{\partial y}$
z-component	p_z	$p_z = -i\hbar \dfrac{\partial}{\partial z} = \dfrac{\hbar}{i}\dfrac{\partial}{\partial z}$
Energy	E	$E = i\hbar \dfrac{\partial}{\partial t} = -\dfrac{\hbar}{i}\dfrac{\partial}{\partial t}$

given by

$$\langle o \rangle = \frac{\int \psi^* O \psi \, dx\, dy\, dz}{\int \psi^* \psi \, dx\, dy\, dz}. \tag{3.31}$$

If the wave function is normalized to unity, then the denominator is equal to unity.

We are now in a position to write the Schrödinger wave equation and start using the above postulates. Then we can compare our results with the experimentally measured values, so as to establish the truth of this theory.

The Schrödinger wave equation

The total energy E of a particle of mass m moving with velocity $\mathbf{v}(v^2 = v_x^2 + v_y^2 + v_z^2)$ and having potential energy $V(x, y, z, t)$ at time t is given by

$$\begin{aligned} E &= \text{kinetic energy} + \text{potential energy} \\ &= \tfrac{1}{2}mv^2 + V(x, y, z, t) \end{aligned}$$

or

$$E = \frac{p^2}{2m} + V(x, y, z, t), \tag{3.32}$$

where $\mathbf{p} = m\mathbf{v}$ is the linear momentum of the particle.

We can make the transition from classical to quantum mechanics by replacing the observable quantities p and E by their corresponding operators. That is,

$$\mathbf{p} \to P \equiv -i\hbar\left(\mathbf{l}\frac{\partial}{\partial x} + \mathbf{m}\frac{\partial}{\partial y} + \mathbf{n}\frac{\partial}{\partial z}\right) \tag{3.33}$$

and

$$E \to E \equiv i\hbar \frac{\partial}{\partial t}, \tag{3.34}$$

where **l**, **m**, and **n** are unit vectors along the X, Y, and Z axes, respectively. The operator corresponding to p^2 is

$$p^2 = \mathbf{p}\cdot\mathbf{p} \to P^2 \equiv -\hbar^2\left(\frac{\partial^2}{\partial x^2} + \frac{\partial^2}{\partial y^2} + \frac{\partial^2}{\partial z^2}\right) \equiv -\hbar^2\nabla^2. \tag{3.35}$$

But these operators must operate on some eigenfunction $\Psi(x, y, z, t)$ of the state of the particle. Thus, in Eq. (3.32), we replace E and p^2 by the operators in Eqs. (3.34) and (3.35), and after rearranging, we get

$$-\frac{\hbar^2}{2m}\left(\frac{\partial^2\Psi}{\partial x^2} + \frac{\partial^2\Psi}{\partial y^2} + \frac{\partial^2\Psi}{\partial z^2}\right) + V(x, y, z, t)\Psi = i\hbar\frac{\partial\Psi}{\partial t}$$

or

$$\boxed{-\frac{\hbar^2}{2m}\nabla^2\Psi + V(x, y, z, t)\Psi = i\hbar\frac{\partial\Psi}{\partial t},} \tag{3.36}$$

which is the famous *time-dependent Schrödinger wave equation.*

When we use the notation

$$\boxed{H \equiv -\frac{\hbar^2}{2m}\nabla^2 + V(x, y, z, t),} \tag{3.37}$$

we can write Eq. (3.36) in the operator form of Eq. (3.30). That is,

$$\boxed{H\Psi = E\Psi,} \tag{3.38}$$

where H is called the *Hamiltonian operator.* We can understand Eq. (3.38) better if we think about the following statement: The operator H (also called the energy operator), when it operates on the eigenfunction Ψ of the state of a particle, yields the eigenvalue energy E of that state times the eigenfunction.

In many cases which we shall come across in this text, the system is stationary; that is, it does not change with time. And also the potential $V(x, y, z, t)$ is independent of time. In such cases, we can use the separation-of-variables technique to obtain a much-easier-to-handle, time-independent Schrödinger wave equation, as shown below.

Let us assume that the function $\Psi(x, y, z, t)$ is a product of two functions $\psi(x, y, z)$, which is a function of space coordinates only, and $T(t)$, which is a function of time only. That is,

$$\Psi(x, y, z, t) = \psi(x, y, z)T(t). \tag{3.39}$$

Substituting this in Eq. (3.36) and also $V(x, y, z, t) = V(x, y, z,)$, we get

$$-\frac{\hbar^2}{2m} T\nabla^2\psi + V(x, y, z)\psi T = i\hbar\psi \frac{\partial T}{\partial t}.$$

Dividing both sides by ψT, we obtain

$$-\frac{\hbar^2}{2m}\frac{\nabla^2\psi(x, y, z)}{\psi(x, y, z)} + V(x, y, z) = i\hbar \frac{1}{T(t)}\frac{\partial T(t)}{\partial t}. \tag{3.40}$$

The left side of this equation is a function of x, y, z only, while the right side is a function of t only. This is possible only if both sides are equal to a constant. We shall see that this constant is the total energy E. Thus the right side of Eq. (3.40) takes the form

$$i\hbar \frac{1}{T}\frac{\partial T}{\partial t} = E. \tag{3.41}$$

Integration of this equation yields

$$T(t) = e^{-i(E/\hbar)t}. \tag{3.42}$$

Note that for the exponent to be dimensionless, E must be identified as energy. That is, $E = \hbar\omega$, as we mentioned earlier.

If we equate the left side of Eq. (3.40) with E, we get

$$-\frac{\hbar^2}{2m}\nabla^2\psi(x, y, z) + V(x, y, z)\psi(x, y, z) = E\psi(x, y, z), \tag{3.43}$$

which is the famous *time-independent Schrödinger wave equation* (*S.W.E.*). It may also be written as

$$\boxed{\nabla^2\psi + \frac{2m}{\hbar^2}[E - V(x, y, z)]\psi = 0,} \tag{3.44}$$

where

$$\nabla^2\psi = \frac{\partial^2\psi}{\partial x^2} + \frac{\partial^2\psi}{\partial y^2} + \frac{\partial^2\psi}{\partial z^2}. \tag{3.45}$$

The complete wave function for the time-dependent S.W.E., obtained by combining Eqs. (3.39) and (3.42), is

$$\Psi(x, y, z, t) = \psi(x, y, z)e^{-i(E/\hbar)t}. \tag{3.46}$$

In the rest of this chapter we shall discuss some applications of the time-independent S.W.E. for the motion of a particle in one dimension, in which case Eq. (3.44) takes the form

$$\frac{d^2\psi(x)}{dx^2} + \frac{2m}{\hbar^2}[E - V(x)]\psi(x) = 0. \tag{3.47}$$

For future use, let us write the solutions of Eq. (3.47) for three special cases.

i) For a particle moving in a potential-free region, that is, $V(x) = V_0 = 0$, the S.W.E. for a free particle is

$$\frac{d^2\psi}{dx^2} + \frac{2mE}{\hbar^2}\psi = 0$$

and has the solution

$$\psi(x) = N \exp\left(\pm i x\sqrt{\frac{2mE}{\hbar^2}}\right) \quad \text{for } V_0 = 0, \tag{3.48}$$

where N is the normalization constant.

ii) For a particle moving in a constant potential V_0 such that $E > V_0$, the solution of Eq. (3.47) is

$$\psi(x) = A \sin kx + B \cos kx \quad \text{for } E > V_0 \tag{3.49a}$$

or

$$\psi(x) = A'e^{ikx} + B'e^{-ikx} \quad \text{for } E > V_0, \tag{3.49b}$$

where $k = \sqrt{2m(E - V_0)}/\hbar$ and A, B, A', and B' are constants.

iii) For a particle moving in a constant potential V_0 such that $E < V_0$, Eq. (3.47) takes the form

$$\frac{d^2\psi}{dx^2} - k'^2\psi = 0$$

where $k' = \sqrt{2m(V_0 - E)}/\hbar$, and the solution is

$$\psi(x) = Ce^{k'x} + De^{-k'x} \quad \text{for } E < V_0, \tag{3.50}$$

where C and D are constants.

In anticipation, let us say that the results obtained by applying the S.W.E. differ from the results obtained by using classical mechanics in two important aspects. (a) Quantum mechanics predicts discrete energy states for bound systems, whereas classical theory predicts a continuous range of energies. (b) In quantum mechanics, we always talk in terms of *probabilities* of things happening instead of certainties. Also, because of the uncertainty principle, a system cannot have a zero value for kinetic energy.

3.6 PARTICLE IN A ONE-DIMENSIONAL POTENTIAL WELL

As a simple application of quantum mechanics, let us consider a single particle whose motion is restricted by the reflecting walls of a one-dimensional infinite potential well (or barrier), as shown in Fig. 3.6, and given by

$$V(x) = 0, \qquad \text{for } 0 < x < a$$
$$V(x) = \infty, \qquad \text{for } x < 0 \text{ and } x > a.$$

The particle inside the well always remains inside because of the infinite walls of the well. Hence the probability of finding the particle outside is zero; i.e., outside the well $\psi(x) = 0$. Inside the well, for x between 0 and a, $V(x) = 0$, and the S.W.E. of the particle in this region is

$$\frac{d^2\psi}{dx^2} + \frac{2mE}{\hbar^2}\psi = 0$$

or

$$\frac{d^2\psi}{dx^2} + k^2\psi = 0, \tag{3.51a}$$

where

$$k = \sqrt{2mE}/\hbar. \tag{3.51b}$$

The general solution of Eq. (3.51) for the particle inside the well is

$$\psi(x) = A \sin kx + B \cos kx, \tag{3.52}$$

where the constants A and B may be evaluated by making use of the boundary conditions. Thus $\psi(x)$ must be zero at $x = 0$ and $x = a$. Since $\psi(x) = 0$ at $x = 0$,

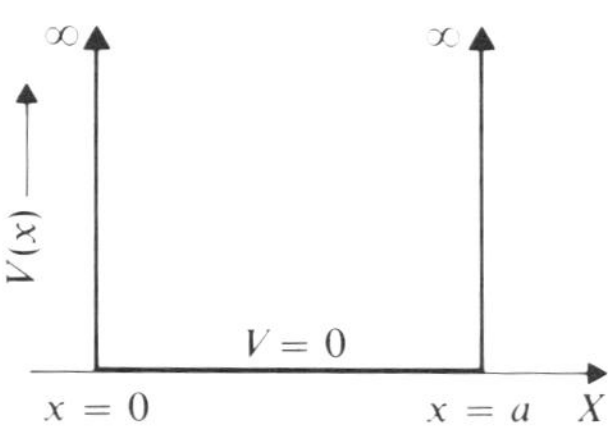

Fig. 3.6 One-dimensional infinite potential well. The motion of a particle inside the well is restricted between $x = 0$ and $x = a$ by the infinite reflecting walls of the well.

from Eq. (3.52) we have

$$\psi(0) = 0 \qquad \text{only if } B = 0. \tag{3.53}$$

Thus

$$\psi(x) = A \sin kx. \tag{3.54}$$

If $\psi(x) = 0$ at $x = a$, then

$$\psi(a) = A \sin ka = 0. \tag{3.55}$$

But A cannot be zero, because then A and B would both be zero, resulting in a trivial solution. Thus

$$\sin ka = 0,$$

which is possible only if

$$ka = n\pi.$$

That is, if

$$k = \frac{n\pi}{a}, \qquad \text{where } n = 1, 2, 3, \ldots. \tag{3.56}$$

Substituting for k from Eq. (3.51b), squaring, and denoting E by E_n, we get

$$E_n = n^2 \frac{\pi^2 \hbar^2}{2ma^2}, \qquad \text{where } n = 1, 2, 3, 4, \ldots \tag{3.57}$$

and n is called the *quantum number*.

Thus, according to Eq. (3.57), a particle inside an infinite potential well can assume only certain discrete energy values given by $E_1 = \pi^2\hbar^2/2ma^2$ for $n = 1$, and $E_2 = 4E_1$, $E_3 = 9E_1$, $E_4 = 16E_1, \ldots,$ for $n = 2, 3, 4, \ldots,$ respectively, as shown in Fig. 3.7. These discrete energy values are the result of applying the boundary conditions. Note that, according to classical mechanics, the particle may take any continuous range of values between zero and infinity.

Example 3.2. Calculate (a) the energy of an electron confined in an atom, and (b) the energy of a proton confined in a nucleus.

a) We can obtain some estimate of the kinetic energy of an electron by using Eq. (3.57),

$$E_n = n^2 \frac{\pi^2 \hbar^2}{2ma^2},$$

where $m = 9.1 \times 10^{-31}$ kg is the mass of the electron and $a \simeq 10^{-8}$ cm $= 10^{-10}$ m is the diameter of the atom. We consider the atom to be an infinitely deep well of width a containing the electron. We assume that the electron is in the ground state, that is, that $n = 1$. Thus

$$E_1 = \frac{(3.142)^2(1.054 \times 10^{-34} \text{ J-sec})^2}{2 \times 9.1 \times 10^{-31} \text{ kg} \times (10^{-10} \text{ m})^2}$$

$$= 0.605 \times 10^{-17} \text{ J} = 37 \text{ eV}.$$

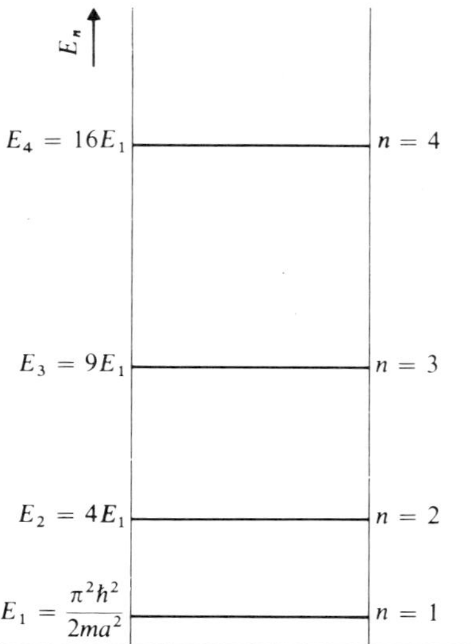

Fig. 3.7 Possible energy levels of a particle inside an infinite potential well according to quantum theory.

b) We assume that the proton is confined to a nucleus. Then, in Eq. (3.57), we substitute for m the mass of the proton, equal to 1.6725×10^{-27} kg, and for a the diameter of the nucleus, which we assume is approximately 10^{-12} cm $= 10^{-14}$ m. Thus we are assuming that the nucleus is providing an infinitely deep well for the proton. Once again let $n = 1$, and we obtain

$$E_1 = \frac{(3.142)^2(1.054 \times 10^{-34}\ \text{J-sec})^2}{2 \times 1.6725 \times 10^{-27}\ \text{kg} \times (10^{-14}\ \text{m})^2}$$

$$\simeq 2 \times 10^6\ \text{eV} = 2\ \text{MeV}.$$

Note that when we calculate the energies of the electron and the proton in this way, they are of the right order of magnitude. On the other hand, if we assume that the electron is confined in a nucleus, we get results that are quite contradictory (see Problem 28).

For a given n, the energy of the particle is given by Eq. (3.57), while we obtain the corresponding wave function by substituting for k from Eq. (3.56) into Eq. (3.54), and replacing $\psi(x)$ by $\psi_n(x)$:

$$\psi_n(x) = A \sin \frac{n\pi}{a} x \qquad \text{for } 0 < x < a. \tag{3.58}$$

The constant A may be calculated by using the normalization condition given by Eq. (3.28). That is,

$$1 = \int_{-\infty}^{\infty} \psi_n^* \psi_n \, dx = \int_0^a A^2 \sin^2 \left(\frac{n\pi x}{a}\right) dx = A^2 \frac{a}{2}$$

or

$$A = \sqrt{(2/a)}.$$

Thus the normalized wave functions of the particle inside the well are

$$\boxed{\psi_n(x) = \sqrt{\frac{2}{a}} \sin \frac{n\pi}{a} x.} \tag{3.59}$$

Figure 3.8 shows plots of the first three wave functions, together with the probability of locating the particle along the X-axis. That is, the distribution function $P_n(x)$ is given by

$$P_n(x) = |\psi_n(x)|^2. \tag{3.60}$$

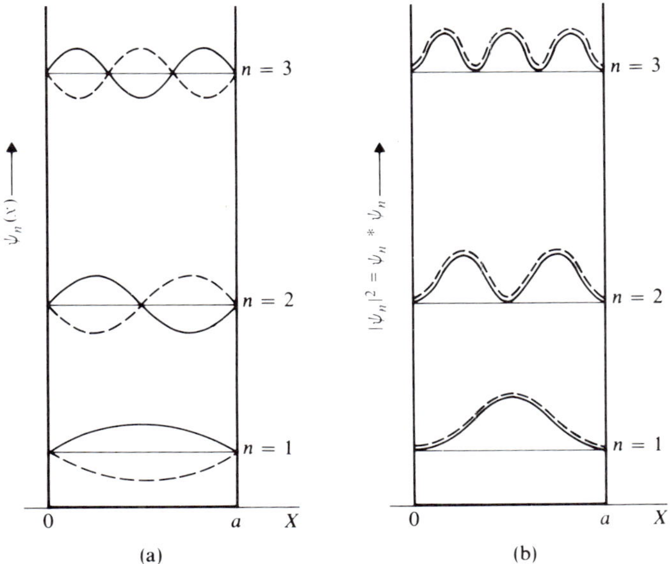

Fig. 3.8 (a) Plots of the first three wave functions of a particle inside an infinite potential well. (b) Plots of the distribution function $P_n(x) = |\psi_n(x)|^2$; that is, the probability of finding the particle along the X-axis inside the infinite well for the first three states.

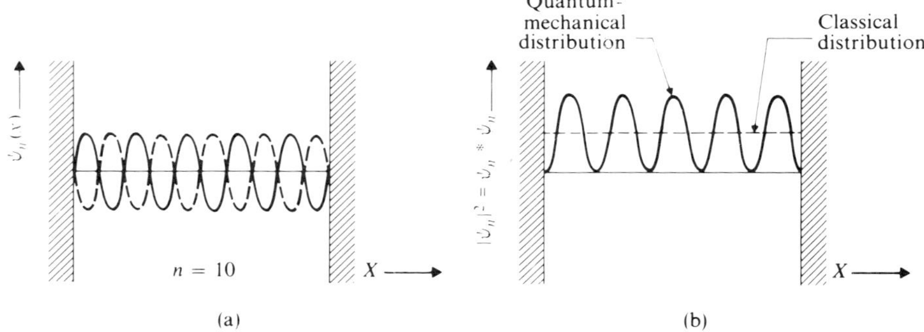

Fig. 3.9 (a) Wave function for $n = 10$. For large values of n, the average quantum-mechanical value of the distribution function $P_n(x) = \psi_n^* \psi_n$ is the same as the classical value, as shown in (b) for $n = 10$, for a particle in an infinite well. The average rapid oscillation in (b) equals the value given by the dashed curve.

Note that these probability distributions are quite different from those expected classically. According to classical mechanics, a particle moving with a velocity v spends a time dx/v in traveling distance dx, and hence the distribution inside the well is constant. Of course, when n is very large, one expects the quantum-mechanical distribution to approach the classical distribution. As shown in Fig. 3.9, for $n = 10$, the peaks are very close, and the average quantum-mechanical value of $P(x)$ is equal to the classical value.

Suppose, in the above example, that the particle of energy E is moving in a potential well of *finite* depth V_0 ($> E$), as shown in Fig. 3.10. Now the problem is more involved, but in principle we can solve it by writing the S.W.E. of the particle in three regions, I, II, III, and using the boundary conditions that $\psi_{\text{I}}(0) = \psi_{\text{II}}(0)$, $\psi'_{\text{I}}(0) = \psi'_{\text{II}}(0)$, $\psi_{\text{II}}(a) = \psi_{\text{III}}(a)$, and $\psi'_{\text{II}}(a) = \psi'_{\text{III}}(a)$.

The resulting wave functions (the first three) and the probability distributions are shown in Fig. 3.11. We still get the same discrete energy levels as in the

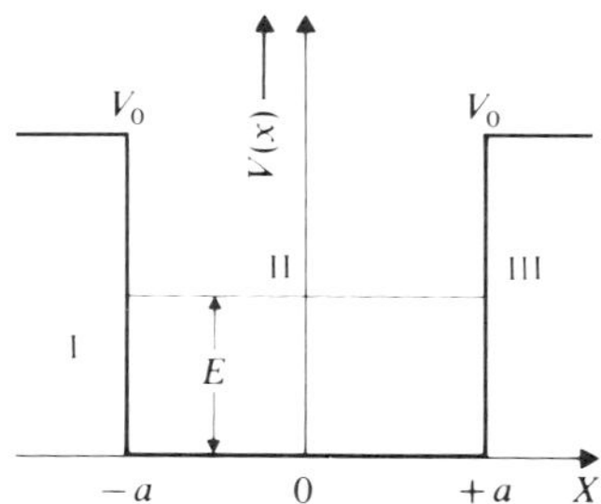

Fig. 3.10 Particle of energy E in a finite potential well such that $E < V_0$.

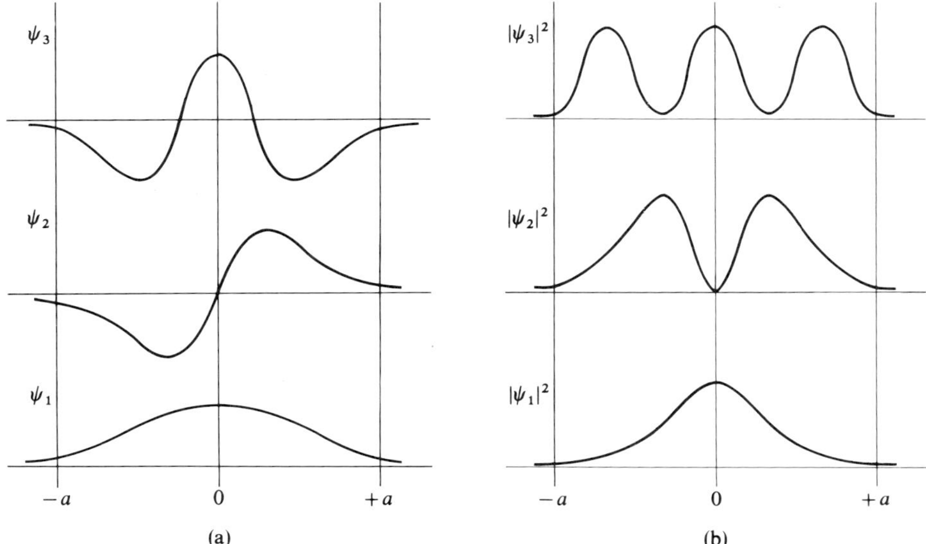

Fig. 3.11 (a) Plots of the first three wave functions of a particle inside a finite potential well. Note the extension of the wave function outside the well. (b) Plots of the probability of finding a particle along the X-axis for the first three states. Note the finite probability of locating the particle outside the well.

infinite-well case, except that their values are shifted to slightly lower levels because the wavelengths are slightly larger (see Fig. 3.11) as compared to the infinite-well case. Also the wave functions and the corresponding probability distributions of particles in different eigenstates (or quantum states) extend beyond the boundaries of the well, as shown in Fig. 3.11. That is, there is a finite (though small) probability of finding a particle outside the well. This situation is unlike classical mechanics, according to which, for $E < V_0$, the particle cannot exist outside the well, because then it would have negative kinetic energy. But quantum mechanics gets around this point by making use of the uncertainty principle. If we are going to measure the negative kinetic energy of a particle outside the well, the uncertainty in energy has to be less than $V_0 - E$. Because of the extremely small amplitude of the wave function outside the well, there is practically no chance of measuring this energy.

3.7 THE SIMPLE HARMONIC OSCILLATOR

The importance of simple harmonic motion lies in classical as well as quantum mechanics. It is used to describe the motion of atoms in molecules and crystals,[6] as well as the quantization of fields.[7,8] We shall briefly discuss this problem of the simple harmonic oscillator from the viewpoint of quantum theory.

From classical mechanics, we know that, when the restoring force acting on a particle is directly proportional to the displacement of the particle, the particle executes *simple harmonic motion* (S.H.M.). One example of this is a mass tied to a spring and then made to vibrate vertically. This is just the kind of motion that is executed by the atoms in molecules and in crystals. That's why it's important to solve this problem by quantum-mechanical methods. Thus, for S.H.M.,

$$F(x) = -kx, \tag{3.61}$$

where k is a constant defined as the force per unit displacement. But according to Newton's law,

$$F(x) = m\frac{d^2x}{dt^2} = -kx;$$

therefore

$$\frac{d^2x}{dt^2} + \frac{k}{m}x = 0, \tag{3.62}$$

which is the classical equation of motion. This equation has an oscillatory solution:

$$x(t) = A\sin(\omega_c t + \phi), \tag{3.63}$$

where A is the amplitude, ϕ is the initial phase angle, and ω_c is the classical angular frequency given by

$$\omega_c = \sqrt{k/m}. \tag{3.64}$$

From Eq. (3.61), for a particle under a potential $V(x)$, we obtain

$$F(x) = -kx = -\frac{dV(x)}{dx}, \qquad dV(x) = kx\,dx$$

which, on integration (when we set the integration constant equal to zero), yields

$$V(x) = \tfrac{1}{2}kx^2 = \tfrac{1}{2}m\omega_c^2x^2. \tag{3.65}$$

Thus, according to classical theory, when a particle is in a potential well $V(x) = \frac{1}{2}kx^2$ (a parabolic potential), as shown in Fig. 3.12, the particle may have any energy from zero to any positive value.

Let us discuss this problem of S.H.M. quantum mechanically. As we expect, we find that the particle can take only certain discrete energy values; we also find that the minimum energy is nonzero. The S.W.E. for a particle moving in a potential well $V(x) = \frac{1}{2}kx^2$ is

$$\frac{d^2\psi}{dx^2} + \frac{2m}{\hbar^2}(E - \tfrac{1}{2}kx^2)\psi = 0. \tag{3.66}$$

Solving such an equation is quite routine in mathematical physics, but is outside the scope of this textbook, so we shall merely state the results. The allowed energy

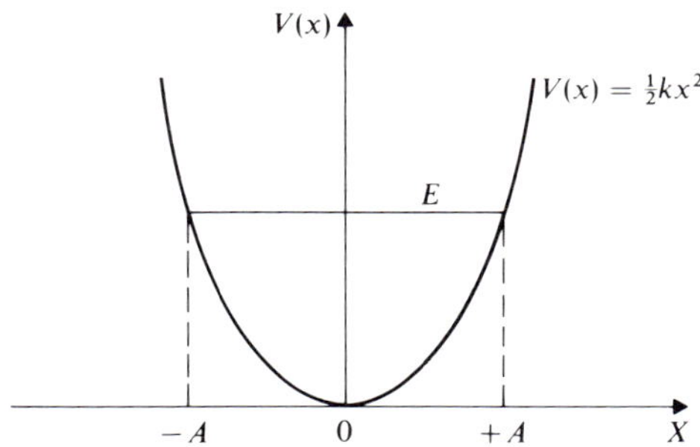

Fig. 3.12 Plot of parabolic potential $V(x) = \frac{1}{2}kx^2$ versus x for simple harmonic oscillator.

levels of the one-dimensional linear harmonic oscillator are found to be discrete, as shown in Fig. 3.13. These energy levels are given by:

$$E_n = (n + \tfrac{1}{2})\hbar\omega, \qquad \text{where } n = 0, 1, 2, 3, \ldots \tag{3.67}$$

The plots of the first three eigenfunctions and relative probabilities, corresponding to $n = 0$, 1, and 2, are shown in Fig. 3.14.

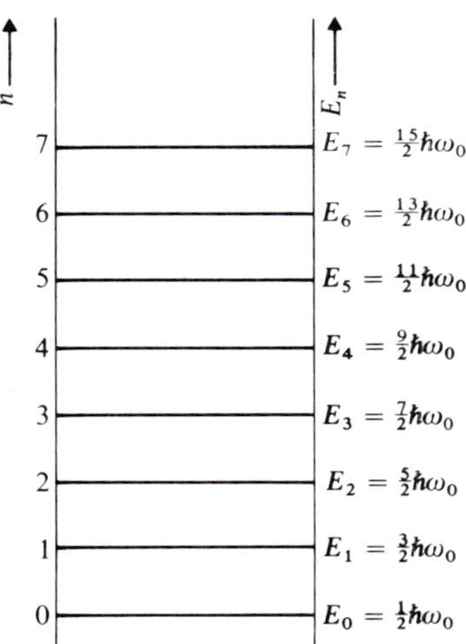

Fig. 3.13 Energy levels in simple-harmonic-oscillator potential $V(x) = \frac{1}{2}kx^2$. Classically, a particle can have any energy, while quantum mechanically, only the discrete energy values $E_n = (n + \frac{1}{2})\hbar\omega$ are possible.

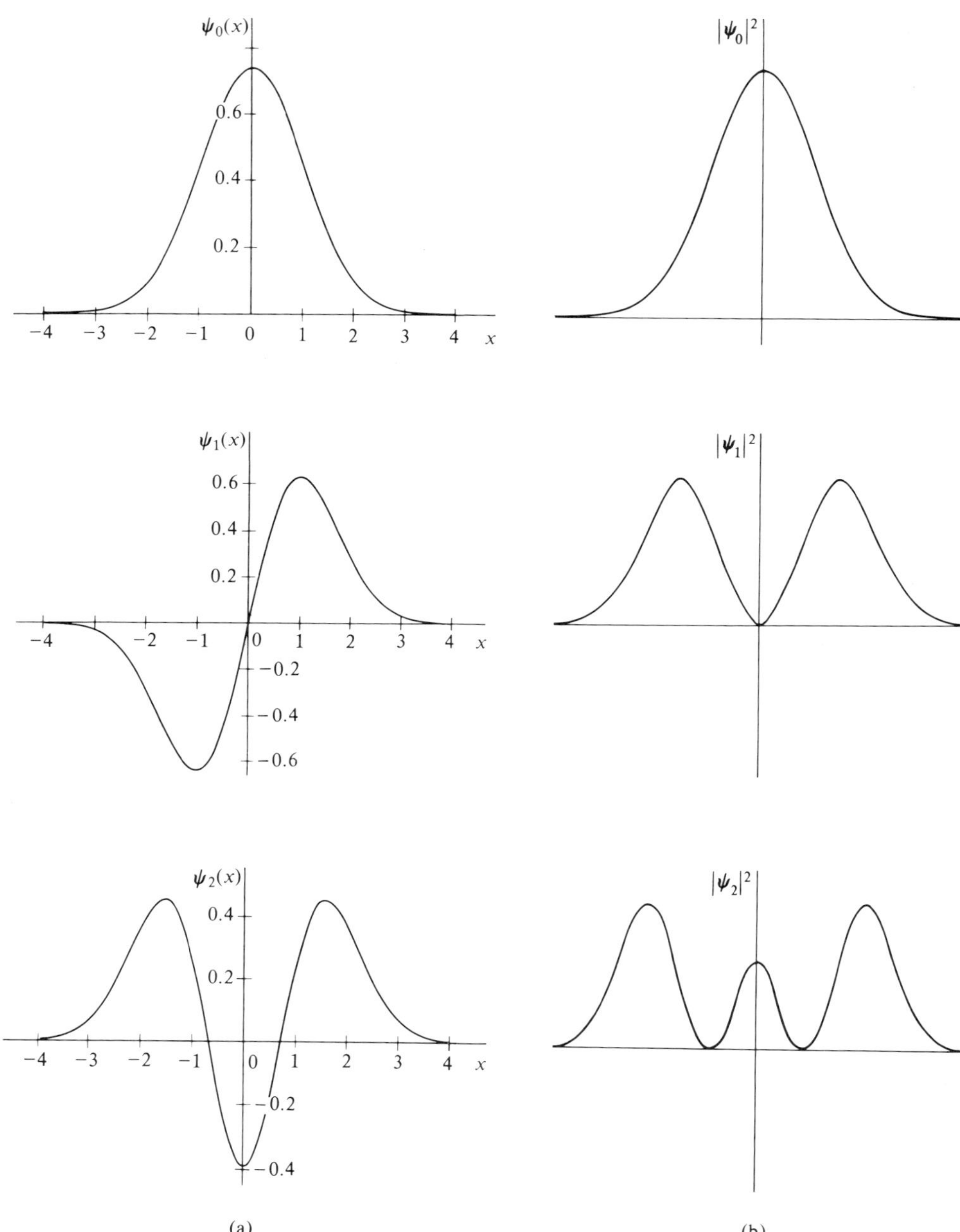

Fig. 3.14 Plots of (a) wave functions of ground state and first two excited states of a harmonic oscillator, and (b) the corresponding relative probabilities.

Thus the results of quantum mechanics differ from those of classical mechanics in the following respects.

a) The particle executing simple harmonic motion can have only discrete energies $\frac{1}{2}\hbar\omega$, $\frac{3}{2}\hbar\omega$, $\frac{5}{2}\hbar\omega$, ...

b) The minimum energy is not zero, but $\frac{1}{2}\hbar\omega$.

c) The wave function extends beyond the potential boundaries.

For n very large, the results of quantum mechanics agree with those of classical mechanics, as we might expect.

★3.8 FINITE POTENTIAL STEPS AND BARRIERS

In Section 3.7 we applied the S.W.E. to a particle confined to a certain potential region, and found the energy eigenvalues and the eigenfunctions of the particle. That is, we were dealing with bound states of the particle. Now let us consider a particle for which $E > V(x)$—that is, an unbound system—and let us see what happens to the particle when it is incident on a finite potential step or a finite potential barrier.

a) Finite potential step

Consider a particle of energy E incident from the left on a potential step, as shown in Fig. 3.15, such that $E > V_0$ and

$$\begin{aligned} V(x) &= 0 \qquad \text{for } x < 0, \qquad \text{region I} \\ V(x) &= V_0 \qquad \text{for } x > 0, \qquad \text{region II} \end{aligned} \tag{3.68}$$

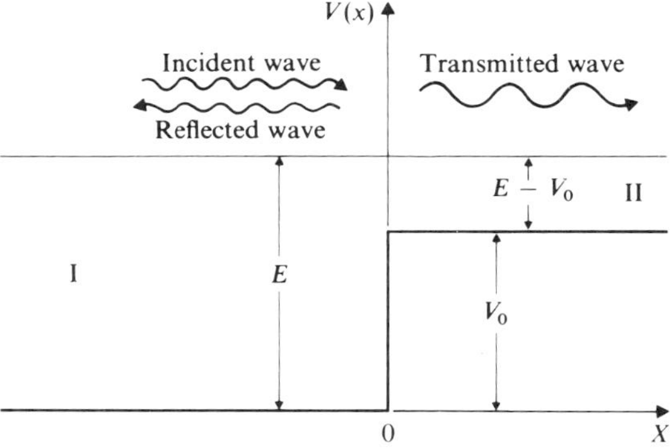

Fig. 3.15 A particle of energy E incident on a step barrier of height V_0 ($E > V_0$). According to classical theory, there should be no reflection at the boundary.

According to classical theory, when this particle encounters the potential step, since $E > V_0$, the particle just keeps on moving and never turns around. Only its velocity, which is $\sqrt{2E/m}$ for $x < 0$, changes at $x = 0$ and becomes $\sqrt{2(E - V_0)/m}$ for $x > 0$.

Quantum-mechanically, the situation is quite different. The wavelength of the particle changes suddenly from

$$\lambda_{\text{I}} = \frac{h}{p_1} = \frac{h}{\sqrt{2mE}}, \qquad \text{region I,} \tag{3.69}$$

to

$$\lambda_{\text{II}} = \frac{h}{p_2} = \frac{h}{\sqrt{2m(E - V_0)}}, \qquad \text{region II.} \tag{3.70}$$

The result of this change in wavelength is that, at the boundary $x = 0$, a small part of the wave is reflected, while the rest is transmitted. We can prove this by solving the Schrödinger wave equations for the two regions. For region I, the S.W.E. is

$$\frac{d^2\psi_{\text{I}}}{dx^2} + \frac{2mE}{\hbar^2}\psi_{\text{I}} = 0 \tag{3.71}$$

with solution

$$\psi_{\text{I}}(x) = Ae^{ik_1x} + Be^{-ik_1x}, \tag{3.72}$$

where A and B are constants, and

$$k_1 = \sqrt{2mE}/\hbar = p_1/\hbar. \tag{3.73}$$

For region II, the S.W.E. is

$$\frac{d^2\psi_{\text{II}}}{dx^2} + \frac{2m}{\hbar^2}(E - V_0)\psi_{\text{II}} = 0, \tag{3.74}$$

the solution of which is

$$\psi_{\text{II}}(x) = Ce^{ik_2x}, \tag{3.75}$$

where C is a constant, and

$$k_2 = \sqrt{2m(E - V_0)}/\hbar = p_2/\hbar \tag{3.76}$$

Just as in optics, we define the *reflection coefficient* R and the *transmission coefficient* T as

$$R = \frac{\text{reflected intensity}}{\text{incident intensity}} = \frac{(\text{reflected amplitude})^2}{(\text{incident amplitude})^2} = \frac{|B|^2}{|A|^2}, \tag{3.77}$$

$$T = \frac{\text{transmitted intensity}}{\text{incident intensity}} = \frac{(\text{transmitted amplitude})^2}{(\text{incident amplitude})^2} = \frac{|C|^2}{|A|^2}. \tag{3.78}$$

According to classical theory, R should be zero and T should be equal to unity. But we can show that, quantum mechanically, this is not so.

First we apply the boundary conditions at $x = 0$, that is,

$$\psi_{\text{I}}(0) = \psi_{\text{II}}(0) \tag{3.79}$$

and

$$\left.\frac{d\psi_{\text{I}}}{dx}\right|_{x=0} = \left.\frac{d\psi_{\text{II}}}{dx}\right|_{x=0}. \tag{3.80}$$

Using Eqs. (3.72) and (3.75) in these two equations, we get

$$A + B = C,$$

$$ik_1(A - B) = ik_2 C.$$

Solving these equations yields

$$\frac{B}{A} = \frac{k_1 - k_2}{k_1 + k_2} \tag{3.81}$$

and

$$\frac{C}{A} = \frac{2k_1}{k_1 + k_2}. \tag{3.82}$$

Thus

$$R = \frac{|B|^2}{|A|^2} = \left(\frac{k_1 - k_2}{k_1 + k_2}\right)^2 \neq 0 \tag{3.83}$$

and

$$T = \frac{|C|^2}{|A|^2}\left(\frac{2k_1}{k_1 + k_2}\right)^2 \neq 1. \tag{3.84}$$

Obviously R is not zero and T is not unity.

b) Finite potential barriers

Suppose that a particle is incident on a barrier of the type shown in Fig. 3.16, such that the energy of the particle $E < V_0$ and

$$\begin{aligned} V(x) &= 0 && \text{for } x < 0, && \text{region I} \\ V(x) &= V_0 && \text{for } 0 \le x \le a, && \text{region II} \\ V(x) &= 0 && \text{for } x > a, && \text{region III} \end{aligned} \tag{3.85}$$

According to classical theory, since $E < V_0$, the particle in region I can never penetrate the potential barrier (region II) and appear in region III. But according to quantum mechanics, the transmission coefficient is not zero; i.e., a fraction of the particles incident from the left will cross the barrier and appear in region III.

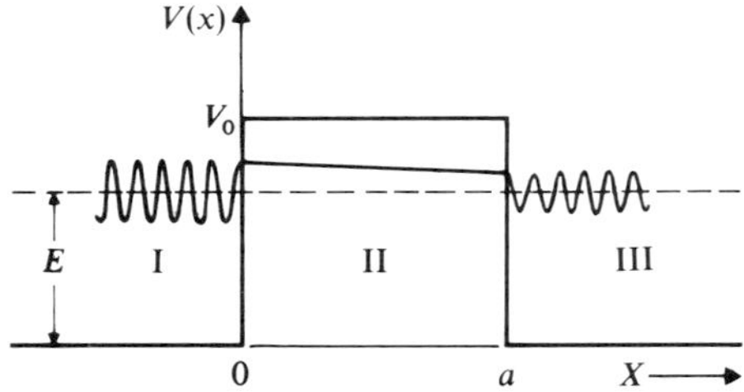

Fig. 3.16 A particle of energy E incident on a square-barrier potential of height V_0 ($E < V_0$) and width a. According to the classical theory, there should be no transmission through the barrier.

Thus the S.W.E. for region I and III is

$$\frac{d^2\psi}{dx^2} + \frac{2mE}{\hbar^2}\psi = 0, \tag{3.86}$$

which gives the solutions

$$\psi_{\text{I}} = Ae^{ik_0x} + Be^{-ik_0x} \tag{3.87}$$

and

$$\psi_{\text{III}} = Ce^{ik_0x}, \tag{3.88}$$

where $k_0 = \sqrt{2mE}/\hbar$, while A, B, and C are constants. Note that the e^{-ik_0x} term (which represents a wave traveling to the left) does not appear in region III because there is no source of particles to the right of the boundary at $x = a$ which would cause a wave to travel toward the left in region III. The S.W.E. in region II is

$$\frac{d^2\psi}{dx^2} - \frac{2m}{\hbar^2}(V_0 - E)\psi = 0. \tag{3.89}$$

Its solution is

$$\psi_{\text{II}} = De^{kx} + Fe^{-kx}, \tag{3.90}$$

where $k = \sqrt{2m(V_0 - E)}/\hbar$, while D and F are constants.

Now we have two boundaries, $x = 0$ and $x = a$, and hence the following four boundary conditions:

$$\begin{aligned} \psi_{\text{I}}(0) &= \psi_{\text{II}}(0), \\ \psi'_{\text{I}}(0) &= \psi'_{\text{II}}(0), \\ \psi_{\text{II}}(a) &= \psi_{\text{III}}(a), \\ \psi'_{\text{II}}(a) &= \psi'_{\text{III}}(a). \end{aligned} \tag{3.91}$$

Using these conditions and ψ_{I}, ψ_{II}, ψ_{III} given by Eqs. (3.87), (3.90), and (3.88), we can calculate the transparency (or the transmission coefficient) $T = |C|^2/|A|^2$

and show that it is definitely not zero, even though E is less than V_0. Figure 3.16 also shows the plots of the wave functions in the three regions.

[The above theory concerning the penetration of a barrier by particles was used by George Gamow in 1926 to explain the theory of alpha decay, as discussed in Chapter 12. The explanation of alpha decay was one of the first successful applications of the quantum theory.]

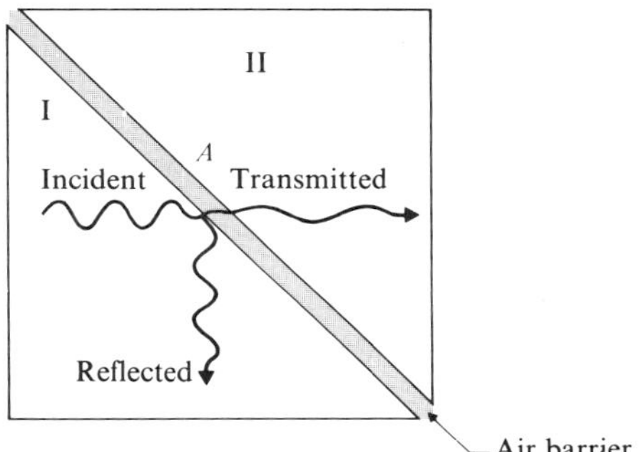

Fig. 3.17 Phenomenon of penetration of an optical barrier. Two 45° prisms are placed as shown, with air as a barrier between the two. When a laser beam is incident on prism I, even though the angle of incidence is greater than the critical angle, not all the light is reflected. Part of the incident beam penetrates the air barrier and appears as a transmitted wave in prism II.

The penetration of a barrier by a matter wave is analogous to the following situation in optics. Suppose that a ray of light is incident on a prism I, as shown in Fig. 3.17. If the angle of incidence is greater than the critical angle, total reflection should occur. If another glass prism II is placed as shown in Fig. 3.17, the situation is similar to that shown in Fig. 3.16. In this case, the air between the prisms is the barrier. Sure enough, a small fraction of the incident light crosses this air barrier and appears as a transmitted wave, as can be shown by using two prisms and a laser beam. [Also, in air, the electric intensity at the outside surface at A is not zero, but decreases exponentially, as is the case with matter waves.]

Figure 3.18 shows water waves crossing a barrier of deep water. However, the gap has to be narrow.

SUMMARY

According to Bohr's principle of complementarity the wave and particle aspects of electromagnetic radiation are complementary.

Any phenomenon concerning energy transport must be explained either by the wave model or the particle model, but not by both simultaneously.

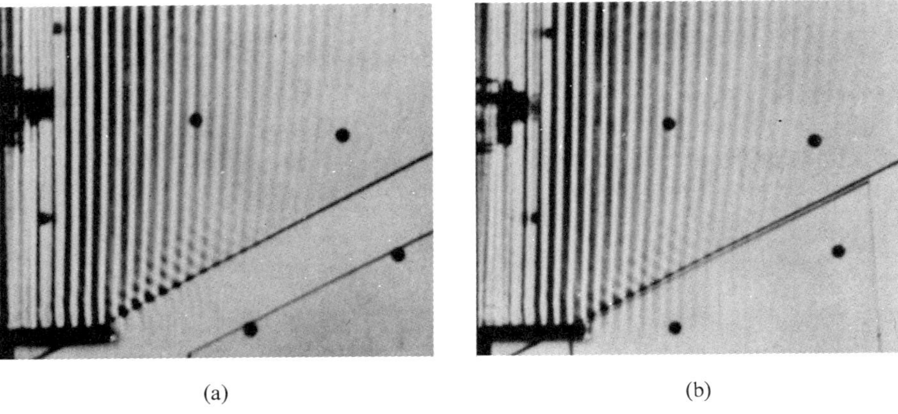

(a) (b)

Fig. 3.18 Penetration of a barrier by water in a ripple tank. (a) Incident waves are totally reflected from a very wide gap of deep water. (b) Incident waves are partly transmitted when deep water gap is very narrow, as shown. (Courtesy Film Studio, Education Development Center, Newton, Mass.)

The phase velocity of an individual wave is given by

$$v_{\text{ph}} = \frac{\omega}{k} = \frac{E}{p} = \frac{c^2}{v}.$$

The modulation in amplitude of a wave packet travels with a group velocity given by

$$v_g = \frac{d\omega}{dk} = \frac{dE}{dp} = v.$$

According to Born, the probability of observing a particle in a volume element dv is

$$|\psi(x, y, z)|^2 \, dv.$$

According to the Heisenberg uncertainty principle,

$$\Delta x \, \Delta p_x \geq \hbar,$$

$$\Delta\theta \, \Delta L_\theta \geq \hbar,$$

$$\Delta t \, \Delta E \geq \hbar.$$

Postulates of quantum mechanics are:

I. ψ and ψ' must be continuous, and the normalization condition is

$$\int_{\text{All space}} \psi^*\psi \, dx \, dy \, dz = 1.$$

II. If an operator O corresponds to an observable o,

$$O\psi = o\psi,$$

where ψ is the eigenfunction and o is the eigenvalue.

*III. The average value of an observable quantity o is

$$\langle o \rangle = \frac{\int \psi^* O \psi \, dx \, dy \, dz}{\int \psi^* \psi \, dx \, dy \, dz}.$$

The time-dependent S.W.E. is $H\Psi = E\Psi$ or

$$-\frac{\hbar^2}{2m} \nabla^2 \Psi + V(x, y, z, t)\Psi = i\hbar \frac{\partial \Psi}{\partial t}.$$

The time-independent S.W.E. in one dimension is

$$\frac{d^2\psi(x)}{dx^2} + \frac{2m}{\hbar^2} [E - V(x)]\psi(x) = 0.$$

For a free particle,

$$\psi(x) = Ne^{\pm i\sqrt{2mE}/\hbar} \qquad \text{for } V = 0.$$

For a particle in regions of step potential,

$$\psi(x) = A \sin kx + B \cos kx \qquad \text{if } E > V_0,$$
$$\psi(x) = Ce^{k'x} + De^{-k'x} \qquad \text{if } E < V_0.$$

For a particle in an infinite potential well,

$$\psi_n(x) = \sqrt{\frac{2}{a}} \sin \frac{n\pi}{a} x,$$

$$E_n = n^2 \frac{\pi^2\hbar^2}{2ma^2} \qquad \text{for } n = 1, 2, 3, 4, \ldots.$$

The allowed energy levels of the one-dimensional linear harmonic oscillator for which $V(x) = \frac{1}{2}kx^2$ are discrete. These levels are given by

$$E_n = (n + \tfrac{1}{2})\hbar\omega, \qquad n = 0, 1, 2, 3, \ldots.$$

*If a particle of energy E is incident on a potential step such that $E > V_0$, the reflection and transmission coefficients are given by

$$R = \left(\frac{k_1 - k_2}{k_1 + k_2}\right)^2 \neq 0, \qquad T = \left(\frac{2k_1}{k_1 + k_2}\right)^2 \neq 1.$$

*If a particle of energy E is incident on a potential barrier such that $E < V_0$, according to classical theory no particles will be transmitted through the barrier. According to quantum mechanics, there is a small probability of penetration.

PROBLEMS

1. Calculate the phase velocity associated with the deBroglie wave of an electron of wavelength 0.1 Å.

2. An electron is moving with a velocity of $0.995c$. Calculate the phase velocity associated with the deBroglie wave of this electron.

3. Calculate the phase velocity of a pi-meson ($m_\pi = 270\ m_e$) which is moving with a velocity of $0.995c$.

4. Calculate the group velocity of the motion of a particle described by the wave function $\psi = e^{i(kx-\omega t)}$.

5. Using the expression $E^2 = p^2c^2 + m_0^2c^4$:

   a) Show that the phase velocity of the electron is greater than c.

   b) Show that the group velocity of the electron wave is equal to the particle velocity of the electron.

   c) If the particle is moving with velocity v, show that the group velocity is equal to the wave velocity.

6. Suppose that the position of a particle which has a mass of 1 g and which is moving with a velocity of 1 cm/sec may be located with an accuracy of 10^{-4} cm. Calculate the uncertainty in the momentum of the particle, and also the fractional uncertainty in the momentum, that is, $\Delta p/p$.

7. The position of an electron moving with a velocity of 10^6 m/sec may be located with an accuracy of 10^{-4} cm. Calculate the uncertainty in the momentum of the electron and also calculate the fractional uncertainty in the momentum, that is, $\Delta p/p$.

8. Using the Heisenberg uncertainty principle, calculate the minimum energy an electron must have to be confined inside an atom of radius 1 Å. (Use $\Delta x = 1$ Å.)

9. Show that for an electron of momentum p,

$$p = mv = \frac{m_0}{\sqrt{1-\beta^2}}\,v,$$

the smallest uncertainty in its position is given by

$$(\Delta x)_{\min} = \frac{\hbar}{2m_0c}\sqrt{1-\beta^2}$$

$$= 0.002 \times 10^{-8}\sqrt{1-\beta^2}\ \text{cm}.$$

10. Derive an expression for the uncertainty in the position of an electron that has been accelerated through a potential difference of $V_0 \pm \Delta V$.

11. Using the Heisenberg uncertainty principle, calculate the minimum energy an electron must have to be confined to a nucleus of radius 10^{-12} cm. (There is no way an electron can stay inside the nucleus because it needs too high an energy, which is not observed experimentally.)

12. Using the Heisenberg uncertainty principle, calculate the minimum energy a proton must have to be confined to a nucleus of radius 10^{-12} cm.

13. A beam of monochromatic visible light of very weak intensity, 1.5×10^{-13} watts per square meter, is incident on a screen. Calculate the number of photons per second falling on an area of 1 cm^2. That is, calculate the photon flux incident on the screen.

14. An electron has been accelerated through a potential difference of (500 ± 0.1) volts. Calculate the uncertainty in the position of the electron. Repeat this for the case of an electron that has been accelerated through a potential difference of $(500{,}000 \pm 200)$ volts. (Write an expression for p, differentiate to find Δp in terms of ΔV, and then use the Heisenberg uncertainty principle to find Δx.)

15. What is the lowest possible energy an electron will have when it is trapped in a vacancy in a crystal lattice such that its motion is restricted to a spherical volume of radius 2.5 Å.

16. When an atom absorbs energy, it retains this energy for 10^{-8} sec before re-emitting it. Calculate the width of the energy state in which the atom exists before re-emitting.

17. Suppose that the angular momentum of a particle moving about an axis can be determined with an accuracy of $0.001\hbar$, $0.01\hbar$, $0.1\hbar$, or $1\hbar$. Calculate the uncertainty in its angular position.

18. Show that the wave functions

$$\Psi(x, t) = A \sin (kx - \omega t) \quad \text{or} \quad \Psi(x, t) = A \cos (kx - \omega t)$$

do not satisfy the time-dependent Schrödinger wave equation.

★19. Assume that a particle is confined in an infinite square potential well. Calculate the average values of the quantities x and x^2 when this particle is in the ground state.

20. Consider a particle in the ground state in an infinite potential well of width a. Calculate the probability of finding this particle:

a) between $x = 0$ and $0.01a$, that is, $\Delta x = 0.01a$

b) between $x = 0.25a$ and $0.26a$

c) between $x = 0.50a$ and $0.51a$

d) between $x = 0.75a$ and $0.76a$

21. Consider an electron confined to a one-dimensional infinite potential well of width 1 Å. Calculate the value of the energy for $n = 1, 10, 100, 101$. Show that, for large values of n, the value of the energy of the electron is the same as that obtained by using classical mechanics.

22. A particle of mass 10^{-4} g is moving with a velocity of 1 cm/sec in a cube whose sides are 1 cm long. Considering this to be a situation similar to a one-dimensional potential well, calculate the value of n.

23. In a manner like that illustrated in Section 3.6, solve for the eigenfunctions and energy eigenvalues of a particle confined to a three-dimensional cubical box whose edges have length a. (Write the S.W.E. in three dimensions, and then use separation of variables and solve separately for $\psi(x)$, $\psi(y)$, $\psi(z)$, E_x, E_y, and E_z.)

24. Solve the problem of an infinite potential well by analogy with the formulation of standing waves on a string. That is, assume $\lambda = 2a/n$, find p, and then E.

25. Using the Heisenberg uncertainty principle, calculate the minimum energy a particle may have when in an infinite potential well, i.e., assume that $\Delta x = a$. How does this value compare with the one found in the text?

26. Consider a particle incident on a step potential, as shown in Fig. 3.15, except that E is less than V_0. Calculate the coefficients of reflection and transmission.

27. Consider a two-dimensional box of length a and width b. Show, by solving the S.W.E., that if a particle is confined to this box, the allowed wave functions and energies are

$$\psi = A \sin\left(\frac{n_x \pi x}{a}\right) \sin\left(\frac{n_y \pi y}{b}\right)$$

and

$$E = \frac{\pi^2 \hbar^2}{2m}\left(\frac{n_x^2}{a^2} + \frac{n_y^2}{b^2}\right),$$

where A is a constant and n_x and n_y are quantum numbers.

28. Following the method used in Example 2, calculate (a) the energy of an electron confined to a nucleus, (b) the energy of a proton confined in an atom. Explain why the results obtained here are contradictory to the experimentally measured values.

29. Calculate the minimum energy, in eV, of a pendulum with a time period of one second. Show that the pendulum has quantized energy states which are so close together that they may be considered continuous.

★30. Consider potential wells of the type shown in the figure below. Write the Schrödinger wave equations and the wave functions for different regions for a particle of energy (a) $E > V_0$, (b) $E < V_0$. Sketch the wave functions in different regions.

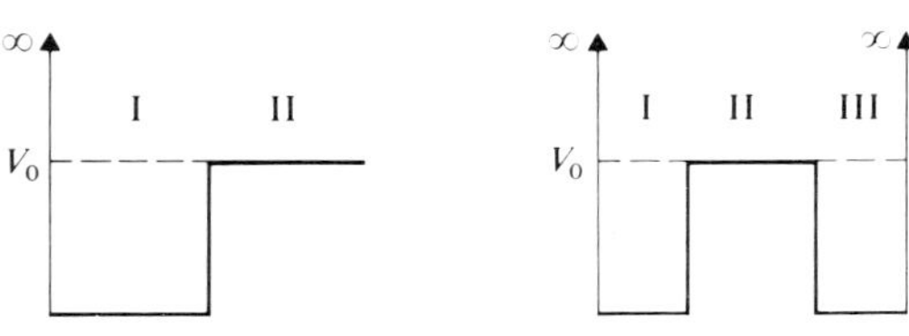

★31. Consider a potential barrier of the form shown in the figure below. Write the Schrödinger wave equations and the wave functions for different regions for a particle of energy (a) $E > V_0$, (b) $0 < E < V_0$. Also sketch the wave functions.

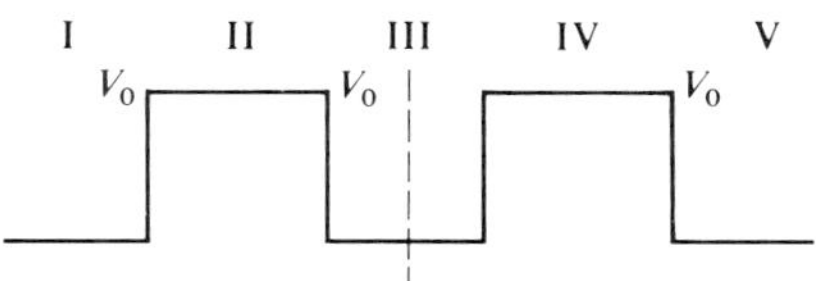

★32. Using Eqs. (3.85) through (3.91), calculate the transparency $T = |C|^2/|A|^2$ for a barrier of the type shown in Fig. 3.16.

REFERENCES

1. Niels Bohr, *Nature* **121,** 580 (1928); *Phys. Rev.* **48,** 696 (1935)
2. Max Born, *Zeits. für Physik* **37,** 863 (1926), and **38,** 803 (1926)
3. Erwin Schrödinger, *Ann. Physik* **79,** 489 (1926)
4. Max Born, W. Heisenberg, and P. Jordan, *Z. Physik*, **35,** 557 (1926)
5. Werner Heisenberg, *Zeits. für Physik* **43,** 172 (1927)
6. G. Herzberg, *Spectra of Diatomic Molecules* and *Molecular Spectra and Molecular Structure*, New York: Dover Publications, 1950
7. Werner Heisenberg and W. Pauli, *Zeits. für Physik* **9,** 338 (1931)
8. Enrico Fermi, *Revs. Mod. Phys.* **1,** 87 (1932)

Suggested Readings

1. F. Hund, "Paths to Quantum Theory Historically Viewed," *Physics Today*, August 1966, page 23
2. George Gamow, "The Principle of Uncertainty," *Sci. Am.* **198,** No. 1, 50 (1958)
3. Richard Feynman, Film of lecture on "Probability and Uncertainty—the Quantum-Mechanical View of Nature," available from Educational Services, Inc., Film Library, Newton, Mass.
4. P. T. Matthews, *Introduction to Quantum Mechanics*, second edition, New York: McGraw-Hill, 1968
5. J. L. Powell and B. Crasemann, *Quantum Mechanics*, Reading, Mass.: Addison-Wesley, 1961
6. C. Sherwin, *Introduction to Quantum Mechanics*, New York: Holt, Rinehart and Winston, 1960
7. K. Ziock, *Basic Quantum Mechanics*, New York: John Wiley, 1969
8. L. T. Schiff, *Quantum Mechanics*, 3rd ed., New York: McGraw-Hill, 1968

CHAPTER 4
ATOMIC STRUCTURE

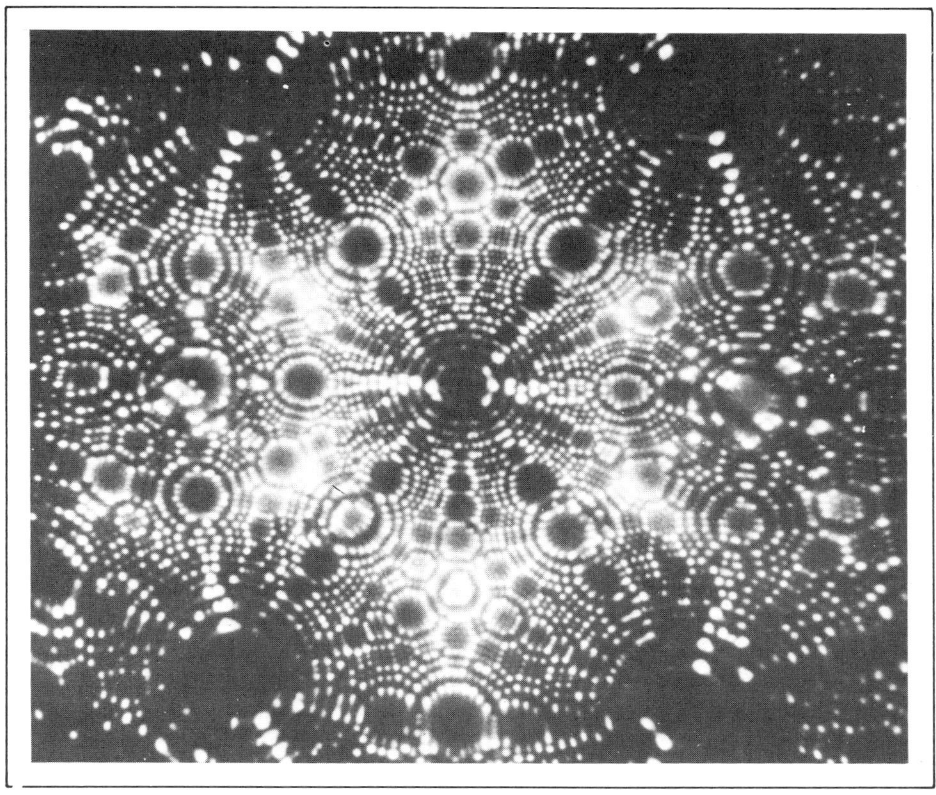

With a field-ion microscope, E. W. Müller used the emission of positive ions from a needlelike specimen to obtain images on a fluorescent screen in which individual atoms can be distinguished. Photograph shows tungsten crystal hemisphere of 400 Å *radius, imaged with helium ions in the field-ion microscope.*
(*Courtesy Dr. Erwin W. Müller, professor of physics, The Pennsylvania State University*)

4.1 THE THOMSON MODEL OF THE ATOM AND ITS FAILURE

By the beginning of the twentieth century it had been well established that the atom was *not* the smallest indivisible unit of matter, as Dalton had stated in 1808. In 1897 J. J. Thomson discovered the electron. Then, too, there were further experiments, by Barkla, concerning the scattering of x-rays from the electrons of the atom. All these pointed to the following conclusion. Since an atom as a whole is neutral, if it contains a negative charge of magnitude Ze (where Z is the atomic number $\simeq A/2$ and e is the charge on the electron), it must contain an equal amount of positive charge. Also, the mass of the electron was found to be so small that it is negligible compared to the mass of the lightest atom. Thus it was concluded that most of the mass of the atom was due to the mass of the particles carrying the positive charge.

The next step in the investigation of atomic structure was to find the arrangement of the negative and the positive charges inside the atom. Of the many models suggested, one suggested by J. J. Thomson, called the *plum-pudding* model (shown in Fig. 4.1), was one of the most attractive. According to this model, the positive charge and the mass of the atom are distributed uniformly over a sphere of radius $\sim 10^{-8}$ cm, with electrons embedded in it to make the atom neutral. (Experimental evidence was available at that time which indicated that the radius of the atom was $\sim 10^{-8}$ cm.) J. J. Thomson tried different configurations of these electrons in the atom (e.g., shells of rotating electrons), but none of these was able to produce a stable equilibrium. Also he could not obtain from this model the frequencies observed in optical spectra. A final blow was given to this model when it was found that it was inadequate to explain the results of the alpha scattering discovered in 1909 by H. Geiger and E. Marsden, in conjunction with their former teacher, E. Rutherford (as discussed below).

One of the ingenious methods of investigating the distributions of charge and mass of microscopic systems which cannot be directly observed is scattering experiments, which involve sending fast-moving charged particles into a microscopic system (which may be treated as a black box, so that nothing is known about its contents) and then looking at their angular distribution as they come out of the system. Geiger and Marsden performed such a scattering experiment, using

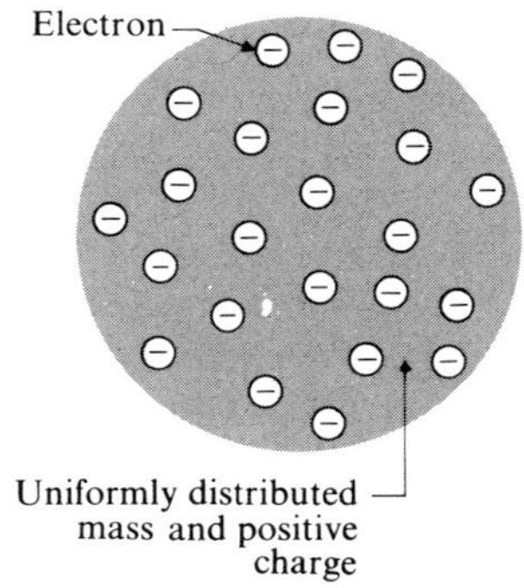

Fig. 4.1 J. J. Thomson's plum-pudding model of the atom.

alpha (or α) particles (α particles are fast-moving, doubly ionized helium atoms, i.e., they have a mass of four units and a positive charge of two units), obtained from radioactive nuclei, as probes for investigating the structure of atoms.

Figure 4.2 shows an experimental arrangement used by Geiger and Marsden.[1,2] They obtained a beam of 5.5-MeV α particles from a radioactive source (radon) placed in a lead block at R. A well-collimated beam of these particles falls on a thin metallic foil F (gold foil about 10^{-4} cm thick). Most of the particles go through the foil undeviated, as if there were empty spaces in the foil. But some of them collide with the atoms of the foil, and are scattered at different angles. This particular scattering has been given the name *Rutherford scattering*. The scattered α particles strike a screen coated with a thin layer of the fluorescent compound ZnS, placed in such a way that the α particles received on the screen make an angle θ with the incident beam. Each α particle produces a small flash of light, called a *scintillation*, on the screen, and may be observed through a magnifying glass or a microscope M. The whole apparatus is placed in a vacuum chamber to avoid the absorption of α particles by air.

Measurements revealed that most of the α particles, as expected, passed through the foil undeviated. But one out of every ~8000 α particles were scattered through an angle greater than 90°, that is, it was back-scattered. This back-scattering of the α particles could not be explained by the Thomson model of the atom. Let's enumerate some possible explanations.

a) The deflection may be due to a collision between the α particle and an electron. Since the mass of the α particle is ~7300 times the mass of the electron, a large-angle scattering could not result from such a collision. Actually the maximum deflection one would expect is ~0.01°.

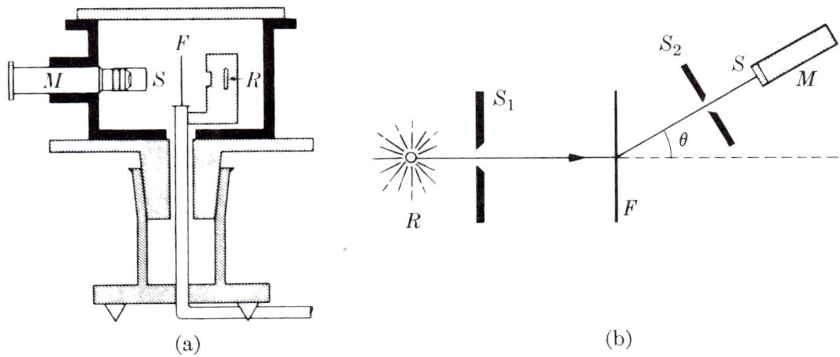

Fig. 4.2 (a) An experimental arrangement used by Geiger and Marsden to investigate the scattering of α particles by thin metallic foils. [From H. Geiger and E. Marsden, *Phil. Mag.* **25,** 607 (1913)] (b) An outline of the experimental arrangement in (a). R is the radioactive source, S_1 and S_2 are slits, F is the metallic foil, S is a ZnS (zinc sulphide) screen, and M is a microscope.

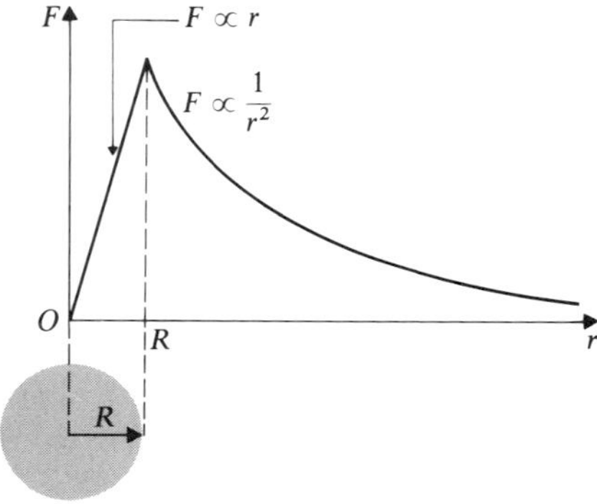

Fig. 4.3 Plot of force on a point charge (an alpha particle) versus distance r from a uniformly charged sphere of radius R (an atom). The variation in the force for $r > R$ is proportional to Q/r^2, while for $r < R$, it is proportional to r $(= Qr/R^3)$.

b) The deflection may be due to the repulsion between the positive charge of the α particle and the positive charge of the atom. According to the Thomson model, the distribution of charge Q (shaded area in Fig. 4.3) and the plot of force F versus r is as shown in Fig. 4.3. But the electrical repulsion is just not strong enough to produce back-scattering. The maximum deflection expected in this case would be $\sim 0.025°$.

Even if we take into account the multiple scattering of α particles by the thin gold foil, the calculated deflection is very small as compared to the back-scattering angle. Thus the Thomson plum-pudding model is inadequate. Ernest Rutherford[3] suggested the *nuclear model of the atom*, which predicted back-scattering due to the collision between the α particle and a heavy nucleus at the center of the atom. Rutherford's model is important, and we shall discuss it in detail.

4.2 RUTHERFORD'S NUCLEAR MODEL OF THE ATOM

Ernest Rutherford[3] suggested that the experimentally observed large-angle scattering (and also the back-scattering) of alpha particles could be explained if we assume the *nuclear model of the atom*.

All the positive charge—and hence almost all the mass of the atom—is confined to a very small volume at the center of the atom, called the nucleus.

The diameter of the nucleus, as we shall see later, is $\sim 10^{-12}$ cm. The electrons surrounding the nucleus are distributed around it like planets around the sun, and fill up the space so as to account for the radius of the atom ($\sim 10^{-8}$ cm). Electrons are so tiny that we may assume that the space in the atom surrounding the nucleus is almost empty.

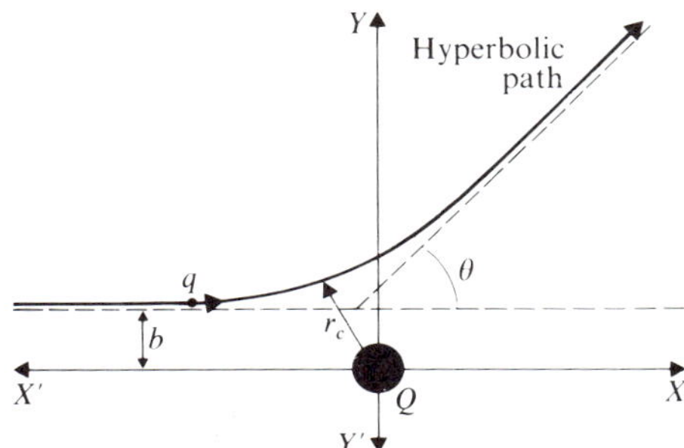

Fig. 4.4 Coulomb repulsive force between incident α particle and stationary nucleus of atom causes the α particle to adopt a hyperbolic orbit.

Thus when a beam of α particles is incident on a thin foil, collisions take place between individual α particles and the heavy charged nuclei of the atoms that make up the gold foil. Since the mass of the α particle is much smaller than the atomic nucleus of any heavy element, one can explain not only large-angle scattering but also back-scattering in a single collision. The force that exists between the α particle and the nucleus is a coulomb repulsive force. Hence the path of the α particle is a hyperbola, with the nucleus at the outside focus, as shown in Fig. 4.4. The dashed lines are the asymptotes of the hyperbola. The angle θ between the two asymptotes—i.e., the angle between the initial and final directions of the α particle—is the *scattering angle*. The distance by which the α particle would have missed the nucleus if there had not been a repulsive force is the *impact parameter b*, as shown in Fig. 4.4. The distance r_c, the closest distance the α particle comes to the nucleus, is called the *distance of closest approach*.

Keeping in mind the above definitions, let us derive a formula which correctly predicts the experimental results, and hence justifies the Rutherford nuclear model of the atom. Because of its simplicity, we shall follow the derivation given by Gordon[4] [M. M. Gordon, *Am. J. Phys.* **23,** 247 (1955)].

Let us assume that (1) the α particle and the nucleus are point charges and (2) the coulomb repulsive force is the only force affecting the path of the α particle, and (3) Newton's law may be used in determining the path of the α particle. Let us also neglect the small amount of recoil energy of the nucleus.

Figure 4.5(a) shows the scattering of an α particle of mass m, velocity v_0, and charge q from a repulsive center O, which is the nucleus, of charge Q. Suppose that $\mathbf{p}_i$ is the momentum of the α particle long before the collision and $\mathbf{p}_f$ is the momentum long after the collision. Then $|\mathbf{p}_i| = |\mathbf{p}_f| = mv_0$, and, from the impulse-momentum theorem, the change in momentum $\Delta\mathbf{p}$ is

$$\Delta\mathbf{p} = \mathbf{p}_f - \mathbf{p}_i = \int_{-\infty}^{\infty} \mathbf{F}\, dt, \tag{4.1}$$

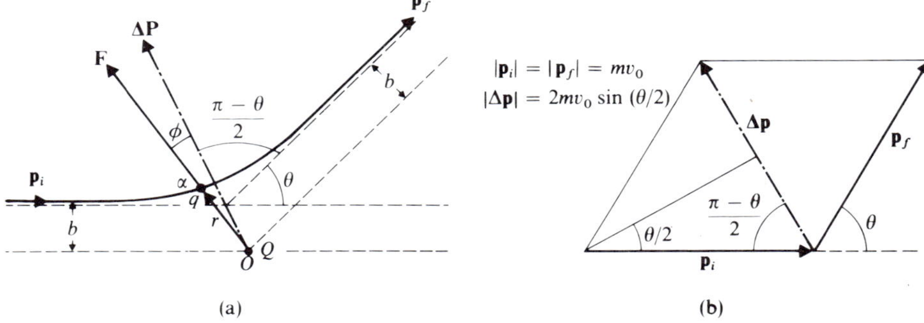

Fig. 4.5 (a) Rutherford scattering, showing the scattering of an α particle by a stationary nucleus of charge Q at O. (b) Momentum vector diagram for Rutherford scattering, showing change in momentum during scattering; that is, $\Delta p = 2mv_0 \sin(\theta/2)$.

where **F** is the coulomb force. From the geometry of Fig. 4.5(b), we obtain

$$\Delta \mathbf{p} = 2m\mathbf{v}_0 \sin(\theta/2). \tag{4.2}$$

From Fig. 4.5(a), we see that **F** is directed outward and makes an angle ϕ with the direction of $\Delta\mathbf{p}$. $\mathbf{F}\,dt$ may be resolved into two components, one parallel to $\Delta\mathbf{p}$ and the other perpendicular. The parallel components add over the whole time from $t = -\infty$ to $t = +\infty$, while the perpendicular components cancel out. Thus, combining Eqs. (4.1) and (4.2), we have

$$2mv_0 \sin(\theta/2) = \int_{-\infty}^{\infty} F \cos\phi \, dt \tag{4.3}$$

and

$$0 = \int_{-\infty}^{\infty} F \sin\phi \, dt. \tag{4.4}$$

Long before the collision, the angular momentum of the α particle about the nucleus is $mv_0 b$, where b is the impact parameter. Also, at any particular r, the angular momentum of the α particle about O (as shown in Fig. 4.5a) is $mr^2\omega$, where ω is the angular velocity. Thus the conservation of angular momentum requires

$$mv_0 b = mr^2\omega. \tag{4.5}$$

By changing variables from t to ϕ, we may write Eq. (4.3) as

$$2mv_0 \sin(\theta/2) = \int_{-(\pi-\theta)/2}^{+(\pi-\theta)/2} F \cos\phi \frac{dt}{d\phi} \, d\phi. \tag{4.6}$$

From Eq. (4.5), we have

$$\frac{dt}{d\phi} = \frac{1}{\omega} = \frac{r^2}{v_0 b}.$$

Therefore

$$2mv_0^2 b \sin(\theta/2) = \int_{-(\pi-\theta)/2}^{+(\pi-\theta)/2} r^2 F \cos\phi \, d\phi. \tag{4.7}$$

But from Coulomb's law, $r^2 F = kqQ$ = constant, and hence

$$2mv_0^2 b \sin(\theta/2) = kqQ \int_{-(\pi-\theta)/2}^{+(\pi-\theta)/2} \cos\phi \, d\phi = kqQ 2 \cos\frac{\theta}{2}, \tag{4.8}$$

which may be written as

$$b = \left(\frac{kqQ}{mv_0^2}\right) \cot\left(\frac{\theta}{2}\right) \tag{4.9}$$

or

$$\boxed{\cot\left(\frac{\theta}{2}\right) = \left(\frac{mv_0^2}{kqQ}\right) b} \tag{4.10}$$

or

$$\theta = 2 \operatorname{arccot}\left[\left(\frac{mv_0^2}{kqQ}\right) b\right]. \tag{4.11}$$

Equation (4.9) implies that if the incident α particle has an impact parameter b, the particle will be scattered at an angle θ given by this relation. If b decreases, the scattering angle increases, as shown in Fig. 4.6(a). Thus all the particles with impact parameters between 0 and b will be scattered through angles greater than θ. This means that if the α particles are incident in an area πb^2 surrounding the nucleus, they will be scattered through an angle greater than θ.

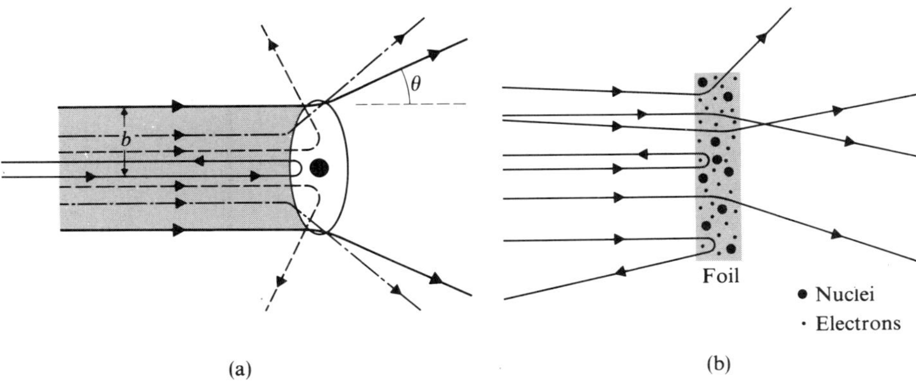

Fig. 4.6 (a) Alpha particles with impact parameter b are scattered through an angle θ, while those with impact parameter less than b are scattered through angles greater than θ. (b) A foil containing many nuclei; each dark circle surrounding a nucleus represents a cross-sectional area of πb^2.

This area πb^2, shown by a circle around the nucleus in Fig. 4.6(a), is called the *cross section* σ for scattering. A practical situation is shown in Fig. 4.6(b), in which each dark circle represents an area πb^2, surrounding each nucleus in a foil. It is assumed that the foil is so thin that no nucleus is hiding another nucleus. Still another way of interpreting the relation between b and θ is: Those particles with impact parameters between b and $b + db$ are scattered through an angle between θ and $\theta + d\theta$, as shown in Fig. 4.7.

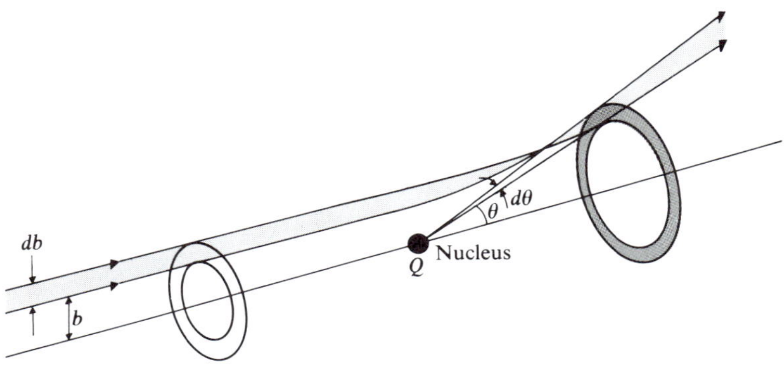

Fig. 4.7 Alpha particles with impact parameters between b and $b + db$ are scattered into a cone of angle $d\theta$ between θ and $\theta + d\theta$.

There is no way of measuring the impact parameter b directly. Therefore, to verify the above theory, we must put these relationships in some other form. We wish to find the number of particles scattered at an angle θ if there are N_i particles incident on a foil. Let the area of the foil shown in Fig. 4.6(b) be A, the thickness of the foil t, and the number of nuclei per unit volume of the foil n. Let πb^2 be the cross-sectional area surrounding each nucleus. The total number of nuclei in the foil is ntA. With each nucleus there is associated an area πb^2 which is effective in causing scattering; we call this the *sensitive area*. The total sensitive area is $\pi b^2 ntA$. Thus the fraction f of the incident α particles that will be scattered through an angle $\geq \theta$ should be equal to the ratio of the total sensitive area to the total area of the foil. That is,

$$f = \frac{nt\pi b^2 A}{A} = nt\pi b^2. \tag{4.12}$$

Substituting the value of b from Eq. (4.9), we get

$$\boxed{f = \pi b^2 nt = \pi nt \left(\frac{kqQ}{mv_0^2}\right)^2 \cot^2 \frac{\theta}{2}.} \tag{4.13}$$

In other words, the number of particles that will be scattered within a cone between angles θ and $\theta + d\theta$, as shown in Fig. 4.7, is given by

$$df = -\pi nt \left(\frac{kqQ}{mv_0^2}\right)^2 \cot\frac{\theta}{2}\csc^2\frac{\theta}{2}\,d\theta, \tag{4.14}$$

where the minus sign implies that f decreases as θ increases. Thus, of the N_i incident particles, only $N_i|df|$ are scattered between θ and $\theta + d\theta$ and received on the shaded area dA between the circles. Usually the size of the detector is such that it does not cover the whole shaded area dA (see Fig. 4.8). Hence $N(\theta)$, the number of particles scattered into a unit area at angle θ, is

$$N(\theta) = \frac{N_i|df|}{dA}. \tag{4.15}$$

We can calculate dA with the help of Fig. 4.8. It is given by

$$dA = (2\pi r\sin\theta)r\,d\theta = 2\pi r^2\sin\theta\,d\theta. \tag{4.16}$$

Combining Eqs. (4.14), (4.15), and (4.16), we obtain

$$N(\theta) = \frac{N_i\pi nt(kqQ/mv_0^2)^2\cot(\theta/2)\csc^2(\theta/2)\,d\theta}{2\pi r^2\sin\theta\,d\theta}. \tag{4.17}$$

Note that, for the α particle, $q = 2e$, and its kinetic energy $K_\alpha = \frac{1}{2}mv_0^2$. (For an α particle with energy less than 10 MeV, relativistic correction is negligible.) The charge on the nucleus is $Q = Ze$, so we may write

$$N(\theta) = \frac{N_i ntk^2Z^2e^4}{4r^2K_\alpha^2\sin^4(\theta/2)}. \tag{4.18}$$

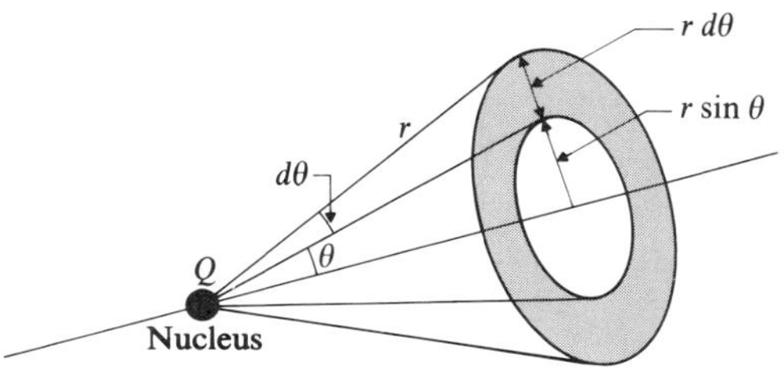

dA = Shaded area = circumference × width
$= 2\pi\,(r\sin\theta) \times r\,d\theta$

Fig. 4.8 Particles with impact parameters between b and $b + db$ are scattered into a cone of angle $d\theta$ between θ and $\theta + d\theta$ and are received in shaded zone.

Thus, according to this relation, the number of scattered α particles $N(\theta)$ received in a unit area at a distance r from the target and at an angle θ should have the following dependence:

$$N(\theta) \propto t$$
$$N(\theta) \propto Z^2$$
$$N(\theta) \propto \frac{1}{K_\alpha^2} \tag{4.19}$$
$$N(\theta) \propto \frac{1}{\sin^4(\theta/2)}$$

Geiger and Marsden verified these relationships while they were working in Rutherford's laboratory. Figure 4.9 shows the results of one of those experiments, clearly demonstrating the agreement between theory and experiment.

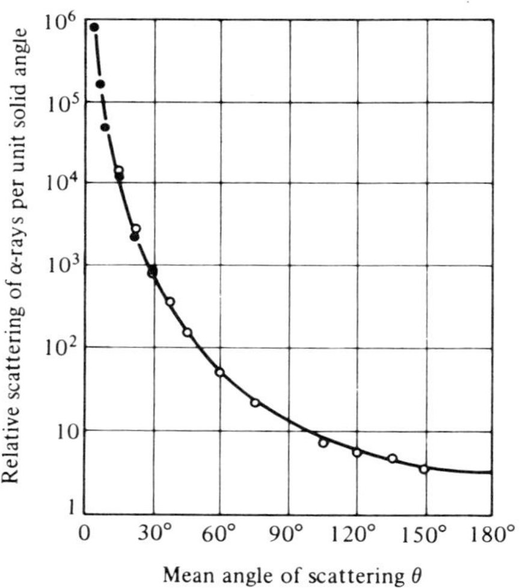

Fig. 4.9 Rutherford formula for scattering of α particles is verified by plotting the relative number of scattered α particles $N(\theta)$ versus θ. The continuous curve is $1/\sin^4(\theta/2)$, as predicted theoretically, while the experimental points are those obtained by Geiger and Marsden. [From R. D. Evans, *The Atomic Nucleus*, New York: McGraw-Hill, 1955)]

Thus the nuclear model of the atom was well established, i.e., it was understood that at the center of the atom there is a very small volume called the nucleus, where almost all the mass of the atom and its positive charge is confined.

Example 4.1. What fraction of 8.8-MeV alpha particles incident on a 2×10^{-5} cm thick gold foil will be scattered through an angle 90° or greater?

According to Eq. (4.14), the fraction of α particles scattered between angles θ and $\theta + d\theta$ is

$$|df| = \pi nt \left(\frac{kqQ}{mv_0^2}\right)^2 \cot\frac{\theta}{2}\csc^2\frac{\theta}{2}\,d\theta.$$

Hence the fraction scattered through an angle θ or greater is

$$|df|_\theta^\pi = \int_\theta^\pi df = \pi nt \left(\frac{kqQ}{mv_0^2}\right)^2 \int_\theta^\pi \cot\frac{\theta}{2}\csc^2\left(\frac{\theta}{2}\right)d\theta$$

$$= \pi nt \left(\frac{kqQ}{mv_0^2}\right)^2 \cot^2\left(\frac{\theta}{2}\right).$$

For $\theta = 90°$,

$$|df|_{90°}^\pi = \pi nt \left(\frac{kqQ}{mv_0^2}\right)^2 = \pi nt \left(\frac{k2e^2Z}{mv_0^2}\right)^2$$

$$= \pi nt \left(\frac{kZe^2}{K_\alpha}\right)^2.$$

In this problem, $Z = 79$, and the number of nuclei per unit volume n is given by

$$n = \frac{\rho N_A}{M} = \frac{19.3\ \text{g/cm}^3 \times 6.02 \times 10^{23}\ \text{atoms/mole}}{197\ \text{g/mole}}$$

$$= 5.9 \times 10^{22}\ \text{atoms/cm}^3.$$

$$\left[\frac{kZe^2}{K_\alpha}\right]^2 = \left[\frac{8.98 \times 10^9\ \text{N-m}^2/\text{coul}^2 \times 79 \times (1.602 \times 10^{-19}\ \text{coul})^2}{8.8 \times 10^6\ \text{eV} \times 1.602 \times 10^{-19}\ \text{joule/eV}}\right]^2$$

$$= \left(\frac{8.98 \times 10^9 \times 79 \times 1.602 \times 10^{-19}}{8.8 \times 10^6}\ \text{m}\right)^2$$

$$= \left(\frac{8.98 \times 7.9 \times 1.602}{8.8} \times 10^{-15}\ \text{m}\right)^2$$

$$= (1.3 \times 10^{-14}\ \text{m})^2 = 1.69 \times 10^{-24}\ \text{cm}^2.$$

Substituting in the above expression, we obtain

$$|df|_{90°}^\pi = 3.142 \times 5.9 \times 10^{22}\ \text{atoms/cm}^3 \times 2 \times 10^{-5}\ \text{cm} \times 1.69 \times 10^{-24}\ \text{cm}^2$$
$$= 6.25 \times 10^{-6}.$$

Example 4.2. In Example 4.1, suppose that a scintillation detector of area 0.1 cm^2 is placed a distance of 5 cm from the scattering center, and that it makes an angle of 45° with the direction of the incident beam. Calculate the number of α particles that would be detected if 10^8 α particles were incident on the gold-foil target.

According to Eq. (4.18), the number of particles $N(\theta)$ received in a unit area at a distance r from the scattering center is

$$N(\theta) = \frac{N_i n t}{4r^2}\left(\frac{kZe^2}{K_\alpha}\right)^2 \frac{1}{\sin^4(\theta/2)},$$

where $N_i = 10^8$/sec, $n = 5.9 \times 10^{22}$ atoms/cm^3 (from Example 4.1), $r = 5$ cm, $t = 2 \times 10^{-5}$ cm, and $\theta = 45°$. Also, from Example 4.1,

$$\left(\frac{kZe^2}{K_\alpha}\right)^2 = 1.69 \times 10^{-24}\ \text{cm}^2.$$

Therefore

$$N(\theta) = \frac{10^8\ \text{sec} \times 5.9 \times 10^{22}\ \text{atoms/cm}^3 \times 2 \times 10^{-5}\ \text{cm}}{4 \times 25\ \text{cm}^2} \times \frac{1.69 \times 10^{-24}\ \text{cm}^2}{\sin^4(22°\ 30')}$$

$$= 107\ \text{cm}^{-2}\ \text{sec}^{-1}.$$

Since $dA = 0.1$ cm^2, then

$$N(\theta)\,dA = 107 \times 0.1\ \text{sec} = 10.7/\text{sec}.$$

That is, 10.7 alpha particles per second will be scattered into the detector.

4.3 SIZE AND STRUCTURE OF THE NUCLEUS

The radius of an atom is $\sim 10^{-8}$ cm, while the nucleus is a great deal smaller: only about 1/10,000 as large as the whole atom.

In the scattering experiment, the α particles reach closest to the nucleus if the collision is head-on, i.e., if the α particle is scattered through an angle of 180°. Imagine that, as the α particle of kinetic energy K_α approaches the target nucleus, the force of the electrostatic repulsion increases. At the distance r_c, the particle's closest approach to the nucleus, the kinetic energy of the particle equals the electrostatic repulsion. So the α particle comes to rest momentarily, and then turns back, as shown in Fig. 4.10. Visualize the α particle climbing the electrostatic hill around the nucleus and coming to a distance r_c from the nucleus, at which point all its kinetic energy has been converted into electrostatic potential energy, which makes it come to rest momentarily. It then starts sliding down the hill and regains its kinetic energy. In this situation we may write kinetic energy equal to potential energy; that is, at $r = r_c$

$$\boxed{K_\alpha = \tfrac{1}{2}mv_0^2 = \frac{kqQ}{r_c}.} \tag{4.20}$$

Since $q = 2e$ and $Q = Ze$, we have

$$r_c = \frac{2kZe^2}{K_\alpha} = \frac{4kZe^2}{mv_0^2}. \tag{4.21}$$

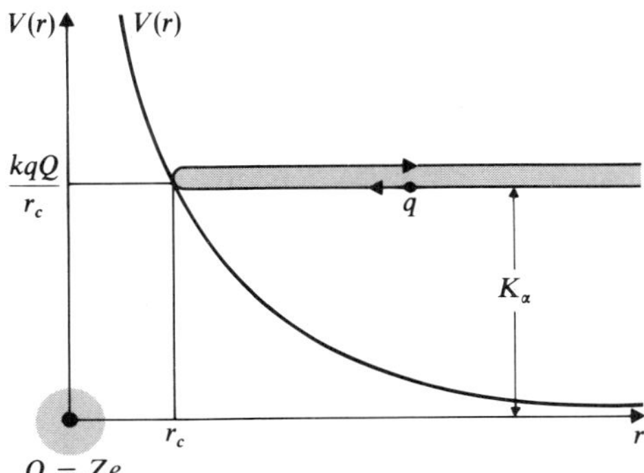

Fig. 4.10 Plot of coulomb potential versus r between nucleus and α particle. The distance of closest approach r_c corresponds to the turning point of the incident α particle.

Rutherford used 7.68-MeV α particles from radium, scattered them from a target of gold foil, and obtained a value for r_c of $\simeq 3 \times 10^{-12}$ cm. Similarly, for an aluminum target, he obtained $r_c = 0.5 \times 10^{-12}$ cm.

Example 4.3. Radium gives out α particles of energy 7.68 MeV. Suppose that we scatter these α particles from foils made of gold and of aluminum. Calculate the distance of closest approach in these two cases.

From Eq. (4.21), we have

$$r_c = \frac{2kZe^2}{K_\alpha},$$

where $k = 8.987 \times 10^9$ N-m^2/coul^2, $e = 1.602 \times 10^{-19}$ coul, $K_\alpha = 7.68 \times 10^6$ eV, and $Z = 79$ for gold and $Z = 13$ for aluminum. Thus

$$r_c \text{ (for Au)} = \frac{2 \times 8.987 \times 10^9 \text{ (N-m}^2\text{/coul}^2) \times 79 \times (1.602 \times 10^{-19} \text{ coul})^2}{7.68 \times 10^6 \text{ eV} \times 1.602 \times 10^{-19} \text{ joule/eV}}$$

$$= 2.98 \times 10^{-14} \text{ m} = 2.98 \times 10^{-12} \text{ cm}$$

$$r_c \text{ (for Al)} = \frac{2 \times 8.987 \times 10^9 \text{ (N-m}^2\text{/coul}^2) \times 13 \times (1.602 \times 10^{-19} \text{ coul})^2}{7.68 \times 10^6 \text{ eV} \times 1.602 \times 10^{-19} \text{ joule/eV}}$$

$$= 0.49 \times 10^{-14} \text{ m} = 0.49 \times 10^{-12} \text{ cm}$$

Suppose that R is the radius of the nucleus, and that the kinetic energy K_α of the α particle is such that $r_c \leq R$. In such cases, it is found that the number of particles scattered is smaller than expected from Rutherford's theory, for the following reasons.

At distances $r \leq R$, the force between the incident α particle and the nucleus is attractive. This attractive force predominates over the coulomb repulsive force, and is represented by the potential

$$V(r) = -V_0, \quad \text{for } r \leq R, \qquad (4.22)$$
$$= 0, \quad \text{for } r > R.$$

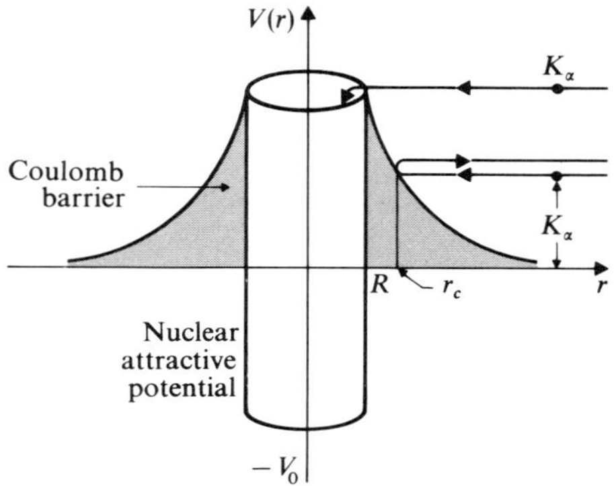

Fig. 4.11 For $r < R$, the nuclear force is attractive and is represented by a constant negative potential $V(r) = -V_0$. For $r > R$, the coulomb potential predominates.

Figure 4.11 shows this relationship, and satisfactorily explains most of the characteristics exhibited by the nuclear force. If K_α is such that $r_c \leq R$, not all the alpha particles are scattered. A fraction of them are absorbed by the nuclei, which is what causes the observed decrease in the number of alpha particles scattered (see Fig. 4.12). By noting the kinetic energy of the alpha particles at which the experimental results of scattering deviate from the Rutherford formula, we can calculate the radius of a given nucleus.

The values of r_c are very significant. Even though these values are not equal to the radius of the nucleus, they indicate that the radius of a nucleus is of the order of 10^{-12} cm. In order to measure this radius accurately, we can proceed as follows: When we use incident α particles of higher energy, the distance r_c decreases. But as soon as the energy of the α particles is such that r_c is less than the radius of the nucleus, the assumption that the nucleus is a point charge (which we made in our discussion of the Rutherford scattering theory) will not hold good, and the measured values of $N(\theta)$ will disagree.

In 1954, Farwell *et al.*[5] reported the scattering of 10- to 45-MeV α particles (obtained from a cyclotron) from a lead (Pb) target, with results as shown in Fig. 4.12. It is obvious that for α particles with energies greater than 27.5 MeV, the experimental results do not agree with the Rutherford theory (coulomb scattering). We assume that at this energy the α particles are very close to the nuclear surface, and by using Eq. (4.21) we can calculate r_c, which in this case is equal to the nuclear radius R. Experiments performed using different targets[6] result in the following expression for the radius of any nucleus:

$$\boxed{R = r_0 A^{1/3},} \tag{4.23}$$

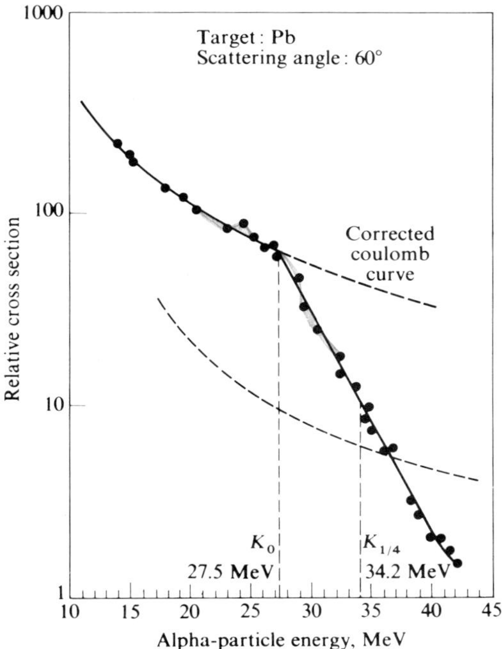

Fig. 4.12 Relative number of α particles scattered at 60° from a lead target versus α particle energy. The critical energy K_0 corresponds to the α particle energy for which one can observe deviations from the formula of classical theory (Rutherford scattering). [From Farwell *et al.*, *Phys. Rev.* **95,** 1212 (1954)]

where A is the mass number of the nucleus and

$$r_0 = 1.414 \times 10^{-13} \text{ cm}. \tag{4.24}$$

Sometimes nuclear sizes are expressed in *fermi*, F (named after the Italian physicist Enrico Fermi), where $1 \text{ F} = 10^{-13} \text{ cm} = 10^{-15} \text{ m}$. Thus $R = 1.414A^{1/3}$ F. The value of r_0 depends on the type of experiment.

Today we know that any nucleus consists of Z protons and $A - Z$ neutrons [that is, there are $Z + (A - Z) = A$ nucleons; a *nucleon* is a generic name for both proton and neutron]. It was not till 1932, however, when James Chadwick[7–9] discovered the neutron, that this *proton–neutron model* of the nucleus was established. To make up a neutral atom of size $\sim 10^{-8}$ cm, the nucleus is surrounded by Z electrons. [Before the discovery of the neutron, it was generally accepted that the nucleus consisted of A protons and $A - Z$ electrons, which would leave the nucleus with a net positive charge Z. The nucleus was assumed to be surrounded by Z electrons to make a neutral atom. This was the so-called *proton–electron* *model* of the nucleus. This model did not survive very long, because of its failure* to explain many facts which became known after 1920.]

Thus we conclude the following about the size and structure of the nucleus.

An atom with a radius of $\sim 10^{-8}$ cm *consists of a nucleus with a radius of* $1.414 \times 10^{-13} A^{1/3}$ cm, *which is located at the center of the atom. This nucleus consists of A nucleons of which Z are protons and A − Z neutrons, and is surrounded by Z electrons, which fill up the rest of the space in the atom.*

For most of the next few chapters, we're going to talk about the arrangement of the electrons in the atom. We shall devote the rest of this chapter, however, to the measurement of atomic masses.

4.4 THE DISCOVERY OF ISOTOPES

The English chemist John Dalton, in 1808, proposed an atomic theory according to which the atoms of any one element were identical in nature. This was contrary to the views held by other scientists at that time. According to Joseph Louis Prout, in 1815, all the elements were made up of hydrogen atoms, and hence the mass of any element would be an integral multiple of the mass of a hydrogen atom. However, experiments revealed that the masses of most elements were *not* integral multiples of the mass of a hydrogen atom, and hence Prout's hypothesis was not correct.

But Sir William Crookes, in 1886, revived Prout's idea. He suggested that all elements must have integral atomic masses. Any elements that had *non*integral atomic masses would have to be mixtures of two or more types of atoms. For example, he postulated that chlorine (atomic mass 35.46 u) was a mixture of two kinds of atoms, having masses of 35 u and 37 u, mixed in such a proportion as to give the observed atomic mass of chlorine. Nowadays we know that those proportions are 75.53% and 24.47%, respectively, resulting in the value,

$$\text{Mass of chlorine} = \frac{35 \times 75.53 + 37 \times 24.47}{75.53 + 24.47} = 35.46 \text{ u}.$$

In addition to Crookes's idea, there were others along the same line. After the discovery of radioactivity, experiment revealed that the atoms of a given element need not be identical in mass. It was also found that there were many elements that were chemically identical but had different masses. The chemist Frederick Soddy suggested the name *isotopes* (*isos* = "equal" and *topes* = "place") for elements that were identical in chemical properties but had different masses, and hence occupied the same place in the periodic table.

In 1910, J. J. Thomson[10] established the fact that isotopes did indeed

* See Atam P. Arya, *Fundamentals of Nuclear Physics*, Chapter I, Boston, Allyn and Bacon, 1966.

exist. The first element he investigated was neon. He found it to be a mixture of two isotopes, of atomic masses 20 and 22, mixed in proportions of 90% and 10%, respectively.

All subsequent measurements showed that the atomic masses of all the isotopes of different elements are very close to being whole numbers. So now we are ready for a formal definition of *isotopes*:

Atoms which have the same atomic number Z, but which have different atomic masses, are called isotopes of the element of given Z.

4.5 MEASUREMENT OF ISOTOPIC MASSES

Once the existence of isotopes had been established, the next task was to measure their masses accurately. Many methods have been developed, all based on the same principle as the original, of J. J. Thomson:[10] the deflection of positive ions by electric and magnetic fields. The method involves measuring the value of q/M for positive ions in a manner similar to that used for cathode rays. If we know the value of the charge q on the ion, we can calculate the mass M of the ion. Thomson's method is called the *positive-ray-analysis* or *parabola method.*

The parabola method (a) is not accurate, (b) is limited to gases, and (c) does not yield abundances of the isotopes. In order to overcome these limitations, F. W. Aston[11] took over the work started by J. J. Thomson, and accomplished more in this field than any other person. The principle of the Aston mass spectrograph[11,12] is to apply electric and magnetic fields one after the other, in different directions. The electric field disperses the positive rays with respect to velocity, and the magnetic field, applied at right angles to the direction of the electric field, brings to a common focus the dispersed rays having the same value of q/M. This is the *velocity focusing method.* Using Aston's mass spectrograph, one can determine isotopic masses with an accuracy of one part in 10,000.

Using the above principle, instruments were built by Arthur J. Dempster,[13] Kenneth Bainbridge and E. Jordan,[14] J. Mattauch,[15] and Alfred Nier.[16] Some have achieved an accuracy of the order of one part in 100,000. These instruments also enable one to measure the abundance of the isotopes. The most modern techniques for measuring masses are (a) the time-of-flight method,[17] and (b) the mass synchrometer.[18] Just to give you an idea, we shall briefly describe the Dempster mass spectrometer.

The Dempster Mass Spectrometer

Figure 4.13 shows the outline of Dempster's mass spectrometer. Positive rays are obtained either by heating salts on platinum strips or by bombarding salts with electrons. By applying an electric field between plates P and Q, one allows the positive ions produced at F to fall through a potential difference V. The passage of these rays through slit S_1 results in a narrow beam. A strong magnetic field B

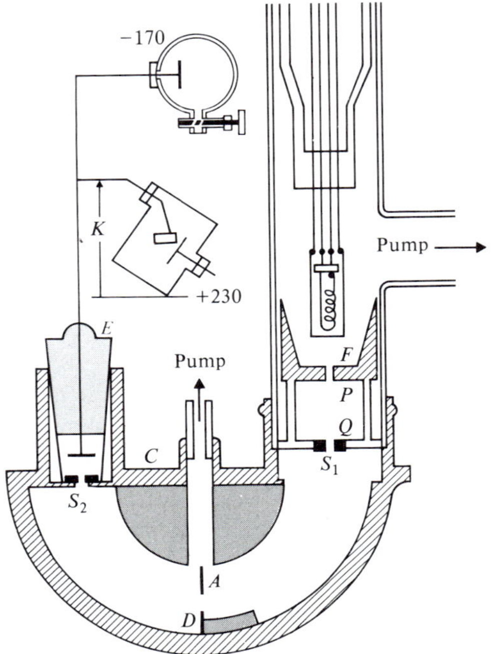

Fig. 4.13 Outline of experimental arrangement of Dempster's mass spectrometer. [From A. J. Dempster, *Phys. Rev.* **22,** 631 (1922)]

applied in a direction perpendicular to the plane of the paper bends these rays into a semicircle. After they pass through slit S_2, they are detected by the electrometer E.

The kinetic (nonrelativistic) energy acquired by a positive ion of mass M and charge q, after falling through a potential difference V, is

$$\tfrac{1}{2}Mv^2 = qV, \tag{4.25}$$

where v is the velocity of the ions coming out of S_1. The radius R of the circular path of the positive ion is related to the magnetic field B by

$$\frac{Mv^2}{R} = qvB. \tag{4.26}$$

Eliminating v from Eqs. (4.25) and (4.26), we get

$$\boxed{q/M = 2V/B^2R^2.} \tag{4.27}$$

Thus for fixed values of V and B, the radius R depends on the value of q/M. Figure 4.14 shows this for three different values of q/M. By changing the electric

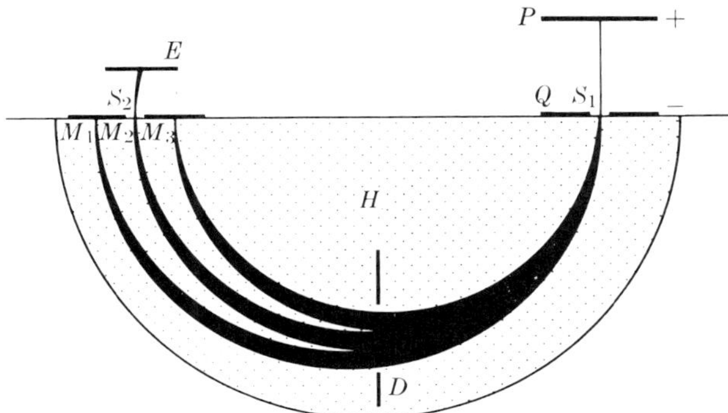

Fig. 4.14 Dispersion and focusing of a beam of ions of three isotopes.

(or magnetic) field, one can bring to a focus at slit S_2 rays having ions with different values of q/M (this means that R is fixed and q/M depends on the value of E), and then bring them to the collector plate of the electrometer E. The current recorded by the electrometer is proportional to the number of positive ions. That is, the current is proportional to the intensity (see Fig. 4.15). Each accelerating voltage V (equal to the potential difference) corresponds to a definite value of M of the positive ions reaching the collector plate.

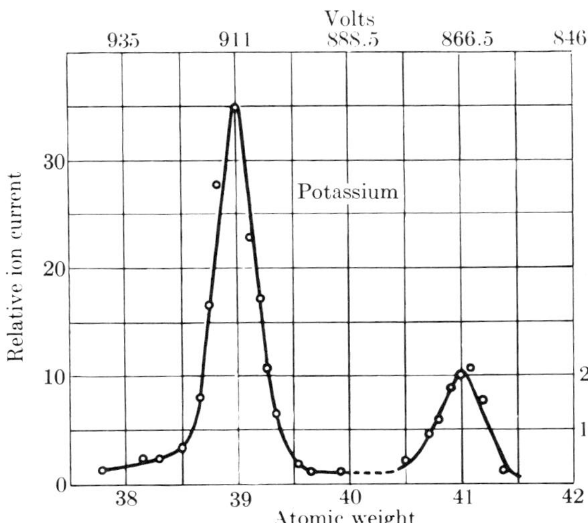

Fig. 4.15 Results obtained by Dempster mass spectrometer for two isotopes of potassium. The areas under the two peaks are proportional to the relative abundances of the isotopes. The small peak has been multiplied by a factor of 10. [From A. J. Dempster, *Phys. Rev.* **22,** 631 (1922)]

One can find the unknown atomic masses of the isotopes by comparing them with some standard atomic masses. When one plots intensity (which is proportional to current) versus atomic weight (which is inversely proportional to the accelerating voltage) one can find the areas under the peaks, which in turn yields the abundance of the isotopes.

Figure 4.15 shows this for the two isotopes of potassium. The atomic weights of these isotopes are 39 and 41 units and the relative intensities, proportional to the areas under the peaks, are 18 and 1, respectively.

SUMMARY

The "plum-pudding" model of atom suggested by J. J. Thomson failed because it could not explain the large-angle alpha scattering observed by Geiger and Marsden.

According to Rutherford's nuclear model of the atom, "all the positive charge and hence almost all the mass of the atom is confined to a very small volume at the center of the atom, called the 'nucleus.'"

According to Rutherford, the relation between the impact parameter b and the scattering angle θ is

$$\theta = 2 \operatorname{arccot}\left[\left(\frac{mv_0^2}{kqQ}\right) b\right]$$

and the cross section of a nucleus is

$$\sigma = \pi b^2.$$

Rutherford's formula for the scattering of α particles is

$$N(\theta) = \frac{N_i n t k^2 Z^2 e^4}{4r^2 K_\alpha^2 \sin^4(\theta/2)}.$$

The distance of closest approach r_c is given by

$$K_\alpha = \tfrac{1}{2} m v_0^2 = \frac{kqQ}{r_c},$$

while the size of a nucleus of mass number A is given by

$$R = r_0 A^{1/3}, \qquad \text{where } r_0 = 1.414 \times 10^{-13} \text{ cm}.$$

According to the proton–neutron model of the atom, the nucleus consists of Z protons and $(A - Z)$ neutrons.

Isotopes are atoms with the same Z but different A.

In a Dempster mass spectrometer, $\frac{1}{2}mv^2 = qV$, $Mv^2/R = qvB$, and hence $q/M = 2V/B^2R^2$.

PROBLEMS

1. The measured value of the radius of an atom is 10^{-8} cm. A nucleon (proton or neutron) is a sphere of radius 1.35×10^{-13} cm, while the size of the electron is almost negligible. What fraction of the total volume of the atom is occupied by a nucleus containing A nucleons?
2. An α particle moving with velocity v strikes an electron at rest. Show that the maximum velocity the electron can gain is $2v$. (Assume that the mass of the electron is very small compared to that of the α particle.) Calculate the fraction of the incident energy lost to the electron.
3. Calculate the fraction of the α-particle energy lost when it makes a head-on collision with a gold ($Z = 79$ and $A = 197$) nucleus. How does this compare with the results of Problem 2?
4. Calculate the maximum energy lost by an α particle of incident kinetic energy 5 MeV in a head-on collision with (a) an electron at rest, (b) a gold nucleus at rest.
5. A particle of charge q and kinetic energy K_i is incident on a target of charge Q. Prove the following relation between the impact parameter b and the distance of closest approach r_c:

$$b = r_c \sqrt{\left(1 - \frac{kqQ}{r_c K_i}\right)}.$$

[*Hint:* Use conservation of angular momentum ($mvb = mr^2\omega$) and conservation of energy.]

6. A particle of mass m, charge ze, and kinetic energy K_i is scattered through an angle θ by a nucleus of charge Ze. Prove the following relation between the distance of closest approach r_c and the scattering angle θ:

$$r_c = \frac{kZze^2}{2K_i}\left(1 + \frac{1}{\sin(\theta/2)}\right)$$

7. $^{210}_{84}$Po emits α particles of kinetic energy 5.3 MeV, which are scattered by a silver foil of thickness 10^{-7} m. Calculate the magnitude of the change of momentum Δp, given that these particles are scattered through (a) $\theta = 30°$, (b) $\theta = 45°$, (c) $\theta = 90°$.
8. $^{210}_{84}$Po emits α particles of kinetic energy 5.3 MeV, which are scattered by a silver foil of thickness 10^{-8} m. Calculate the fraction of the α particles scattered through angles greater than (a) 45°, (b) 90°. Calculate the distance of closest approach in both cases.
9. What is the thickness of a gold ($Z = 79$) foil that will scatter one out of every 1000 α particles of 5 MeV energy through an angle greater than 90°?
10. What is the impact parameter of an α particle of 7 MeV energy which is scattered through an angle of (a) 30°, (b) 45°, by a silver ($Z = 47$) foil of thickness 10^{-7} m?
11. Alpha particles of energy 5 MeV are incident on a copper ($Z = 29$) foil of thickness 10^{-7} m. Calculate the scattering angle for an impact parameter of 10^{-11} cm. What is the distance of closest approach in this case? (See Problem 5.)
12. What fraction of incident 7-MeV α particles are scattered by a gold foil 10^{-5} cm thick into a detector whose sensitive area is 10 cm², and which is placed 40 cm from the target and makes an angle of 30° with the incident beam?

13. Suppose that 10^8 α particles per second, of kinetic energy 5 MeV, are incident on a silver foil 10^{-5} cm thick. What is the number of particles per second scattered into a detector of area 5 cm^2 placed 10 cm from the target and at an angle of 90°?

14. Consider 5 MeV, 8 MeV, 10 MeV, and 15 MeV α particles incident on a target of gold ($A = 197$) foil. Calculate the distance of closest approach in each case. Can you estimate the size of the nucleus from such calculation? How do these values compare with the radius of the nucleus given by $R = 1.35 \times 10^{-13} A^{1/3}$ cm?

15. Consider Compton scattering between an incident photon of energy E_γ ($= h\nu$) and a nucleus which has a rest mass M. Show that:

   a) The change in wavelength of the incident photon is

$$(\Delta\lambda)_{\max} = \frac{2hc}{Mc^2}.$$

   b) The change in energy of the incident photon is

$$\Delta E \simeq -\frac{hc\,\Delta\lambda}{\lambda^2}.$$

   c) The energy E_N of the recoil nucleus is related to the incident photon energy E_γ by the following relation:

$$E_N = \frac{2E_\gamma^2}{Mc^2}.$$

16. By means of the uncertainty principle, show that a proton or a neutron can reside inside the nucleus. That is, the kinetic energy of a nucleon inside the nucleus is a few MeV.

17. Two positive ions, of masses M_1 and M_2, but of the same charge q, are accelerated along the X-axis through a potential difference V. They then pass through a uniform magnetic field of magnetic induction B normal to the plane of the paper (along the Z-axis).

   a) Show that, after the ions enter the magnetic field the value of the y coordinate (perpendicular to the X-axis in the plane of the paper) at any time t is given by

$$y = Bx^2 \frac{q}{2Mv}.$$

   b) What is the ratio of the y deflections of the two ions?
   c) What is the ratio of the radii of the paths of the two ions in the magnetic field?
   d) Suppose that these ions, after entering the magnetic field, were to travel in circular orbits. Calculate the orbital frequencies.

18. A beam of electrons moving with a velocity $\mathbf{v}$ enters a uniform field of strength $\mathbf{B}$; $\mathbf{v}$ makes an angle θ with the direction of $\mathbf{B}$.

   a) Show that the path of these electrons is a helix.
   b) Show that the electrons cross the axis again at time $t = 2\pi m/Be$, and at the distance $x = 2\pi mv \cos\theta/Be$. (Note that this is the so-called magnetic lens.)

19. The stable isotopes of zinc, their percentage abundances, and their isotopic masses on the ^{12}C scale are as follows.

Isotopes	Abundance	Isotopic masses
^{64}Zn	48.89	63.9291
^{66}Zn	27.81	65.9260
^{67}Zn	4.11	66.9271
^{68}Zn	18.57	67.9249
^{70}Zn	0.62	69.9253

Calculate the atomic weight of zinc.

20. The stable isotopes of iron, their percentage abundances, and their isotopic masses on the ^{12}C scale are as follows.

Isotopes	Abundance	Isotopic masses
^{54}Fe	5.82	53.9396
^{56}Fe	91.66	55.9349
^{57}Fe	2.19	56.9354
^{58}Fe	0.33	57.9333

Calculate the atomic weight of iron.

21. How will the behavior of a doubly ionized atom differ from that of a singly ionized atom with half the actual mass?

22. Locate the position of the collector plates for singly ionized ions with mass numbers 20, 21, and 22 that have been accelerated through a potential difference of 3000 volts, after which they enter a uniform magnetic field of 0.3 weber/m^2 perpendicular to the beam of ions.

23. A beam of singly ionized boron atoms, after being accelerated through a potential difference of 2000 volts, enters a uniform magnetic field of 0.3 weber/m^2 perpendicular to the beam. The boron ions then impinge on a photographic plate placed at S_2 in Fig. 4.14. What is the absolute separation of $^{10}B^+$ and $^{11}B^+$ ions on the photographic plate?

24. Suppose that you are using a Dempster mass spectrograph for the purpose of measuring the mass difference of the doublet $(^1H_2)^+ - (^2H)^+$. Find an expression for the mass difference in terms of the absolute separation $\Delta r = r_1 - r_2$ and the average distance $r = (r_1 + r_2)/2$, where r_1 and r_2 are the radii of the paths of the ions in the magnetic field.

REFERENCES

1. H. Geiger and E. Marsden, *Proc. Roy. Soc.* **A82,** 495 (1909)
2. H. Geiger and E. Marsden, *Phil. Mag.* **25,** 607 (1913)
3. E. Rutherford, *Phil. Mag.* **21,** 669 (1911)

4. M. M. Gordon, *Am. J. Phys.* **23,** 247 (1955)
5. G. W. Farwell and H. E. Wegner, *Phys. Rev.* **95,** 1212 (1954)
6. D. D. Kerles, J. S. Blair, and G. W. Farwell, *Phys. Rev.* **107,** 1343 (1957)
7. W. Bothe and H. Becker, *Z. Physik* (Berlin) **66,** 289 (1930)
8. I. Curie and F. Joliot, *Compt. Rend.* **194,** 273 (1932)
9. J. Chadwick, "The Existence of a Neutron," *Proc. Roy. Soc.* (London) **A-136,** 692 (1932)
10. J. J. Thomson, *Phil. Mag.* **24,** 209, 669 (1912), **21,** 225 (1911)
11. F. W. Aston, *Phil. Mag.* **38,** 707 (1909); **39,** 449 (1920)
12. F. W. Aston, *Proc. Roy. Soc.* **A115,** 484 (1927); **163,** 391 (1937)
13. A. J. Dempster, *Phys. Rev.* **11,** 316 (1918); **18,** 415 (1921); **20,** 631 (1922)
14. K. T. Bainbridge and E. B. Jordan, *Phys. Rev.* **50,** 282 (1936)
15. J. Mattauch, *Phys. Rev.* **50,** 617 (1936)
16. A. O. Nier, *Rev. Sci. Instr.* **18,** 398 (1947)
17. S. A. Goudsmit, *Phys. Rev.* **74,** 622 (1948)
18. L. G. Smith and C. C. Damm, *Rev. Sci. Instr.* **27,** 638 (1956)

Suggested Readings

1. Atam P. Arya, *Fundamentals of Atomic Physics*, Chapter VII; Boston: Allyn and Bacon, 1971
2. M. H. Shamos, editor, *Great Experiments in Physics*, New York: Holt, Rinehart and Winston, 1962
3. H. Boorse and L. Motz, editors, *The World of the Atom*, New York: Basic Books, 1966
4. M. M. Gordon, *Am. J. Phys.* **23,** 247 (1955)
5. William F. Magie, *A Source Book in Physics*, Cambridge, Mass.: Harvard University Press, 1963
6. Atam P. Arya, *Fundamentals of Nuclear Physics*, Chapter V; Boston: Allyn and Bacon, 1966
7. J. J. Thomson, *Rays of Positive Electricity and Their Application to Chemical Analysis*, second edition, London: Longmans Green, 1921
8. L. T. Bainbridge, "Charged Particle Dynamics and Optics, Relative Isotope Abundances of the Elements, Atomic Masses," in *Experimental Nuclear Physics*, Vol. I, edited by E. Segre; New York: John Wiley, 1953

CHAPTER 5
BOHR AND SOMMERFELD THEORIES OF THE HYDROGEN ATOM

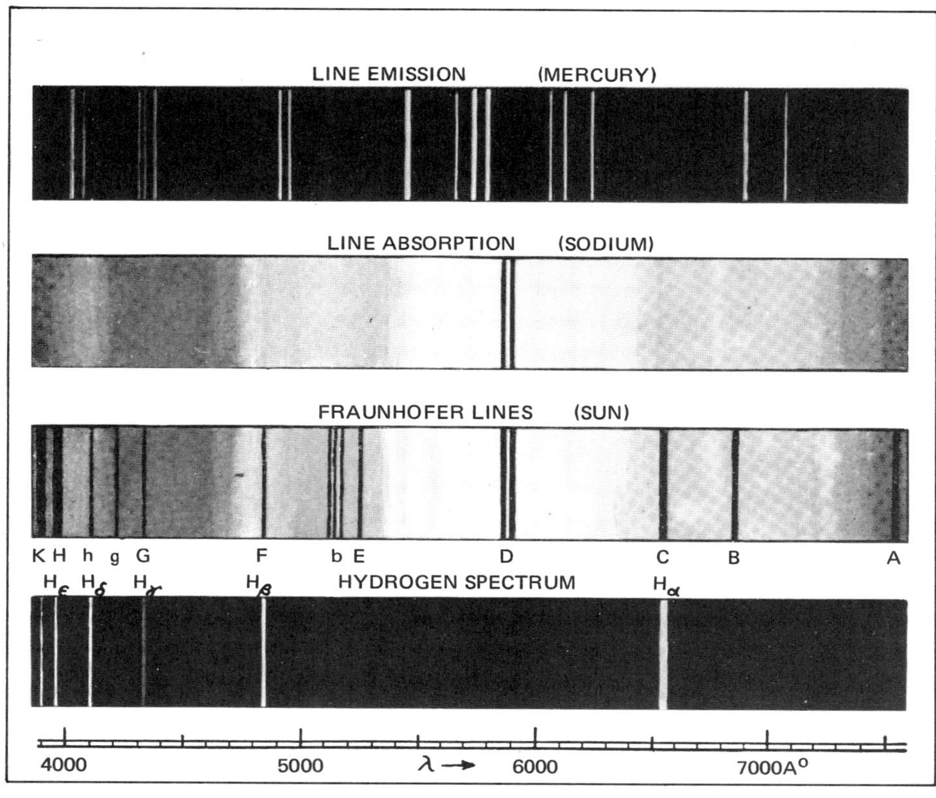

Continuous and line spectra, both emission and absorption.
(*From H. E. White,* Atomic and Nuclear Physics, *D. Van Nostrand, New York, 1964*)

In this and the next few chapters, we're going to investigate the arrangement of the electrons surrounding the nucleus inside the atom. Here we shall limit ourselves to the study of the simplest of all the atoms—hydrogen. We shall also limit ourselves to the so-called semiclassical or old quantum-mechanical theories of the hydrogen atom, developed before the introduction of quantum (or wave) mechanics. To help us in our work, we shall use the most useful experimental tool for studying atomic structure: *spectroscopy*.

5.1 ATOMIC SPECTRA

The study of atomic spectra is concerned with the measurement of the wavelengths and intensities of the electromagnetic radiation emitted or absorbed by atoms. Using the results of these experiments as guidelines, scientists have constructed theories to explain the arrangement of the electrons in the atoms. Figure 5.1 shows a typical arrangement for observing atomic spectra. The source S may be in the form of an electric discharge passing through a monatomic gas. Or it may be an electric arc or spark, or one may heat the salts of the material one is studying. In any of these processes, the atoms are put into (or raised to) those states in which their total energy is greater—the *excited states*—than when they are in their normal or ground state. Atoms in a higher or excited energy state do not remain there for more than 10^{-8} second before they return to the normal state. During their transition from the higher to the lower state, the atoms emit excess energy in the form of electromagnetic radiation. This electromagnetic radiation, after it passes through a system of lenses and collimating slits, falls on a prism (or diffraction grating). The prism disperses the radiation and brings different wavelengths to focus at different points on the photographic plate. The impressions on the photographic plate appear as lines (because of the rectangular slit in front of the source) corresponding to different wavelengths. Hence the spectrum obtained is called a *line spectrum*.

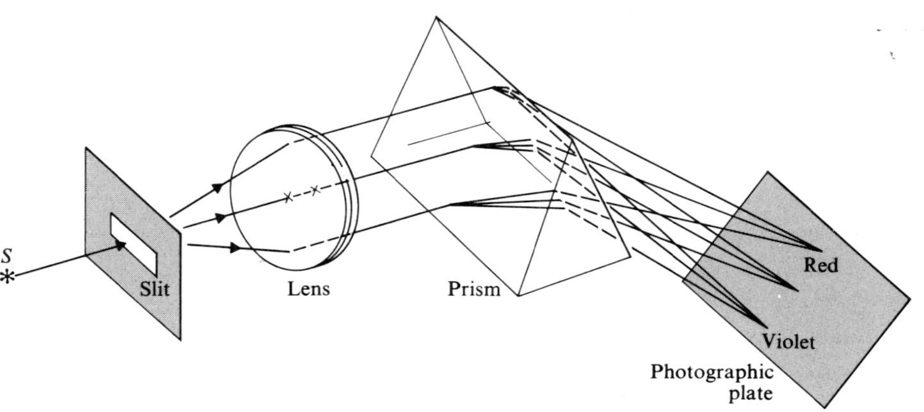

Fig. 5.1 Experimental arrangement for observing spectra of different atoms. Instead of the prism, one often uses a diffraction grating.

Any material's atomic spectrum consists of discrete lines, and is due to the transitions of individual atoms; thus the spectrum is characteristic of the type of atom. [This is quite different from the continuous spectrum of electromagnetic radiation obtained from the surface of a heated solid, as in the case of blackbody radiation, discussed in Chapter 2.] The arrangement used for making visual observations of spectra is called a *spectroscope*; if it is arranged for measuring wavelength, it is called a *spectrometer*.

In the second half of the nineteenth century, scientists expended great efforts in measuring the wavelelengths of line spectra of different elements. Atomic spectra became a useful tool because scientists could identify elements by their characteristic line spectra. In addition the regularities they observed in such spectra helped them formulate theories about the structure of the atom. They found that the spectrum of any element consists of wavelengths which exhibit definite regularities, and could be classified into groups called the *spectral series*. The first such spectral series was observed by J. J. Balmer[1] in 1885 in the spectrum of the hydrogen atom. The hydrogen atom consists of a single proton—the nucleus—and a single electron outside the nucleus. The frequencies of the lines in this series, called the *Balmer series*, are in the visible region (see Fig. 5.2). Balmer expressed the wavelengths of the lines in this series by the formula

$$\lambda = (3645.6)\,\frac{n^2}{n^2 - 4}\ \text{Å}, \tag{5.1}$$

where $n = 3, 4, 5, 6, \ldots$, and λ is the wavelength in angstrom units ($1\ \text{Å} = 10^{-10}$ m $= 10^{-8}$ cm). For $n = 3$, $\lambda = 6562.08$ Å ($\lambda_{\text{expt}} = 6562.8$ Å), and is called the H_α line. For $n = 4$, $\lambda = 4860.80$ Å ($\lambda_{\text{expt}} = 4861.3$ Å) and is called the H_β line. Similarly, $n = 5, 6, \ldots$, give the H_γ, $H_\delta, \ldots$ lines, respectively, as shown in Fig. 5.2. As n increases, the wavelengths become closer and closer, merging into one at $n = \infty$, with $\lambda = 3645.6$ Å, which is called the *series limit*.

J. R. Rydberg,[2] in 1896, expressed the Balmer series given by Eq. (5.1) in a more convenient form by using the reciprocal wavelength, called the wave number

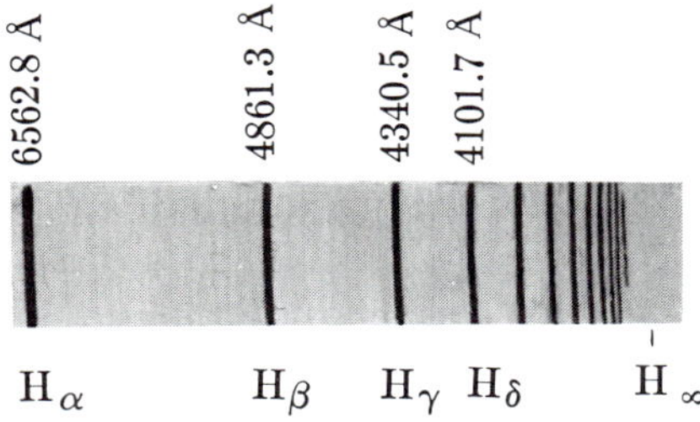

Fig. 5.2 A portion of the emission spectrum of hydrogen, corresponding to visible lines of the Balmer series and the near-ultraviolet region. [From *Atomic Spectra and Atomic Structure*, by G. Herzberg, New York: Dover Publications, 1944]

$\bar{\nu} = 1/\lambda$. The number of wavelengths per centimeter is given by

$$\frac{1}{\lambda} = \bar{\nu} = R_H\left(\frac{1}{2^2} - \frac{1}{n^2}\right) \tag{5.2}$$

where $n = 3, 4, 5, 6, \ldots$. R_H is the Rydberg constant, the most recent value of which (with an accuracy for wavelength measurement of one part in 100,000 as compared to one part in 1000 in about 1900) is

$$R_H = 109677.576 \pm 0.012 \text{ cm}^{-1}. \tag{5.3}$$

Many more series were observed after Bohr formulated the theory of the hydrogen atom in 1913, as we shall see later.

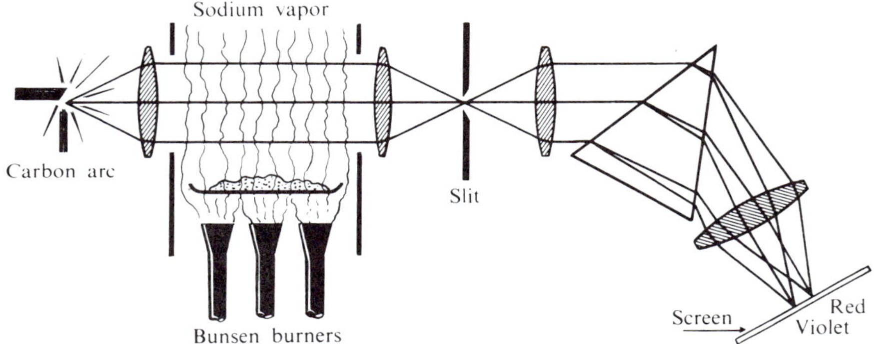

Fig. 5.3 Experimental arrangement for observing absorption spectra of different atoms.

So far we have been discussing emission spectra of atoms. One may observe *absorption spectra* of atoms by using the arrangement shown in Fig. 5.3. In this case, light containing *all* wavelengths is made to pass through a tube containing atoms, in the gaseous state, of the element under investigation. The light coming out is analyzed as before. The spectrum now consists of dark lines against a bright background. The frequencies of these dark lines are the same as those of the lines in the emission spectrum of the same element. For every line in the absorption spectrum, there is a corresponding line in the emission spectrum, *but the reverse is not true*, because, in absorption, the probability of an atom making a transition from one excited state to another is negligible (the atoms are in the ground state).

5.2 INADEQUACY OF CLASSICAL PLANETARY MODEL OF HYDROGEN ATOM

According to the classical planetary model of the hydrogen atom, the nucleus (in this case the proton, which has 1836 times the mass of an electron) is at rest, while the electron is going around it in a circular or elliptical orbit. This is like the planetary model of the solar system, with the nucleus the sun, the electron the planet, and the coulomb force between the electron and the nucleus like the gravitational force between the planet and the sun.

For simplicity, let us assume that the orbit of the electron in the hydrogen atom is a circular one, as shown in Fig. 5.4. The centripetal force F_c which keeps the electron of mass m moving with a velocity v in a circular orbit of radius r is provided by the electrostatic attractive force F_e between the electron and the proton. Hence we may write

$$F_c = F_e, \qquad \frac{mv^2}{r} = k\frac{e^2}{r^2}, \tag{5.4}$$

where $k = \dfrac{1}{4\pi\epsilon_0} = 8.99 \times 10^9$ N-m²/coul².

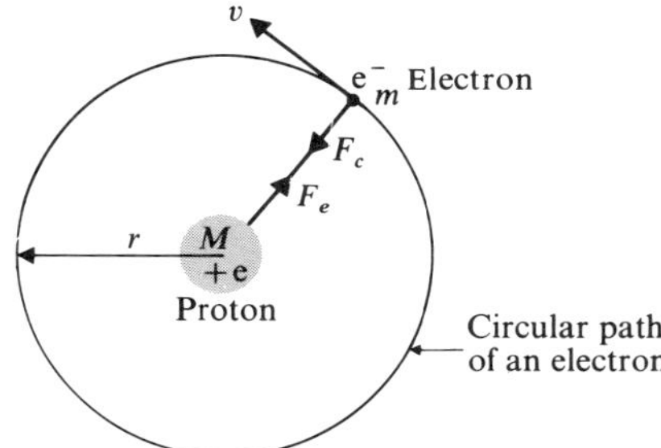

Fig. 5.4 For an electron moving in a circular orbit with the proton at the center, the centripetal force F_c is provided by the coulomb force F_e; that is, $mv^2/r = ke^2/r^2$.

The total energy E of the hydrogen atom is equal to the sum of its kinetic energy K and its potential energy P. That is,

$$E = K + P = \tfrac{1}{2}mv^2 + \left(-\frac{ke^2}{r}\right). \tag{5.5}$$

Substituting for mv^2 from Eq. (5.4) into Eq. (5.5), we obtain

$$\boxed{E = -\frac{1}{2}\frac{ke^2}{r}.} \tag{5.6}$$

The fact that the total energy is negative means that the electron is bound to the nucleus. Also, if we substitute $E = -13.58$ eV, which is the binding energy of the electron in the hydrogen atom, we get $r = 0.53$ Å (as shown in Example 5.1). This agrees with the value of the radius of the hydrogen atom as measured by other methods. So far, Eq. (5.6) is in excellent agreement with experiment, but there are some serious shortcomings, as we shall soon see.

Example 5.1. Calculate the radius of the orbit of the electron in the hydrogen atom. Also calculate the frequency of the electromagnetic radiation if radiation is emitted by this circulating electron.

The binding energy of an electron in a hydrogen atom is -13.58 eV, and from Eq. (5.6), we have

$$E = -\frac{ke^2}{2r}$$

or

$$r = -\frac{ke^2}{2E} = -\frac{(8.99 \times 10^9 \text{ N-m}^2/\text{coul}^2)(1.602 \times 10^{-19} \text{ coul})^2}{2(-13.58 \text{ eV} \times 1.602 \times 10^{-19} \text{ joule/eV})}$$

$$= 0.53 \times 10^{-10} \text{ m} = 0.53 \text{ Å}.$$

The orbital frequency f of the electron is

$$f = \frac{\omega}{2\pi} = \frac{v}{2\pi r},$$

where v is calculated from Eq. (5.4). That is,

$$\frac{mv^2}{r} = \frac{ke^2}{r^2},$$

or

$$v = \sqrt{\frac{ke^2}{mr}}.$$

Thus

$$f = \frac{1}{2\pi}\sqrt{\frac{ke^2}{mr^3}},$$

which, when we substitute for k, e, m, and r, yields

$$f = \frac{1}{2\pi}\sqrt{\frac{8.99 \times 10^9 \text{ N-m}^2/\text{coul}^2 \times (1.602 \times 10^{-19} \text{ coul})^2}{9.108 \times 10^{-31} \text{ kg} \times (0.53 \times 10^{-10} \text{ m})^3}}$$

$$= 0.66 \times 10^{16} \text{ cycles/sec},$$

which is in the ultraviolet frequency region.

According to classical electromagnetic theory, an accelerated charge must radiate electromagnetic waves. An electron in a circular orbit in the hydrogen atom has an acceleration v^2/r, and hence should radiate electromagnetic waves. In this process of radiation, the electron naturally loses energy, and hence E decreases, i.e., becomes more negative. This is possible only if r decreases. Since the electron accelerates constantly, it radiates continuously, and hence E and r decrease continuously. This means that the electron ought eventually to collapse into the nucleus (it should take only 10^{-8} sec to collapse), as shown in Fig. 5.5.

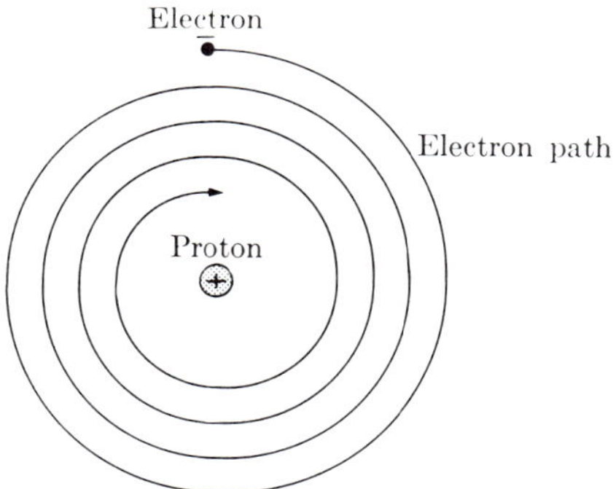

Fig. 5.5 According to classical theory, the radiation of energy by the electron would lead to a decrease in radius of the orbit and eventually (within $\sim 10^{-8}$ sec) to a collapse of the electron into the nucleus.

But this isn't what happens! The hydrogen atom does not itself radiate electromagnetic waves. Furthermore, it is stable. The classical model also predicts a continuous spectrum of emitted radiation, but this is contrary to the discrete line spectrum observed in the laboratory. Hence we must reject the classical planetary model.

5.3 THE BOHR THEORY OF THE HYDROGEN ATOM

In 1913, Niels Bohr[3] formulated a theory of the hydrogen atom which satisfactorily explained the observed spectrum. Bohr's theory is a mixture of classical physics and the energy-quantization ideas introduced by Planck. That's why his theory is also referred to as the *semiclassical theory*, or as the old quantum theory of the hydrogen atom. The Bohr theory is based on the following three postulates.

a) The basic postulates

The three basic postulates, or the quantization rules, may be stated as follows.

Postulate I. *An electron in an atom can move about the nucleus in certain circular orbits without radiating. These are called the discrete stationary states of the atom (or the system).*

Postulate II. *The allowed stationary states are those for which the orbital angular momentum, $L = mvr$, of the electron is equal to an integral multiple of $\hbar$ (where $\hbar = h/2\pi$, and h is Planck's constant). That is,*

$$L = mvr = n\hbar, \tag{5.7}$$

where $n = 1, 2, 3, 4, \ldots$, and n is called the principal quantum number.

[*This quantization condition may also be written as*

$$\int_0^{2\pi} p_\theta \, d\theta = nh,$$

where $p_\theta = mvr$ *is constant. Hence*

$$p_\theta \int_0^{2\pi} d\theta = nh$$

or

$$p_\theta 2\pi = nh, \qquad mvr = n\hbar.]$$

Postulate III. *Whenever an electron jumps from an initial higher energy orbit* E_i *to a final lower energy orbit* E_f *(i.e., whenever a transition takes place from the initial state of energy* E_i *to the final state of energy* E_f*), an electromagnetic radiation (or photon) of energy hv is emitted so that*

$$hv = E_i - E_f. \tag{5.8}$$

Postulate I is completely arbitrary; but postulate III does have some justification from Planck's quantum hypothesis. The justification for postulate II came later, with the deBroglie hypothesis (1926).

Suppose there is a string of length l tied at both ends. Let there be n waves of wavelength λ, such that $l = n\lambda$. These waves are standing (or stationary) waves. Let us bend this string into a circle of radius r, as shown in Fig. 5.6 for $n = 4$. Thus $l = 2\pi r = n\lambda$. According to the deBroglie hypothesis, $\lambda = h/p = h/mv$, and therefore

$$n\lambda = n\frac{h}{mv} = 2\pi r \qquad \text{or} \qquad mvr = nh/2\pi,$$

which is postulate II, given above. That is, the electron in a stationary state can be represented by a standing matter wave.

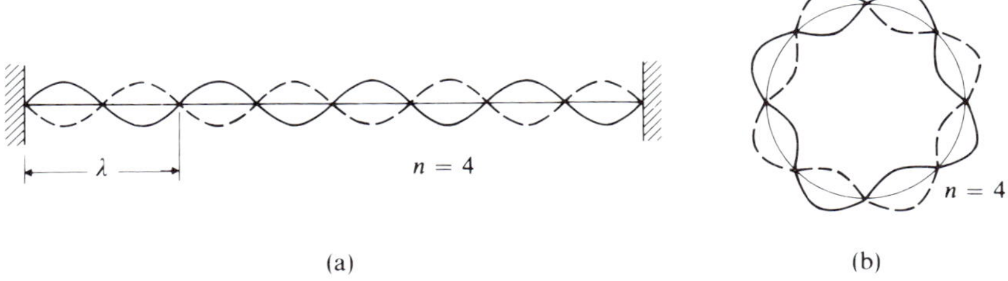

Fig. 5.6 (a) Formation of standing or stationary resonance vibrations in a string for $n = 4$. (b) Formation of standing or stationary resonance vibrations on a circular loop for $n = 4$.

b) Energy levels

Suppose that the electron is in a stationary circular orbit of radius r_n with velocity v_n, so that the coulomb force provides the necessary centripetal force (recall Fig. 5.4). Thus

$$\frac{mv_n^2}{r_n} = k\frac{e^2}{r_n^2}. \tag{5.9}$$

But v_n, from postulate II above, or from Eq. (5.7), is

$$v_n = \frac{n\hbar}{mr_n}. \tag{5.10}$$

Substituting for v_n in Eq. (5.9) and solving for r_n, we obtain

$$\boxed{r_n = n^2\frac{\hbar^2}{kme^2} = n^2 r_1,} \tag{5.11}$$

where $n = 1, 2, 3, 4, \ldots$ and $r_1 = \hbar^2/kme^2$ is the radius of the first Bohr orbit of the hydrogen atom. If we substitute the values of $\hbar$, k, m, and e^2, we get $r_1 = 0.528$ Å, which agrees with the experimentally measured value. Thus, according to the Bohr theory, the electron can only be in those orbits which are given by

$$r_n = r_1, 4r_1, 9r_1, 16r_1, \ldots. \tag{5.12}$$

We can obtain the velocity of the electron in these orbits by substituting for r_n from Eq. (5.11) into Eq. (5.10). That is,

$$\boxed{v_n = \frac{1}{n}\frac{ke^2}{\hbar} = \frac{v_1}{n},} \tag{5.13a}$$

$$v_n = v_1, \frac{v_1}{2}, \frac{v_1}{3}, \frac{v_1}{4}, \qquad \text{where } v_1 = \frac{k}{mr_1} = \frac{ke^2}{\hbar}. \tag{5.13b}$$

The total energy E_n of the electron in the nth orbit is the sum of the kinetic energy and the potential energy. That is,

$$E_n = \tfrac{1}{2}mv_n^2 + \left(-\frac{ke^2}{r_n}\right). \tag{5.14}$$

Substituting for v_n from Eq. (5.10) and then for r_n from Eq. (5.11) into Eq. (5.14), we obtain

$$E_n = \frac{1}{2}m\left(\frac{n^2\hbar^2}{m^2r_n^2}\right) - \frac{ke^2}{r_n} = \frac{1}{2}\frac{n^2\hbar^2}{m}\left(\frac{n^2\hbar^2}{kme^2}\right)^{-2} - ke^2\left(\frac{n^2\hbar^2}{kme^2}\right)^{-1}.$$

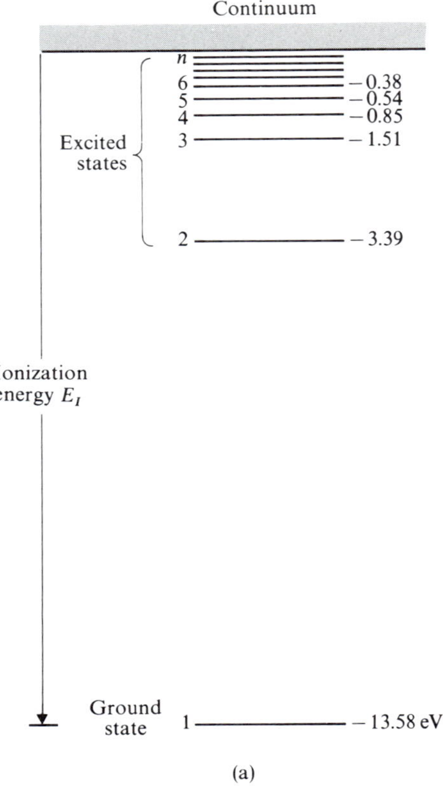

Fig. 5.7 (a) Possible energy states of the hydrogen atom corresponding to different values of the quantum number n.

Therefore

$$E_n = -\frac{1}{n^2}\left(\frac{k^2e^4m}{2\hbar^2}\right) \tag{5.15}$$

or

$$\boxed{E_n = -\frac{E_I}{n^2},} \tag{5.16}$$

where

$$\boxed{E_I = \frac{k^2e^4m}{2\hbar^2} = 13.58 \text{ eV}} \tag{5.17}$$

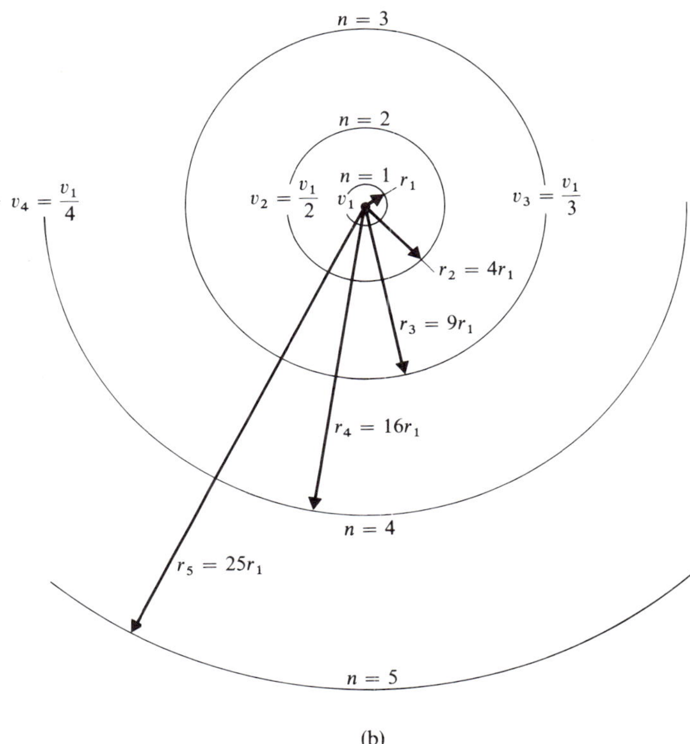

(b)

Fig. 5.7 (b) Relative sizes of circular orbits of the electron when in different states.

is the magnitude of the energy of the electron in the first Bohr orbit of the hydrogen atom. Thus, for $n = 1$, $E_1 = -E_I = -13.58$ eV, which is in perfect agreement with the measured value of the binding energy of the electron in the hydrogen atom. Thus, for $n = 1, 2, 3, 4, \ldots$, the allowed energy levels of the hydrogen atom are given by

$$E_n = -E_I, -\frac{E_I}{4}, -\frac{E_I}{9}, -\frac{E_I}{16}, \ldots. \tag{5.18}$$

Figure 5.7 shows these levels schematically. Normally the electron in the hydrogen atom is found in the lowest energy state, which is the first Bohr orbit. This lowest energy state, corresponding to $n = 1$, is called the *ground state* or the *normal state* of the hydrogen atom. The energy needed to remove the electron from the atom is equal to the magnitude, E_I, of the ground-state energy, and is called the *ionization energy*. If the electron is in a state other than the ground state, we say that the electron, and hence the atom, is in the *excited state*. The energy needed to transfer an electron from the ground state to any excited state E_n is $(E_n - E_I)$, and is called the *excitation energy*.

c) The hydrogen spectrum

We can now use postulate III to calculate the energies and frequencies of the possible transitions. If the electron is in the initial state with energy E_i (other than the ground state) and makes a transition to a final state E_f, the energy $h\nu$ of the photon emitted is given by

$$h\nu = E_i - E_f. \tag{5.19}$$

From Eq. (5.16), we have

$$E_i = -\frac{E_I}{n_i^2} \quad \text{and} \quad E_f = -\frac{E_I}{n_f^2}. \tag{5.20}$$

Substituting these in Eq. (5.19), we obtain

$$\nu = \frac{E_i}{h} - \frac{E_f}{h} = \frac{E_I}{h}\left(\frac{1}{n_f^2} - \frac{1}{n_i^2}\right). \tag{5.21}$$

Using the relation $\nu = c/\lambda$, the wave number $\bar{\nu}$ $(= 1/\lambda)$ is given by

$$\bar{\nu} = \frac{1}{\lambda} = \frac{E_I}{hc}\left(\frac{1}{n_f^2} - \frac{1}{n_i^2}\right) \tag{5.22a}$$

or

$$\boxed{\bar{\nu} = \frac{1}{\lambda} = R\left(\frac{1}{n_f^2} - \frac{1}{n_i^2}\right),} \tag{5.22b}$$

where R is the *Rydberg constant*, given by (using Eq. 5.17)

$$R = \frac{E_I}{hc} = \frac{k^2e^4m}{4\pi\hbar^3c} = 109740\ \text{cm}^{-1} = 1.0974 \times 10^{-3}\ \text{Å}^{-1} \tag{5.23}$$

while

$$R_{\text{expt}} = 109677.576 \pm 0.012\ \text{cm}^{-1}. \tag{5.24}$$

Hence there is very good agreement between theory and experiment. Also Eq. (5.22) is the same as Rydberg's empirical equation for the Balmer series, as given by Eq. (5.2), provided we substitute $n_f = 2$ and $n_i = n$. By substituting $n_i = 3, 4, 5, 6, \ldots$, we get all the lines in the Balmer series, provided the electron always jumps to the final state, corresponding to $n_f = 2$.

Even a small volume of hydrogen contains many billions of atoms, which are in the ground state at room temperature. By electrical discharge or other means, one can raise the electrons to higher excited states. Since they can stay in the excited states for only a short time, they make transitions to lower energy states and emit electromagnetic radiation, i.e., photons of discrete frequencies.

The Bohr theory predicts not only the Balmer series, but many other series, corresponding to different values of n_f. These series were named after the various scientists who discovered them, as shown in Fig. 5.8. Note that only the Balmer-series lines are in the visible region; this is the reason why the other series were discovered only after they had been predicted by the Bohr theory.

Thus the Bohr theory proved very successful in explaining the known aspects of the hydrogen spectrum.

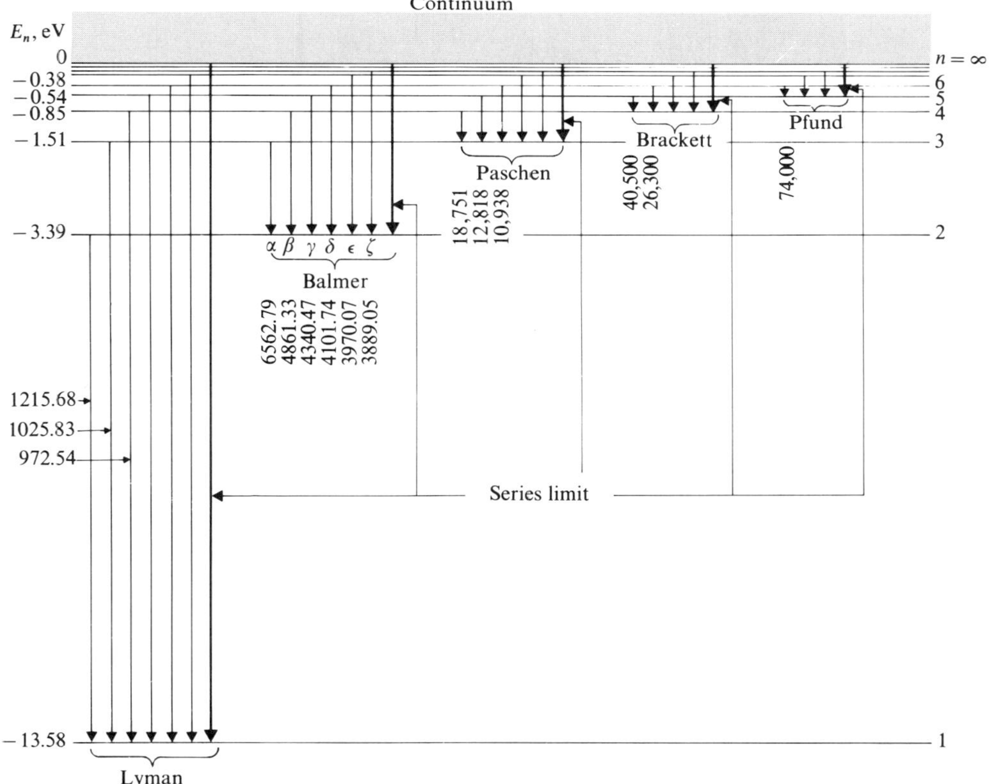

Fig. 5.8 Horizontal lines show the energy-level diagram of the hydrogen atom. Vertical lines show possible transitions corresponding to different series. Numbers with vertical lines are wavelengths of transitions, in angstroms.

5.4 EXTENSION OF THE BOHR THEORY

Now let's discuss the way the correction for nuclear motion, as applied to the Bohr theory, was used in the discovery of deuterium (an isotope of hydrogen), and the way to modify the Bohr theory in order to extend it to hydrogenlike atoms.

a) Correction for nuclear motion

In the above derivations, we assumed that the nucleus was at rest, which would be true only if the mass of the nucleus were infinite. A very small correction (measurable with sensitive instruments) must be made to the previous results if we assume that the mass of the nucleus, M, is finite. In this case the nucleus and the electron move about their common center of mass, as shown in Fig. 5.9. Given that V is the velocity of the nucleus, its momentum is $p = MV$, while the kinetic energy of the nucleus is equal to

$$K_N = \tfrac{1}{2}MV^2 = \frac{M^2V^2}{2M} = \frac{p^2}{2M}.$$

Since the total momentum of the atom is zero, the momentum of the electron is equal to that of the nucleus; that is, $mv = MV = p$, and its kinetic energy will be

$$K_e = \tfrac{1}{2}mv^2 = \frac{m^2v^2}{2m} = \frac{p^2}{2m}.$$

Thus the total kinetic energy of the hydrogen atom is

$$K = K_N + K_e = \frac{p^2}{2M} + \frac{p^2}{2m} = \frac{p^2}{2}\left(\frac{1}{M} + \frac{1}{m}\right) = \frac{p^2}{2\mu}. \tag{5.25}$$

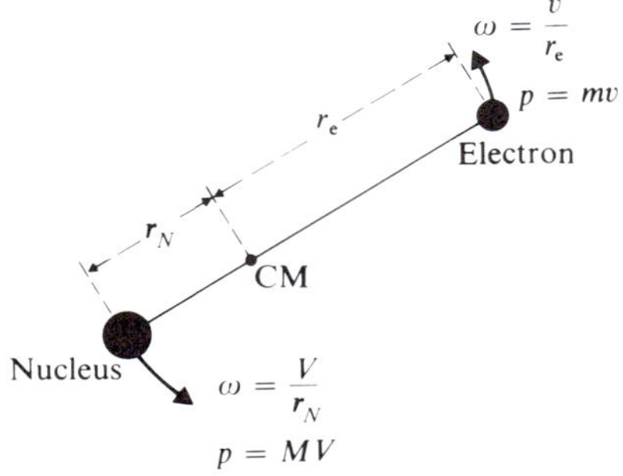

Fig. 5.9 In an atom, the nucleus and the electron move about their common center of mass with the same angular frequency $\omega = v/r_e = V/r_N$ so as to conserve momentum; that is, $mv = MV$.

That is, in replacing K_e by K, we have replaced $p^2/2m$ by $p^2/2\mu$, where

$$\frac{1}{\mu} = \frac{1}{M} + \frac{1}{m} = \frac{M + m}{Mm}$$

or

$$\boxed{\mu = \frac{Mm}{M + m} = \frac{m}{1 + (m/M)}.} \tag{5.26}$$

μ is called the *reduced mass*, and differs from m by a very small factor of $1/(1 + m/M)$. Given that M is the mass of the proton, $M = 1836m$. Therefore $\mu = m \times (1836/1837)$. ($p^2/2m$ differs from $p^2/2\mu$ only in the fact that m is replaced by μ.) Thus, in order to take into account the finite mass of the nucleus, we must replace the electron mass m by the reduced mass μ. This means that Eq. (5.22) takes the form, for the corrected wave number $\bar{\nu}_c$ (corrected for the finite nuclear mass),

$$\boxed{\bar{\nu}_c = \frac{1}{\lambda_c} = R_M\left(\frac{1}{n_f^2} - \frac{1}{n_i^2}\right),} \tag{5.27}$$

where

$$R_M = \frac{k^2e^4\mu}{4\pi\hbar^3c} = 1.0967758 \times 10^{-3}\ \text{Å}^{-1}. \tag{5.28a}$$

The Rydberg constant for the approximation of the infinite mass of an atom is

$$R_\infty = \frac{k^2e^4m}{4\pi\hbar^3c} = 1.0973731 \times 10^{-3}\ \text{Å}^{-1} \tag{5.28b}$$

and

$$\boxed{R_M = \frac{\mu}{m}R_\infty = \left(\frac{M}{M + m}\right)R_\infty.} \tag{5.29}$$

Thus it is obvious that $\bar{\nu}_c$ differs from $\bar{\nu}$ by a very small amount, because, for any atom, $M \gg m$.

The above correction led to the discovery of an isotope of hydrogen called deuterium. Natural hydrogen consists of two isotopes: one has a proton in the nucleus and an electron outside, called hydrogen, ^{1}H; it has a mass of 1.007825 u. The other has a neutron *and* a proton in the nucleus, and an electron outside. It is called heavy hydrogen or deuterium, $^2_1$H; it has a mass of 2.01410 u. From Eq. (5.29), the Rydberg constants for these two isotopes are

$$R_{^1\text{H}} = R_\infty\left(\frac{M_{^1\text{H}}}{M_{^1\text{H}} + m}\right) = 1.09678 \times 10^{-3}\ \text{Å}^{-1} \tag{5.30}$$

and

$$R_{^2\text{H}} = R_\infty\left(\frac{M_{^2\text{H}}}{M_{^2\text{H}} + m}\right) = 1.09707 \times 10^{-3}\ \text{Å}^{-1}. \tag{5.31}$$

Because of this small difference in $R_{^1\text{H}}$ and $R_{^2\text{H}}$, the spectral lines due to these two isotopes are shifted only slightly. Thus the wavelengths of the H_α line for the two isotopes are 6562.80 Å and 6561.01 Å; that is, a difference of 1.79 Å. Similarly, the difference in the wavelengths of the H_β lines in the two isotopes is 1.326 Å. Figure 5.10 shows a transition between the levels $n = 3$ and $n = 2$ (that is, the H_α line) for hydrogen with infinite and finite nuclear masses and deuterium with finite nuclear mass.

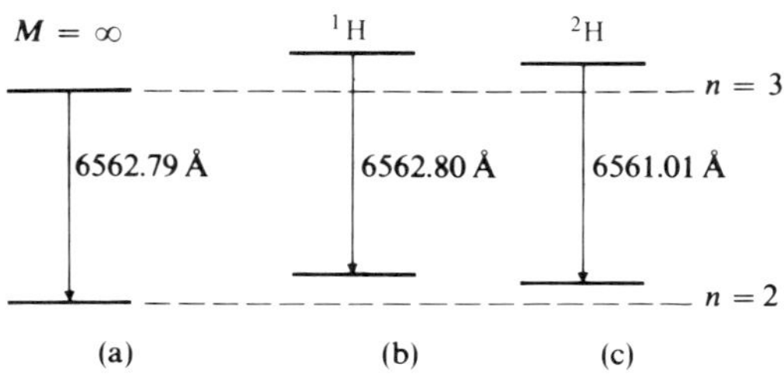

Fig. 5.10 H_α line transition (between $n = 3$ and $n = 2$) for (a) hydrogen atom with infinite nuclear mass, (b) hydrogen atom with finite nuclear mass, (c) deuterium atom with finite nuclear mass. Note that, for infinite nuclear mass, (b) and (c) coincide with (a). (Diagram not to scale.)

Natural hydrogen contains only 0.015% of ^2H isotopes, and hence lines due to ^2H isotopes are weak. Using hydrogen enriched with ^2H isotopes, Urey, Brickwedde, and Murphy[4] were able to identify the lines due to ^2H isotopes slightly shifted from the ^1H isotope. This led to the discovery of deuterium.

Example 5.2. The wavelength of the first Balmer line (H_α line) of hydrogen is 6562.80 Å. From this datum, calculate the wavelength of the H_α line for tritium.

The tritium atom, like the hydrogen atom, has one electron, except that its nucleus consists of 2 neutrons and 1 proton (instead of one proton only, as in hydrogen). Thus $Z = 1$ and $M_\text{T} = 3M_\text{H}$, and for the H_α line, $n_i = 3$ and $n_f = 2$. Equations (5.27) and (5.29), that is,

$$\bar{\nu}_c = \frac{1}{\lambda_c} = R_M\left(\frac{1}{n_f^2} - \frac{1}{n_i^2}\right) \quad \text{and} \quad R_M = \frac{\mu}{m}R_\infty = \left(\frac{M}{M + m}\right)R_\infty = \frac{R_\infty}{1 + (m/M)}$$

take the form

$$\bar{\nu}_\text{H} = \frac{1}{\lambda_\text{H}} = R_{M_\text{H}}\left(\frac{1}{2^2} - \frac{1}{3^2}\right) = \frac{R_\infty}{1 + (m/M_\text{H})}\left(\frac{1}{2^2} - \frac{1}{3^2}\right)$$

and

$$\bar{\nu}_\text{T} = \frac{1}{\lambda_\text{T}} = R_{M_\text{T}}\left(\frac{1}{2^2} - \frac{1}{3^2}\right) = \frac{R_\infty}{1 + (m/M_\text{T})}\left(\frac{1}{2^2} - \frac{1}{3^2}\right)$$

for hydrogen and tritium, respectively. Dividing one by the other, we get

$$\frac{\lambda_T}{\lambda_H} = \frac{1 + (m/M_T)}{1 + (m/M_H)}.$$

Since

$$\frac{m}{M_H} = \frac{1}{1837} \quad \text{and} \quad \frac{m}{M_T} = \frac{m}{3M_H} = \frac{1}{5611},$$

then

$$\frac{\lambda_T}{\lambda_H} = \frac{1 + (1/5611)}{1 + (1/1837)} = \frac{5612}{5514}.$$

Hence

$$\lambda_T = \tfrac{5612}{5514} \times \lambda_H = \tfrac{5612}{5514} \times 6562.80 \text{ Å} = 6680.93 \text{ Å}.$$

---

b) Hydrogenic atom

Ions like singly ionized helium, He^+, doubly ionized lithium, Li^{2+}, and triply ionized beryllium, Be^{3+}, etc., each have one electron outside the nucleus, and hence can be treated like the hydrogen atom. They are called hydrogenlike ions or *hydrogenic atoms.* For these ions, the nuclear charge is $Z = 2$, 3, and 4, respectively, for He^+, Li^{2+}, and Be^{3+}. In the hydrogenlike atoms the coulomb force takes the form $F_e = kZe^2/r^2$ instead of $F_e = ke^2/r^2$, where Z is the atomic number and Ze is the charge on the nucleus. Therefore, in every equation for the hydrogen atom, if we replace e^2 by Ze^2 and m by μ, we get the following equations, which are applicable to hydrogenlike ions:

$$\boxed{r_n = n^2 \frac{\hbar^2}{k\mu(Ze^2)}} \tag{5.32a}$$

$$v_n = \frac{1}{n}\frac{k(Ze^2)}{\hbar}, \tag{5.32b}$$

$$E_n = -\frac{1}{n^2}\frac{\mu k^2(Z^2e^4)}{2\hbar^2}, \tag{5.32c}$$

$$\bar{\nu} = \frac{1}{\lambda} = Z^2 R_M \left(\frac{1}{n_f^2} - \frac{1}{n_i^2}\right). \tag{5.32d}$$

Equation (5.32c) indicates that, compared to hydrogen energy levels, for hydrogenic atoms the energy levels are lowered by Z^2, while the energies of the photons, as indicated by Eq. (5.32d), are increased by a factor of Z^2, as shown in Fig. 5.11.

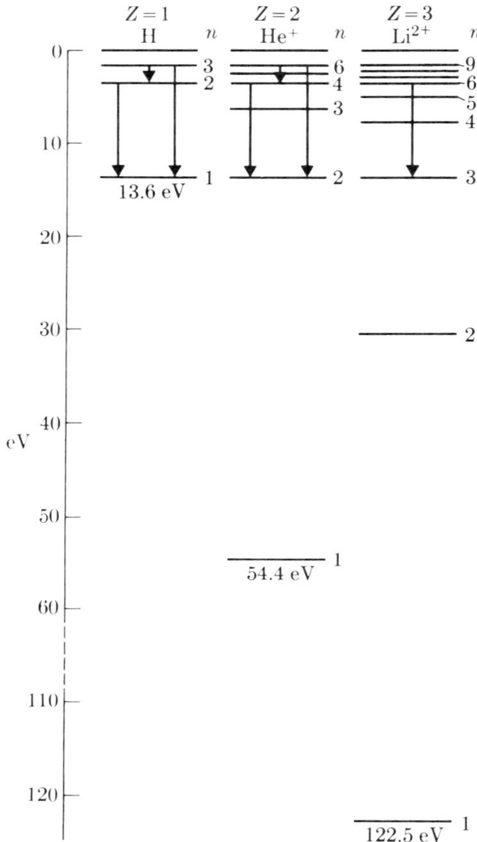

Fig. 5.11 Energy-level diagrams of hydrogen and hydrogenic atoms: singly ionized He (He^+) and doubly ionized Li (Li^{2+}). The energy levels are lowered by a factor of Z^2. Note that the energies of some of the transitions in different atoms are the same.

5.5 EXPERIMENTAL VERIFICATION OF DISCRETE ATOMIC ENERGY LEVELS

In 1914 the existence of discrete stationary states for electrons in atoms, as given by the Bohr theory, was demonstrated directly in experiments carried out by Franck and Hertz.[5] Figure 5.12 shows an outline of their experimental setup.

A tube T is filled at low pressure with the vapors of the element under investigation. Electrons emitted from a heated filament F are accelerated by a potential V_0 toward the grid G, and then, after encountering a small retarding potential V_r between G and the plate P, arrive at P, resulting in a current I_p indicated by the current-measuring device A (an ammeter or galvanometer).

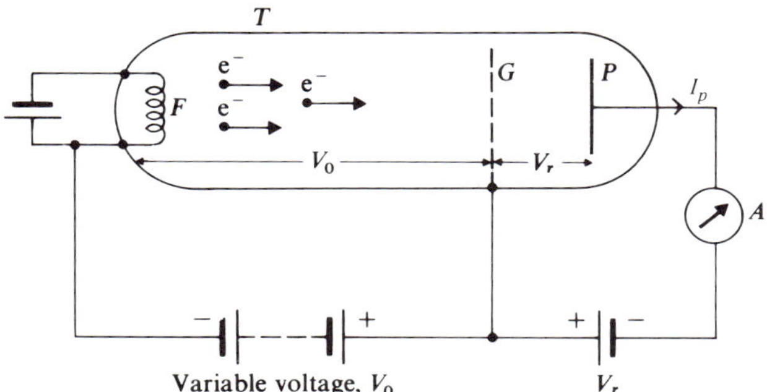

Fig. 5.12 The Franck–Hertz experimental arrangement for demonstrating the existence of discrete energy levels in mercury.

The velocity of an electron after it is emitted from F and arrives at G is

$$\tfrac{1}{2}mv^2 = eV_0. \tag{5.33}$$

As V_0 increases, so does the velocity of the electrons.

There are two types of collisions between the electrons and the atoms of the vapor. For low v, the collisions between the electrons and the atoms are *elastic*; in this case, the energy that is exchanged is energy of translation. That is, the atom is not excited and the electron simply changes its direction of motion. The electrons reaching G have enough energy to overcome the small retarding potential V_r and reach P. As V_0 increases, more and more electrons reach P; this increases the current I_P, as shown in Fig. 5.13.

As V_0 is increased still further, the electrons have enough energy to cause *inelastic* collisions with the atoms. In this case, the electron may lose a part or all of its energy, and at the same time increase the potential energy of the atom, i.e., cause excitation of the atom. Such electrons, when they reach G, don't have enough energy to overcome the small retarding potential V_r. Hence there is a sudden decrease in the current, as shown in Fig. 5.13. A further increase in V_0 increases the current I_P once again.

Figure 5.13 shows a plot of I_P versus V_0 for the case of mercury vapor. The first excited state of mercury above the ground state is 4.9 electron volts. If the electron has enough energy, it can cause more than one elastic collision. Thus, for mercury, we may expect peaks in the current I_P at 4.9 volts, $2 \times 4.9 = 9.8$ volts, $3 \times 4.9 = 14.7$ volts, etc. This is quite obvious, as we see from the results of the Franck and Hertz experiment, shown in Fig. 5.13. The voltages corresponding to the peaks in the current are called *excitation potentials*. When the voltage is very high, the electrons may cause inelastic collisions in which the atom is completely ionized. Thus there is an additional drop in the current at a voltage corresponding to the *ionization potential voltage*.

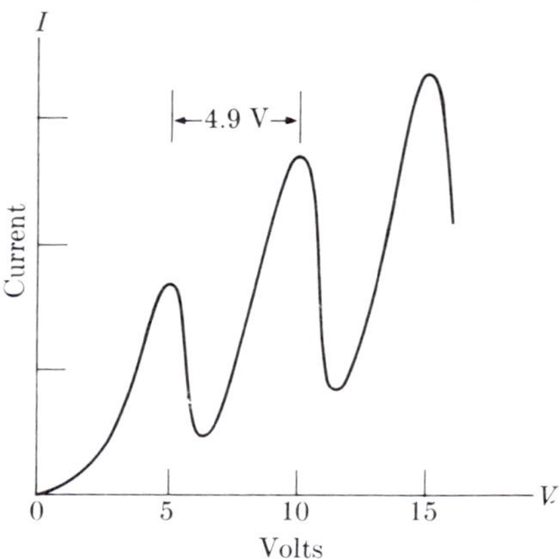

Fig. 5.13 Variation of the plate current, I_P, versus the accelerating voltage, V_0, in the Franck–Hertz experiment, showing peaks corresponding to excitation potentials of mercury.

The above experiment, though crude, was successful beyond any doubt.

Example 5.3. In sodium, the wavelength at which there is transition of an electron from the first excited state to the ground state is 5896 Å. The ionization potential of sodium is 5.1 volts. According to the Franck–Hertz experiment, at what different voltages should we expect drops in the current I_P?

The voltage at which the first drop occurs due to excitation of the sodium atom is

$$\tfrac{1}{2}mv^2 = eV = h\nu = \frac{hc}{\lambda} \tag{5.34}$$

or

$$V = \frac{hc}{e\lambda} = \frac{6.25 \times 10^{-34}\,\text{joule-sec} \times 3 \times 10^8\,\text{m/sec}}{1.602 \times 10^{-19}\,\text{coul} \times 5896 \times 10^{-10}\,\text{m}}$$

$$= 2.1 \text{ volts.}$$

Thus drops occur at 2.1 volts, 4.2 volts, and 6.3 volts, corresponding to excitation potentials; and at 5.1 volts, 7.2 volts (2.1 excitation + 5.1 ionization) corresponding to ionization potentials.

5.6 THE CORRESPONDENCE PRINCIPLE

The quantum theory predicts the behavior of microscopic systems, while the classical theory predicts the behavior of macroscopic systems. When the quantum theory is applied to large-scale phenomena (macroscopic systems), it should always reduce to the classical theory. The Bohr theory proves our point.

Consider an atom, the radius of whose Bohr orbit is ~ 1 mm, which is very large compared to 1 Å $= 10^{-8}$ cm, the diameter of the first Bohr orbit of the hydrogen atom. Hence in this case the Bohr theory should certainly reduce to the classical theory. Note that for a 1-mm diameter, the value of the quantum number n is very large: $n \simeq 5000$. Thus, according to the Bohr theory,

$$\nu = cR\left(\frac{1}{n_f^2} - \frac{1}{n_i^2}\right),$$

which may be written as

$$\nu = cR\left(\frac{n_i^2 - n_f^2}{n_i^2 n_f^2}\right)$$

$$= cR\left[\frac{(n_i + n_f)(n_i - n_f)}{n_i^2 n_f^2}\right]. \tag{5.35}$$

Since n is very large, let $n_i \simeq n_f = n$ and $n_i - n_f = \Delta n$. This reduces Eq. (5.35) to

$$\underset{n\to\infty}{\text{Limit}}\ \nu = \frac{2cR}{n^3}\,\Delta n.$$

Substituting for R from Eq. (5.23) and using Eq. (5.11) for the value of r, we get

$$\underset{n\to\infty}{\text{Limit}}\ \nu = \frac{2c}{n^3}\frac{k^2e^4m}{4\pi\hbar^3c}\,\Delta n = \frac{1}{2\pi}\frac{k^2e^4m}{n^3\hbar^3}\,\Delta n$$

$$= \frac{1}{2\pi}\sqrt{\frac{ke^2}{m}\frac{k^3m^3e^6}{n^6\hbar^6}}\,\Delta n$$

$$= \frac{1}{2\pi}\sqrt{\frac{ke^2}{mr^3}}\,\Delta n. \tag{5.36}$$

For $\Delta n = 1$, this agrees with the frequency predicted by the classical theory as given in Example 5.1. That is,

$$f = \frac{1}{2\pi}\sqrt{\frac{ke^2}{mr^3}}. \tag{5.37}$$

Also for $\Delta n = 2, 3, 4, \ldots$, we get the same harmonies $2f$, $3f$, $4f$, ..., predicted by the classical theory.

Thus there is a complete correspondence between the classical theory and quantum theory. That is,

$$\underset{\text{large size}}{\text{Limit}} \text{ (quantum theory)} \rightarrow \text{(classical theory)}. \tag{5.38}$$

★5.7 BOHR–SOMMERFELD MODEL AND RELATIVISTIC EFFECTS

The assumption in the Bohr theory that an electron moves in a circular orbit is an oversimplification. In general, the path of a particle under the influence of an inverse-square force (a coulomb force in this case) is a conic section. The orbit may be in the shape of a parabola, a hyperbola, or an ellipse; the shape depends on the total energy of the particle.[6,7] We have shown that, in the case of an electron in a hydrogen atom, the total energy is negative, and hence the path of the electron is an ellipse.

According to the Bohr–Sommerfeld[6] model of hydrogenlike atoms, the electron moves in an elliptical orbit around the nucleus at one focus. Hence its position can be located by two polar coordinates r and θ, because both r and θ are constantly changing and are periodic functions of time. (Note that, in the Bohr theory, only θ was changing, while r was constant in time.) This means that the Bohr quantization condition for angular momentum for a circular orbit of the electron must be replaced by two quantum conditions: One, for the quantization of radial momentum L_r $(= p_r)$ leads to the quantization condition

$$\oint p_r \, dr = n_r h, \tag{5.39}$$

where n_r is the *radial quantum number* and $p_r = m \, dr/dt$ is the *radial momentum.* The second condition for the quantization of angular momentum L_θ $(= p_\theta)$ leads to the quantization condition

$$\oint p_\theta \, d\theta = n_\theta h, \tag{5.40}$$

where n_θ is the *azimuthal quantum number* and $p_\theta = mr^2 \, d\theta/dt$ is the *angular momentum.* The integrals in both the above equations are evaluated over one complete period of orbital motion. A complete analysis shows that the energy calculated by imposing the above two quantization conditions depends only on the sum of the two quantum numbers. That is, it depends on the total or principal quantum number n, where

$$\boxed{n = n_r + n_\theta} \tag{5.41}$$

and

$$\begin{aligned} n &= 1, 2, 3, 4, 5, \ldots \\ n_\theta &= 1, 2, 3, 4, 5, \ldots \\ n_r &= 0, 1, 2, 3, 4, \ldots \end{aligned} \tag{5.42}$$

Thus, when $n = 1$, $n_\theta = 1$, and $n_r = 0$, the orbit is circular, as shown in Fig. 5.14(a). For $n = 2$, we can have either $n_\theta = 2$ and $n_r = 0$ or $n_\theta = 1$ and $n_r = 1$. The two orbits corresponding to these two possibilities are shown in Fig. 5.14(b); both these have the same energy. Similarly, for $n = 3$, as shown in Fig. 5.14(c), there are three possible orbits for the electron. All three have the same energy.

Thus the introduction of elliptical orbits does not introduce any new energy levels, and hence no new transitions. Atomic levels that can be described by more than one set of quantum numbers, all giving the same energy, are called *degenerate*.

Sommerfeld showed that degeneracy can be removed (i.e., two or more sets of quantum numbers describing a given level can have different energies) if relativistic effects are taken into consideration. An electron in a circular orbit has constant velocity, but an electron in an elliptical orbit has velocity that is different at different positions, speeding up when the electron is near the nucleus and slowing down when it is far away.

The elliptical path actually becomes a precessing ellipse, i.e., a rosette, as shown in Fig. 5.15. This precession rate is different for different elliptical orbits (i.e., the rate depends on n_θ), even though n may be the same. Thus degeneracy is removed. The energy levels of hydrogenlike atoms are given by

$$E_n = -\frac{\mu k^2 Z^2 e^4}{2n^2\hbar^2}\left[1 + \frac{Z^2\alpha^2}{n}\left(\frac{1}{n_\theta} - \frac{3}{4n}\right)\right], \tag{5.43}$$

where μ is the reduced mass and α is the *fine-structure constant*, given by

$$\alpha = \frac{ke^2}{\hbar c} \simeq \frac{1}{137}. \tag{5.44}$$

Thus, according to Eq. (5.43), for a given n, since n_θ can take n different values, there are n different states with slightly different energies. Figure 5.16 shows the energy-level diagram of the hydrogen atom according to the Bohr–Sommerfeld model. The resulting close levels and corresponding transitions are referred to as *fine structure*.

Recall that, in the Bohr theory, there was no restriction on the change in the

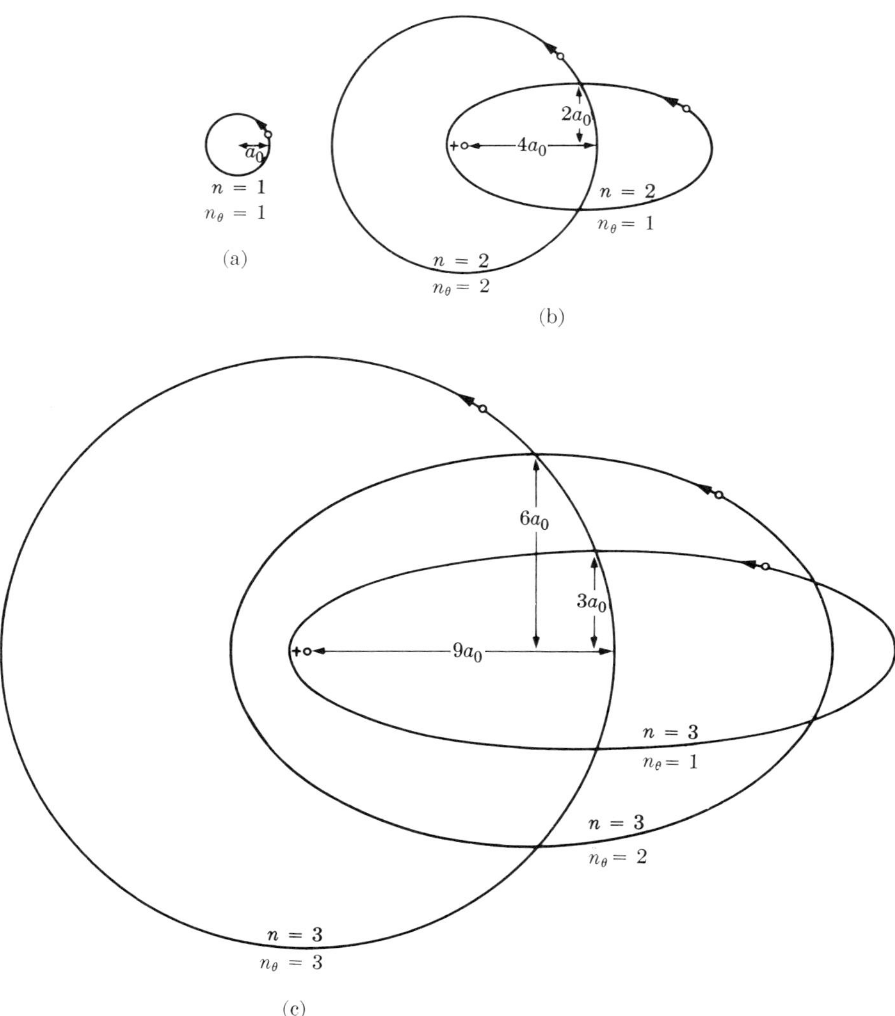

Fig. 5.14 The different possible paths of an electron in the hydrogen atom for $n = 1, 2,$ and 3, according to the Bohr–Sommerfeld model.

value of n for the observed transitions. In the Bohr–Sommerfeld theory, there are still no restrictions on n. Only those transitions are observed for which n_θ changes by unity. That is,

$$\boxed{\Delta n_\theta = n_{\theta_i} - n_{\theta_f} = \pm 1,} \tag{5.45}$$

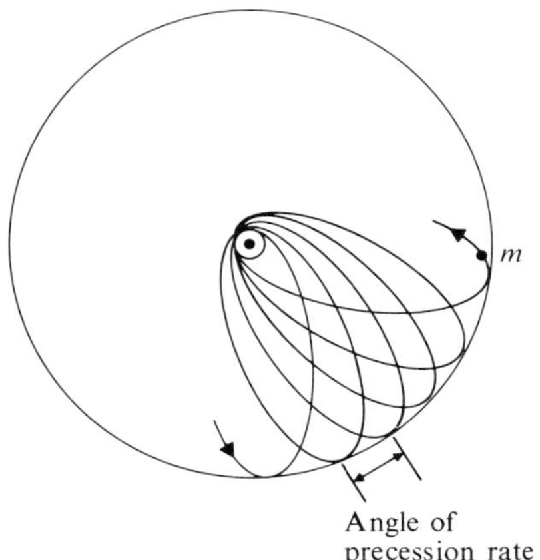

Fig. 5.15 The elliptical path of an electron around a nucleus becomes a rosette-shaped path when a relativistic mass correction (i.e., the mass of an electron changes with its velocity) is applied to the electron.

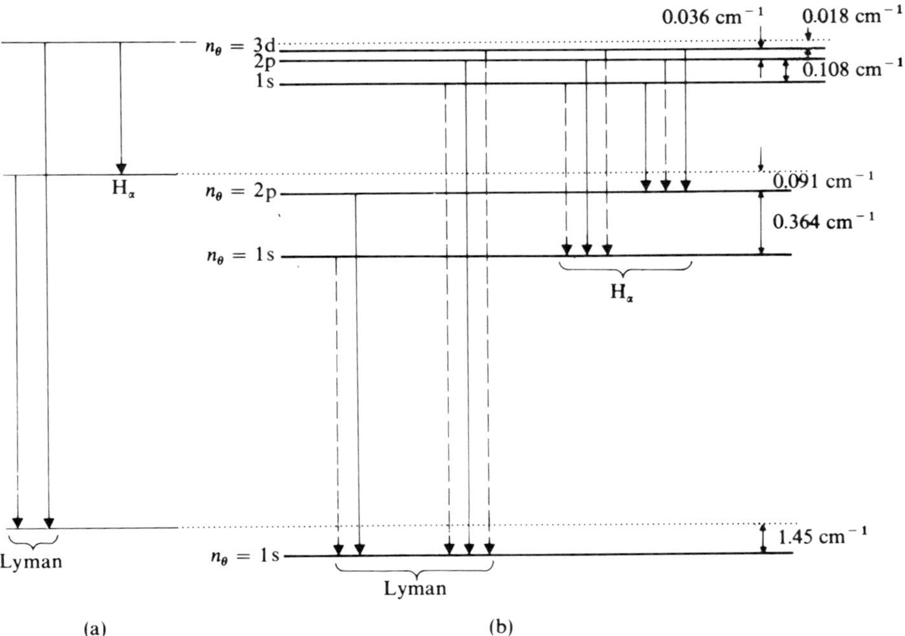

Fig. 5.16 A portion of the energy-level diagram of hydrogen (a) without relativistic correction, and (b) with relativistic correction, showing the fine structure. Note that the dashed transitions are transitions that are not allowed by the selection rules.

which is called the *selection rule* for n_θ for allowed transitions. By the Bohr–Sommerfeld selection rules, the dashed transitions shown in Fig. 5.16 are not allowed. This means, for example, that the H_α line, which was a singlet according to the Bohr theory, is a close doublet according to the Bohr–Sommerfeld model.

5.8 LIMITATIONS OF OLD QUANTUM-MECHANICAL MODELS

We have seen in preceding sections that the old quantum-mechanical models put forth by Bohr and Sommerfeld succeeded in predicting many elementary aspects of the atomic spectra of hydrogen and hydrogenlike atoms. For example, they explained different series of transitions observed in the hydrogen atom, the size of the hydrogen atom, the correct values of excitation and ionization potentials, and many other aspects. But these models weren't completely satisfactory. Many aspects of atomic spectra remained unexplained. Some of these are the following.

1. The relative intensities of different transitions
2. The spectra of atoms having more than one valence electron
3. The fine structure observed in many situations
4. The behavior (except for the simplest case) of atoms under the influence of a magnetic field, i.e., the effect of a magnetic field on atomic spectra
5. The logical reason for assuming different quantum numbers

These difficulties were overcome by the development of quantum mechanics. Next we are going to briefly outline this development, although not all aspects of quantum mechanics will be treated in this book.

SUMMARY

According to the Bohr theory of the hydrogen atom, the radius of the electron in the nth orbit is

$$r_n = n^2 \frac{\hbar^2}{kme^2} = n^2 r_1, \qquad n = 1, 2, 3, \ldots$$

and its energy levels are

$$E_n = -\frac{E_I}{n^2} = -\frac{13.58}{n^2}\ \text{eV},$$

while the spectrum of the hydrogen atom is given by

$$\bar{\nu} = \frac{1}{\lambda} = R\left(\frac{1}{n_f^2} - \frac{1}{n_i^2}\right)$$

For the Balmer series, $n_f = 2$, while $n_i = 3, 4, 5, 6, \ldots$

If a correction for nuclear mass is made,

$$R_M = \frac{\mu}{m} R_\infty,$$

where $\mu = mM/(m + M)$ is the reduced mass, and $R_\infty = (k^2e^4m/4\pi\hbar^3 c)$.

For hydrogenic atoms the energy levels are

$$E_n = -\frac{1}{n^2}\frac{\mu k^2(Z^2e^4)}{2\hbar^2}$$

and the transitions are

$$\bar{\nu} = \frac{1}{\lambda} = Z^2R_M\left(\frac{1}{n_f^2} - \frac{1}{n_i^2}\right).$$

Franck and Hertz, in 1914, carried out experiments which demonstrated the existence of discrete stationary states for electrons in atoms.

According to the correspondence principle, the quantum theory, when applied to large (macroscopic) systems, should always reduce to the classical theory.

★According to the Bohr–Sommerfeld model, the path of an electron in an atom is elliptical; the total quantum number n is given by

$$n = n_r + n_\theta, \quad \text{where } n = 1, 2, 3, \ldots, n_\theta = 1, 2, 3, \ldots, n_r = 0, 1, 2, \ldots.$$

Atomic levels that can be described by more than one set of quantum numbers, all of which yield the same energy, are called *degenerate*.

★For all possible transitions of electrons from one energy level to another, there is no restriction on change of n, but n_θ must change by unity. That is, $\Delta n_\theta = \pm 1$ in the Bohr–Sommerfeld model of hydrogenlike atoms.

PROBLEMS

1. Show that the force of gravitational attraction between an electron and a proton in a hydrogen atom is negligible compared to the force of electrostatic attraction between them. Assume the radius of the hydrogen atom to be 0.53 Å.

2. According to the classical model of the hydrogen atom, an electron moving in a circular orbit of radius 0.53 Å around a proton fixed at the center is unstable, and the electron should eventually collapse into the proton, as shown in Fig. 5.5. Estimate how long it would take for the electron to collapse into the proton. [*Hint:* Start with the classical expression for radiation from an accelerated charge,

$$\frac{dE}{dt} = -\frac{2}{3}\frac{ke^2a^2}{c^3}, \quad \text{where } a \text{ is the acceleration.]}$$

Also

$$E = K + V = -\frac{ke^2}{2r} \quad \text{and} \quad a = ke^2/mr^2,$$

which yields

$$dt = -\frac{3}{4}\frac{m^2c^3}{(ke^2)^2}\, r^2\, dr,$$

which, on integration from $r = a_0$ (the Bohr radius) to $r = 0$, should give the required time.]

3. Consider an electron in the hydrogen atom moving in the first Bohr orbit. Calculate the magnetic field produced by this electron at the center of its orbit.

4. The range of transitions in the visible spectrum lies between the wavelengths 3800 Å and 7000 Å. Calculate this range in terms of energy units.

5. Suppose that the radius of the orbit of an electron in a hydrogen atom is 0.00001 cm, 0.001 cm, 0.1 cm, or 10 cm. Calculate the values of the quantum number n corresponding to these orbits. Which of these orbits must be treated quantum mechanically?

6. Calculate the energies corresponding to the following transitions of electrons from one energy state to another in the hydrogen atom.

   a) $n_i = 2 \rightarrow n_f = 1$
   b) $n_i = 6 \rightarrow n_f = 5$
   c) $n_i = 11 \rightarrow n_f = 10$
   d) $n_i = 51 \rightarrow n_f = 50$
   e) $n_i = 101 \rightarrow n_f = 100$

   For what value of n would you say that the discrete nature of the system is lost?

7. Show that the recoil energy of an atom when a transition from one energy state to another takes place, though small, is not zero, and is given by

$$E_r = (h\nu)^2/2Mc^2,$$

   where M is the mass of the atom and $h\nu$ is the transition energy of the atom when it goes from one state to another. What is the energy of the photon emitted?

8. Calculate (a) the recoil energies of the hydrogen atoms, (b) the momenta of the photons emitted, (c) the energies of the photons emitted corresponding to the first two lines, H_α and H_β, of the Balmer series of the hydrogen atom.

9. Calculate the ionization energy of the hydrogen atom from the following data. The wavelength of the Balmer-series limit is 3645 Å and the wavelength of the first line in the Lyman series is 1215.7 Å. How do these values compare with the one calculated from the wavelength of the series limit, $\lambda = 911.3$ Å, of the Lyman series?

10. Suppose that an electron in the first excited state of the hydrogen atom stays in that state for 10^{-8} sec. What is the width (in eV) of this excited level? In this time interval, how many revolutions does this electron make before making a transition to the ground state?

11. Draw approximate energy-level diagrams for (a) the singly ionized helium, He^+, (b) the doubly ionized lithium, Li^{2+}.

12. Draw an approximate energy-level diagram for triply ionized beryllium, Be^{3+}.

13. Calculate the wavelengths of the first four lines of the Balmer series for the deuterium atom. Find the difference in wavelengths between these and the corresponding wavelengths of the transitions in the hydrogen atom.

14. Calculate the wavelength of the second spectral line of a singly ionized helium atom, He^+, corresponding to the second spectral line of the Balmer series of the hydrogen atom. What is the ionization potential of He^+?

15. Calculate the wavelengths of the following transitions with and without the mass correction.

   a) $n_i = 2$ to $n_f = 1$ in ^1H
   b) $n_i = 4$ to $n_f = 2$ in He^+
   c) $n_i = 6$ to $n_f = 3$ in Li^{2+}

16. The difference between the wavelengths of the first Balmer line of the hydrogen atom and the first Balmer line of the deuterium atom is 1.79 Å, while the wavelength of the first Balmer line in hydrogen is 6564.7 Å. From this information, calculate the ratio of the mass of the deuterium atom to that of the hydrogen atom.

17. Calculate the wavelength of the first line of the Lyman series for a tritium atom (the nucleus of tritium consists of two neutrons and one proton). Compare this with that of hydrogen and deuterium.

18. The μ-meson is a particle with the same charge as the electron. It has a mass 207 times the mass of the electron: $m_\mu = 207\ m_e$. When a μ-meson combines with a proton, it forms a μ-mesonic atom, just as an electron combines with a proton to form a hydrogen atom.

   a) Derive expressions for the velocities, radii, and energies of the different Bohr-type orbits in a μ-mesonic atom.
   b) Compare the values in (a) with those for the hydrogen atom.
   c) What is the ionization energy of the μ-mesonic atom?
   d) Compare the energies of the following transitions in the μ-mesonic atom with those in the hydrogen atom.

      i) $n_i = 2$ to $n_f = 1$ ii) $n_i = 3$ to $n_f = 2$
      iii) $n_i = 3$ to $n_f = 1$ iv) $n_i = 4$ to $n_f = 2$

19. If white light passes through a tube containing hydrogen, what wavelengths can be absorbed? What do you expect to see if you analyze the light which has passed through hydrogen?

20. The fine structure constant α is defined as $\alpha = v_1/c$, where v_1 is the velocity of the electron in the first Bohr orbit of the hydrogen atom.

   a) Show that $v_1 = c\alpha = ke^2/\hbar$.
   b) Show that $\alpha \simeq 1/137$.
   c) Show that, in hydrogenlike atoms, the speed of an electron in the lowest energy state is αcZ.
   d) In case (c), if Z is greater than 137, the velocity of the electron in the atom is greater than c. How do you explain this?

21. The binding energies of the ground state and the first excited state of potassium are 4.3 eV and 2.7 eV, respectively. Suppose that you are using potassium vapor in the Franck–Hertz experiment. At what voltages do you expect to see drops in the plot of current versus voltage?

22. Calculate the minimum potential differences through which electrons must be accelerated so that, when they cause inelastic collisions with hydrogen atoms in the ground state, the hydrogen atoms are left in the first and the second excited states.

23. The average kinetic energy of the atoms in a heated gas is $\frac{3}{2}kT$. These atoms can be used to excite and ionize other atoms by thermally induced collisions. At what temperature should mercury vapor be in order to stimulate emission of light of 2580 Å wavelength?

★24. Calculate the possible energy values for the first and second excited states ($n = 2$ and 3, respectively) of the hydrogen atom according to the relativistically corrected Sommerfeld formula. Calculate the energies and the wavelengths of the allowed transitions between these two levels.

REFERENCES

1. J. J. Balmer, *Wied. Ann.* **25,** 80 (1885)
2. J. R. Rydberg, *Astrophys. J.* **4,** 91 (1896)
3. Niels Bohr, *Phil. Mag.* **26,** 1 (1913); **27,** 506 (1914)
4. H. C. Urey, F. G. Brickwedde, and G. M. Murphy, *Phys. Rev.* **40,** 1 (1932)
5. J. Franck and G. Hertz, *Verhandl. dent. physik Ges.* **16,** 457, 512 (1912)
6. A. Sommerfeld, *Ann. Physik* **51,** 1 (1916); *Atomic Structure and Spectral Lines*, London: Methuen, 1929
7. W. Wilson, *Phil. Mag.* **29,** 795 (1915)

Suggested Readings

1. M. H. Shamos, editor, *Great Experiments in Physics*, New York: Holt, Rinehart and Winston, 1962
2. H. Boorse and L. Motz, editors, *The World of the Atom*, New York: Basic Books, 1966
3. Atam P. Arya, *Fundamentals of Atomic Physics*, Chapter 8; Boston, Mass.: Allyn and Bacon, 1971
4. R. M. Eisberg, *Fundamentals of Modern Physics*, Chapter 5; New York: John Wiley, 1961
5. A. Mellissinos, *Experiments in Modern Physics*, Chapters 2 and 6; New York: Academic Press, 1966
6. H. E. White, *Introduction to Atomic Spectra*, Chapters I, II, III, and IV; New York: McGraw-Hill, 1934
7. G. Herzberg, *Atomic Spectra and Atomic Structure*, Chapter I; New York: Dover Publications, 1944
8. F. K. Richtmyer, E. H. Kennard, and John H. Cooper, *Introduction to Modern Physics*, sixth edition, Chapters 8 and 9; New York: McGraw-Hill, 1969

CHAPTER 6
QUANTUM THEORY OF THE HYDROGEN ATOM

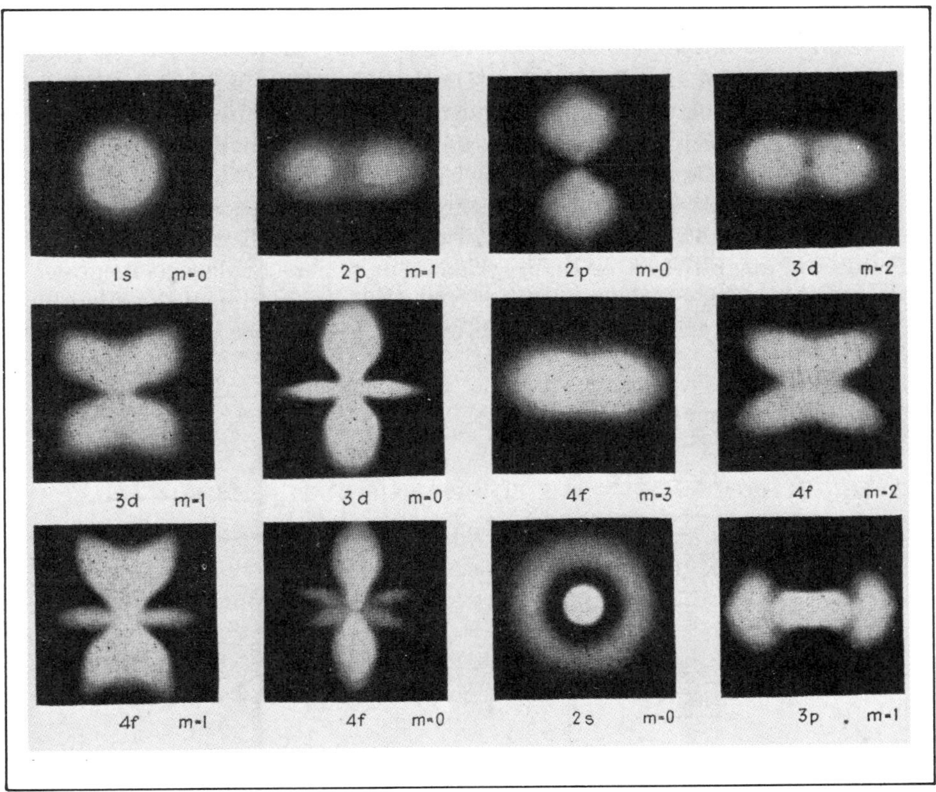

Photographic representation of electron cloud: the probability of electron distribution $|\Psi|^2$ *for several energy states of the hydrogen atom. Figure was made by using a spinning mechanical model; scale varies from figure to figure. (From H. E. White,* Introduction to Atomic Spectra, *McGraw-Hill, New York, 1934*)

6.1 THE NEED FOR A NEW THEORY

It is true that the Bohr and Sommerfeld theories of the hydrogen atom succeeded in explaining many experimentally observed aspects of atomic spectra. But these theories have their shortcomings: First, one has to accept many basic assumptions without proof. Second, the theories are based on the simplest of all the atomic systems, and cannot be extended to more complex atoms. Third, the theories cannot explain certain aspects of atomic spectra, such as the intensities of spectral lines and the selection rules for the transitions of atoms from one energy level to another. Between the Bohr theory's birth in 1913 and the advent of the quantum theory in 1925, many empirical rules were formulated to account for the regularities observed in the experimental data.

The invention of quantum mechanics by Schrödinger in 1925 changed the situation altogether. Starting with one basic assumption—the Schrödinger wave equation—he solved the problem of the hydrogen atom. The Schrödinger wave equation for the atomic system led not only to the explanation of discrete energy levels, but also to the postulation of different quantum numbers. Its results agreed with the experimental results for intensities of spectral lines, and was capable of being extended to many complex systems. It also could be used to clarify many until-then-unexplained features of atomic spectra.

Because the Schrödinger wave equation demands considerable mathematical sophistication and because of the limitations of the scope of this text, we cannot go into the details of the quantum theory of the hydrogen atom. Therefore, in the following sections, we shan't solve the Schrödinger wave equation, but merely outline the steps and summarize the results of it.

6.2 THE SCHRÖDINGER WAVE EQUATION FOR THE HYDROGEN ATOM

Let us assume that the nucleus of the hydrogen atom (a proton in this case) has a mass M, located at the origin. The electron has mass m and charge e^-, and is somewhere in space at (x, y, z). The nonrelativistic Schrödinger wave equation in this case is

$$\frac{\partial^2\psi}{\partial x^2} + \frac{\partial^2\psi}{\partial y^2} + \frac{\partial^2\psi}{\partial z^2} + \frac{2\mu}{\hbar^2}(E - V)\psi = 0. \tag{6.1}$$

Here $\psi = \psi(x, y, z)$ is the wave function of the electron, μ is the reduced mass of the electron and the nucleus, given by

$$\mu = \frac{Mm}{M + m}, \tag{6.2}$$

and $V = V(x, y, z)$ is the potential energy of the electron in the hydrogen atom,

$$V(x, y, z) = -\frac{ke^2}{\sqrt{x^2 + y^2 + z^2}} = -\frac{ke^2}{r}, \tag{6.3}$$

where r is the distance between the electron and the nucleus. [Note that, unlike the previous cases, in which we solved the S.W.E. for a single particle, here, in the case of the hydrogen atom, we have a two-body problem: the electron and the nucleus. However, we can reduce this to a one-body problem, as we have done above, provided $V(r)$ depends on r only.]

Our aim is to solve Eq. (6.1) for the wave function $\psi(x, y, z)$, which has a definite value at each point of the space surrounding the nucleus. The value of $|\psi(x, y, z)|^2$ gives the probability of finding the electron at any of the points (x, y, z). Where the probability is large, the electron is likely to be found; where the probability is small, it is less likely to be found. It is much more convenient to solve and discuss Eq. (6.1) in terms of the spherical polar coordinates r, θ, and ϕ instead of the rectangular coordinates x, y, and z.

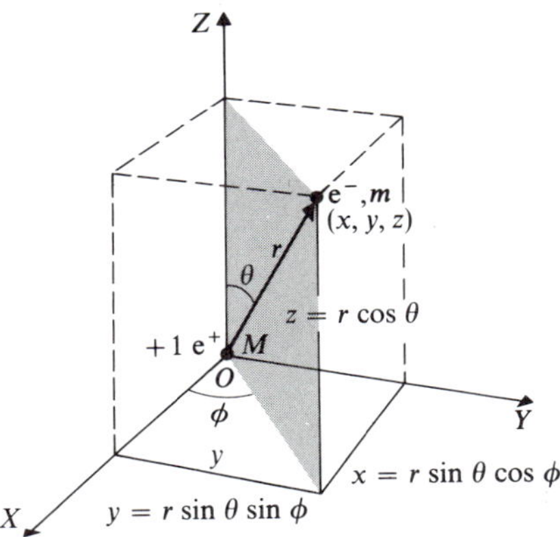

Fig. 6.1 Geometrical relation between rectangular coordinates (x, y, z) and spherical polar coordinates (r, θ, ϕ) of an electron. The nucleus is assumed to be at the origin.

Figure 6.1 shows the location of the electron by the two sets of coordinates (x, y, z) and (r, θ, ϕ), which are related as follows.

$$r = \text{distance of electron from nucleus (origin)} = \sqrt{x^2 + y^2 + z^2}$$

$$\theta = \text{angle of } r \text{ with } Z\text{-axis} = \arccos \frac{z}{\sqrt{x^2 + y^2 + z^2}}$$

$$\phi = \text{angle of } r \text{ around } Z\text{-axis in } xy \text{ plane measured from } +X\text{-axis} = \arctan (y/x) \tag{6.4}$$

or

$$x = r \sin \theta \cos \phi, \qquad y = r \sin \theta \sin \phi, \qquad z = r \cos \theta. \tag{6.5}$$

Substituting the relations given by (6.5) into Eq. (6.1), we get the Schrödinger wave equation for spherical coordinates. Since this transformation involves a great deal of algebra, we shall omit the details and simply state the final results.

Thus, replacing $\psi(x, y, z)$ by $\psi(r, \theta, \phi)$, we find the S.W.E. of the hydrogen atom:

$$\frac{1}{r^2}\frac{\partial}{\partial r}\left(r^2\frac{\partial\psi}{\partial r}\right) + \frac{1}{r^2\sin\theta}\frac{\partial}{\partial\theta}\left(\sin\theta\frac{\partial\psi}{\partial\theta}\right) + \frac{1}{r^2\sin^2\theta}\frac{\partial^2\psi}{\partial\phi^2} + \frac{2\mu}{\hbar^2}[E - V(r)]\psi = 0. \tag{6.6}$$

To further simplify the problem, we use the technique of separation of variables and separate the partial differential equation, Eq. (6.6), into three ordinary differential equations. Also we remember that the potential energy $V(r)$ is a function of the radial distance r only, and not of θ and ϕ. Let $\psi(r, \theta, \phi)$ be the product of three wave functions,

$$\psi(r, \theta, \phi) = R(r)\Theta(\theta)\Phi(\phi), \tag{6.7}$$

where R is a function of r only, Θ is a function of θ only, and Φ is a function of ϕ only. Also from this equation,

$$\frac{\partial\psi}{\partial r} = \Theta\Phi\frac{\partial R}{\partial r}, \qquad \frac{\partial\psi}{\partial\theta} = R\Phi\frac{\partial\Theta}{\partial\theta}, \qquad \frac{\partial^2\psi}{\partial\phi^2} = R\Theta\frac{\partial^2\Phi}{\partial\phi^2}.$$

Substituting this in Eq. (6.6) and dividing by $R\Theta\Phi$, we get

$$\left[\frac{1}{r^2}\frac{1}{R}\frac{\partial}{\partial r}\left(r^2\frac{\partial R}{\partial r}\right)\right] + \left[\frac{1}{r^2\sin\theta}\frac{1}{\Theta}\frac{\partial}{\partial\theta}\left(\sin\theta\frac{\partial\Theta}{\partial\theta}\right)\right] + \left[\frac{1}{r^2\sin^2\theta}\frac{1}{\Phi}\frac{\partial^2\Phi}{\partial\phi^2}\right] + \frac{2\mu}{\hbar^2}[E - V(r)] = 0. \tag{6.8}$$

Multiplying both sides by $r^2\sin^2\theta$ and rearranging so that the terms containing r and θ are on the left side, while those containing ϕ are on the right side, we obtain

$$-\frac{\sin^2\theta}{R}\frac{\partial}{\partial r}\left(r^2\frac{\partial R}{\partial r}\right) - \frac{2\mu}{\hbar^2}r^2\sin^2\theta[E - V(r)] - \frac{\sin\theta}{\Theta}\frac{\partial}{\partial\theta}\left(\sin\theta\frac{\partial\Theta}{\partial\theta}\right) = \frac{1}{\Phi}\frac{\partial^2\Phi}{\partial\phi^2}. \tag{6.9}$$

The left side of this equation is a function of r and θ only, while the right side is a function of ϕ only, which is possible only if each side is equal to a constant, say $-m_l^2$. Thus we obtain the following two equations:

$$\frac{1}{\Phi}\frac{d^2\Phi}{d\phi^2} = -m_l^2 \tag{6.10}$$

and

$$-\frac{\sin^2\theta}{R}\frac{d}{dr}\left(r^2\frac{dR}{dr}\right)-\frac{2\mu}{\hbar^2}r^2\sin^2\theta[E-V(r)]$$
$$-\frac{\sin\theta}{\Theta}\frac{d}{d\theta}\left(\sin\theta\frac{d\Theta}{d\theta}\right)=-m_l^2. \tag{6.11}$$

Dividing both sides of this equation by $\sin^2\theta$ and rearranging, we obtain the following:

$$\frac{1}{R}\frac{d}{dr}\left(r^2\frac{dR}{dr}\right)+\frac{2\mu r^2}{\hbar^2}[E-V(r)]=\frac{m_l^2}{\sin^2\theta}-\frac{1}{\Theta}\frac{1}{\sin\theta}\frac{d}{d\theta}\left(\sin\theta\frac{d\Theta}{d\theta}\right). \tag{6.12}$$

The left side of this equation is a function of r only, while the right side is a function of θ only. This is possible only if each side is equal to a constant, say $l(l+1)$. Therefore

$$\frac{m_l^2}{\sin^2\theta}-\frac{1}{\Theta}\frac{1}{\sin\theta}\frac{d}{d\theta}\left(\sin\theta\frac{d\Theta}{d\theta}\right)=l(l+1) \tag{6.13}$$

and

$$\frac{1}{R}\frac{d}{dr}\left(r^2\frac{dR}{dr}\right)+\frac{2\mu r^2}{\hbar^2}[E-V(r)]=l(l+1). \tag{6.14}$$

Equations (6.10), (6.13), and (6.14), after rearranging, may be written as

$$\frac{d^2\Phi}{d\phi^2}+m_l^2\Phi=0, \tag{6.15}$$

$$\frac{1}{\sin\theta}\frac{d}{d\theta}\left(\sin\theta\frac{d\Theta}{d\theta}\right)+\left[l(l+1)-\frac{m_l^2}{\sin^2\theta}\right]\Theta=0, \tag{6.16}$$

$$\frac{1}{r^2}\frac{d}{dr}\left(r^2\frac{dR}{dr}\right)+\frac{2\mu}{\hbar^2}\left[E-V(r)-\frac{\hbar^2}{2\mu}\frac{l(l+1)}{r^2}\right]R=0. \tag{6.17}$$

Thus solving the S.W.E. of the hydrogen atom means solving the above three equations. The first two are the angular equations and are independent of the potential $V(r)$; the third is the radial equation. Such a separation is always possible if the potential energy is a function of the radial distance only.

a) Solution of the radial equation

The radial equation for the hydrogen atom is Eq. (6.17). Solving it is mathematically quite involved. However, let's just take a look at some general characteristics of the solution corresponding to the lowest energy state, called the *ground state.*

For this case, as we shall soon see, $l = 0$, and hence Eq. (6.17) takes the form

$$\frac{d^2R}{dr^2} + \frac{2}{r}\frac{dR}{dr} + \frac{2\mu}{\hbar^2}\left(E + \frac{ke^2}{r}\right)R = 0. \tag{6.18}$$

Figure 6.2 shows the form of the potential $V(r) = -ke^2/r$. If the total energy is greater than zero, say E_2, the electron is not bound to the atom. But if E is less than zero, say E_1, the potential energy is greater in magnitude than the total energy. As always in such situations, the S.W.E. gives discrete values of the bound energies of the electron.

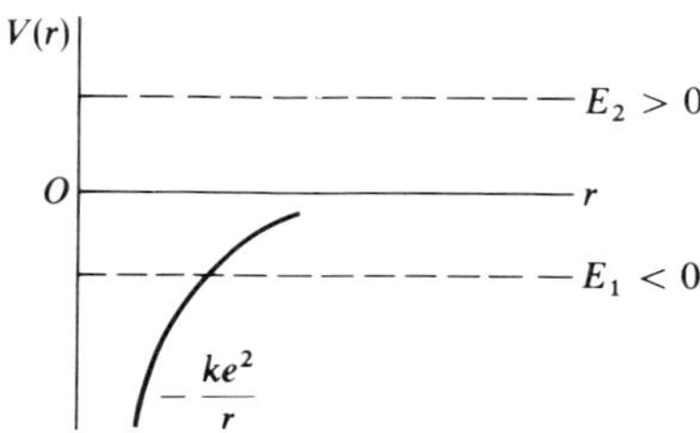

Fig. 6.2 Plot of potential energy $V(r)$ of an electron in a hydrogen atom versus r. If the total energy is greater than zero, say E_2, the electron is not bound. If the total energy is less than zero, say E_1, the electron is bound and can take discrete energy values.

Whenever the electron is close to the nucleus, the magnitude of the wave function $R(r)$ describing the electron will be large, because of high probability of finding the electron near the nucleus. The probability of finding an electron an infinite distance away from the nucleus is zero, and hence the wave function at infinity is zero. The wave function that satisfies these requirements may be written as

$$R(r) = C_n e^{-r/a_0}, \tag{6.19}$$

where C_n is a constant, and since the exponent must be dimensionless, a_0 is another constant, and has the dimensions of length.

Figure 6.3 shows the plot of this wave function, $R(r)$ versus r, for $C_n = 1$. The value of a_0 determines how fast the wave function goes to zero. From Eq. (6.19), we have

$$\frac{dR}{dr} = -\frac{C_n}{a_0}e^{-r/a_0} = -\frac{1}{a_0}R$$

and

$$\frac{d^2R}{dr^2} = \frac{C_n}{a_0^2}e^{-r/a_0} = \frac{1}{a_0^2}R.$$

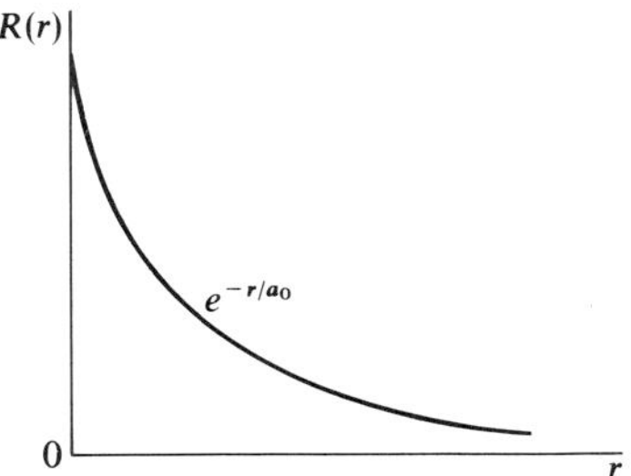

Fig. 6.3 Wave function $R(r) = C_n e^{-r/a_0}$ for the ground state of the hydrogen atom. Note that in this case $\psi(r) = R(r)$.

Substituting for dR/dr and d^2R/dr^2 in Eq. (6.18), we obtain

$$\left[\frac{1}{a_0^2} - \frac{2}{r}\frac{1}{a_0} + \frac{2\mu}{\hbar^2}\left(E + \frac{ke^2}{r}\right)\right] R(r) = 0. \tag{6.20}$$

Since $R(r)$ cannot be zero, the terms in the brackets are zero, and after rearranging we may write them as

$$\left(\frac{1}{a_0^2} + \frac{2\mu}{\hbar^2} E\right) + \left(\frac{2\mu ke^2}{\hbar^2} - \frac{2}{a_0}\right)\frac{1}{r} = 0. \tag{6.21}$$

For this equation to hold for all values of r between 0 and ∞, both the enclosed terms must be zero. Equating the second enclosed term to zero, we get

$$\boxed{a_0 = \frac{\hbar^2}{\mu ke^2} = 0.53\ \text{Å} = r_1.} \tag{6.22}$$

That is, the constant a_0 is nothing but the radius of the first Bohr orbit, r_1. Equating the first enclosed term to zero, we obtain

$$E = -\frac{\hbar^2}{2\mu a_0^2}.$$

Substituting for a_0, we have

$$\boxed{E = -\frac{\mu k^2 e^4}{2\hbar^2} = -13.6\ \text{eV} = E_I,} \tag{6.23}$$

which is the ground-state energy of the hydrogen atom, exactly equal to the value obtained by the Bohr theory. Thus the wave function $R(r)$ given by Eq. (6.19) does represent the ground state of the hydrogen atom.

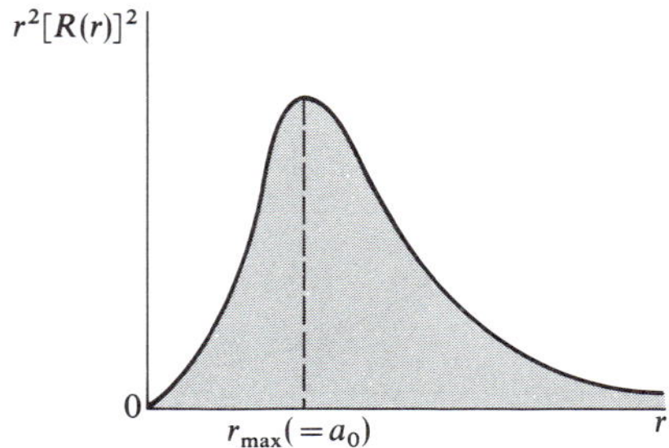

Fig. 6.4 Plot of the probability of finding an electron in the hydrogen ground state ($\propto r^2R^2$) between r and $r + dr$ versus r, the distance from the nucleus; that is, the radial probability distribution for the ground state of hydrogen. Note that r_{max} corresponds to the first Bohr radius.

As we see from Fig. 6.3, the wave function at $r = 0$ is larger than it is for any other value of r. But this does not mean that the probability of finding the electron is a maximum at $r = 0$. The probability $P(r)$ of finding the electron in a small volume element dv of the shell of radius between r and $r + dr$ is given by

$$|R(r)|^2\,dv = |R(r)|^2 4\pi r^2\,dr \propto e^{-2r/a_0}r^2\,dr. \tag{6.24}$$

Figure 6.4 shows the plot of $r^2[R(r)]^2$ versus r. Even though $R(r)$ is largest at $r = 0$, the shell is very small, the volume element dv is almost zero, and hence $R^2(r)r^2$ is zero. Similarly, as $r \to \infty$, $dv \to \infty$, but $R(\infty) \to 0$, and hence $R^2(r)r^2 \to 0$. Thus, somewhere between $r = 0$ and $r = \infty$, the probability of finding the electron must be maximum. We can find the position of r_{max}, corresponding to the maximum value of $[R(r)]^2r^2$, by differentiating this with respect to r and equating it to zero. That is, after substituting from Eq. (6.19),

$$\frac{d}{dr}\,(C_n^2 e^{-2r/a_0}r^2) = 0,$$

$$2re^{-2r/a_0} - \frac{2}{a_0}\,r^2e^{-2r/a_0} = 0.$$

Or, substituting $r = r_{max}$, we obtain

$$r_{max} = a_0 = r_1. \tag{6.25}$$

Thus quantum mechanics differs from the Bohr theory, in that the electron in the hydrogen atom can be found anywhere between $r = 0$ and $r = \infty$, but it is most likely to be found at $r = a_0 = r_1$, as illustrated in Fig. 6.5 (and where its binding energy corresponds to the ground-state energy of -13.6 eV).

There, you see! We have succeeded in discussing the ground state of the hydrogen atom without getting involved in any detailed mathematics. (Well, not *very* detailed.)

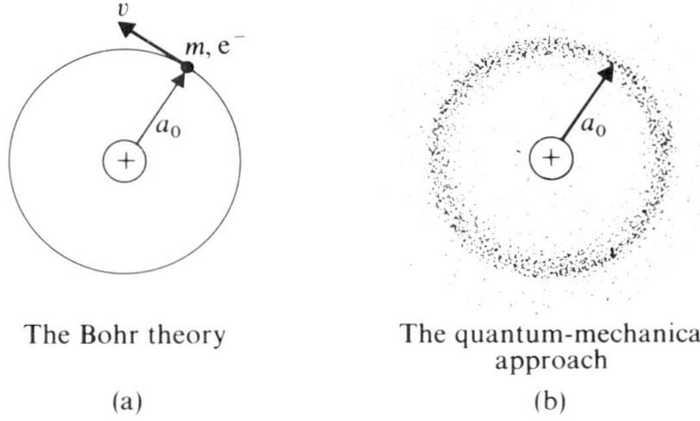

Fig. 6.5 Comparison of results of (a) Bohr theory, and (b) quantum theory of the hydrogen atom. Density of dots indicates relative probability of locating an electron. It is most likely to be found at $r = a_0$.

The general solution of Eq. (6.17) gives the following expression for the energy levels of the hydrogen atom. That is, the energy eigenvalues are

$$E_n = -\frac{1}{n^2}\left(\frac{\mu k^2 e^4}{2\hbar^2}\right) = -\frac{E_I}{n^2}, \tag{6.26}$$

where n is the *principal quantum number*. It can assume the values

$$n = 1, 2, 3, 4, 5, \ldots, \tag{6.27}$$

which is in agreement with the Bohr theory of the hydrogen atom.

Similarly, for hydrogenlike atoms,

$$E_n = -\frac{Z^2 E_I}{n^2} \tag{6.28}$$

where $n = 1, 2, 3, 4, 5, \ldots$.

b) Solutions of the angular equations

We can solve Eq. (6.16) only if l is an integer or zero, and the maximum value of l is less than n. For a given n, l can assume the values

$$l = 0, 1, 2, 3, 4, \ldots, (n - 1). \tag{6.29}$$

That is, for a given energy level, we get n solutions of Θ corresponding to one radial solution. Thus, for each value of n, the Bohr theory gives us only one orbit, while the quantum theory gives us n states. The value l is called the *orbital quantum number* (or angular momentum quantum number). It determines how fast ψ changes with θ for fixed values of r and ϕ. For large values of l, ψ charges rapidly with θ, and for small values of l, ψ charges slowly with θ. $l = 0$ means that ψ does not change with θ, and hence at a given r, $\Theta(\theta)$ has the same value in all directions.

The solution of the differential equation (6.15) is simple and straightforward. That is,

$$\frac{d^2\Phi}{d\phi^2} + m_l^2\Phi = 0$$

has a solution

$$\Phi(\phi) = Ae^{im_l\phi}, \tag{6.30}$$

where A is a constant and (keeping in mind that in order for Eq. (6.16) to be satisfied as well, $|m_l|$ must be less than or equal to l) m_l can assume the values

$$\boxed{m_l = -l, -(l-1), -(l-2), \ldots, -1, 0, 1, \ldots, (l-1), l.} \tag{6.31}$$

That is, for a given value of l, m_l can take $(2l + 1)$ values given by Eq. (6.31). m_l is called the *magnetic quantum number*. For fixed values of r and θ, the wave function $\Phi(\phi)$ determines the variation of ψ with ϕ. For large positive or negative values of m_l, ψ changes rapidly with ϕ, and small values of m_l yield slow variations of ψ with ϕ; while $m_l = 0$ means no variation of ψ with ϕ.

Thus the complete or total wave function of an electron in any state is given by

$$\psi_{n,l,m_l}(r, \theta, \phi) = R_{nl}(r)\Theta_{l,m_l}(\theta)\Phi_{m_l}(\phi).$$

c) Summary of possible solutions

Solving the problem of the hydrogen atom by means of quantum mechanics has meant that we've had to introduce three quantum numbers, n, l, and m_l instead of just n, as in the Bohr theory. For a given n, there are n values of l, and for a given l, there are $(2l + 1)$ values of m_l, as given by Eqs. (6.29) and (6.31), respectively. Thus, for a given n, there are

$$\sum_{l=0}^{n-1} (2l + 1) = n^2$$

possible solutions; see Table 6.1, page 196. Since the energy of the hydrogen atom is determined by n only, this equation means that a given energy state has n^2

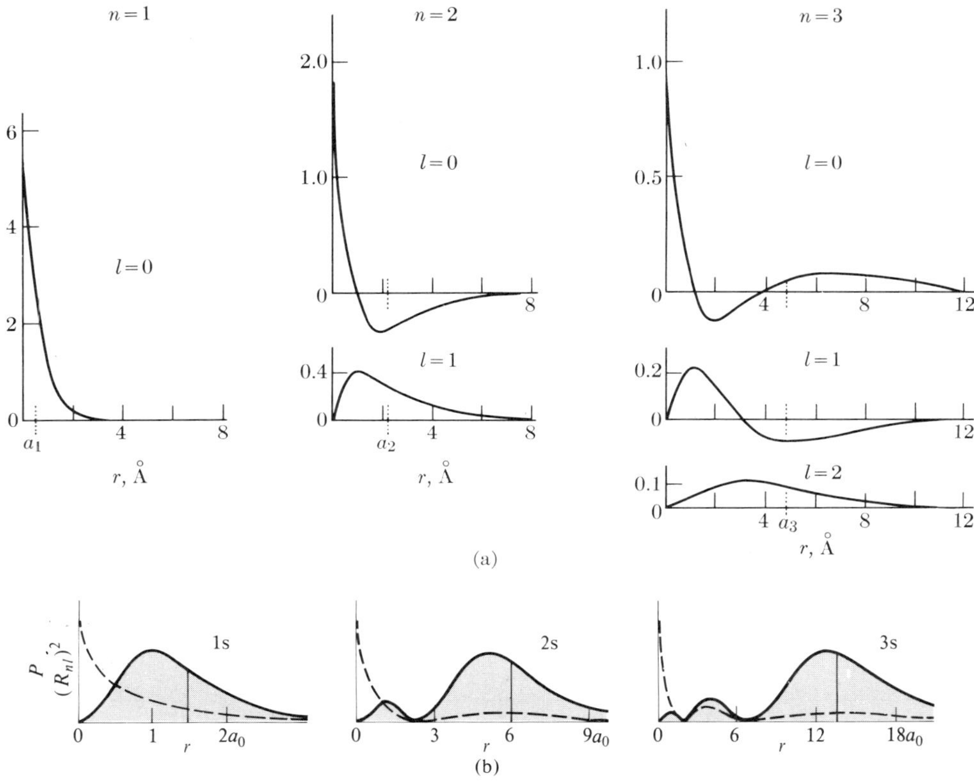

Fig. 6.6 (a) Plots of radial wave function $R_{nl}(r)$ versus r for $n = 1$, 2, and 3. (b) Probability of finding electron (shaded area) $P = \{4\pi r^2 [R_{nl}(r)]^2\}$ in a hydrogen atom at a distance between r and $r + dr$ from the nucleus for $n = 1$, 2, and 3. Dashed lines are the plots of $[R_{nl}(r)]^2$ versus r. [From H. E. White, *Introduction to Atomic Spectra*, New York: McGraw-Hill, 1934]

eigenfunctions and n^2 corresponding eigenvalue solutions, but all these have the same energy. The system is said to be *degenerate* if there are more than one eigenfunction (or eigenstate), all of which have the same energy. Thus, for example, $n = 2$ is 4-fold degenerate, $n = 3$ is 9-fold degenerate, etc.

Because of the probability interpretation of the wave function in quantum mechanics, electrons do not move in certain definite orbits. To give some idea of the probability of finding an electron orbiting around the nucleus, Fig. 6.6 shows the plot of R_{nl}, $[R_{nl}(r)]^2$, and $P = (4\pi r^2[R_{nl}(r)]^2)$ versus r for $n = 1$, 2, and 3. Figure 6.7 shows the plot of $r^2\psi^2_{nlm_l}$ for $n = 2$. The density of the dots in Figure 6.7 indicates the relative probability of finding electrons around the nucleus.

Table 6.1

Number of Possible States of a Hydrogenlike Atom

n	l	m_l	Number of solutions for a given $n = n^2$. Each solution denoted by (n, l, m_l)	
1	0	0	(1, 0, 0)	1
2	0	0	(2, 0, 0)	4
	1	−1, 0, 1	(2, 1, −1), (2, 1, 0), (2, 1, 1)	
3	0	0	(3, 0, 0)	9
	1	−1, 0, 1	(3, 1, −1), (3, 1, 0), (3, 1, 1)	
	2	−2, −1, 0, 1, 2	(3, 2, −2), (3, 2, −1), (3, 2, 0), (3, 2, 1) (3, 2, 2)	
4	0	0	(4, 0, 0)	16
	1	−1, 0, 1	(4, 1, −1), (4, 1, 0), (4, 1, 1)	
	2	−2, −1, 0, 1, 2	(4, 2, −2), (4, 2, −1), (4, 2, 0), (4, 2, 1) (4, 2, 2)	
	3	−3, −2, −1, 0, 1, 2, 3	(4, 3, −3), (4, 3, −2), (4, 3, −1), (4, 3, 0) (4, 3, 1), (4, 3, 2), (4, 3, 3)	

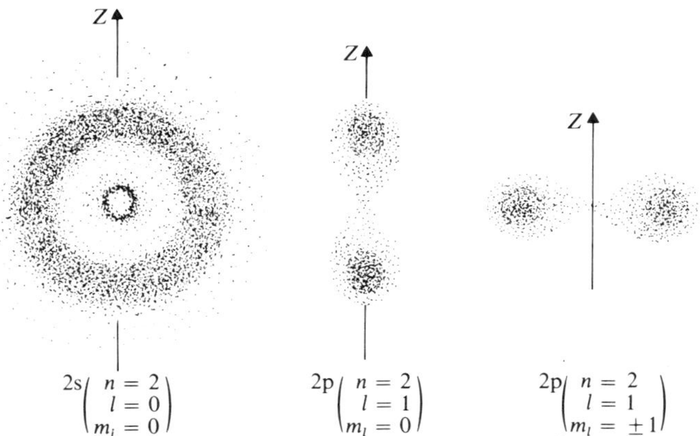

Fig. 6.7 Density of points indicates the probability $r^2\psi^2$ of the hydrogen atom being in quantum state $n = 2$, according to quantum-mechanical model.

6.3 PHYSICAL INTERPRETATION OF QUANTUM NUMBERS

Handling the hydrogen atom according to the rules of quantum mechanics has revealed that the motion of the electron is governed by three quantum numbers.

a) *The principal quantum number n* given by

$$n = 1, 2, 3, 4, \ldots. \tag{6.32}$$

b) *The angular momentum (or orbital momentum) quantum number l.* For a given n, l can take the following n different values:

$$l = 0, 1, 2, 3, \ldots, (n - 1) \tag{6.33}$$

c) *The magnetic quantum number m_l.* For a given l, m_l can take the following $(2l + 1)$ values:

$$m_l = -l, -(l - 1), \ldots, -1, 0, 1, \ldots, (l - 1), l \tag{6.34}$$

The quantum number n determines the energy of the electron in the atom. The orbital quantum number l determines the magnitude of the angular moment, and l and the magnetic quantum number m_l together determine the magnitude and orientation in space of the angular momentum vector **L** of the electron in the atom. Let's look at how all this is put together.

Each possible set of values (n, l, m_l) corresponds to a different quantum state, and hence a different wave function and a different probability distribution for the electron around the nucleus. Table 6.1 clearly shows that, for a given n, there are n^2 possible states (n, l, m_l). The energy of a quantum state is determined by n, and is the same according to the quantum theory or Bohr theory. That is,

$$E_n = -\frac{1}{n^2}\left(\frac{\mu k^2 e^4}{2\hbar^2}\right) = -\frac{E_I}{n^2}. \tag{6.35}$$

Therefore all the n^2 states for a given n have the same energy. For example, $n = 2$ has four states with different eigenfunctions, but each of the four states has the same energy $-E_I/2^2$. Thus, for a given n, different values of l and m_l denote the same energy state; and such states, as we said earlier, are called *degenerate*. We say that $n = 2$ is $2^2 = 4$-fold degenerate, and $n = 3$ is $3^2 = 9$-fold degenerate. And so forth. However, degeneracy *can* be avoided under certain conditions. We can do this, for example, by applying a magnetic field, in which case each (n, l, m_l) state will have different energy, thereby justifying the existence of the three quantum numbers, and hence the quantum theory.

The orbital quantum number l determines the variation of the wave function of the electron as the angle θ is changed. Also, from quantum mechanics, the magnitude of the angular momentum L of the electron in the atom is given by

$$L = \sqrt{l(l+1)}\,\hbar. \tag{6.36}$$

This is different from the value given by the Bohr theory, $L = n\hbar$. According to quantum mechanics, the magnitude of L is determined by l and not by n. In the Bohr theory, there is only one value of L for each given value of n. But according to quantum theory, there are n values of L for each given value of n. These values are given by combining Eqs. (6.33) and (6.36). Of course, n determines the energy and puts an upper limit to the angular momentum. This means that the maximum value of l for a given n is $n - 1$. Hence $L_{\text{max}} = \sqrt{(n-1)n}\,\hbar$, and this is less than the Bohr value of $L = n\hbar$. The maximum value of L corresponding to $l = n - 1$ corresponds to the Bohr circular orbit. A common notation for denoting different l states is the following.

$$\begin{aligned} l &= 0, 1, 2, 3, 4, \ldots \\ \text{Electron state notation} &= \text{s, p, d, f, g}, \ldots \\ \text{Atomic state notation} &= \text{S, P, D, F, G}, \ldots \end{aligned} \tag{6.37}$$

That is, "s state" means that $l = 0$, "p state" means that $l = 1$, "d state" means that $l = 2$, etc. Similarly, 1s stands for a state with $n = 1$, $l = 0$, 2s stands for a state with $n = 2$, $l = 0$, 2p stands for a state with $n = 2$ and $l = 1$, 3d stands for a state with $n = 3$ and $l = 2$, and so on.

Let us consider the special case of a 1s state, that is, $n = 1$ and $l = 0$. According to quantum mechanics, $L = \sqrt{0(0+1)}\,\hbar = 0$. This disagrees with the Bohr theory, in which you can't have an orbit with zero angular momentum, because such an orbit would mean that the electron would be traveling on a straight line passing through the nucleus! No such difficulty arises in quantum mechanics because the states having zero angular momentum have a probability distribution of zero at the nucleus.

The z-component of the angular momentum $\mathbf{L}$ of the electron in the atom is given by

$$L_z = m_l\hbar, \tag{6.38}$$

where m_l determines the variation of ψ with ϕ and also determines the orientation of the angular momentum vector $\mathbf{L}$ in space, i.e., the *space quantization of the angular momentum.* Thus the angle θ which the vector **L** rotating around the Z-axis makes with the Z-axis is given by (see Fig. 6.8a):

$$\cos\theta = \frac{L_z}{L} = \frac{m_l\hbar}{\sqrt{l(l+1)}\,\hbar} = \frac{m_l}{\sqrt{l(l+1)}}. \tag{6.39}$$

Since m_l can take the $(2l + 1)$ values given by Eq. (6.34), the angular momentum vector **L** can have only $(2l + 1)$ orientations in space, the angle of orientation being given by Eq. (6.39), so that the projection of **L** on the Z-axis gives values of L_z according to Eq. (6.39). Figure 6.8(b) illustrates this for $l = 2$.

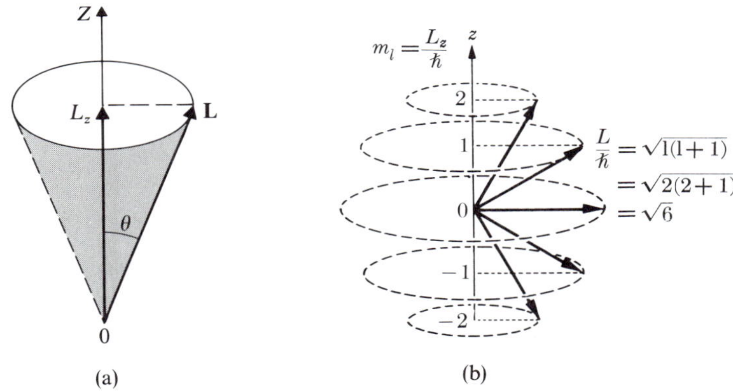

Fig. 6.8 (a) Rotation of angular momentum vector **L** about Z-axis and its relation to L_z. (b) According to classical theory, θ can have any value, but according to quantum mechanics, only those values of θ (and hence only certain orientations of L) are possible which satisfy the condition $\cos\theta = L_z/L$, where $L = \sqrt{l(l+1)}\,\hbar$ and $L_z = m_l\hbar$, as illustrated for $l = 2$.

As shown in Fig. 6.9, for $l = 1$, $m_l = 1, 0$, and -1, corresponding to $L = \sqrt{2}\,\hbar$; $L_z = \hbar, 0, -\hbar$. The three orientations in space are given by $\cos\theta = 1/\sqrt{2}$, 0, $-1/\sqrt{2}$. Similarly, for $l = 2$, $m_l = 2, 1, 0, -1, -2$ gives $L = \sqrt{6}\,\hbar$; $L_z = 2\hbar$, $\hbar, 0, -\hbar, -2\hbar$. The corresponding five orientations in space are

$$\cos\theta = 2/\sqrt{6},\ 1/\sqrt{6},\ 0,\ -1/\sqrt{6},\ -2/\sqrt{6}.$$

But there is only one orientation which corresponds to $l = 0$ and for which $m_l = 0$.

Note that θ has a maximum value when $m_l = l$. From Eq. (6.39), $\cos\theta = l/\sqrt{l(l+1)}$, which is always less than l, and implies (as shown in Fig. 6.8), that **L**

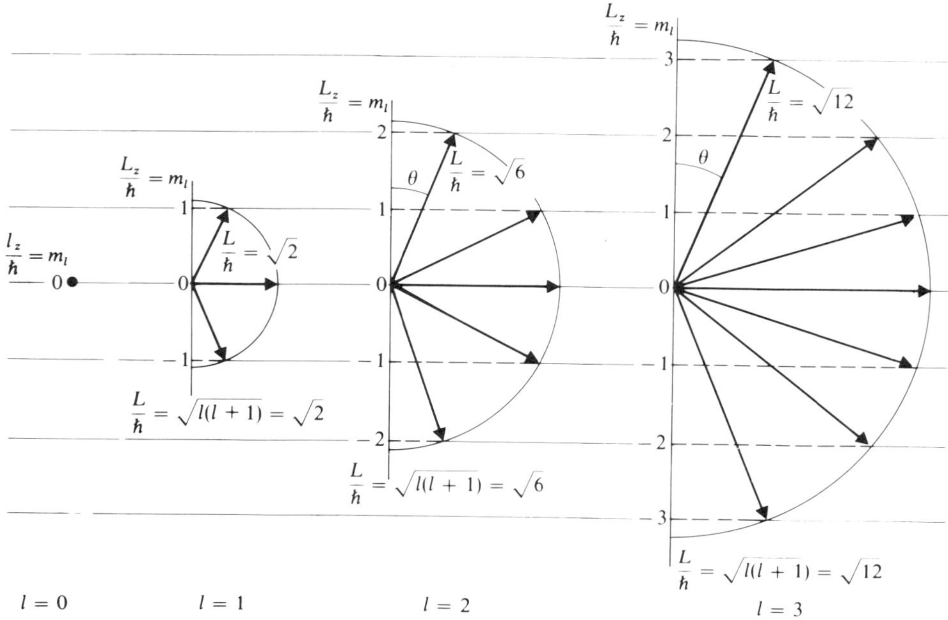

Fig. 6.9 The possible orientations of the angular momentum vectors for different values of l, according to quantum mechanics.

can never align itself completely in the direction of the Z-axis. Classically there is no such restriction on alignment. In classical physics, complete alignment *is* possible, because $|\mathbf{L}|$ is equal to $l\hbar$ ($\simeq n\hbar$) and not $\sqrt{l(l+1)}\,\hbar$. Of course, experimental measurements agree with the predictions of the quantum theory.

It is important to note that the space quantization of $\mathbf{L}$, as discussed above, is not a unique property of the coulomb potential. Space quantization is a characteristic of any potential which is a function of the radial distance r only. In Chapter 7 we are going to prove space quantization experimentally by placing the atom in a magnetic field. The direction of the magnetic field provides the Z-axis needed for space orientation.

Before leaving this section, let's ask ourselves this question: How does the introduction of these quantum numbers modify the energy-level diagram and the resulting transitions in the hydrogen atom? Figure 6.10 shows the energy-level diagram of the hydrogen atom according to quantum theory. Even though, for a given n, the different l states have the same energy, we have shown them separated. Because when we consider atoms other than hydrogen, the l degeneracy is removed, and different l states for a given n have slightly different energies. Recall that the l degeneracy was removed by the introduction of elliptical orbits together

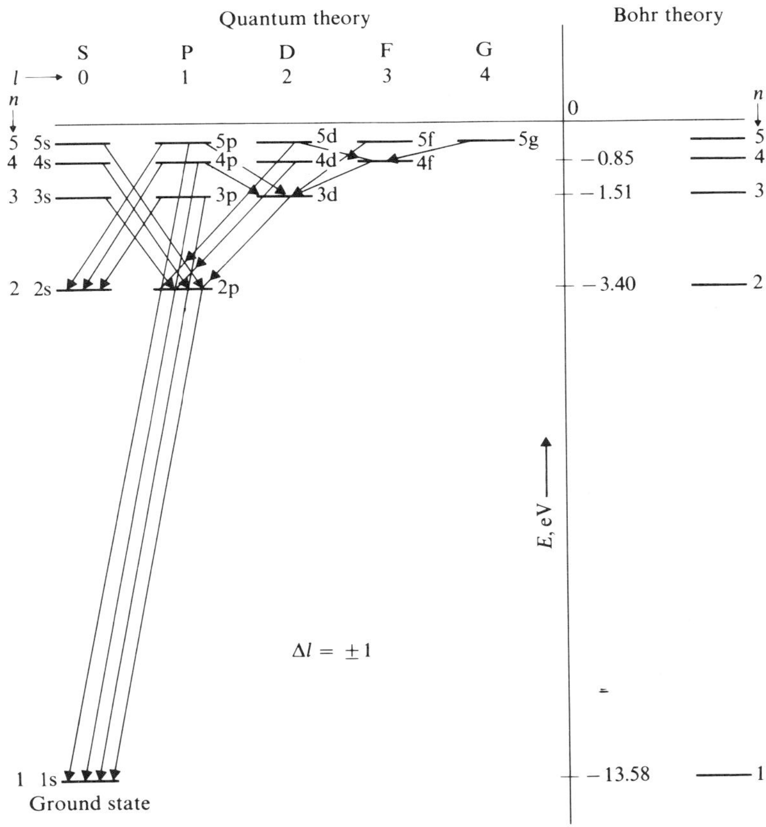

Fig. 6.10 Energy-level diagram for hydrogen atom according to quantum mechanics. Transitions shown follow the selection rule $\Delta l = \pm 1$. For comparison, energy levels predicted by the Bohr theory are also shown.

with relativistic correction by Sommerfeld, but this did not always provide the correct explanation of the observed spectrum.

In Bohr's theory there was no restriction on changes in n for the transitions of atoms from one energy level to another. According to quantum theory, there is also no restriction on changes in n. However, when it comes to l, only changes in l given by $l_i - l_f = \Delta l = \pm 1$ are allowed. The corresponding transitions are called *allowed transitions*, while the transitions which do *not* follow this rule are very much less probable, and hence are called *forbidden transitions*. That $\Delta l = \pm 1$ is allowed. This is called the *selection rule for l*, and may be derived from quantum theory.[1,2,3] (All through the text we shall merely state the selection rules. But if we wanted to, we could actually derive them from quantum theory.)

6.4 THE SPINNING ELECTRON AND THE VECTOR MODEL OF THE ATOM

As scientists continued to investigate atomic spectra, they found that the three quantum numbers n, l, and m_l were not sufficient to explain all the experimentally observed transitions. For example, many transitions believed to be singlets, like the transition from 2p to 1s in Fig. 6.10, were actually found to be doublets when investigators observed them with spectroscopes that had a high resolving power. That is, some of the transitions exhibited fine structure. Also, the number of transitions observed when the atoms were placed in an external magnetic field could not be accounted for.

These new experimentally determined facts led, in 1925, to the assignment of a new angular momentum to the electron by Uhlenbeck and Goudsmit,[4] and independently by Bichowsky and Urey.[5] According to them, the electron in any state spins about its own mechanical axis, as shown in Fig. 6.11(a). That is, the electron has an intrinsic angular momentum called the *spin angular momentum*, denoted by **S**. This spin angular momentum **S** is in addition to the orbital angular momentum **L** which the electron has because of its motion around the nucleus, and is similar to the motion of a planet (say earth) spinning or rotating about its own axis. (Note that, like the spinning motion of the electron, the spinning motion of the earth is *in addition to* its orbital motion around the sun.) To explain the experimentally observed results, these scientists had to assign to the electron a spin angular momentum of only one-half a quantum unit of angular momentum. That is,

$$S = |\mathbf{S}| = s\hbar = \tfrac{1}{2}\hbar, \tag{6.40}$$

where s is the *spin quantum number*. It can take on only the one value of $\frac{1}{2}$. As quantum mechanics developed, scientists found that the correct magnitude of **S** is not $\frac{1}{2}\hbar$, but is given by

$$\boxed{S = \sqrt{s(s+1)}\,\hbar = \sqrt{\tfrac{1}{2}(\tfrac{1}{2}+1)}\,\hbar = \frac{\sqrt{3}}{2}\,\hbar,} \tag{6.41}$$

where $s = \frac{1}{2}$ is still the spin quantum number. Furthermore, like the orbital angular momentum vector **L**, the spin angular momentum vector **S** is also space-quantized. Its z components, S_z, can take the following two values:

$$\boxed{S_z = m_s\hbar,} \tag{6.42}$$

where

$$\boxed{m_s = +\tfrac{1}{2} \quad \text{or} \quad -\tfrac{1}{2},} \tag{6.43}$$

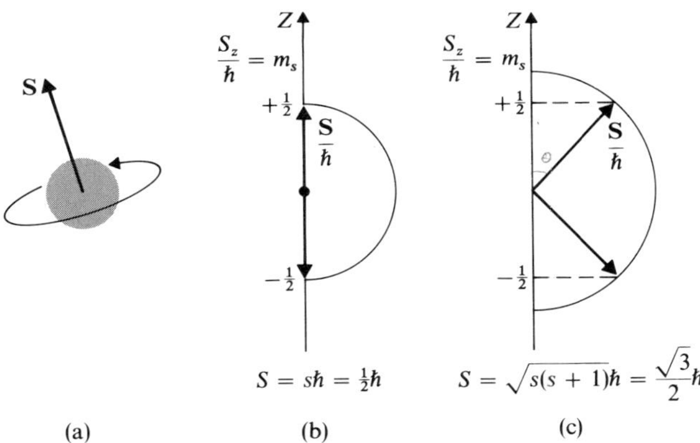

Fig. 6.11 (a) Electron in an atom spinning about its own axis with an intrinsic angular momentum **S**. (b) Space quantization of **S** according to the Sommerfeld model. (c) Space quantization of **S** according to the quantum-mechanical model.

and m_s is called the *spin magnetic quantum number*. Figure 6.11(b) shows the space quantization of **S** according to the Sommerfeld model (or semiclassical model). Here it is assumed that $S = s\hbar$ and that **S** is in complete alignment with the Z-axis, i.e., the axis of the externally applied magnetic field. Figure 6.11(c) shows the space quantization of **S** according to quantum mechanics. It is obvious that because $S = \sqrt{s(s+1)}\,\hbar$, you can't have complete alignment of **S** with the externally applied magnetic field. A direct experimental proof of the existence of electron spin angular momentum and its space quantization was given by Stern and Gerlach.[6]. We'll talk about this in Chapter 7.

Example 6.1. Calculate the two possible orientations of the spin vector **S** with respect to a magnetic field direction.

$$S = \sqrt{s(s+1)}\,\hbar, \qquad \text{where } s = \tfrac{1}{2}$$

while

$$S_z = m_s\hbar, \qquad \text{where } m_s = \pm\tfrac{1}{2}.$$

Therefore

$$\cos\theta = \frac{S_z}{S} = \frac{m_s\hbar}{\sqrt{s(s+1)}\,\hbar} = \frac{m_s}{\sqrt{s(s+1)}};$$

for $m_s = +\frac{1}{2}$,

$$\cos\theta_1 = \frac{\frac{1}{2}}{\sqrt{\frac{1}{2}(\frac{1}{2}+1)}} = \frac{1}{\sqrt{3}} = 0.577 \qquad \text{or} \qquad \theta_1 = 54^\circ\,45'$$

and for $m_s = -\frac{1}{2}$,

$$\cos\theta_2 = \frac{-\frac{1}{2}}{\sqrt{\frac{1}{2}(\frac{1}{2}+1)}} = -\frac{1}{\sqrt{3}} = -0.577 \qquad \text{or} \qquad \theta_2 = 125°\ 15'.$$

These two orientations are as shown in the accompanying figure.

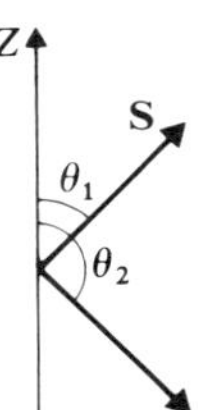

[The theoretical justification for assigning spin angular momentum to the electron came about through the work of P. A. M. Dirac.[7] Because the Schrödinger wave equation which we have been using so far is nonrelativistic, it applies only to particles moving with velocities small compared to the velocity of light. Dirac solved the relativistic form of the Schrödinger wave equation (known as the Klein–Gordon equation) for the case of the electron, and came up with the following two conclusions.

a) Each electron must be assigned a spin angular momentum of $\frac{1}{2}\hbar$, and the component of the spin along any direction in space is $\pm\frac{1}{2}\hbar$.

b) For every particle there exists its counterpart, called an *antiparticle*. Thus, corresponding to the electron, there must be an antielectron, called a *positron*, which has the same mass as the electron, *but opposite charge*. (We'll discuss electron–positron pairs in Chapter 12.)]

When we talk about angular momentum of an atom, we mean that the electron going around the nucleus in an atom has two angular momenta: the orbital angular momentum **L** and the spin angular momentum **S**. These two angular momenta together give a *total angular momentum vector*, **J**, which is the vector sum of **L** and **S**:

$$\boxed{\mathbf{J} = \mathbf{L} + \mathbf{S}.} \tag{6.44}$$

The magnitude of **J**, in analogy with the magnitudes of **L** and **S**, is given by

$$\boxed{J = |\mathbf{J}| = \sqrt{j(j+1)}\,\hbar,} \tag{6.45}$$

where j is called the *total angular momentum quantum number* or simply the *total quantum number*. Since both **L** and **S** are quantized, **J** must also be quantized. That

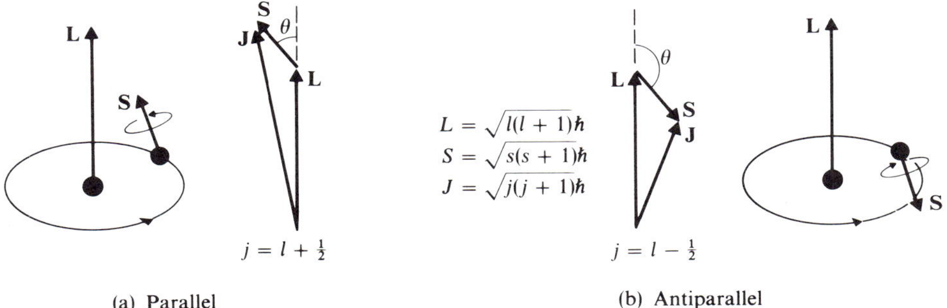

Fig. 6.12 Orbital angular momentum vector **L** and spin angular momentum vector **S** can add either parallel [as shown in (a)] or antiparallel [as shown in (b)]. The total angular momentum vector $\mathbf{J} = \mathbf{L} + \mathbf{S}$ has a magnitude $J = \sqrt{j(j+1)}\,\hbar$, where $j = l + \frac{1}{2}$ for (a) and $j = l - \frac{1}{2}$ for (b).

is, **J** can take on only certain definite values. When one adds the vectors **L** and **S**, the vector **L** provides the arbitrary axis to which **S** can be added parallel or antiparallel, as shown in Fig. 6.12. This adding of **S** to **L** is a coupling scheme with the purpose of obtaining **J**. (Note that, according to quantum theory, perfect alignment or antialignment with an axis is not possible.) Thus, if the quantum numbers are added, we get the following two values of j:

$$\boxed{\begin{aligned} j &= l + \tfrac{1}{2}, \\ j &= l - \tfrac{1}{2}. \end{aligned}} \tag{6.46}$$

For example, for $l = 1$, $j = \frac{3}{2}$ or $\frac{1}{2}$; for $l = 2$, $j = \frac{5}{2}$ or $\frac{3}{2}$. However, for $l = 0$, we get $j = \frac{1}{2}$ only. This implies that all the l states which were supposed to be singlets are actually doublets except for the $l = 0$ state (s state), which is still a singlet. According to spectroscopic notation, the $n = 1$, $l = 0$, $s = \frac{1}{2}$, $j = \frac{1}{2}$ state is denoted by $1S_{1/2}$; the $n = 2$, $l = 0$, $s = \frac{1}{2}$, $j = \frac{1}{2}$ state is denoted by $2S_{1/2}$, while the $n = 2$, $l = 1$, $s = \frac{1}{2}$, $j = \frac{1}{2}$ or $\frac{3}{2}$ states are denoted by $2P_{1/2}$ and $2P_{3/2}$, and so on.

The magnitudes of the angular momenta **L**, **S**, and **J** in units of $\hbar$, according to quantum mechanics, are given by

$$\frac{L}{\hbar} = \sqrt{l(l+1)}, \qquad \frac{S}{\hbar} = \sqrt{s(s+1)}, \qquad \frac{J}{\hbar} = \sqrt{j(j+1)}. \tag{6.47}$$

For an s electron, $l = 0$, $s = \frac{1}{2}$, $j = \frac{1}{2}$, which gives $L/\hbar = 0$, $S/\hbar = \sqrt{\frac{1}{2}(\frac{1}{2}+1)} = \sqrt{3}/2$, and $J/\hbar = \sqrt{\frac{1}{2}(\frac{1}{2}+1)} = \sqrt{3}/2$. For a p electron, $l = 1$, $s = \frac{1}{2}$, $j = \frac{3}{2}$ or $\frac{1}{2}$, which gives $L/\hbar = \sqrt{1(1+1)} = \sqrt{2}$, $S/\hbar = \sqrt{\frac{1}{2}(\frac{1}{2}+1)} = \sqrt{3}/2$, and $J/\hbar$ has two values: $\sqrt{\frac{3}{2}(\frac{3}{2}+1)} = \sqrt{15}/2$ and $\sqrt{\frac{1}{2}(\frac{1}{2}+1)} = \sqrt{3}/2$. For a d electron, $l = 2$, $s = \frac{1}{2}$, $j = \frac{5}{2}$ or $\frac{3}{2}$, which yields $L/\hbar = \sqrt{6}$, $S/\hbar = \sqrt{3}/2$, and $J/\hbar$ has two values: $\sqrt{35}/2$ and $\sqrt{15}/2$. Similarly, we can obtain values of $J/\hbar$ for higher values of l.

Our next step will be to calculate the angle between the vectors **L** and **S**. Since **L** and **S** are quantized, θ can take on only certain definite values, not a continuous range. Applying the law of cosines to Fig. 6.12, we get

$$J^2 = L^2 + S^2 + 2LS\cos\theta. \tag{6.48}$$

That is,

$$j(j+1)\hbar^2 = l(l+1)\hbar^2 + s(s+1)\hbar^2 + 2\sqrt{l(l+1)}\,\hbar\sqrt{s(s+1)}\,\hbar\cos\theta. \tag{6.49}$$

Or we can rewrite it as

$$\cos\theta = \frac{j(j+1) - l(l+1) - s(s+1)}{2\sqrt{l(l+1)}\sqrt{s(s+1)}}. \tag{6.50}$$

For example, if $l = 1$, $s = \frac{1}{2}$, $j = \frac{3}{2}$ or $\frac{1}{2}$. If $j = \frac{3}{2}$, $\cos\theta = 1/\sqrt{6}$ or $\theta \simeq 66°$; and if $j = \frac{1}{2}$, $\cos\theta = -\frac{2}{6}$ or $\theta \simeq 145°$.

Example 6.2. Consider a d electron in a one-electron atomic system. Calculate the values of (a) l, s, and j, (b) L, S, and J, (c) possible angles between **L** and **S**.

a) For a d electron, $l = 2$ and $s = \frac{1}{2}$. Therefore $j = 2 + \frac{1}{2} = \frac{5}{2}$ or $j = 2 - \frac{1}{2} = \frac{3}{2}$. That is, there are two states available, corresponding to $j = \frac{5}{2}$ and $\frac{1}{2}$. These two states are denoted by $^2D_{5/2}$ and $^2D_{3/2}$, or simply $^2D_{5/2,\,3/2}$.

b) Since $L = \sqrt{l(l+1)}\,\hbar$, $S = \sqrt{s(s+1)}\,\hbar$, $J = \sqrt{j(j+1)}\,\hbar$,

$$\text{for } l = 2, \qquad L = \sqrt{2(2+1)}\,\hbar = \sqrt{6}\,\hbar$$

$$\text{for } s = \tfrac{1}{2}, \qquad S = \sqrt{\tfrac{1}{2}(\tfrac{1}{2}+1)}\,\hbar = \frac{\sqrt{3}}{2}\hbar$$

$$\text{for } j = \tfrac{5}{2}, \qquad J = \sqrt{\tfrac{5}{2}(\tfrac{5}{2}+1)}\,\hbar = \frac{\sqrt{35}}{2}\hbar$$

$$\text{for } j = \tfrac{3}{2}, \qquad J = \sqrt{\tfrac{3}{2}(\tfrac{3}{2}+1)}\,\hbar = \frac{\sqrt{15}}{2}\hbar$$

c) According to Eq. (6.50),

$$\cos\theta = \frac{j(j+1) - l(l+1) - s(s+1)}{2\sqrt{l(l+1)}\sqrt{s(s+1)}}.$$

Thus, for $l = 2$, $s = \frac{1}{2}$, $j = \frac{5}{2}$ (that is, the $^2\mathrm{D}_{5/2}$ state),

$$\cos\theta = \frac{\frac{5}{2}(\frac{5}{2}+1) - 2(2+1) - \frac{1}{2}(\frac{1}{2}+1)}{2\sqrt{2(2+1)}\sqrt{\frac{1}{2}(\frac{1}{2}+1)}} = 0.47.$$

That is,

$$\theta = 42°.$$

For $l = 2$, $s = \frac{1}{2}$, $j = \frac{3}{2}$ (that is, the $^2\mathrm{D}_{3/2}$ state)

$$\cos\theta = \frac{\frac{3}{2}(\frac{3}{2}+1) - 2(2+1) - \frac{1}{2}(\frac{1}{2}+1)}{2\sqrt{(2+1)}\sqrt{\frac{1}{2}(\frac{1}{2}+1)}} = -\frac{1}{\sqrt{2}}.$$

That is, $\theta = 135°$.

Since both **L** and **S** show space quantization, it is natural to conclude that **J** will also show space quantization. The picture is as follows. The vectors **L** and **S** precess about **J**, as shown in Fig. 6.13. In the presence of an arbitrary Z axis, such as one obtained by placing the atom in an external magnetic field, the vector **J** will precess about this Z axis, as shown in Fig. 6.14. The rules of space quantization require that **J** precess around the Z axis only at those angles for which the Z component of **J**—that is, J_z—is given by

$$\boxed{J_z = m_j\hbar,} \tag{6.51}$$

where m_j is called the *total magnetic quantum number*. It can take the values

$$\boxed{m_j = -j, -(j-1), \ldots, -1, 0, 1, \ldots, (j-1), j.} \tag{6.52}$$

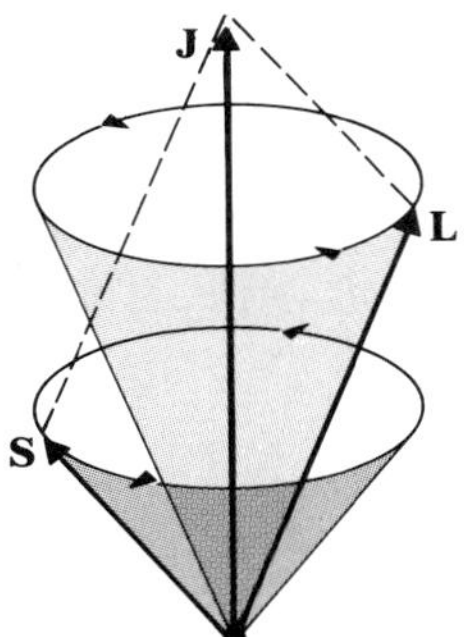

Fig. 6.13 Precession of angular momentum vectors **L** and **S** about their resultant total angular momentum vector **J**.

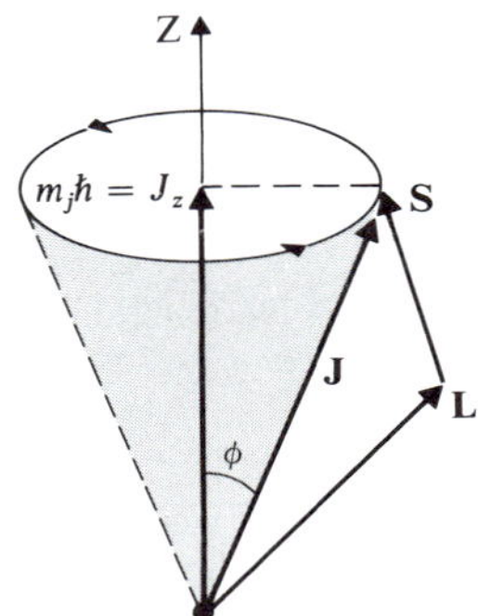

Fig. 6.14 Space quantization of total angular momentum vector **J**. Vector **J** precesses about an arbitrary Z axis, and only those orientations of **J** are possible for which $J_z = m_j\hbar$.

That is, m_j can take $(2j + 1)$ values. Thus, for example, $j = \frac{1}{2}$, $m_j = -\frac{1}{2}$ or $+\frac{1}{2}$; for $j = \frac{3}{2}$, $m_j = -\frac{3}{2}, -\frac{1}{2}, +\frac{1}{2}$, or $+\frac{3}{2}$. The angle between J_z and J is given by

$$\cos\phi = \frac{J_z}{J} = \frac{m_j\hbar}{\sqrt{j(j+1)}\,\hbar} = \frac{m_j}{\sqrt{(j(j+1)}}. \tag{6.53}$$

Note that once again **J** can never align itself completely along the Z axis. Figure 6.15 shows the relation between m_j and j for different values of j. We shall demonstrate the space quantization of **J** and that of **L** and **S** experimentally in Chapter 7.

Example 6.3. Consider the $D_{5/2}$ state of the electron discussed in Example 6.2. Calculate (a) the possible values of m_j and J_z, (b) the different possible orientations of the **J** vector in space. What are the possible orientations if $s = 0$?

a) For this state, $l = 2$, $s = \frac{1}{2}$, and $j = \frac{5}{2}$. Therefore $m_j = \frac{5}{2}, \frac{3}{2}, \frac{1}{2}, -\frac{1}{2}, -\frac{3}{2} -\frac{5}{2}$, and $J_z = m_j\hbar = \frac{5}{2}\hbar, \frac{3}{2}\hbar, \frac{1}{2}\hbar, -\frac{1}{2}\hbar, -\frac{3}{2}\hbar, -\frac{5}{2}\hbar$. That is, there are six different possible values of m_j for $j = \frac{5}{2}$.

b) According to Eq. (6.53),

$$\cos\phi = \frac{m_j}{\sqrt{j(j+1)}}.$$

Let us therefore substitute $\sqrt{j(j+1)} = \dfrac{\sqrt{35}}{2}$ and different values of m_j,

m_j	$\cos\phi$	ϕ
$\frac{5}{2}$	0.85	32°
$\frac{3}{2}$	0.51	59.5°
$\frac{1}{2}$	0.17	79.8°
$-\frac{1}{2}$	-0.17	100.2°
$-\frac{3}{2}$	-0.51	120.5°
$-\frac{5}{2}$	-0.85	148°

where ϕ is the angle which **J** makes with the z-axis.

c) If $s = 0, j = l = 2$, and j and m_j and J and J_z have no meaning. So we must talk in terms of l, L, and L_z, and m_l. Thus $l = 2, m_l = 2, 1, 0, -1, -2$; $L = \sqrt{2(2+1)}\,\hbar = 6\hbar$; $L_z = 2\hbar, 1\hbar, 0\hbar, -1\hbar, -2\hbar$; and $\cos\theta = L_z/L = 2/\sqrt{6}, 1/\sqrt{6}, 0, -1/\sqrt{6}, -2\sqrt{6}$.

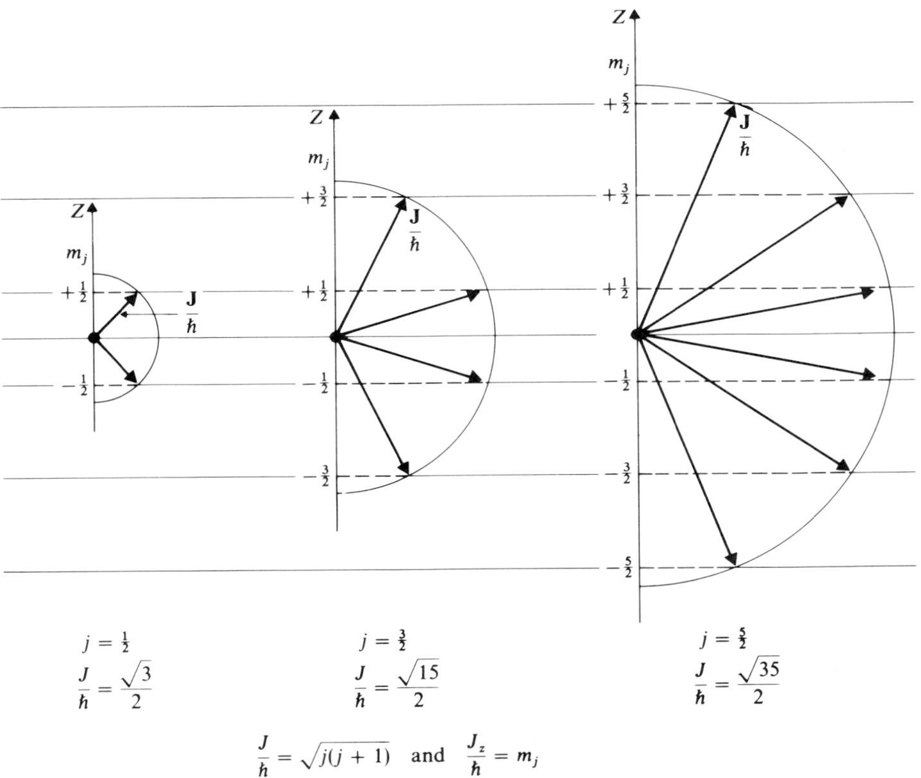

Fig. 6.15 Possible orientations of the total angular momentum vector **J** for different values of j.

We have now introduced all the quantum numbers necessary for our study of atomic spectra in this text. But this does not mean that there are no more quantum numbers! Far from it. We need to introduce new quantum numbers to explain hyperfine spectra and molecular spectra. For the present, let us summarize the quantum numbers we've talked about so far.

a) The total quantum number n determines the energy of the state, which is given by

$$E_n = -Z^2 \frac{E_I}{n^2}, \qquad \text{where } n = 1, 2, 3, 4, \ldots, \infty.$$

b) The orbital quantum number l determines the magnitude of L, and for a given n,

$$l = 0, 1, 2, 3, 4, \ldots, (n-1), \qquad \text{where } L = \sqrt{l(l+1)}\,\hbar.$$

c) The orbital magnetic quantum number m_l determines the z component L_z of L. For a given l,

$$m_l = -l, -(l-1), \ldots, -1, 0, 1, \ldots, (l-1), l, \qquad \text{where } L_z = m_l\hbar.$$

d) The spin quantum number s determines the magnitude of the intrinsic angular momentum S of the electron, which is given by

$$S = \sqrt{s(s+1)}\,\hbar \qquad \text{where } s = \tfrac{1}{2}.$$

e) The magnetic spin quantum number m_s determines the z components, S_z, of S. That is,

$$S_z = m_s\hbar, \qquad \text{where } m_s = +\tfrac{1}{2} \quad \text{or} \quad -\tfrac{1}{2}.$$

f) The total angular momentum quantum number j determines the total angular momentum J, which is given by

$$J = \sqrt{j(j+1)}\,\hbar, \qquad \text{where } j = l + \tfrac{1}{2} \quad \text{or} \quad l - \tfrac{1}{2}.$$

g) The total magnetic quantum number m_j determines the z components, J_z, of J. That is,

$$J_z = m_j\hbar, \qquad \text{where } m_j = -j, -(j-1), \ldots, -1, 0, +1, \ldots, (j-1), j.$$

★6.5 SPIN–ORBIT INTERACTION AND FINE STRUCTURE OF HYDROGEN

The ground state of the hydrogen atom corresponds to $n = 1$, $l = 0$, and $s = \frac{1}{2}$, thereby resulting in the total angular momentum quantum number $j = \frac{1}{2}$. According to spectroscopic notation, this state is written as $1S_{1/2}$ in which the first number indicates the n-value, followed by a letter which stands for the l-value. The subscript to this letter indicates the j values. Thus $1S_{1/2}$ means $n = 1$, $l = 0$, and

$j = \frac{1}{2}$. For the first excited level, $n = 2$, which means either $l = 0$ or $l = 1$. For $n = 2$, $l = 0$, $s = \frac{1}{2}$ gives $j = \frac{1}{2}$, corresponding to a $2S_{1/2}$ state. For $n = 2$, $l = 1$, $s = \frac{1}{2}$, we get $j = \frac{3}{2}$ or $\frac{1}{2}$, resulting in two states: $2P_{3/2}$ and $2P_{1/2}$. Thus the first excited level of hydrogen has three states, $2S_{1/2}$, $2P_{3/2}$, $2P_{1/2}$. Similarly, for the second excited level, $n = 3$, we get the following five states: $3S_{1/2}$, $3P_{3/2}$, $3P_{1/2}$, $3D_{5/2}$, $3D_{3/2}$. Thus the introduction of spin has led to the removal of degeneracy. If there were no spin, the 2S and 2P states would have the same energy $E = -E_I/2^2 = -13.6/4 = -3.4$ eV. But the introduction of spin and the subsequent coupling between spin angular momentum and orbital angular momentum splits the 2P level into $2P_{3/2}$ and $2P_{1/2}$ states, which have energies slightly different from the unsplit energy level, and hence leads to the removal of the degeneracy. Except for the S states such as 1S, 2S, 3S, . . . , etc., which are singlets, all other states such as 1P, 2P, 3P, . . . , 2D, 3D, 4D, . . . , 3F, 4F, 5F, . . . , etc., should be doublet levels. These doublet levels are what constitute the fine structure of the hydrogen atom.

The amount of splitting depends on the amount of interaction between the spin angular momentum **S** and the orbital angular momentum **L**, and hence is called the *spin–orbit effect*. We shall not go through the actual calculations of the magnitude of this energy change, but will briefly outline how it is done and then quote the results. Calculations can be carried out by complete quantum-mechanical methods, such as Dirac and others used, or by semiclassical methods.[8–11] The results are the same in both cases.

Consider an electron moving with a speed v in a circle of radius r around a nucleus of mass M and charge Ze^+ as in Fig. 6.16(a). Since the motion is relative, we may consider the electron to be at rest, while the nucleus is going around the electron in a circle of radius r. The nucleus going in a circular loop produces a magnetic field at the center of the loop, i.e., at the position where the electron is at rest, as in Fig. 6.16(b). If the electron had no spin, there would be no change in the energy of the system. But due to the spin angular momentum, the electron behaves like a small magnetic dipole placed in a magnetic field produced by the nucleus, and this interaction leads to the orientational potential energy denoted by $\Delta E_{S \cdot L}$. Since $\Delta E_{S \cdot L}$ for electrons with $j = +\frac{1}{2}$ is different from the value for electrons with $j = -\frac{1}{2}$ (i.e., for electrons with spins parallel

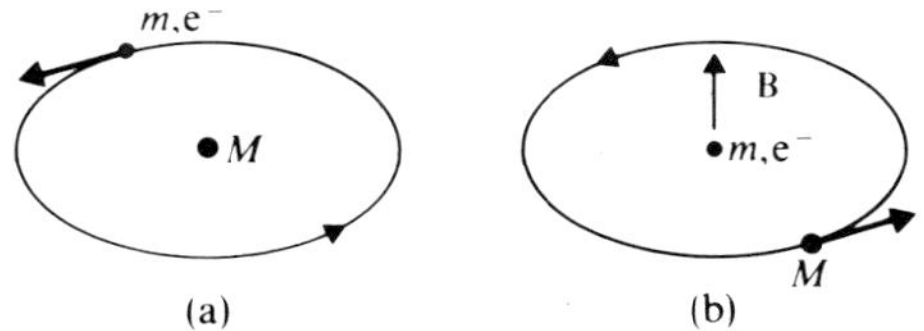

Fig. 6.16 (a) Electron moving in a Bohr-type circular orbit around a stationary nucleus located at the center of the atom. (b) Nucleus as seen by electron is equivalent to the nucleus going around stationary electron.

and antiparallel to L), this leads to the removal of the degeneracy. The detailed calculation shows that

$$\boxed{\Delta E_{\mathbf{S}\cdot\mathbf{L}} = a\,\frac{J^2 - L^2 - S^2}{2} = a\mathbf{L}\cdot\mathbf{S},} \tag{6.63}$$

where a is a constant (except for its dependence on r, that is, $a = \text{constant}/\langle r^3\rangle$); while the energy levels of the hydrogenlike atoms are given by

$$\boxed{E_n = -\frac{\mu k^2 Z^2 e^4}{2n^2\hbar^2}\left[1 + \frac{Z^2\alpha^2}{n}\left(\frac{1}{j+\frac{1}{2}} - \frac{3}{4n}\right)\right],} \tag{6.64}$$

where μ is the reduced mass and α is the *fine-structure constant*, given by

$$\alpha = ke^2/\hbar c \simeq 1/137 \tag{6.65}$$

For comparison, the energy levels of hydrogenlike atoms according to the Bohr theory are

$$E_n = -\frac{\mu k^2 Z^2 e^4}{2n^2\hbar^2} \tag{6.66}$$

and those according to the Bohr–Sommerfeld model with relativistic corrections are

$$E_n = -\frac{\mu k^2 Z^2 e^4}{2n^2\hbar^2}\left[1 + \frac{Z^2\alpha^2}{n}\left(\frac{1}{n_\theta} - \frac{3}{4n}\right)\right]. \tag{6.67}$$

For comparison, Fig. 6.17 shows the energy levels of the hydrogen atom as predicted by all three models—the Bohr model, the Bohr–Sommerfeld model, and the quantum-mechanical model. The separation between the $2P_{3/2}$ level and $2P_{1/2}$ level as calculated from Eq. (6.64) is $\sim 10^{-4}$ eV. This and other splittings of the levels, as calculated from the spin–orbit interaction, are in agreement with the experimentally measured values. Note that not every transition between any two states is possible. According to quantum theory, only those transitions are allowed which obey the following selection rules:

a) l must change by unity. That is,

$$l_i - l_f = \Delta l = \pm 1.$$

b) There is no change in s. That is,

$$s_i - s_f = \Delta s = 0.$$

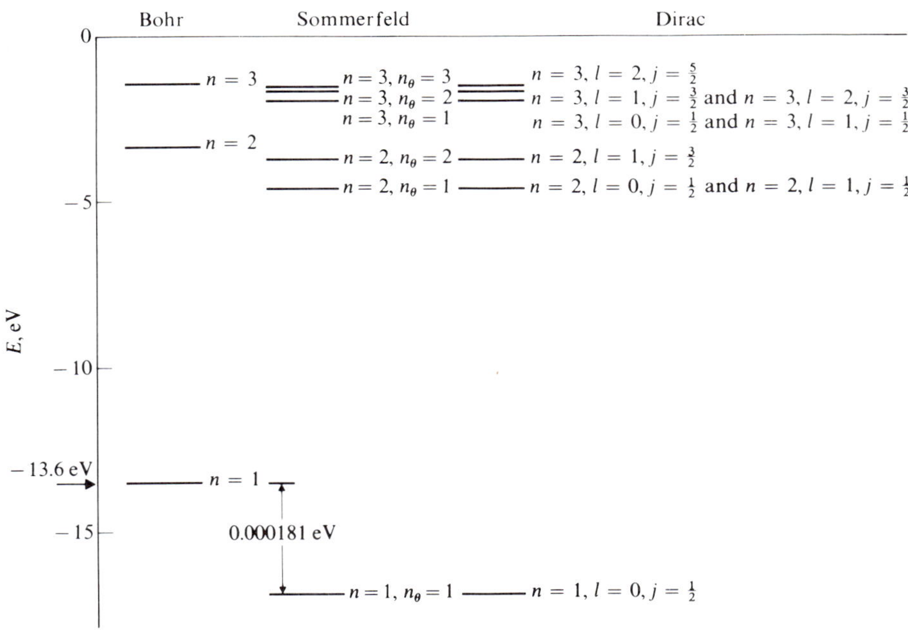

Fig. 6.17 Energy-level diagrams of the hydrogen atom according to three models: (a) Bohr, (b) Sommerfeld, (c) Dirac. Sommerfeld and Dirac models exhibit displacements of split levels from Bohr energy levels which are exaggerated by a factor of $(137)^2$. [From R. M. Eisberg, *Fundamentals of Modern Physics*, New York: John Wiley, 1961]

c) j must change by zero or unity. That is,

$$j_i - j_f = \Delta j = 0,\ \pm 1.$$

But $j_i = 0$ to $j_f = 0$ is not allowed.

d) There is no restriction on n.

The above selection rules are not very rigid, and we do observe transitions which are not allowed by the above selection rules, but those are usually very weak.

It appears that the Bohr–Sommerfeld model is as good as the quantum-mechanical model, at least for the hydrogen atom. But this is just a coincidence, and the Bohr–Sommerfeld model fails completely when applied to multielectron atoms. For example, in multielectron atoms, the electrons move so slowly that the relativistic correction is negligibly small. The correct answer lies in the quantum theory.

Example 6.4. Using the Bohr theory, calculate the flux density produced at the site of the electron in the first Bohr orbit of the hydrogen atom due to the motion of the proton (the nucleus).

Let us assume the electron to be at rest and the proton to be moving in a circle of radius equal to that of the first Bohr orbit. The current i due to the circulating proton is

$$i = \frac{ev}{2\pi r_1},$$

where e is the proton charge, v its velocity, and r_1 is the Bohr orbit radius $= 0.52$ Å.

Current in a loop produces a magnetic field, resulting in a magnetic flux density at the center, given by

$$B = \frac{\mu_0 i}{2r_1}$$

where $\mu_0 = 4\pi \times 10^{-7}$ weber/amp-m. Thus, combining the above two equations, we have, after submitting for μ_0,

$$B = \frac{2\pi i}{r_1} \times 10^{-7} \text{ weber/amp-m.}$$

That is,

$$B = \frac{ev}{r_1^2} \times 10^{-7} \text{ weber/amp-m,}$$

or substituting for e, $v = v_1 = 2.18 \times 10^6$ m/sec and r_1, we get

$$B = \frac{1.602 \times 10^{-19} \text{ coul} \times 2.18 \times 10^6 \text{ m/sec} \times 10^{-7} \text{ weber/amp-m}}{(0.52 \times 10^{-10} \text{ m})^2}$$

or $B = 6.72$ weber/m^2.

SUMMARY

Solving the S.W.E. for the hydrogen atom, that is,

$$\frac{\partial^2 \psi}{\partial x^2} + \frac{\partial^2 \psi}{\partial y^2} + \frac{\partial^2 \psi}{\partial z^2} + \frac{2\mu}{\hbar^2}(E - V)\psi = 0,$$

where $V(r) = -ke^2/r$ and $\mu = mM/(m + M)$, leads to the following results. The wave function is

$$\psi(r, \theta, \phi) = R_{nl}(r)\Theta_l(\theta)\Phi_{m_l}(\phi)$$

and the ground-state radial wave function is

$$R(r) = C_n e^{-r/a_0}$$

The energies of the different states are

$$E_n = -\frac{E_I}{n^2}.$$

The principal quantum number n can take values

$$n = 1, 2, 3, 4, \ldots.$$

For a given n, the angular (or orbital) quantum number l takes the values

$$l = 0, 1, 2, 3, \ldots, (n - 1).$$

For a given l, the magnetic quantum number m_l takes the values

$$m_l = -l, -(l - 1), \ldots, -1, 0, 1, \ldots, (l - 1), l.$$

The magnitude of the angular momentum is

$$L = \sqrt{l(l + 1)}\,\hbar,$$

while the z component of the angular momentum is

$$L_z = m_l\hbar$$

and the angle between L and L_z is

$$\cos\theta = \frac{L_z}{L} = \frac{m_l}{\sqrt{l(l + 1)}}.$$

In spectroscopic notation

$$l = 0, 1, 2, 3, 4, \ldots$$

Electron state = s, p, d, f, g, . . .

Atomic state = S, P, D, F, G, ...

The electron has an intrinsic spin angular momentum

$$S = \sqrt{s(s + 1)}\,\hbar,$$

where $s = \frac{1}{2}$ is the spin quantum number, while the z component of S is

$$S_z = m_s\hbar, \qquad \text{where } m_s = \pm\tfrac{1}{2}.$$

The total angular momentum vector **J** is the vector sum of **L** and **S**,

$$\mathbf{J} = \mathbf{L} + \mathbf{S} \qquad \text{and} \qquad J = \sqrt{j(j + 1)}\,\hbar,$$

where j is the total quantum number, with possible values of $j = l + \frac{1}{2}$ and $j = l - \frac{1}{2}$.

The angle between **L** and **S** is

$$\cos\theta = \frac{j(j+1) - l(l+1) - s(s+1)}{2\sqrt{l(l+1)}\sqrt{s(s+1)}}$$

The z component of J is

$$J_z = m_j\hbar, \qquad \text{where } m_j = -j, -(j-1), \ldots, -1, 0, 1, \ldots, (j-1), j,$$

and

$$\cos\phi = J_z/J = m_j/\sqrt{j(j+1)}.$$

★ The spin–orbit interaction correction $\Delta E_{\mathrm{S\cdot L}} \simeq a\mathbf{L}\cdot\mathbf{S}$ leads to the removal of degeneracy, resulting in two levels corresponding to $j = l + \frac{1}{2}$ and $j = l - \frac{1}{2}$, with slightly different energies for a given n.

PROBLEMS

1. According to Eq. (6.19), the ground-state radial wave function of the hydrogen atom is $R(r) = C_n e^{-r/a_0}$. By using the normalization condition, calculate the value of the constant C_n.

2. Show by direct substitution into Eq. (6.18) that

$$R_{10}(r) = \frac{2}{a_0^{3/2}}\, e^{-r/a_0}$$

is a solution of the radial part of the Schrödinger wave equation for the hydrogen atom.

3. Show by direct substitution into Eq. (6.18) that $R = e^{\alpha r}$ (where α is a positive constant) is *not* a solution of the radial part of the Schrödinger wave equation for the hydrogen atom.

4. The ground-state normalized wave function for the hydrogen atom ($n = 1$, $l = 0$, $m_l = 0$) is given by

$$\psi = \frac{1}{\sqrt{4\pi}}\,\frac{2}{a_0^{3/2}}\, e^{-r/a_0}.$$

Show that the average distance of the electron from the nucleus for the ground state of the hydrogen atom is

$$\langle r\rangle = \int \psi^* r\psi\, dv = \tfrac{3}{2}a_0,$$

where $dv = r^2 \sin\theta\, d\theta\, d\phi\, dr$ and a_0 is the radius of the first Bohr orbit.

5. Using the ground-state normalized wave function given in Problem 4, calculate the average value of $1/r$. That is, show that

$$\left\langle\frac{1}{r}\right\rangle = \int \psi^* \frac{1}{r}\,\psi\, dv = \frac{1}{a_0}.$$

What do you conclude from the results that

$$\langle r \rangle = \tfrac{3}{2}a_0, \qquad \text{while} \qquad \left\langle \frac{1}{r} \right\rangle = \frac{1}{a_0}?$$

6. The average or expectation value for the radius of a hydrogenlike atom obtained from quantum mechanics is

$$\langle r \rangle = \frac{n^2 a_0}{Z}\left\{1 + \tfrac{1}{2}\left[1 - \frac{l(l+1)}{n^2}\right]\right\}.$$

Calculate $\langle r \rangle$ for $n = 1, 2, 3$, and for different l, and compare these values with the corresponding Bohr orbit radii.

7. Show that for large values of l, the quantum-mechanical value of the angular momentum approaches the classical value of the angular momentum.

8. We have shown that an electron in any atom has a well-defined angular momentum, but its angular position at any time is not always well-defined. Explain why.

9. Using Fig. 6.8, calculate the angles between the angular momentum vector **L** and its components L_z along the Z axis for $l = 1$, $l = 2$, and $l = 3$.

10. What are the values of the orbital angular momentum for f and g electrons according to (a) the quantum-mechanical model, (b) the Bohr–Sommerfeld model?

11. What are the possible values of j for $l = 2$, that is, for a d electron, in an atom? Also calculate the possible values of the total angular momentum according to (a) the quantum-mechanical model, (b) the Bohr–Sommerfeld model. Draw diagrams illustrating the above calculations.

12. What are the possible values of j for $l = 3$, that is, for a g electron, in an atom? Also calculate the possible values of total angular momentum according to (a) the quantum-mechanical model, (b) the Bohr–Sommerfeld model.

13. Calculate the possible angles between the **L** and **S** vectors for the f electron in an atom according to the quantum-mechanical model.

14. Calculate the possible angles between the **L** and **S** vectors for the g electron in an atom according to the quantum-mechanical model.

15. Calculate the angles between the total and the orbital angular momentum vectors for the following states: (a) $^2D_{3/2}$, (b) $^2D_{5/2}$, (c) $^2F_{5/2}$, (d) $^2F_{7/2}$. Assume that the systems are single-electron ones.

16. Using Fig. 6.14, calculate the angles between the total angular momentum vector **J** and its components J_z along the Z axis for $j = \frac{3}{2}$ and $j = \frac{5}{2}$.

★17. Using Eq. (6.64), calculate the energies corresponding to the following three states in the hydrogen atom: $n = 2$, $l = 1$, $j = \frac{3}{2}$; $n = 2$, $l = 1$, $j = \frac{1}{2}$; $n = 2$, $l = 0$, $j = \frac{1}{2}$. Are all the states degenerate?

★18. Using Eq. (6.64), calculate the energies corresponding to the following five states in the hydrogen atom: $n = 3$, $l = 0$, $j = \frac{1}{2}$; $n = 3$, $l = 1$, $j = \frac{1}{2}$; $n = 3$, $l = 1$, $j = \frac{3}{2}$; $n = 3$, $l = 2$, $j = \frac{3}{2}$; $n = 3$, $l = 2$, $j = \frac{5}{2}$. Are all the five states degenerate? If not, how many levels are observed?

★19. After you have applied the spin–orbit correction to the levels in the hydrogen atom, draw, with the help of Fig. 6.17, the possible transitions between $n = 3$ and $n = 2$.

★20. In sodium, the two levels $3\ ^2P_{3/2}$ and $3\ ^2P_{1/2}$ (that is, the states for $j = \frac{3}{2}$ and $j = \frac{1}{2}$) are separated by 5.97 Å. Calculate the value of a in the expression $\Delta E_{\mathbf{S}\cdot\mathbf{L}} = a\mathbf{S}\cdot\mathbf{L}$.

★21. Calculate the magnitude of the spin–orbit interactions for $l = 4$, 10, and 15. Show that for large values of l, the spin and the orbital angular momentum vectors **S** and **L** are almost at right angles to each other for both values of j: $j = l + \frac{1}{2}$ and $j = l - \frac{1}{2}$.

REFERENCES

1. E. Schrödinger, *Ann. Physik* **80,** 437 (1926)
2. Albert Einstein, *Zeits. für Physik* **18,** 121 (1917)
3. P. A. M. Dirac, *Proc. Roy. Soc.* **A114,** 243 (1927)
4. G. E. Uhlenbeck and S. Goudsmit, *Physics* **5,** 266 (1925); *Nature* **107,** 264 (1920)
5. F. R. Bichowsky and H. C. Urey, *Proc. Nat. Acad. Sci.* **12,** 80 (1926)
6. O. Stern and W. Gerlach, *Zeits. für Physik* **8,** 110 (1921); **9,** 349, 353 (1922); *Ann. Physik* **74,** 673 (1924)
7. P. A. M. Dirac, *Proc. Roy. Soc.* **A117,** 610 (1927)
8. P. A. M. Dirac, *Proc. Roy. Soc.* **A117,** 616 (1927); **A118,** 351 (1928)
9. W. Pauli, *Zeits. für Physik* **43,** 601 (1927)
10. C. G. Darwin, *Proc. Roy. Soc.* **A118,** 654 (1928)
11. W. Gordon, *Zeits. für Physik* **48,** 11 (1929)

Suggested Readings

1. Atam P. Arya, *Fundamentals of Atomic Physics*, Chapter IX; Boston, Mass.: Allyn and Bacon, 1971
2. R. M. Eisberg, *Fundamentals of Modern Physics*, Chapter 10; New York: John Wiley, 1961
3. H. E. White, *Introduction to Atomic Spectra*, Chapters IV and IX; New York: McGraw-Hill, 1934
4. G. Herzberg, *Atomic Spectra and Atomic Structure*, New York: Dover Publications, 1944
5. B. W. Shore and D. H. Menzel, *Principles of Atomic Spectra*, New York: John Wiley, 1968
6. L. I. Schiff, *Quantum Mechanics*, third edition, Chapters IV, X, and XII; New York: McGraw-Hill, 1968
7. Also see Suggested Readings for Chapter 4, References 5, 6, 7.
8. H. G. Kuhn, *Atomic Spectra*, New York: Academic Press, 1962

CHAPTER 7
STRUCTURE AND SPECTRA OF MANY-ELECTRON ATOMS

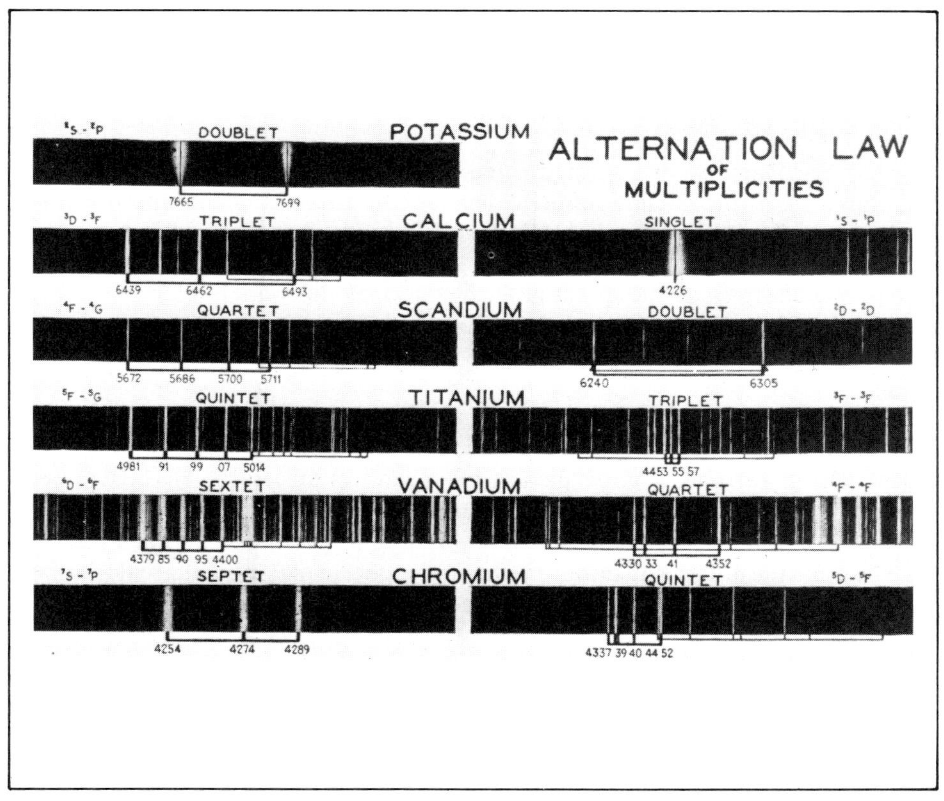

The spectra of the first six elements in the fourth period of the periodic table. (From H. E. White, Introduction to Atomic Spectra, *McGraw-Hill, New York, 1934*)

In this chapter we shall investigate the structure of a number of atoms and consider their energy-level diagrams. Then we shall discuss the experimental proofs of space quantization. Finally, to shed further light on the arrangement of electrons in atoms, we shall look at the characteristic spectra of x-rays. For convenience, we shall divide this chapter into two parts.

I. Optical spectra and atomic structure

II. Characteristic x-ray spectra and atomic structure

I. OPTICAL SPECTRA AND ATOMIC STRUCTURE

7.1 INTRODUCTION

In Chapters 5 and 6 we discussed three models of the hydrogen atom: (a) the Bohr model, (b) the Bohr–Sommerfeld model, and (c) the quantum-mechanical model. We showed that the Bohr–Sommerfeld model was quite an improvement over the Bohr model, but that when it was applied to atoms other than hydrogen, it failed. Actually, since the quantum-mechanical model was developed, it is the only model used. The quantum-mechanical model is the one that most frequently predicts accurate answers concerning the structure and spectra of many-electron atoms.

However, one of the practical difficulties with the quantum-mechanical model is that all results must be interpreted in terms of probabilities, which is not much help in forming a good picture of an atom's structure. In this respect, Sommerfeld's old quantum-mechanical model, in which electrons are assumed to be moving in different elliptical orbits, is very useful. Therefore, even though we shall be using the modern quantum-mechanical model in making actual calculations, we'll make an occasional reference to the Sommerfeld model whenever we are interested in obtaining a visual picture. Note that the energy of a given level is determined by the quantum number n in both models, but the magnitude of the orbital angular momentum according to the Sommerfeld model is $l\hbar$ (where $l = k - 1 = n_\theta - 1$), while according to the quantum-mechanical model it is $\sqrt{l(l + 1)}\,\hbar$.

The Schrödinger wave equation has been solved accurately for the hydrogen and the helium ($Z = 2$) atoms, but for many-electron atoms, the S.W.E. cannot be solved exactly because of the involved mathematics. Hence the methods that have been used have been only approximate. However, even approximate methods enable one to predict general features of the atomic spectra.

7.2 PAULI'S EXCLUSION PRINCIPLE AND ELECTRON CONFIGURATION

Before we talk about the atomic spectra of many-electron atoms, we need to understand the arrangement of electrons in these atoms. All electrons are identical in charge, mass, and spin. So an atom is equivalent to a system containing many indistinguishable, identical particles. The position of an electron in an atom can be specified by three coordinates, x, y, z (or r, θ, ϕ), and its spin state m_s ($m_s = +\frac{1}{2}$ corresponds to pointing up and $m_s = -\frac{1}{2}$ to pointing down). We can give the same information if, instead of stating r, θ, and ϕ, we specify the corresponding quantum numbers n, l, and m_l, respectively. Thus the quantum state of an electron in an atom is completely specified by a set of four quantum numbers: n, l, m_l, and m_s. (Another set of four quantum numbers which serves the same purpose is n, l, j, m_j.)

Now we ask: Is there any restriction on the number of electrons that can occupy the same quantum state? Quantum mechanics did not provide an exact answer to this question. However, the Austrian-born physicist Wolfgang Pauli concluded from experimental observations that a given quantum state can*not* be occupied by more than one electron. Thus, in 1925, Pauli[1] stated *Pauli's exclusion principle* in the following form.

No two electrons in any atom can be in the same quantum state, which is the same thing as saying that

No two electrons in an atom can have identical values for a set of four quantum numbers (n, l, m_l, m_s) or (n, l, j, m_j).

To show you the type of experimental observations from which Pauli derived his principle, let's consider the helium atom. It has two electrons, and its ground state may be represented by four quantum numbers, n, l, m_l, m_s. Three of the four quantum numbers are the same for both electrons, that is, $n = 1$, $l = 0$, $m_l = 0$. The fourth quantum number, m_s, can be either $+\frac{1}{2}$ or $-\frac{1}{2}$. If both the electrons were in the same m_s state, the total spin of the ground state of the helium atom would be unity. But if one electron is in the state $m_s = +\frac{1}{2}$ and the other in the state $m_s = -\frac{1}{2}$, the total spin of the ground state is zero. The observed transitions of electrons to and from the ground state in the helium atom indicate that the ground-state spin must be zero, and not unity. Hence the two electrons in the helium atom are in two different quantum states, $(1, 0, 0, \frac{1}{2})$ and $(1, 0, 0, -\frac{1}{2})$, where ( ) stands for (n, l, m_l, m_s).

★Let us see what the Pauli exclusion principle means in quantum mechanics. For simplicity, consider a system of two noninteracting, identical, indistinguishable particles. Let particle 1 be in a quantum state denoted by α, so that its wave function is $\psi_\alpha(1)$, and let particle 2 be in quantum state β, so that its wave function is $\psi_\beta(2)$. The total wave function $\psi_{\alpha\beta}(1, 2)$ of the system is

$$\psi_{\alpha\beta}(1, 2) = \psi_\alpha(1)\psi_\beta(2). \tag{7.1}$$

If the positions of these two particles are interchanged, the new wave function is

$$\psi_{\alpha\beta}(2, 1) = \psi_\alpha(2)\psi_\beta(1). \tag{7.2}$$

Since the two particles are identical, exchanging their positions shouldn't make any difference in their probabilities of being in a given position. Thus

$$|\psi_{\alpha\beta}(2, 1)|^2 = |\psi_{\alpha\beta}(1, 2)|^2, \tag{7.3}$$

which means that either the wave function is *symmetric* if

$$\psi_{\alpha\beta}(2, 1) = +\psi_{\alpha\beta}(1, 2), \tag{7.4}$$

or the wave function is *antisymmetric* if

$$\psi_{\alpha\beta}(2, 1) = -\psi_{\alpha\beta}(1, 2). \tag{7.5}$$

That is, the symmetric wave function does not change sign when the two particles exchange places, but the antisymmetric wave function does change sign.

Thus both $\psi_{\alpha\beta}(1, 2)$ and $\psi_{\alpha\beta}(2, 1)$ are the eigenfunctions of the system, and so are their two linear combinations, resulting in the *symmetric wave function* ψ_S given by

$$\psi_S = \frac{1}{\sqrt{2}}[\psi_{\alpha\beta}(1, 2) + \psi_{\alpha\beta}(2, 1)] \tag{7.6}$$

and the *antisymmetric wave function* ψ_A, given by

$$\psi_A = \frac{1}{\sqrt{2}}[\psi_{\alpha\beta}(1, 2) - \psi_{\alpha\beta}(2, 1)]. \tag{7.7}$$

Both ψ_S and ψ_A are solutions of the Schrödinger wave equation. Also the total energy of the system corresponding to the eigenfunction $\psi_{\alpha\beta}(1, 2)$ is the same as that corresponding to the exchanged eigenfunction $\psi_{\alpha\beta}(2, 1)$. Hence the system is degenerate. This type of degeneracy is called *exchange degeneracy*.

The quantum theory cannot tell us whether the total wave function of the system is symmetric, ψ_S, or antisymmetric, ψ_A. Let us see what happens if we try to place both the particles in the same state, i.e., $\alpha = \beta$. Then Eq. (7.7) takes the form

$$\begin{aligned}\psi_A &= \frac{1}{\sqrt{2}}[\psi_{\alpha\alpha}(1, 2) - \psi_{\alpha\alpha}(2, 1)] \\ &= \frac{1}{\sqrt{2}}[\psi_\alpha(1)\psi_\alpha(2) - \psi_\alpha(2)\psi_\alpha(1)] = 0.\end{aligned} \tag{7.8}$$

Thus for ψ_A to be nonzero, $\alpha \neq \beta$. That is, the two electrons cannot be placed in the same quantum state. And there we have Pauli's exclusion principle! Thus another way of stating Pauli's exclusion principle is the following.

The eigenfunction of a system containing several electrons must be antisymmetric.

Or, to be more general,

The eigenfunction of a system containing identical, indistinguishable particles, each having a half-integer spin, must be antisymmetric.

With the help of Pauli's exclusion principle, we can now assign different quantum states to the electrons in a given atom. In assigning quantum states, we shall use four quantum numbers, n, l, m_l, and m_s. In a given atom, the electrons that have the same principal quantum number n are said to be in the same *group*, or *shell*. For a given n, those electrons that have the same value of l are said to be in the same *subgroup*, *subshell*, or *sublevel*. Also, according to x-ray notation, $n = 1$ is called the K shell, $n = 2$ the L shell, $n = 3$ the M shell, $n = 4$ the N shell, and so on. Remember that $l = 0, 1, 2, 3, 4, \ldots$ states are denoted by S, P, D, F, G, ..., respectively.

Now we can calculate the number of electrons in any shell or subshell. For a given l, m_l can take $(2l + 1)$ different values, and for each m_l, m_s can take two different values, $+\frac{1}{2}$ or $-\frac{1}{2}$. Thus, for a given l, there are $2(2l + 1)$ different combinations; i.e., any subshell has $2(2l + 1)$ different quantum states. Thus there are 2 quantum states in the s-subshell, 6 in the p-subshell, 10 in the d-subshell, ..., and $2(2l + 1)$ in the l-subshell.

For a given n, l can take n values; that is, $0, 1, 2, 3, \ldots, n - 1$. Thus the total number of quantum states, N_t, for a given n is

$$N_t = \sum_{l=0}^{l=n-1} 2(2l + 1) = 2[1 + 3 + 5 + \cdots (2n - 1)]$$

$$= 2\left[n\left\{\frac{1 + (2n - 1)}{2}\right\}\right] = 2n^2,$$

$$\boxed{N_t = 2n^2.} \tag{7.9}$$

That is, in any shell there are $2n^2$ different quantum states, each denoted by (n, l, m_l, m_s). Table 7.1 shows the possible number of different shells and subshells. Each horizontal line represents a quantum state, and no two horizontal lines have all four quantum numbers n, l, m_l, m_s the same. Note that, for any subgroup $\sum m_s = 0$, $\sum m_l = 0$. This means that, in any complete shell, $S = 0$, $L = 0$ and hence $J = 0$. (Note that, in multielectron atoms, $\mathbf{S}$ and $\mathbf{L}$ are the vector sums of orbital and spin momenta of individual electrons, and $\mathbf{J} = \mathbf{L} + \mathbf{S}$.)

Assigning quantum states to electrons in any atom is now a simple matter. We just start from the top of the table and assign quantum numbers in each horizontal line to different electrons. Thus hydrogen ($Z = 1$), which as we know has one electron, has ground state $(1, 0, 0, \frac{1}{2})$. We denote the ground-state configuration of the hydrogen atom as $1s^1$, which means $n = 1$, $l = 0$; the superscript means that there is one electron in the s state. Similarly, the two electrons in

Table 7.1

Quantum States Available in Different Subshells and Shells

Numbers of electrons in an atom	Quantum states of electron n	l	m_l	m_s	Number of electrons in different l subshells	Electron configuration at closed subshells and shells
1	1	0	0	$+\frac{1}{2}$		
2	1	0	0	$-\frac{1}{2}$	2	$1s^2$
3	2	0	0	$+\frac{1}{2}$		
4	2	0	0	$-\frac{1}{2}$	2	$1s^22s^2$
5	2	1	-1	$+\frac{1}{2}$		
6	2	1	-1	$-\frac{1}{2}$		
7	2	1	0	$+\frac{1}{2}$		
8	2	1	0	$-\frac{1}{2}$		
9	2	1	$+1$	$+\frac{1}{2}$		
10	2	1	$+1$	$-\frac{1}{2}$	6	$1s^22s^22p^6$
11	3	0	0	$+\frac{1}{2}$		
12	3	0	0	$-\frac{1}{2}$	2	$1s^22s^22p^63s^2$
13	3	1	-1	$+\frac{1}{2}$		
14	3	1	-1	$-\frac{1}{2}$		
15	3	1	0	$+\frac{1}{2}$		
16	3	1	0	$-\frac{1}{2}$		
17	3	1	$+1$	$+\frac{1}{2}$		
18	3	1	$+1$	$-\frac{1}{2}$	6	$1s^22s^22p^63s^23p^6$
19	3	2	-2	$+\frac{1}{2}$		
20	3	2	-2	$-\frac{1}{2}$		
21	3	2	-1	$+\frac{1}{2}$		
22	3	2	-1	$-\frac{1}{2}$		
23	3	2	0	$+\frac{1}{2}$		
24	3	2	0	$-\frac{1}{2}$		
25	3	2	$+1$	$+\frac{1}{2}$		
26	3	2	$+1$	$-\frac{1}{2}$		
27	3	2	$+2$	$+\frac{1}{2}$		
28	3	2	$+2$	$-\frac{1}{2}$	10	$1s^22s^22p^63s^23p^63d^{10}$

the $_2$He ($Z = 2$) atom have quantum states $(1, 0, 0, \frac{1}{2})$ and $(1, 0, 0, -\frac{1}{2})$. The ground-state electron configuration is $1s^2$, which means that $n = 1$, $l = 0$, and that there are two electrons in the s state.

For nitrogen, $Z = 7$, therefore the ground-state electron configuration of nitrogen is $1s^22s^22p^3$, which means that there are two electrons in the 1s state,

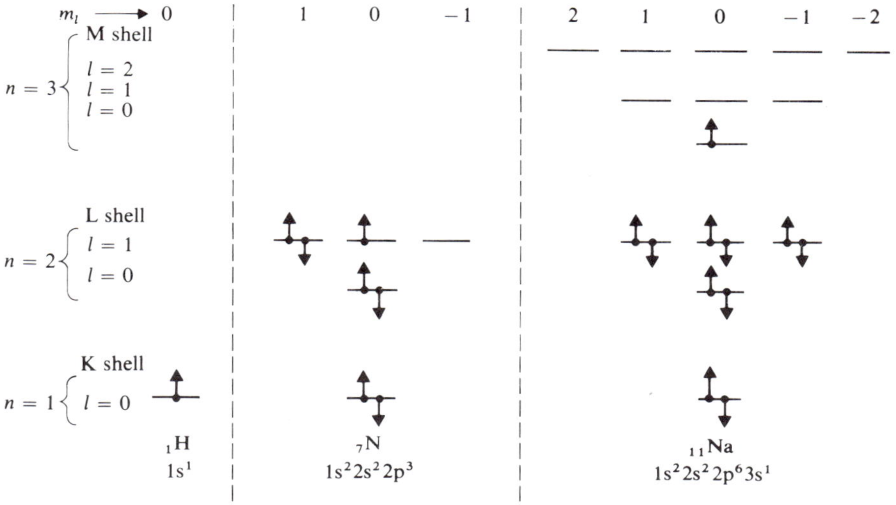

Fig. 7.1 Schematic representation of configurations of the electrons of H ($Z = 1$), N ($Z = 7$), and Na ($Z = 11$).

2 electrons in the 2s state, and 3 electrons in the 2p state. The ground state of nitrogen is given by the quantum numbers in the seventh ($Z = 7$) horizontal row in Table 7.1. That is, $(2, 1, 0, \frac{1}{2})$. Similarly, for sodium ($Z = 11$), the electron configuration is $1s^2 2s^2 2p^6 3s^1$, while its ground state is, from the eleventh horizontal line, $(3, 0, 0, \frac{1}{2})$. That is, the last electron is in the 3s state.

Figure 7.1 shows another convenient way of displaying electron configuration for $_1$H, $_7$N, and $_{11}$Na. Note that the arrow pointing up corresponds to $m_s = +\frac{1}{2}$, and the arrow pointing down corresponds to $m_s = -\frac{1}{2}$.

We may be tempted to say that the electron configuration of any atom follows the general rule

$$1s^2 2s^2 2p^6 3s^2 3p^6 3d^{10} 4s^2 4p^6 4d^{10} 4f^{14} \cdots$$

But this is not true. The actual order in which the levels must be filled so that the resulting energy is minimum (corresponding to a stable atom) is as follows.

$$\boxed{1s^2 2s^2 2p^6 3s^2 3p^6 4s^2 3d^{10} 4p^6 5s^2 4d^{10} 5p^6 6s^2 4f^{14} 5d^{10} 6p^6 7s^2 6d^{10}} \tag{7.10}$$

7.3 THE SPECTRA OF ATOMS WITH ONE VALENCE ELECTRON

Atoms with one valence electron are those which have one electron outside the closed shells of electrons. These atoms belong to the alkali metals, which, together with the hydrogen atom, have the following electron configurations.

Hydrogen	H	$Z = 1$	$\mathbf{1s^1}$
Lithium	Li	$Z = 3$	$1s^2\mathbf{2s^1}$
Sodium	Na	$Z = 11$	$1s^22s^22p^6\mathbf{3s^1}$
Potassium	K	$Z = 19$	$1s^22s^22p^63s^23p^6\mathbf{4s^1}$
Rubidium	Rb	$Z = 37$	$1s^22s^22p^63s^23p^64s^23d^{10}4p^6\mathbf{5s^1}$
Cesium	Cs	$Z = 55$	$1s^22s^22p^63s^23p^64s^23d^{10}4p^65s^24d^{10}5p^6\mathbf{6s^1}$

One characteristic common to all these atoms is that, in each atom, there is only one electron outside the closed shells, and this single electron is in the s state ($l = 0$ state) when the atom is in the normal or ground state. Chemists call the electron in the ground state (the s electron in this case) the *valence electron*, because it determines the chemical characteristics of the atom. In spectroscopy, this electron is called the *optical electron*, because it is responsible for the optical spectrum of a given material. The optical spectra of all these atoms must be very similar because the optical spectrum is caused by the single optical electron in the s state outside the closed shell. Let's take sodium, Na ($Z = 11$), as an example, and investigate its energy-level diagram and optical spectrum.

We can consider the sodium atom as consisting of three parts (see Fig. 7.2a): (a) the nucleus, (b) the ten electrons in states $1s^22s^22p^6$—for which L = 0, S = 0, and hence J = 0—form an inert core of electrons, and do not contribute to the angular momentum, (c) the optical electron in the 3s state.

Now the sodium atom, like the hydrogen atom, also has one optical electron,

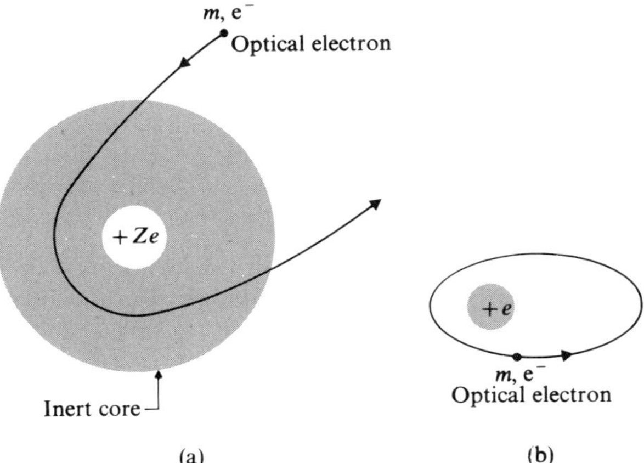

Fig. 7.2 (a) The structure of the sodium atom may be divided into three parts: the nucleus, the inert core of electrons shielding the nucleus, and the optical electron. (b) The hydrogen atom is similar to the sodium atom, except that it does not have a core of inert electrons.

but there is one significant difference, as shown in Fig. 7.2(b). In the sodium atom, the optical electron moves in the net field of the nucleus of charge $+Ze$ and the core of electrons with charge $(Z - 1)e^-$ surrounding the nucleus. This core of electrons acts as a shield between the optical electron and the nucleus. There isn't any such shielding in the case of the hydrogen atom. Shielding by the core electrons has a very important effect, in that it removes the l-degeneracy, i.e., it makes the energy of the levels dependent on l, and hence causes shifts in the energy levels corresponding to the same n, but different l. Let's examine the reasons why.

In sodium, the first two shells are completely filled with ten electrons, while the eleventh electron is in the 3S state, the ground state. When the sodium atom is excited (by heat or electrical discharge), it results in the transformation of the 3s electron to the higher energy states, such as 3P, 4S, 3D, 4P, 5S, 4D, ... According to the quantum-mechanical analysis of the hydrogen atom, the 3S, 3P, and 3D states should all have the same energy; the 4S, 4P, 4D, and 4F states should all have the same energy, and so on. But this is not so in the case of sodium, as we illustrate for the 3S, 3P, and 3D states. The net effective charge of the sodium nucleus and its shielding electrons is quite different for a 3s electron and for a 3p or 3d electron.

We can see this both from the quantum-mechanical picture[2–5] in Fig. 7.3 and from the equivalent classical picture in Fig. 7.4. The shaded area in Fig. 7.3 shows the plot of probability charge density distribution $4\pi r^2\psi^*\psi$ of the core electrons versus the radial length r, plus the plots of the distributions for the valence electron in the 3S, 3P, and 3D states. The bumps in the core charge distribution correspond to the location of K, L, M, . . . shell electrons. It is obvious from these plots that

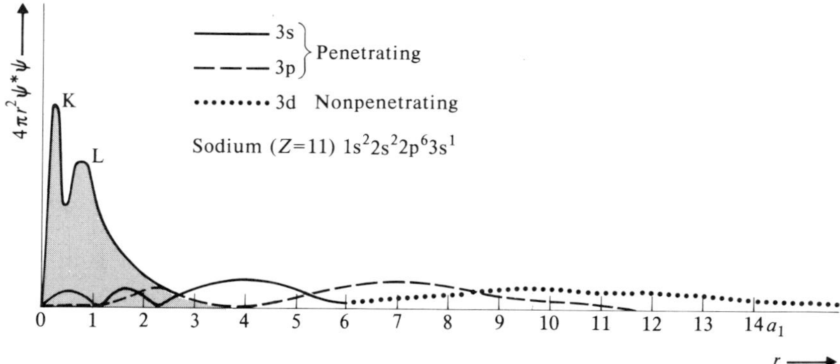

Fig. 7.3 The shaded area is the radial charge-probability-density distribution of the inert core electrons of sodium. Unshaded areas indicate the probability of the valence electron of sodium being in the 3S, 3P, and 3D states. Note the penetration of 3s and 3p electrons into the core distribution. [From H. E. White, *Introduction to Atomic Spectra*, page 103, New York: McGraw-Hill, 1934]

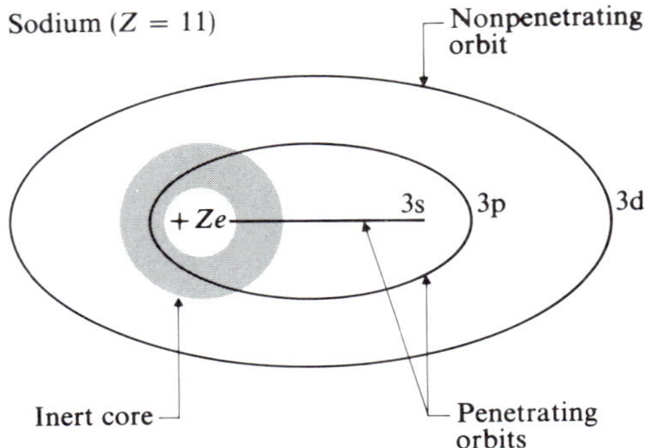

Fig. 7.4 Classical picture of core electrons and valence electron of sodium (equivalent of quantum-mechanical picture in Fig. 7.3), showing penetrating (3s, 3p) and non-penetrating (3d) orbits.

the 3s electron penetrates more deeply into the core electron charge distribution than the 3p electron, while the 3d electron is almost outside the core. We call the 3s and 3p electrons *penetrating electrons* and the 3d electron *nonpenetrating*. Or, in terms of the classical picture in Fig. 7.4, the 3s and 3p electrons have *penetrating orbits*, while the 3d one has a *nonpenetrating orbit*.

For a nonpenetrating orbit like the 3d one in Fig. 7.4, the net effective charge "seen" by this electron is $11e^{+} - 10e^{-} = 1e^{+}$. Hence the system is hydrogenlike, and the energy of the 3D state of sodium is the same as that of the 3D (or $n = 3$) state of hydrogen, as shown in Fig. 7.5. The net effective charge for the 3P electron is more than $1e^{+}$ because of the penetration of this electron into the core. Therefore the energy of the 3P state is depressed below 3D; that is, it is more negative because

$$E_n = -Z_{\text{eff}}^2 \frac{Rch}{n^2}, \tag{7.11}$$

where Z_{eff} is the net effective Z for the penetrating orbit and is larger for 3P than for 3D. Since the 3S state is highly penetrating, its Z_{eff} is even larger than that of the 3P state, and hence the 3S state is depressed much more than the 3P state, as shown in Fig. 7.5.

This type of argument can be carried out for higher n values, resulting in the energy-level diagram of sodium (without taking into account the spin–orbit coupling) which we see in Fig. 7.5. For comparison, we show the energy-level diagram of hydrogen also. Equation (7.11) may also be written as

$$E_n = -(Z - \sigma)^2 \frac{Rch}{n^2}, \tag{7.12}$$

where σ is the screening constant and is different for different l-states[6] (for a

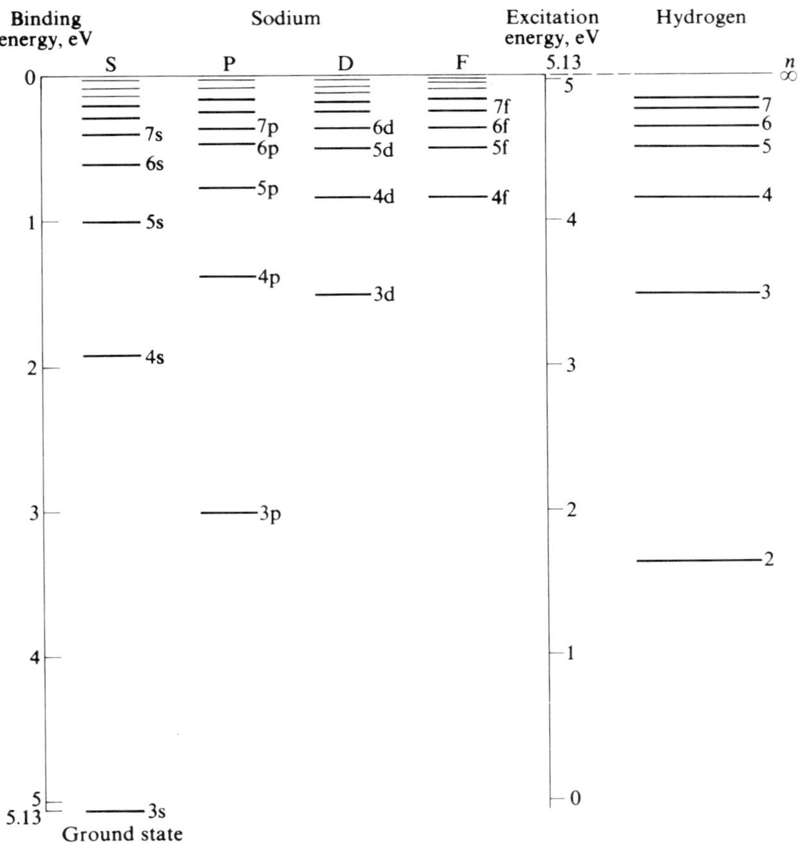

Fig. 7.5 Energy-level diagram of sodium (neglecting intrinsic spin of electron). Short horizontal lines indicate the possible energy states to which the valence electron of sodium may be excited. For comparison, look at the energy-level diagram of hydrogen.

given n); hence it removes the l-degeneracy. Some of the features we can see in the energy-level diagram of sodium in Fig. 7.5 are the following.

i) For a given n, the S state has less energy than the P state, the P state less than the D state, the D state less than the F state, and so on.

ii) With increasing n, the corresponding diminution of energy becomes less and less. Actually, for $n = 5$, the depressions in energy are so small that the 5D and 5F levels of sodium coincide with the $n = 5$ level of hydrogen.

In sketching the energy-level diagram of sodium in Fig. 7.5, we neglected the intrinsic spin of the electron. The single valence electron outside the closed shell has a spin angular momentum $|\mathbf{S}| = \sqrt{s(s+1)}\,\hbar$, where $s = \frac{1}{2}$, which couples with its orbital angular momentum $|\mathbf{L}| = \sqrt{l(l+1)}\,\hbar$, resulting in a total angular momentum $|\mathbf{J}| = \sqrt{j(j+1)}\,\hbar$. The total angular momentum quantum number j can take two values, if $l \neq 0$,

$$j = l + \tfrac{1}{2} \text{ parallel coupling}, \qquad j = l - \tfrac{1}{2} \text{ antiparallel coupling}, \tag{7.13}$$

and, if $l = 0$, takes only one value, $j = \frac{1}{2}$. Thus each of the l levels in Fig. 7.5 is split into a doublet, except for the S level, corresponding to $l = 0$, which is still a singlet (degeneracy has not been removed in this case).

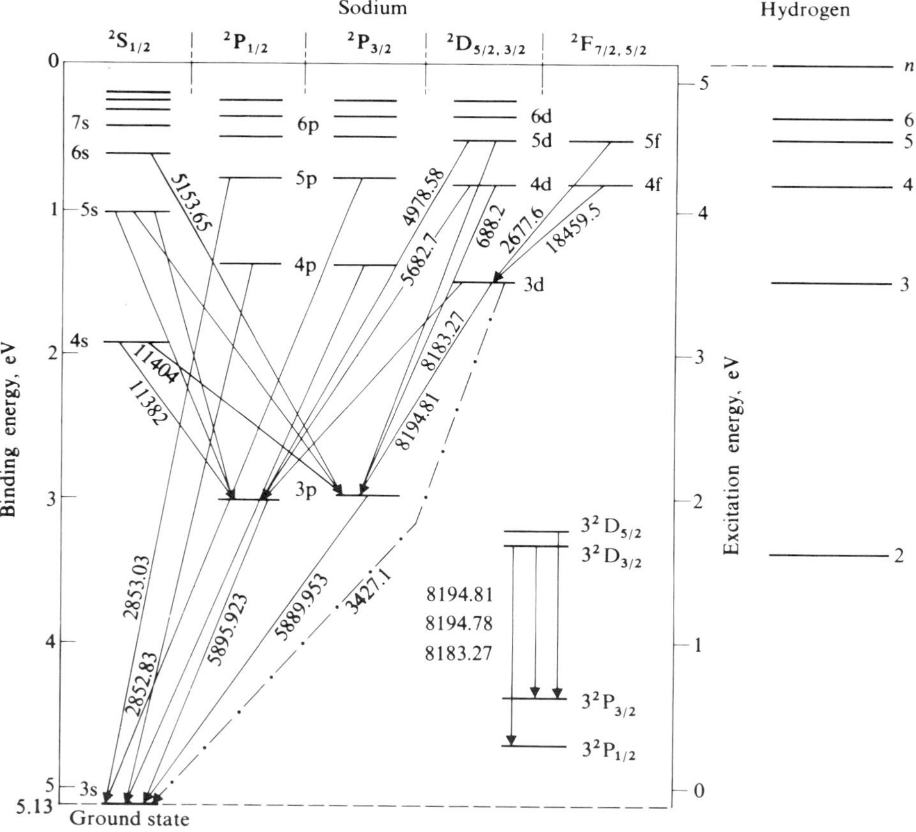

Fig. 7.6 A complete energy-level diagram (including spin–orbit coupling) and possible transitions of electrons in the sodium atom. For comparison the hydrogen energy levels are also shown. [From A. P. Arya, *Fundamentals of Atomic Physics*, page 370, Boston, Mass.: Allyn and Bacon, 1971]

After we take into account spin–orbit splitting, the new energy-level diagram of sodium is as shown in Fig. 7.6. Each P level ($l = 1$) is split into two levels, corresponding to $j = \frac{3}{2}$ and $\frac{1}{2}$; each D level ($l = 2$) is split into two levels, corresponding to $j = \frac{5}{2}$ and $\frac{3}{2}$; and similarly, F levels are $j = \frac{7}{2}$ and $\frac{5}{2}$; while the S level is $j = \frac{1}{2}$. A particular state is denoted by the spectroscopic notation

$$\boxed{n^{2s+1}L_j,}$$

where s is the electron's intrinsic spin quantum number and $(2s + 1)$ is the *multiplicity* of the level. This multiplicity, for example, in the case of a single optical electron, $s = \frac{1}{2}$, is $(2s + 1) = (2 \times \frac{1}{2} + 1) = 2$. That is, the levels are doublets. n stands for the principal quantum number, L for the orbital state S, P, D, F, . . . , and j is the total angular quantum number. Thus $3^2P_{1/2}$ means $n = 3$, $l = 1$, $j = \frac{1}{2}$, and 2 is the multiplicity. That is,

$$\boxed{2 \times \text{spin} + 1 = \text{multiplicity},}$$

$$2 \times \tfrac{1}{2} + 1 = 2.$$

Thus $3^2P_{1/2}$ is read as "three doublet P one-half state". The other member of this doublet is $3^2P_{3/2}$. Even though S is a singlet state, we denote it as a doublet, so as to be consistent. (We can show that, when we apply a magnetic field, we can remove the degeneracy and then S is a doublet level.)

Note that in Fig. 7.6, for a given n, the doublet separation decreases with increasing l as well as increasing n. Thus the 3P doublet is wider than the 3D, and the 3D wider than the 3F; also the 3P doublet is wider than the 4P, the 4P wider than the 5P, and so on.

Figure 7.6 shows the possible transitions resulting from electrons in the excited states returning to their lower energy states (or ground states). These transitions may be classified into the following four series.

1)	Sharp series	$n^2S_{1/2} \to 3^2P_{3/2,1/2}$	$n = 4, 5, 6$	doublets
2)	Principal series	$n^2P_{3/2,1/2} \to 3^2S_{1/2}$	$n = 3, 4, 5$	doublets
3)	Diffuse series	$n^2D_{5/2,3/2} \to 3^2P_{3/2,1/2}$	$n = 3, 4, 5$	triplets
4)	Fundamental series	$n^2F_{7/2,5/2} \to 3^2D_{5/2,3/2}$	$n = 4, 5, 6$	triplets

The reason we label the levels as S, P, D, and F and classify the transitions as principal, sharp, diffuse, and fundamental is purely historical. The lines of the S series were found to be relatively sharp; hence the name "sharp series." The lines of the P series were observed both in the emission and absorption spectra of sodium for very small excitations; hence the name "principal series." The diffuse series lines were found to be somewhat diffused; hence the name "diffuse." The frequencies of the fundamental series are in the infrared region and come very close to the frequencies of the "fundamental" hydrogen atom; and hence that name. The two transitions $3^2P_{3/2} \rightarrow 3^2S_{1/2}$ and $3^2P_{1/2} \rightarrow 3^2S_{1/2}$ are the first members of the principal series, and are the prominent yellow lines of sodium referred to as the sodium D lines doublet.

The diffuse and fundamental series start from doublet levels and end on doublet levels, so we'd expect to find quartets and not triplets. The reason they are triplets is that one of these transitions is not allowed by the so-called *selection rules*. The transitions between different states must obey the following selection rules.

a) The orbital quantum number must change by unity:

$$\boxed{\Delta l = l_i - l_f = \pm 1.}$$

In such transitions one unit of angular momentum, $1\hbar$, is carried away by the emitted photon, so as to conserve angular momentum.

b) The total quantum number j should change by zero or unity. That is,

$$\boxed{\Delta j = j_i - j_f = 0, \pm 1.}$$

But $j_i = 0$ to $j_f = 0$ is forbidden.

c) There is no restriction on change in n.

The general features—energy-level diagrams as well as transitions—of other one-valence-electron atoms are similar to that of the sodium atom shown in Fig. 7.6.

7.4 MAGNETIC DIPOLE MOMENT AND LARMOR'S PRECESSION

We can verify the rules of space quantization for **L**, **S**, and **J** (that is, these quantities can take only certain definite positions with respect to some axis in space) by placing atoms in an external magnetic field which is in the direction of the arbitrarily chosen Z axis. We don't measure **L** directly, but by measuring the interaction of $\boldsymbol{\mu}$ (the magnetic dipole moment associated with the atom) with the applied external field, we can show the space quantization of **L**. To investigate these interactions, we must know the magnetic dipole moment resulting from the motion of an electron in an atom. Thus we divide our discussion into the following parts.

A) Magnetic dipole moment due to orbital motion of the electron
B) Magnetic dipole moment due to spinning motion of the electron
C) Magnetic dipole moment due to total angular momentum
D) Effect of an external magnetic field on the atom

The results derived below make use of the Bohr theory, the electromagnetic theory, and quantum theory. Though we could derive all these results directly from quantum-mechanical treatment, due to the limitations of the scope of this text, we shall follow the semiclassical treatment, which has the advantage of being easily visualized.

A) Magnetic dipole moment due to orbital motion of the electron

Consider an electron of mass m and charge $-e$ moving with a velocity v in a circular Bohr orbit of radius r, as shown in Fig. 7.7(a). This circulating charge is equivalent to a loop of wire carrying a current i, given by

$$i = \frac{e}{T} = \frac{e}{2\pi r/v} = \frac{ev}{2\pi r}, \tag{7.14}$$

where T is the period of time the electron takes to make its circular motion, which is equal to $2\pi r/v$. From our knowledge of elementary electromagnetic theory, we know that, at large distances from the loop, the magnetic field due to the loop is the same as that of a magnetic dipole located at the center of the loop, as shown in Fig. 7.7(b). Actually, the loop behaves like a small permanent bar magnet, as shown in Fig. 7.7(c). According to Ampère's theorem,[7] the magnetic dipole moment μ_l resulting from the current in this loop is

$$\boxed{\mu_l = \text{current} \times \text{area enclosed by the loop} = iA.} \tag{7.15}$$

The direction of this dipole is perpendicular to the plane of the orbit, as shown.

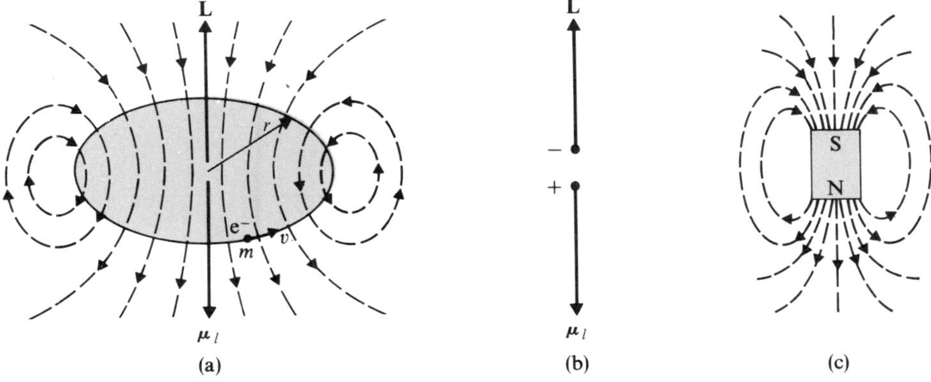

Fig. 7.7 (a) The orbital motion of an electron in an atom is equivalent to a current-carrying loop. The resulting magnetic lines of force are as shown. (b) Magnetic dipole equivalent of a current-carrying loop. (c) A permanent magnetic dipole, which is also the equivalent of a current-carrying loop.

The direction of $\boldsymbol{\mu}_l$ is opposite to that of the angular momentum **L** because of the negative charge of the electron. The area A of the loop is the area of the Bohr circular orbit. That is,

$$A = \pi r^2. \tag{7.16}$$

Substituting for i and A from Eq. (7.14) and (7.16) into Eq. (7.15), we get

$$\mu_l = iA = \frac{ev}{2\pi r}\,\pi r^2 = \frac{evr}{2},$$

which may also be written as

$$\mu_l = \frac{e}{2m}\,mvr. \tag{7.17}$$

But the angular momentum L of the electron moving in a circular orbit is

$$L = mvr. \tag{7.18}$$

Therefore

$$\mu_l = \frac{e}{2m}\,L. \tag{7.19}$$

Since L is quantized and is equal to $\sqrt{l(l+1)}\,\hbar$, hence

$$\mu_l = \frac{e\hbar}{2m}\sqrt{l(l+1)}. \tag{7.20}$$

According to this equation, for $l = 0$, $\mu_l = 0$; for $l = 1$, $\mu_l = \sqrt{2}\,e\hbar/2m$; for $l = 2$, $\mu_l = \sqrt{6}\,e\hbar/2m, \ldots$. The quantity $e\hbar/2m$ consists of only universal constants and is used as a unit for magnetic moment. It is called the *Bohr magneton*, and is equal to[8]

$$\begin{aligned}\mu_\beta = \frac{e\hbar}{2m} = \frac{eh}{4\pi m} &= 0.927 \times 10^{-23}\ \text{joule/weber/m}^2\\ &= 0.927 \times 10^{-20}\ \text{erg/gauss}.\end{aligned} \tag{7.21}$$

Equations (7.19) or (7.20) may also be written as

$$\boldsymbol{\mu}_l = -\frac{e}{2m}\,\mathbf{L} \tag{7.22}$$

or

$$\boldsymbol{\mu}_l = -\frac{\mu_\beta}{\hbar}\,\mathbf{L}. \tag{7.23}$$

Because of the negative charge of the electron, the direction of $\boldsymbol{\mu}_l$ is opposite to that of **L**. According to these equations, $\boldsymbol{\mu}_l/\mathbf{L}$ is independent of the shape of the orbit of the electron, so these results hold good for any shape orbit, even though we assumed a circular orbit. We get the same results if we assume that the electron is moving in an elliptical orbit. We can also write Eq. (7.23) as

$$\frac{\boldsymbol{\mu}_l/\mu_\beta}{\mathbf{L}/\hbar} = -1. \tag{7.24}$$

The left-hand side of this equation is *the ratio of the magnetic dipole moment in units of Bohr magnetons to the orbital angular momentum in units of* $\hbar$, and is equal to minus one in this case. This ratio is called the *gyromagnetic ratio.* To make it more general, we may define this ratio to be equal to g_l instead of -1. That is,

$$\boxed{\frac{\boldsymbol{\mu}_l/\mu_\beta}{\mathbf{L}/\hbar} = g_l,} \tag{7.25}$$

where $g_l = -1$, in this case of orbital motion of an electron in an atom, and may be different for different types of motions. g_l is called the *Landé g factor* or the *g-factor* or the *gyromagnetic ratio.* Thus Eq. (7.25) may be written as

$$\boldsymbol{\mu}_l = \frac{g_l\mu_\beta}{\hbar}\mathbf{L} \tag{7.26}$$

or

$$\boxed{\mu_l = g_l\mu_\beta\sqrt{l(l+1)},} \tag{7.27}$$

with $g_l = -1$.

B) Magnetic dipole moment due to spinning motion of the electron

An electron spinning about its own axis with an angular momentum $S = \sqrt{s(s+1)}\,\hbar = \sqrt{\frac{1}{2}(\frac{1}{2}+1)}\,\hbar = (\sqrt{3}/2)\hbar$ should behave like a tiny magnet. Also the electron spinning about its own axis may be replaced by an equivalent current-carrying loop, shown by the dashed line in Fig. 7.8. Hence the atom should have a magnetic dipole moment $\boldsymbol{\mu}_s$. Again because of the negative charge, the direction of $\boldsymbol{\mu}_s$ should be opposite to that of S. When we don't know the structure and the charge distribution of an electron, this prevents us from calculating $\boldsymbol{\mu}_s$ the same way we calculated $\boldsymbol{\mu}_l$. But from the definition of gyromagnetic ratio given in Eq. (7.25) for orbital motion, we may write a similar equation for the spinning motion:

$$\boxed{\frac{\boldsymbol{\mu}_s/\mu_\beta}{\mathbf{S}/\hbar} = g_s,} \tag{7.28}$$

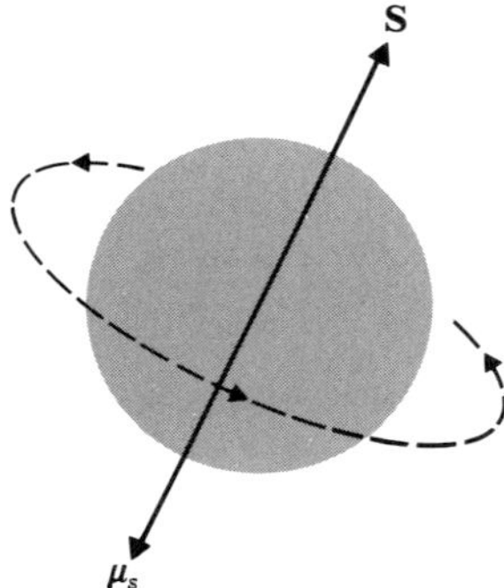

Fig. 7.8 An electron spinning about its own axis may be visualized as being equivalent to a current-carrying loop, shown by a dashed curve.

where g_s is the *spin g factor*. We may write this as

$$\boldsymbol{\mu}_s = \frac{g_s \mu_\beta \mathbf{S}}{\hbar} \tag{7.29}$$

or

$$\boxed{\mu_s = g_s \mu_\beta \sqrt{s(s+1)},} \tag{7.30}$$

where $\mu_\beta = e\hbar/2m = eh/4\pi m$. Since μ_s can be measured experimentally, the only unknown is g_s. All experiments indicate that, for an electron, g_s must be assigned a value[8] of -2 (or, to be more precise, $g_s = -2.002319230$).

Thus for $s = \frac{1}{2}$, $g_s = -2$, from Eq. (7.30), we get

$$\boxed{\mu_s = -2\mu_\beta \sqrt{\tfrac{1}{2}(\tfrac{1}{2}+1)} = -\sqrt{3}\,\mu_\beta = -1.73\mu_\beta.} \tag{7.31}$$

⋆C) Magnetic dipole moment due to total angular momentum

If an electron in an atom has both orbital angular momentum **L** and spin angular momentum **S**, we must consider the total dipole moment $\boldsymbol{\mu}_j$ (instead of the individual $\boldsymbol{\mu}_l$ and $\boldsymbol{\mu}_s$) of the atom, corresponding to the total angular momentum **J** of the valence electron, where

$$\mathbf{J} = \mathbf{L} + \mathbf{S}. \tag{7.32}$$

Once again, we shan't go into the detailed derivation of $\boldsymbol{\mu}_j$, but will make use of the definition of gyromagnetic ratio given by Eq. (7.25). That is, the ratio of the magnetic dipole moment $\boldsymbol{\mu}_j$ in units of μ_β to the total angular momentum **J** in units of $\hbar$

is equal to the Landé g factor,

$$\frac{\boldsymbol{\mu}_j/\mu_\beta}{\mathbf{J}/\hbar} = g_j. \tag{7.33}$$

That is,

$$\boldsymbol{\mu}_j = \frac{g_j\mu_\beta}{\hbar}\mathbf{J} \tag{7.34}$$

or

$$\boxed{\mu_j = g_j\mu_\beta\sqrt{j(j+1)},} \tag{7.35}$$

where the Landé g_j factor can be calculated, and is given by[9]

$$g_j = 1 + \frac{j(j+1) + s(s+1) - l(l+1)}{2j(j+1)}. \tag{7.36}$$

Thus, in summary, we have three different cases.

a) If $l \neq 0$, but $s = 0$,

$$\mu_l = g_l\mu_\beta\sqrt{l(l+1)}, \qquad \text{where } g_l = -1. \tag{7.37a}$$

b) If $l = 0$, but $s \neq 0$,

$$\mu_s = g_s\mu_\beta\sqrt{s(s+1)}, \qquad \text{where } g_s = -2. \tag{7.37b}$$

c) If $l \neq 0$, $s \neq 0$, and $j \neq 0$,

$$\mu_j = g_j\mu_\beta\sqrt{j(j+1)}, \tag{7.37c}$$

where g_j is given by Eq. (7.36).

D) Effect of an external magnetic field on the atom

When we say that there is interaction between an atom and a magnetic field, we mean that there is a change in energy, ΔE, of the electron (or atom) placed in the magnetic field. Let us now calculate this change in energy. An electron which is in an orbit around a nucleus is equivalent to an electric dipole of magnetic moment $\boldsymbol{\mu}_l$ which interacts with an applied magnetic field $\mathbf{B}$. Since $\mathbf{B}$ is uniform, there are no net forces acting on the dipole, and hence no translational motion. Figure 7.9 shows the lines of action of an equivalent dipole of a bar magnet. Note that they do not pass through the same point. Therefore there is a net torque $\boldsymbol{\tau}$ acting on the dipole (which leads to rotational motion). This torque is defined as the rate of change of angular momentum

$$\boldsymbol{\tau} = \frac{d\mathbf{L}}{dt}, \tag{7.38}$$

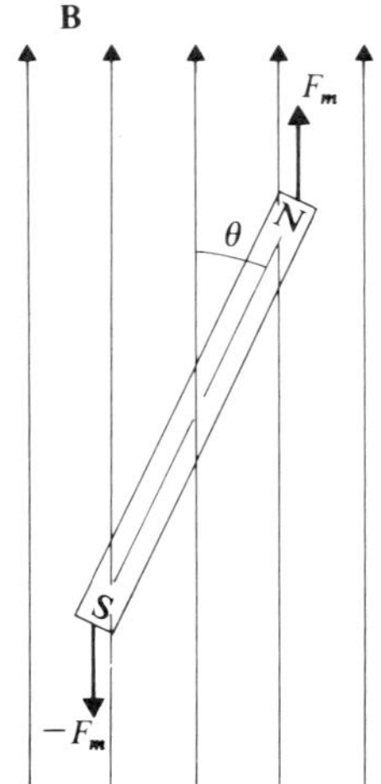

Fig. 7.9 A magnetic bar, equivalent to an atom, placed in an external magnetic field **B**. Two equal and opposite forces, F_m and $-F_m$, give rise to a torque, resulting in a twist of the magnet by an angle θ.

where

$$\tau = \boldsymbol{\mu}_l \times \mathbf{B} \tag{7.39}$$

or

$$\tau = \mu_l B \sin \theta, \tag{7.40}$$

where θ is as shown in Fig. 7.9. Because of this net torque, the change in energy ΔE which results from the rotation is a completely orientational potential energy. It is usual to assume that when $\boldsymbol{\mu}_l$ is at 90° to **B**, $\Delta E = 0$, and to determine the potential energy at any angle θ by calculating the amount of external work which must be done to rotate the dipole from $\theta = 90°$ (corresponding to $\Delta E = 0$) to some angle θ. That is,

$$\Delta E = \int_{90°}^{\theta} \tau \, d\theta. \tag{7.41}$$

Substituting for τ from Eq. (7.40), we obtain

$$\Delta E = \int_{90°}^{\theta} \mu_l B \sin \theta \, d\theta = \mu_l B \int_{90°}^{\theta} \sin \theta \, d\theta$$

$$= -\mu_l B \cos \theta \tag{7.42}$$

or

$$\Delta E = -\boldsymbol{\mu}_l \cdot \mathbf{B}. \tag{7.43}$$

Thus for $\theta = 90°$, $\Delta E = 0$; for $\theta = 0°$, $\Delta E = -\mu_l B$; and for $\theta = 180°$, $\Delta E = \mu_l B$. That is, $\theta = 0°$ corresponds to a position of minimum energy while

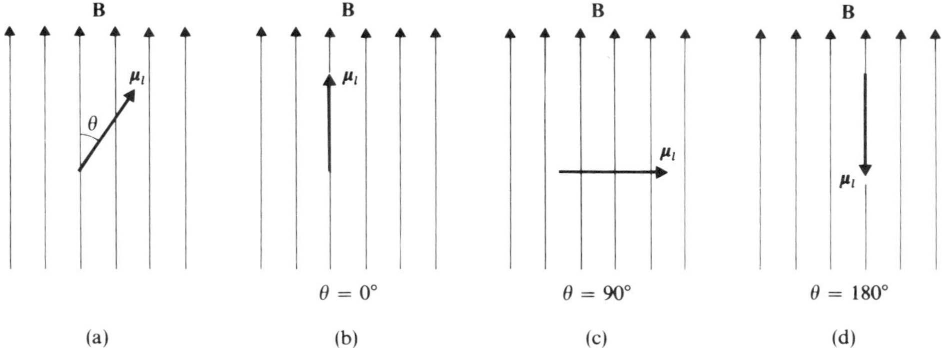

Fig. 7.10 The orientational potential energy ΔE of an atom placed in a magnetic field **B** is equal to: (a) $-\mu_l B \cos\theta$ when μ_l makes an angle θ with the magnetic field, (b) $-\mu_l B$ when $\theta = 0$, (c) 0 when $\theta = 90°$, (d) $+\mu_l B$ when $\theta = 180°$. [Note that this is a classical picture, because, according to quantum mechanics, perfect alignment at $\theta = 0°$ and $\theta = 180°$ is not possible.]

$\theta = 180°$ corresponds to a position of maximum energy. Figure 7.10 shows the different situations. When a uniform magnetic field is applied to an atom, the dipole of magnetic moment $\boldsymbol{\mu}_l$ tries to align itself along the applied magnetic field **B**, so that the system may have minimum energy. This tendency to align results in rotational energy, which must be dissipated. But there is no process available by which it can be, so the tendency of $\boldsymbol{\mu}_l$ to align itself with **B** results in the precession of $\boldsymbol{\mu}_l$ around **B**, keeping θ and ΔE fixed. Since **L** is quantized, so is $\boldsymbol{\mu}_l$. Therefore $\boldsymbol{\mu}_l$ precesses around **B** only at certain specific angles.

Thus when a charged body which has a given angular momentum is placed in a uniform external magnetic field, the angular momentum vector $\mathbf{L}(\propto \boldsymbol{\mu}_l)$ starts precessing with a uniform angular frequency around the axis of the magnetic field, just the way a mechanical top precesses in a gravitational field.

Figure 7.11 shows the precession of an angular momentum vector **L** of an electron in an atom when the atom is placed in a uniform magnetic field **B**. This is called *Larmor precession.* The relation expressing the frequency of precession is called *Larmor's theorem.*[10] According to this theorem, the angular frequency of precession, or the Larmor precessional angular frequency ω_L, is related to different quantities by the equation

$$\omega_L = \frac{d\Omega}{dt} = g_l \frac{e}{2m} B, \tag{7.44}$$

where $d\Omega$ is as defined in Fig. 7.11. For the orbital motion of an electron, $g_l = 1$. Therefore (ignoring the negative sign),

$$\omega_L = \frac{e}{2m} B. \tag{7.45a}$$

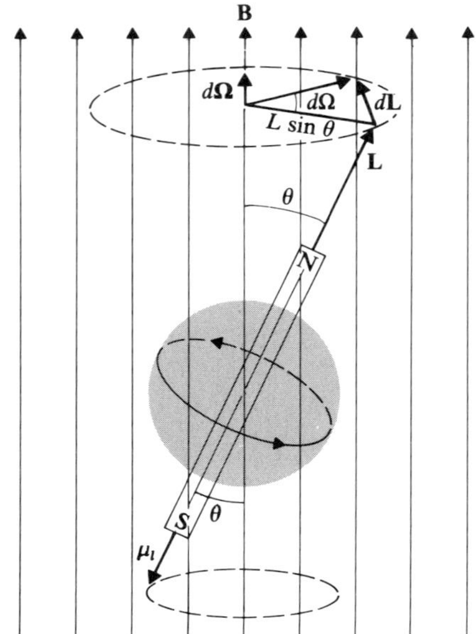

Fig. 7.11 The angular motion of the electron in an atom in a magnetic field is equivalent to a bar magnet in which the torque causes the angular momentum vector to precess with an angular frequency $\omega_L = d\Omega/dt$ about the axis at an angle θ with the externally applied magnetic field.

Or the frequency ν_L, where $\omega_L = 2\pi\nu_L$, is

$$\nu_L = \frac{e}{4\pi m} B. \tag{7.45b}$$

Equation (7.44) is the statement of Larmor's theorem. Note that ω_L or ν_L is independent of $\mathbf{L}$, and is a function of the charge and mass of the particle and of the magnitude of the applied field B. If $B = 0$, $\omega_L = 0$; also, if $e = 0$, $\omega_L = 0$.

Similar arguments may be extended to $\boldsymbol{\mu}_s$ and $\boldsymbol{\mu}_j$. Thus, in summary, from Eqs. (7.42) and (7.43), we have

$$\Delta E_L = -\boldsymbol{\mu}_l \cdot \mathbf{B} = -\mu_l B \cos\theta \tag{7.46a}$$

$$\Delta E_S = -\boldsymbol{\mu}_s \cdot \mathbf{B} = -\mu_s B \cos\theta \tag{7.46b}$$

$$\Delta E_J = -\boldsymbol{\mu}_j \cdot \mathbf{B} = -\mu_j B \cos\theta \tag{7.46c}$$

The corresponding frequencies of precession around **B** are

$$\omega_L = g_l \frac{e}{2m} B \tag{7.47a}$$

$$\omega_S = g_s \frac{e}{2m} B \tag{7.47b}$$

$$\omega_J = g_j \frac{e}{2m} B \tag{7.47c}$$

Since $g_l = 1$ and $g_s = 2$, $\omega_S = 2\omega_L$. That is, **S** precesses around **B** twice as fast as **L** does, while ω_J depends on g_j, given by Eq. (7.36).

We shall make use of these results to prove the space quantization of **L**, **S**, and **J**.

7.5 THE ZEEMAN EFFECT

Now we can prove the space quantization of **L**, **S**, and **J** directly, by showing what happens when we apply magnetic fields.

a) Experimental observation of the quantization of orbital angular momentum **L** by placing atoms in a uniform magnetic field is called the *normal Zeeman effect*.

b) Experimental observation of the quantization of total angular momentum **J** by placing atoms in a uniform magnetic field is called the *anomalous Zeeman effect*.

c) Experimental observation of the quantization of spin angular momentum **S** by placing atoms in a *non*uniform magnetic field is called the *Stern–Gerlach experiment*.

We shall discuss the first of these effects in this section, and the Stern–Gerlach experiment in Section 7.6. The anomalous Zeeman effect is more complex and will not be discussed here.

The normal Zeeman effect[11]

Let us consider an atom which has only orbital angular momentum **L**, but no spin angular momentum. That is, $\mathbf{S} = 0$ or $s = 0$. [In the case of atoms having only one valence electron, s is $\frac{1}{2}$; hence $S \neq 0$. But for atoms with two valence electrons, each electron has $s = \frac{1}{2}$; hence the sum of the two will be either 0 or 1.] If such an atom is placed in a uniform magnetic field **B** applied along the Z axis, the change in energy ΔE_L according to Eq. (7.46a) is

$$\Delta E_L = -\boldsymbol{\mu}_l \cdot \mathbf{B} = -\mu_l B \cos\theta, \tag{7.48}$$

where, from Eq. (7.22),

$$\boldsymbol{\mu}_l = -\frac{e}{2m}\mathbf{L}. \tag{7.22}$$

Combining these two equations, we get

$$\Delta E_L = \frac{e}{2m}\mathbf{L}\cdot\mathbf{B} = \frac{e}{2m}LB\cos\theta. \tag{7.49}$$

But $L\cos\theta = L_z$, where L_z is the z component of L, so we get

$$\Delta E_L = \frac{e}{2m}L_z B. \tag{7.50}$$

Or, since $L_z = m_l\hbar$ and $e\hbar/2m = \mu_\beta$, we may write Eq. (7.50) as

$$\Delta E_L = m_l\mu_\beta B. \tag{7.51}$$

Thus, if an atom is in a quantum state corresponding to orbital quantum number l and energy E_0, when it is placed in a magnetic field B, it has an energy E given by

$$\boxed{E = E_0 + \Delta E_L = E_0 + m_l\mu_\beta B.} \tag{7.52}$$

If l has $(2l + 1)$-fold degeneracy—that is, if for a given l, m_l can take $(2l + 1)$ values: $-l, -(l - 1), \ldots, -1, 0, 1, \ldots, (l - 1), l$—then, from Eq. (7.52), E can also take $(2l + 1)$ values. Thus, if $l = 0$, $m_l = 0$, and $E = E_0$. For $l = 1$, $m_l = -1, 0, +1$, and E can take three values:

$$E = E_0 - \mu_\beta B, \qquad E = E_0, \qquad E = E_0 + \mu_\beta B.$$

Similarly, $l = 2$ results in five values of E. Figure 7.12 shows the splittings for the $l = 0$ (S state), $l = 1$ (P state), and $l = 2$ (D state).

When we are observing the orbital angular momentum $\mathbf{L}$, using the normal Zeeman effect, we get our information about the splitting of these levels by looking at the transitions between the split levels. Thus if, in the absence of a magnetic field, we observe a transition of frequency ν_0 between the initial energy level E_0^i and the final energy level E_0^f, then

$$h\nu_0 = E_0^i - E_0^f. \tag{7.53}$$

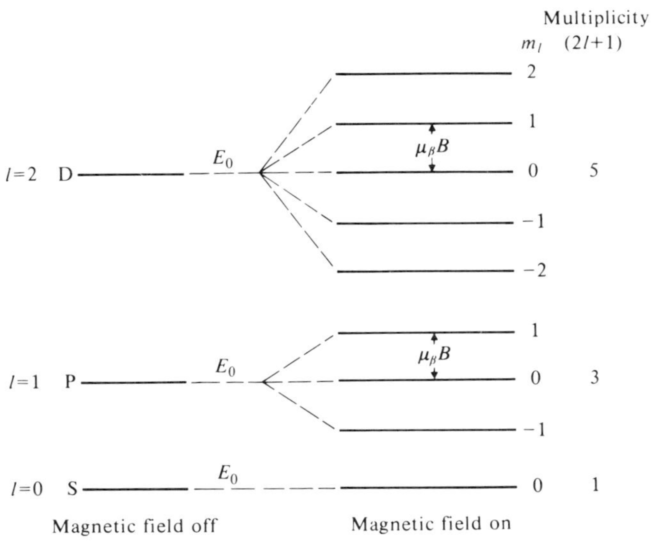

Fig. 7.12 On the left are energy levels of an atom in the absence of a magnetic field. In the *presence* of a magnetic field, as on the right, each level splits into $(2l + 1)$ levels.

Now suppose we apply a uniform magnetic field **B**; it will cause splitting of the initial and final levels, and according to Eq. (7.52),

$$E^i = E_0^i + \Delta E_L^i = E_0^i + m_l^i \mu_\beta B, \tag{7.54}$$

$$E^f = E_0^f + \Delta E_L^f = E_0^f + m_l^f \mu_\beta B. \tag{7.55}$$

Subtracting and denoting by ν the frequency of any transition between E^i and E^f, we get

$$h\nu = E^i - E^f = (E_0^i - E_0^f) + (m_l^i - m_l^f)\mu_\beta B. \tag{7.56}$$

Let

$$\Delta m_l = m_l^i - m_l^f. \tag{7.57}$$

Therefore

$$h\nu = h\nu_0 + \Delta m_l \mu_\beta B$$

or

$$\boxed{\nu = \nu_0 + \Delta m_l \frac{eB}{4\pi m}.} \tag{7.58}$$

According to the selection rules, however, not all the transitions between E^i and E^f are allowed—only those for which Δm_l changes by 0 or ± 1. That is,

$$\boxed{m_l^i - m_l^f = \Delta m_l = 0, \pm 1.} \tag{7.59}$$

From Eqs. (7.58) and (7.59), we may conclude that, in the presence of the magnetic field **B**, the original transition (the field-free transition of frequency ν_0) is replaced by three transitions of the following frequencies:

$$\nu = \nu_0 - \left(\frac{eB}{4\pi m}\right), \qquad \nu = \nu_0, \qquad \nu = \nu_0 + \left(\frac{eB}{4\pi m}\right). \tag{7.60}$$

Figure 7.13(a) shows the transitions between the P and the S states brought about by the normal Zeeman effect. The values determined by experiment agree with the values calculated from Eq. (7.60), thereby verifying the space quantization of **L**. Figure 7.13(b) shows the normal Zeeman transitions between the D ($l = 2$) state and the P ($l = 1$) state. The selection rules of Eq. (7.59) allow nine transitions, but the magnitudes of the separations are such that we can observe only three groups (each group containing three transitions) of three different frequencies. We would expect this from Eq. (7.60). Figure 7.13(c) shows a photograph of such a normal triplet. In this case we can see that degeneracy is partially removed. The anomalous Zeeman effect, a more sophisticated test of space quantization, which removes this partial degeneracy as well, will not be discussed.

Example 7.1. One of the most prominent spectral lines of calcium is the one with wavelength $\lambda = 4226.73$ Å [from P ($l = 1$) to S ($l = 0$)]. Calcium atoms exhibit normal Zeeman patterns when placed in a uniform magnetic field of 4 webers/m^2. Calculate the wavelengths of the three components of a normal Zeeman pattern and the separation between them.

According to Eq. (7.60), the frequencies ν of the three components are

$$\nu_1 = \nu_0 - \frac{eB}{4\pi m}, \qquad \nu_2 = \nu_0, \qquad \nu_3 = \nu_0 + \frac{eB}{4\pi m}.$$

Thus the separation among the three components is

$$\Delta\nu = \nu_2 - \nu_1 = \nu_3 - \nu_2 = \frac{eB}{4\pi m}$$

or

$$\Delta\nu = \frac{e}{m}\frac{B}{4\pi} = 1.7588 \times 10^{11}\,\frac{\text{coul}}{\text{kg}}\,\frac{1}{4\pi}\,4\,\frac{\text{webers}}{\text{m}^2}$$

$$= 0.56 \times 10^{11}\ \text{cycles/sec.}$$

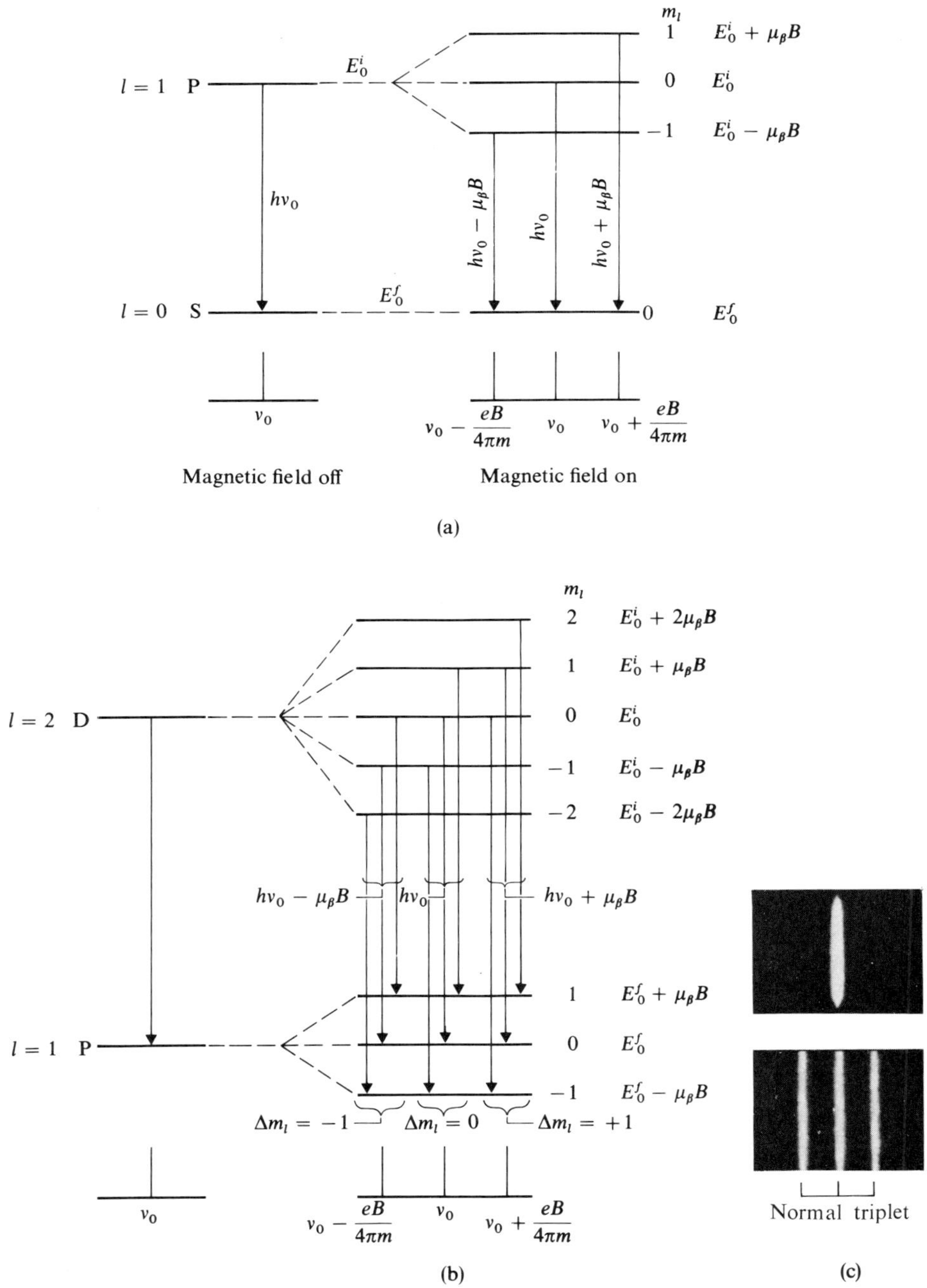

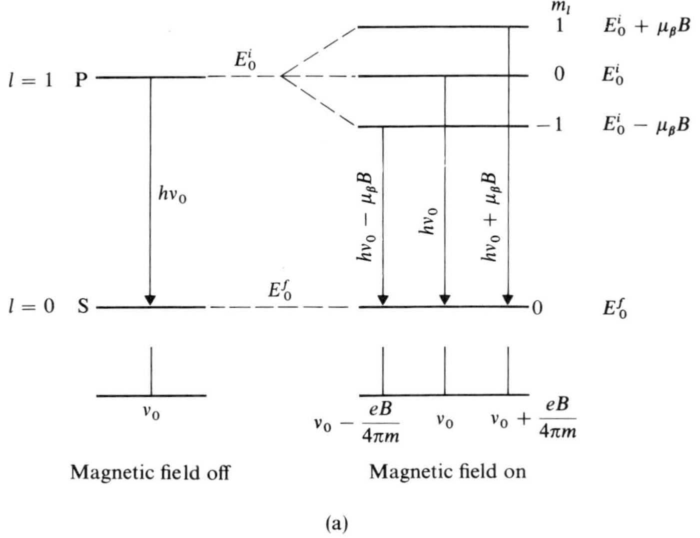

Fig. 7.13 (a) The normal Zeeman splitting of a transition between the P ($l = 1$) state and the S ($l = 0$) state. (b) The normal Zeeman splitting of a transition between the D ($l = 2$) state and the P ($l = 1$) state. Note that the degeneracy is partially removed, and we observe only three transitions instead of nine. (c) Photograph of normal Zeeman effect in a zinc singlet when observed in a magnetic field results in a triplet. [From H. E. White, *Introduction to Atomic Spectra*, page 152, New York: McGraw-Hill, 1934]

Since

$$\nu = \frac{c}{\lambda} \qquad \text{or} \qquad d\nu = -\frac{c}{\lambda^2}\, d\lambda,$$

$$|d\lambda| = \frac{\lambda^2}{c}\, d\nu = \frac{(4226.73 \times 10^{-10}\ \text{m})^2}{3 \times 10^8\ \text{m/sec}}\, 0.56 \times 10^{11}\ \text{cycles/sec},$$

$$|d\lambda| = 0.33 \times 10^{-10}\ \text{m} = 0.33\ \text{Å}.$$

Thus the wavelengths of the three normal Zeeman components are

$$6226.40\ \text{Å}, \qquad 6226.73\ \text{Å}, \qquad 6227.06\ \text{Å}.$$

Note that when we measure $\Delta\nu$, knowing B, to calculate e/m we can use the expression

$$\Delta\nu = \frac{e}{m}\frac{B}{4\pi}.$$

7.6 THE STERN–GERLACH EXPERIMENT

In 1921, Otto Stern and Walter Gerlach[12] devised an experiment which directly proved the space quantization of spin angular momentum. They did this without taking into consideration the transitions of electrons from one energy level to another, which we encounter in the Zeeman effect. Their method consisted of placing an atom for which $\mathbf{L} = 0$ but $\mathbf{S} \neq 0$ in an inhomogeneous magnetic field. Let's look at how this works.

An electron that has an intrinsic angular momentum also has an intrinsic magnetic moment. So we can treat an atom having such an electron as a small bar magnet (see Fig. 7.14). If we place such a dipole magnet in a uniform magnetic field, as in Fig. 7.14(a), the forces at the two poles are equal and opposite, and constitute a couple. The result is a rotational motion of the atomic magnet, but no translational motion. To produce a translational motion, we place the dipole magnet in an inhomogeneous magnetic field, as shown in Fig. 7.14(b) or (c). Some of these atomic dipole magnets find themselves in a situation like that in Fig. 7.14(b), in which the upward force is greater than the downward force (since $B_1 > B_2$). This makes the dipole drift along the $+Z$ axis. At the same time, those atoms finding themselves in a situation like that in Fig. 7.14(c) drift along the $-Z$ axis. This is shown mathematically as follows: Let us calculate the force acting on an atomic dipole for which $\mathbf{L} = 0$, but $\mathbf{S} \neq 0$. From Eq. (7.46b), we know that the change in energy ΔE_S when such an atom is at a point at which the field is $\mathbf{B}$ is

$$\Delta E_S = -\boldsymbol{\mu}_s \cdot \mathbf{B} = -\mu_s B \cos\theta,$$

where

$$\boldsymbol{\mu}_s = -g_s \frac{e}{2m}\mathbf{S}.$$

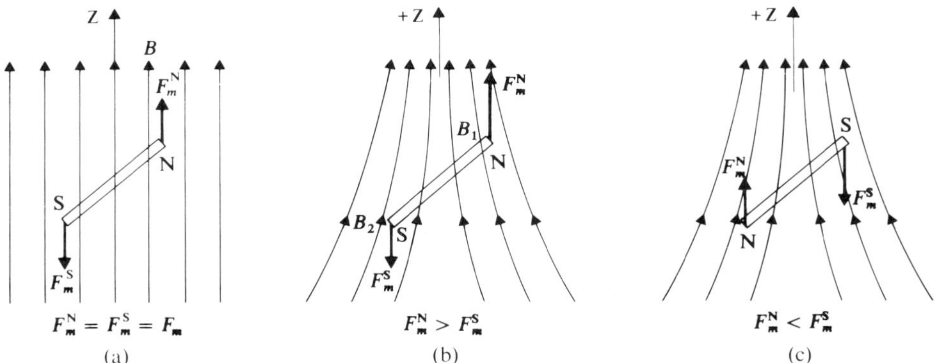

Fig. 7.14 A magnetic-bar equivalent of an atom (with $L = 0$ and $S \neq 0$) in a magnetic field. (a) In a uniform magnetic field there is only rotational motion, no translational motion. In an inhomogeneous magnetic field there is translational motion. In (b), the bar magnet moves along $+Z$-axis. In (c), it moves along $-Z$-axis.

Combining the above equations, we get

$$\Delta E_S = g_s \frac{e}{2m} \mathbf{S} \cdot \mathbf{B} = g_s \frac{e}{2m} SB \cos\theta. \tag{7.61}$$

But $S \cos\theta = S_z = m_s\hbar$; therefore

$$\boxed{\Delta E_S = g_s m_s \frac{e\hbar}{2m} B.} \tag{7.62}$$

Remember that this atom is in an inhomogeneous field, so the net force is equal to the derivative of ΔE_S with respect to z. The Z axis is the direction of symmetry of the inhomogeneous field and provides the axis for space quantization of spin. That is,

$$F_z = -\frac{d}{dz}(\Delta E_S) \tag{7.63}$$

or

$$\boxed{F_z = -g_s m_s \frac{e\hbar}{2m} \frac{dB}{dz},} \tag{7.64}$$

where dB/dz is the inhomogeneity of the magnetic field. Thus, assuming that $dB/dz > 0$, $s = \frac{1}{2}$, $m_s = +\frac{1}{2}$ will make the atom drift along the $-Z$ axis and $m_s = -\frac{1}{2}$ will make the atom drift along the $+Z$ axis. Since the atomic dipoles in a beam of atoms are randomly oriented, such a beam, after passing through an inhomogeneous magnetic field, is split in two.

Figure 7.15 shows the experimental arrangement. A beam of atoms of silver (Ag) obtained from oven O is collimated by slit system S. It passes between the poles of a magnet and is received on a photographic plate P. The shape of the pole faces is such that the magnet provides the required inhomogeneous field. The figure shows the difference between the impression on the photographic plate with the applied magnetic field and without it.

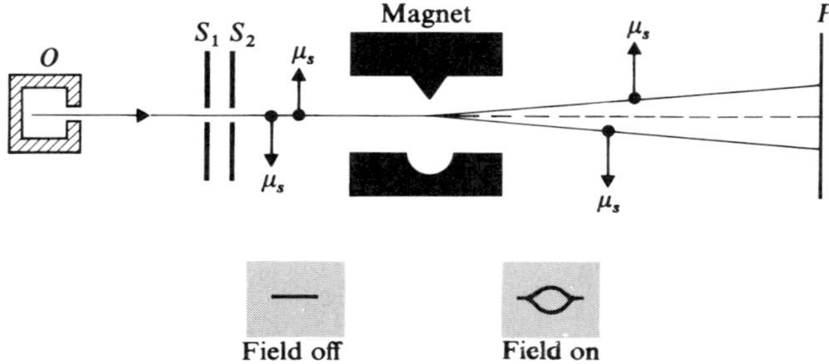

Field off

Field on

Fig. 7.15 The experimental arrangement of Stern and Gerlach for observing space quantization of intrinsic spin angular momentum of an atom. In the presence of an inhomogeneous magnetic field, the beam of silver atoms splits into two, as shown.

For silver, the ground-state configuration is $5^2S_{1/2}$, that is, $l = 0, j = s = \frac{1}{2}$, and hence $m_s = \pm\frac{1}{2}$. According to the results in the figure, the multiplicity is 2. Thus $2s + 1 = 2$ or $s = \frac{1}{2}$, the assumed spin quantum number for the electron. This verifies not only the space quantization of **S**, but also proves that $s = \frac{1}{2}$.

Even though the original experiment was used to prove the space quantization of **S**, scientists can (and often do) use it to prove the space quantization of **L** and **J**. Actually, by counting the number of splittings, we can find the values of l and j for unknown cases.

Example 7.3. Suppose that, in a Stern–Gerlach experiment, a beam of silver atoms is obtained from an oven heated to a temperature of 1500°K. This beam passes through an inhomogeneous magnetic field having a field gradient of 20,000 gauss/cm (10^4 gauss $=$ 1 Wb/m^2) perpendicular to the beam. The pole faces are 10 cm long. What is the separation between the two components of the split beam on a photographic plate placed at a distance of 50 cm? (See the accompanying figure.)

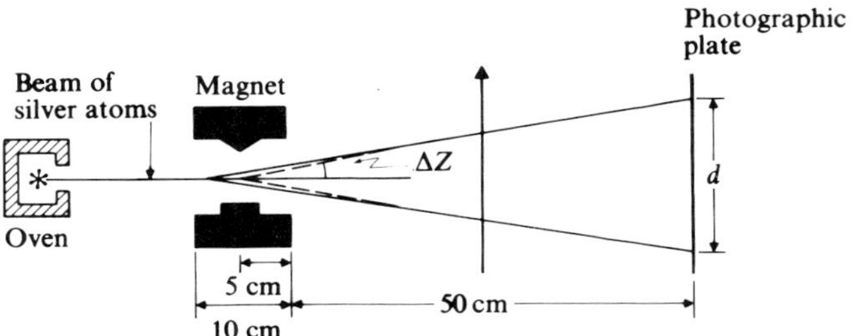

According to Eq. (7.64), (ignoring the minus sign),

$$F_z = g_s m_s \frac{eh}{4\pi m}\frac{dB}{dZ} = g_s m_s \mu_\beta \frac{dB}{dZ}. \tag{7.65}$$

Therefore the acceleration a of the silver atom of mass M in the Z direction is

$$a = \frac{F_z}{M} = \frac{g_s m_s \mu_\beta (dB/dZ)}{M}. \tag{7.66}$$

Thus the deflection ΔZ in the Z direction is

$$\Delta Z = \tfrac{1}{2}at^2, \tag{7.67}$$

where t is the time the silver atoms take to go through the magnetic field produced by pole faces of length l. That is,

$$t = \frac{l}{v}, \tag{7.68}$$

where v is the velocity of the silver atoms (assumed constant). Substituting for a and t from Eqs. (7.66) and (7.68) into (7.67), we get

$$\Delta Z = \frac{1}{2}\frac{g_s m_s \mu_\beta}{M}\frac{dB}{dZ}\frac{l^2}{v^2}. \tag{7.69}$$

According to the kinetic theory of gases, atoms at T°K have kinetic energy $\frac{3}{2}kT$, where k is Boltzmann's constant. Thus

$$\tfrac{1}{2}Mv^2 = \tfrac{3}{2}kT. \tag{7.70}$$

Substituting for Mv^2 in Eq. (7.69), we obtain

$$\begin{aligned}\Delta Z &= \frac{1}{2}\frac{g_s m_s \mu_\beta l^2}{3kT}\frac{dB}{dZ}\\ &= \frac{1}{2}\frac{2(\pm\frac{1}{2})(0.93\times 10^{-23}\ \text{J/Wb/m}^2)\ (10\ \text{cm})^2}{3(1.38\times 10^{-23}\ \text{J/°K})(1500\text{°K})}\ (2\text{Wb/m}^2/\text{cm})\\ &= 0.015\ \text{cm}.\end{aligned}$$

ΔZ, as shown in the figure, is the deflection the beam undergoes as it leaves the magnet. The deflection d at the photographic plate is given by

$$\frac{d}{2\,\Delta Z} = \frac{55 \text{ cm}}{5 \text{ cm}} = 11,$$

$$d = 11 \times 2Z = 11 \times 2 \times 0.015 \text{ cm} = 0.330 \text{ cm}.$$

II. CHARACTERISTIC X-RAY SPECTRA AND ATOMIC STRUCTURE

7.7 PRODUCTION OF CHARACTERISTIC X-RAY SPECTRA

For some time now we've been talking about optical spectra. In Chapter 2 we discussed continuous x-ray spectra and briefly mentioned characteristic x-ray spectra. These three types of spectra—continuous, optical, and characteristic—are interrelated in many ways. Let us now discuss them briefly before we go into the details of the production and properties of characteristic x-rays.

Figure 7.16 shows a typical energy-level diagram of an atom of cadmium, for which $Z = 48$. Note that we are using the x-ray notation according to which $n = 1, 2, 3, 4, 5, \ldots$, corresponding to K, L, M, N shells, respectively. The three spectra are produced in the following ways.

a) *Optical spectra* result when the valence electrons, like those in the outermost shell (such as the O shell in Fig. 7.16) are removed, or are excited to higher energy states. Then, when these valence electrons are de-excited, or fall back to a lower energy state, optical spectra result. The wavelengths of optical transitions are between $\sim$4000 Å to 7000 Å.

b) *Characteristic x-ray spectra* are produced whenever electrons from innermost shells, say the K or L shell of a heavy element (see Fig. 7.16) are removed or excited to higher energy states. The characteristic x-ray spectra consist of discrete lines. These lines form a definite pattern which is *characteristic* of the element under investigation.

c) *Continuous x-ray spectra* are produced when fast-moving electrons in the coulomb field of heavy atoms are slowed down. The range of wavelengths of continuous spectra is between $\sim$0.01 Å and 10 Å, the same as that of the characteristic x-rays. The minimum wavelength (or maximum frequency) depends only on the operating voltage of the x-ray tube.

Two other commonly encountered phenomena with which we ought to be familiar are the following.

i) *Resonance radiation.* An atom absorbs a photon of energy $h\nu$ and then is de-excited by emitting a photon of the same energy. This process is called

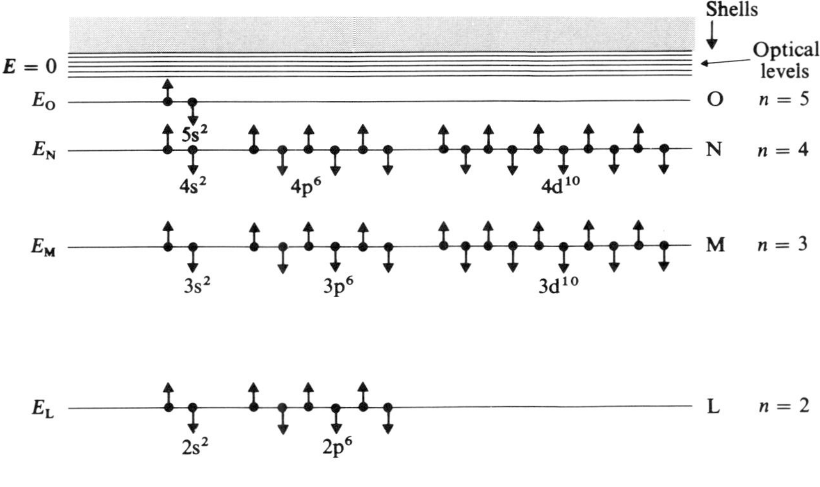

E_K K $n = 1$

1s²

(a)

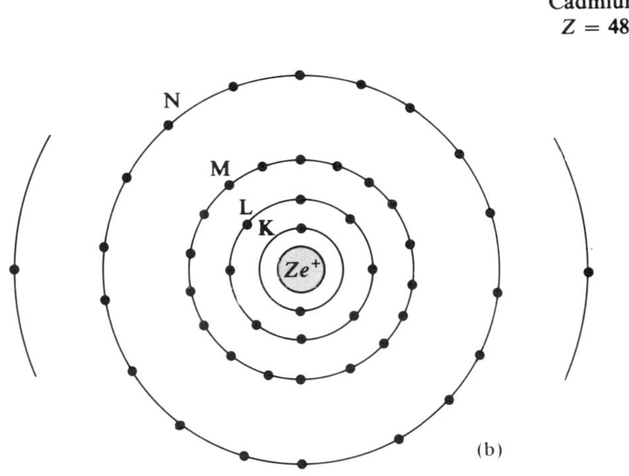

(b)

Fig. 7.16 (a) Schematic energy-level diagram of cadmium ($Z = 48$). The K, L, M, N, O levels are x-ray levels, while the levels near the continuum are optical levels. Excitation of outer electrons, such as ones from the O level, leads to the optical spectrum, while excitation or removal of inner electrons, such as ones from the K, L, M, or N levels, leads to x-ray spectra. (b) Simplified Bohr-type representation of orbiting electrons in cadmium.

resonance. It takes place when the energy of the incident photon is such that the atom can jump up only to the first excited state. In this case, when the atom is in the process of de-excitation, it has no other choice but to emit a photon of the same energy.

ii) *Fluorescence radiation.* The energy of the incident photon is such that the atom that absorbs it jumps up to the second or third excited state. Then, when it is in the process of de-excitation, the atom decays by emitting two or more photons. The process by which an atom absorbs one very energetic photon and then emits two or more photons in a short interval is called *fluorescence.*

Now let's get back to characteristic x-ray spectra, x-ray energy-level diagrams, and selection rules.

In order to remove one of the electrons in the K shell (see Fig. 7.16), we must supply enough energy to raise the electron to a continuum or to one of the optical levels near the continuum, because the other levels—such as L, M, and N—are filled by electrons in heavy atoms. Since in heavy atoms the binding energies of the K or L shells are of the order of a few keV, the incident electrons or photons trying to cause the ionization of K-shell electrons must have high energies. The incident photon (or electron) must come close to the electron that is to be raised to another energy level. Therefore the probability of an electron being removed from the K shell is greater than the probability of an electron being removed from the L shell, and the probability of an electron being removed from the L shell is greater than the probability of one being removed from the M shell, and so on. The reason is that K electrons are mostly found in a small volume (or in an orbit with a small radius, according to the Bohr theory) surrounding the nucleus, and are more likely to be hit by a photon than N or O electrons, which are spread over a larger volume (Fig. 7.16b). Let us call E_K the binding energy of the K electron. If we supply energy E_K and remove the K electron, the atom has an excess energy of E_K. Similarly, if the electrons from the L, M, N, or O shells are removed, the atom has an excess energy E_L, E_M, E_N, respectively.

Figure 7.17 shows the energy levels corresponding to the removal of K, L, M, N, . . . , electrons from an atom. This diagram is called an x-ray energy-level diagram of an atom.

How do characteristic x-ray transitions take place? Consider an atom from which a K electron has been removed. We know that this leaves the atom with an excess energy E_K. That is, the atom is in the E_K level in Fig. 7.17. Now an electron from the L shell may jump to occupy the vacancy in the K shell. Since a K-shell electron is more tightly bound to the nucleus (i.e., it has more negative energy) than an L-shell electron, it emits excess energy in the form of a photon of energy $h\nu_{K\alpha}$, given by

$$h\nu_{K\alpha} = E_K - E_L.$$

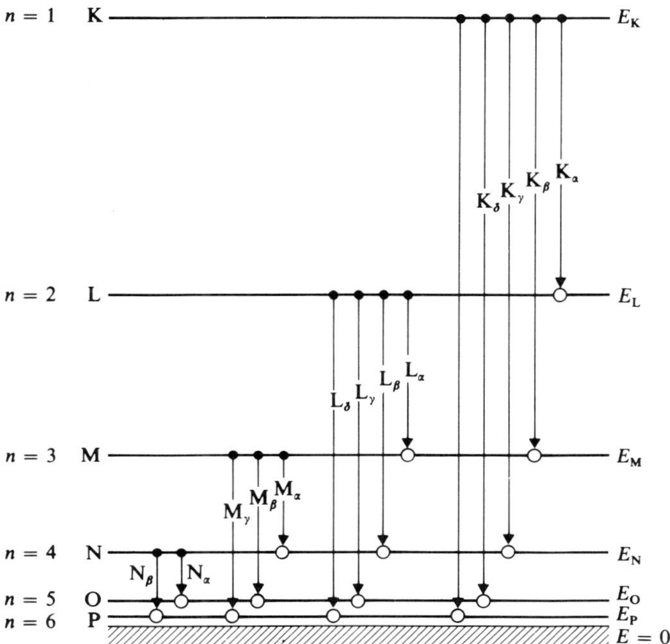

Fig. 7.17 An x-ray energy-level diagram of an atom. Unlike the optical energy-level diagram, here $E = 0$ corresponds to the ground state (instead of $E =$ the binding energy of the valence electron, as in optical representation). When an electron departs, it leaves a hole. Moving of the hole from the K to the L level results in an emission of a photon of energy, accompanied by a transition of an electron from the L to the K shell. Shown here are transitions corresponding to the K, L, M, and N series.

A more convenient way of describing the transition is as follows. Instead of saying that an *electron* jumps from the L shell to the K shell (i.e., closer to the nucleus), we say that a *hole* (i.e., a vacancy created by the K-shell electron leaving) moves from the K shell to the L shell. This transition of the hole gives rise to the emission of a photon of energy $h\nu_{K\alpha}$. If the hole moves from the K shell to the M shell, it emits a photon of energy $h\nu_{K\beta}$. Thus the holes created in the K shells of the atoms move to the lower energy levels L, M, N, O, . . . , and each time they move there is an emission of a photon, or a characteristic x-ray, as shown in Fig. 7.17.

The K series consists of the following transitions:

$$\begin{aligned} h\nu_{K\alpha} &= E_K - E_L \\ h\nu_{K\beta} &= E_K - E_M \\ h\nu_{K\gamma} &= E_K - E_N \end{aligned} \tag{7.71}$$

Similarly, if the electron was originally removed from the L shell, the hole moves to the M, N, or O shell, resulting in x-rays which constitute the L series:

L series
$$\begin{aligned} h\nu_{L\alpha} &= E_L - E_M \\ h\nu_{L\beta} &= E_L - E_N \\ h\nu_{L\gamma} &= E_L - E_O \end{aligned} \tag{7.72}$$

As shown in Fig. 7.17, we can define the M and N series in just this manner. Since the binding energies of the K, L, M shells in heavy atoms are of the order of a few keV, most characteristic x-ray transitions also naturally have energies of a few keV, as compared to a few eV in optical spectra. [Note that L consists of two subshells, M consists of three subshells, and so on. We have ignored the fine structure in the above discussion.]

Transitions of electrons (or holes) from one energy state to another must obey the following selection rules:

$$\Delta l = \pm 1, \qquad \Delta j = 0, \pm 1. \tag{7.73}$$

There is no restriction on n. The transitions for $\Delta n = 0$ are very weak, which is undoubtedly why they have not been observed.

7.8 THE MOSELEY LAW AND ATOMIC STRUCTURE

In 1913, H. G. B. Moseley[13] realized that the wavelengths of characteristic x-ray lines change smoothly as one goes from element to element. He made a systematic study of K x-ray radiation from the elements Ca, Ti, V, Cr, Mn, Fe, Co, Ni, Cu, and Zn, with $Z = 20, 22, 23, 24, 25, 26, 27, 28, 29$, and 30, respectively. He used these elements as targets in an x-ray tube of the type shown in Fig. 2.12. Then, using a single-crystal diffraction spectrometer, he analyzed the characteristic x-radiation, which appeared as discrete lines superimposed on the continuous x-ray spectrum. He found that with increasing Z, the wavelength of the x-rays decreased, and hence the frequency of the x-rays increased. Moseley made a plot of the square root of the frequency of K radiation versus Z—that is, $\sqrt{\nu_{K\alpha}}$ versus Z—and found it to be straight line, as shown in Fig. 7.18. This figure also shows the plot of $\sqrt{\nu_{K\alpha}}$ versus atomic weight. Obviously the plot of $\sqrt{\nu_{K\alpha}}$ versus Z gives a perfect fit to a straight line, so Moseley concluded that it is the atomic number Z, not the mass number A, which is the definitive quantity in classifying elements and investigating atomic structure.

Mathematically, Moseley expressed the frequencies of the K_α lines $\nu_{K\alpha}$ of different elements as

$$\boxed{\sqrt{\nu_{K\alpha}} \propto Z} \tag{7.74}$$

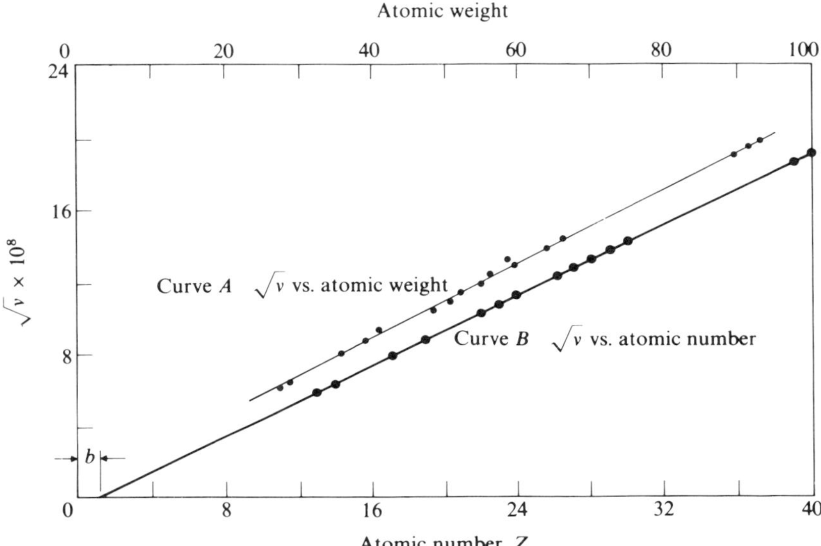

Fig. 7.18 The solid curve *B* represents Moseley's law. [From F. K. Richtmyer, E. H. Kennard, and J. Cooper, *Introduction to Modern Physics*, sixth edition, New York: McGraw-Hill, 1969]

or

$$\nu_{K\alpha} = \tfrac{3}{4}cR(Z - 1)^2, \tag{7.75}$$

where R is the Rydberg constant and c is the speed of light. From these observations he drew the following important conclusions.

a) Elements in the periodic table should be arranged according to their increasing atomic number Z, and not their atomic weight A, as had been done previously; see Table 7.2.

b) Moseley's work removed many discrepancies in the arrangement of the elements in the periodic table. For example, cobalt, with $Z = 27$ and atomic weight 58.9, should come before nickel ($Z = 28$ and atomic weight 58.7), even though the atomic weight of cobalt is more than that of nickel. (See Table 7.2.)

c) Moseley predicted many elements that were then missing from the periodic table, such as scandium (Sc, $Z = 21$) and promethium (Pm, $Z = 61$); these elements were discovered later. (See Table 7.2.)

With the invention of spectrometers of improved resolution, scientists were able to measure the frequencies of the fine structure of doublets, like $K_{\alpha 1}$, $K_{\alpha 2}$, and $K_{\beta 1}$, $K_{\beta 2}$. The Moseley law applied to these four lines also yielded straight lines when $\sqrt{\nu_K}$ was plotted against Z, but the slopes and intercepts were different. Actually, Eq. (7.75) may be written for the general case as

$$\boxed{\nu = C_n(Z - S)^2,} \tag{7.76}$$

where C_n and S are constants, and are different for different series. S is also called the *screening constant*. We can derive this relation by making use of the Bohr theory. For hydrogenlike atoms,

$$\begin{aligned} \nu &= cRZ_{\text{eff}}^2\left(\frac{1}{n_f^2} - \frac{1}{n_i^2}\right) \\ &= cR(Z - S)^2\left(\frac{1}{n_f^2} - \frac{1}{n_i^2}\right), \end{aligned} \tag{7.77}$$

where Z_{eff} has been replaced by $(Z - S)$. Now consider the special case of K x-rays, in which $n_i = 2$ and $n_f = 1$:

$$\nu_K = cR(Z - S)^2\left(\frac{1}{1^2} - \frac{1}{2^2}\right)$$

or

$$\nu_K = \tfrac{3}{4}cR(Z - S)^2. \tag{7.78}$$

When an electron from the L shell jumps to the K shell, the effective charge "seen" by the L-shell electron is $+Ze$ on the nucleus and $-1e$ on the K electron. Hence $Z_{\text{eff}} = (Z - 1)$. This leads to

$$\boxed{\nu_K = \tfrac{3}{4}cR(Z - 1)^2,} \tag{7.79}$$

which is the same as the Moseley law stated by Eq. (7.75).

The screening constants for other series have also been calculated,[4] but they are not so simple. For example, for the L series, $n_f = 2$ and $n_i = 3$, but when an electron jumps from the M shell to the L shell, even though there are 9 electrons (2 in the K shell and 7 in the L shell), the effective charge is not $(Z - 9)e$, but only $(Z - 7.4)e$. This is because electrons in the L shell are more spread out than they are in the K shell, and not all of them are screening the nucleus from the M-shell electrons as effectively as the K-shell electrons screen it. Thus, for the L series,

$$\nu_L = cR\left(\frac{1}{2^2} - \frac{1}{3^2}\right)(Z - 7.4)^2 = \tfrac{5}{36}cR(Z - 7.4)^2. \tag{7.80}$$

That is, for the L series, $S = 7.4$.

Table 7.2

Periodic Table of the Elements

Period ↓	Group →	I	II													III	IV	V	VI	VII	VIII
1	1s	1 **H** 1.00 $1s^1$																			2 **He** 4.00 $1s^2$
2	2s	3 **Li** 6.94 $2s^1$	4 **Be** 9.01 $2s^2$												2p	5 **B** 10.81 $2p^1$	6 **C** 12.01 $2p^2$	7 **N** 14.01 $2p^3$	8 **O** 16.00 $2p^4$	9 **F** 19.00 $2p^5$	10 **Ne** 20.18 $2p^6$
3	3s	11 **Na** 22.99 $3s^1$	12 **Mg** 24.31 $3s^2$												3p	13 **Al** 26.98 $3p^1$	14 **Si** 28.09 $3p^2$	15 **P** 30.98 $3p^3$	16 **S** 32.07 $3p^4$	17 **Cl** 35.46 $3p^5$	18 **Ar** 39.94 $3p^6$
4	4s	19 **K** 39.10 $4s^1$	20 **Ca** 40.08 $4s^2$	3d	21 **Sc** 44.96 $3d^1$	22 **Ti** 47.90 $3d^2$	23 **V** 50.94 $3d^3$	24 **Cr** 52.00 $4s^1 3d^5$	25 **Mn** 54.9 $4s^2 3d^5$	26 **Fe** 55.85 $4s^2 3d^6$	27 **Co** 58.93 $4s^2 3d^7$	28 **Ni** 58.71 $4s^2 3d^8$	29 **Cu** 63.54 $4s3d^{10}$	30 **Zn** 65.37 $4s^2 3d^{10}$	4p	31 **Ga** 69.72 $4p^1$	32 **Ge** 72.59 $4p^2$	33 **As** 74.92 $4p^3$	34 **Se** 78.96 $4p^4$	35 **Br** 79.91 $4p^5$	36 **Kr** 83.8 $4p^6$
5	5s	37 **Rb** 85.47 $5s^1$	38 **Sr** 87.66 $5s^2$	4d	39 **Y** 88.91 $5s^2 4d^1$	40 **Zr** 91.22 $5s^1 4d^2$	41 **Nb** 92.91 $5s^1 4d^4$	42 **Mo** 95.94 $5s^1 4d^5$	43 **Tc** (99) $5s^2 4d^5$	44 **Ru** 101.1 $5s^1 4d^7$	45 **Rh** 102.91 $5s^1 4d^8$	46 **Pd** 106.4 $5s^0 4d^{10}$	47 **Ag** 107.87 $5s^1 4d^{10}$	48 **Cd** 112.40 $5s^2 4d^{10}$	5p	49 **In** 114.82 $5p^1$	50 **Sn** 118.69 $5p^2$	51 **Sb** 121.75 $5p^3$	52 **Te** 127.60 $5p^4$	53 **I** 126.90 $5p^5$	54 **Xe** 131.30 $5p^6$
6	6s	55 **Cs** 132.91 $6s^1$	56 **Ba** 137.34 $6s^2$	5d	57–71 *	72 **Hf** 178.49 $6s^2 5d^2$	73 **Ta** 180.95 $6s^2 5d^3$	74 **W** 183.85 $6s^2 5d^4$	75 **Re** 186.2 $6s^2 5d^5$	76 **Os** 190.2 $6s^2 5d^6$	77 **Ir** 192.2 $6s^2 5d^7$	78 **Pt** 195.09 $6s^1 5d^9$	79 **Au** 197.0 $6s5d^{10}$	80 **Hg** 200.59 $6s^2 5d^{10}$	6p	81 **Tl** 204.37 $6p^1$	82 **Pb** 207.19 $6p^2$	83 **Bi** 208.98 $6p^3$	84 **Po** (210) $6p^4$	85 **At** (210) $6p^5$	86 **Rn** 222 $6p^6$
7	7s	87 **Fr** (223) $7s^1$	88 **Ra** 226.05 $7s^2$	6d	89–103 †																

4f	57 **La** 138.91 $6s^2 5d^1$	58 **Ce** 140.12 $5d^1 5f^2$	59 **Pr** 140.91 $5d^0 4f^3$	60 **Nd** 144.24 $5d^0 4f^4$	61 **Pm** (145)	62 **Sm** 150.35 $5d^0 4f^6$	63 **Eu** 152.0 $5d^0 4f^7$	64 **Gd** 157.25 $5d^1 4f^7$	65 **Tb** 158.92 $5d^1 4f^8$	66 **Dy** 162.50	67 **Ho** 164.92	68 **Er** 167.26	69 **Tm** 168.93 $5d^0 4f^{13}$	70 **Yb** 173.04 $5d^0 4f^{14}$	71 **Lu** 174.97 $5d^1 4f^{14}$
5f	89 **Ac** 227 $7s^2 6d^1$	90 **Th** 232.04 $6d^2 5f^0$	91 **Pa** 231	92 **U** 238.03 $5d^1 5f^3$	93 **Np** (237)	94 **Pu** (242)	95 **Am** (243) $6d^0 5f^7$	96 **Cm** (247)	97 **Bk** (249)	98 **Cf** (251)	99 **Es** (254)	100 **Fm** (253)	101 **Md** (256)	102 **No** (254)	103 **Lr** (257)

*Lanthanides (rare earths).

†Actinides.

7.9 THE PERIODIC TABLE

When we study the physical and chemical properties of the various elements, we realize that elements with similar properties occur at regular intervals. The tabular arrangement of these elements is called the *periodic table.* Dmitri Mendeléev (Russia, 1834–1907), and Lothar Meyer (Germany, 1830–1895) suggested the form of the periodic table as it is known today. As we have said, the elements were first arranged in order of increasing mass, which resulted in minor discrepancies. But with the establishment of Moseley's law, elements in Mendeléev's table were rearranged in order of increasing atomic number, as shown in Table 7.2.

Elements with similar chemical and physical properties form a *group.* There are eight groups, I through VIII, shown in vertical columns in Table 7.2. For example, Group I consists of hydrogen (H) plus the alkali metals (Li, Na, K, Rb, Cs, and Fr), all of which have valences of $+1$ and are chemically very active. Similarly, Group VIII consists of the so-called rare gases (He, Ne, Ar, Kr, Xe, and Rn), all of which have 0 valence. This means that they are chemically very inert, naturally.

The horizontal rows in the periodic table are called *periods.* Properties of the elements in a given period vary steadily from one element to the next, so that the properties of the elements at the two extremes of the periods (left and right) are drastically different. Thus, in any one group, the properties of the elements do not change appreciably with the atomic number, but in a given period, the properties of the elements change rapidly with atomic number.

In each box in Table 7.2, the atomic number is at the top, followed by the symbol of the element, the atomic weight of the element, and the quantum states of the outermost electrons in the last two subshells. Quantum states are assigned according to the Pauli exclusion principle, as we have said. We explained the formation of shells and subshells in Section 7.2, and Table 7.1 showed how the system works.

The first period consists of only two elements, $_1H$ and $_2He$, belonging to Group I and Group VIII, respectively. The second period consists of eight elements—$_3Li$, $_4Be$, $_5B$, $_6C$, $_7N$, $_8O$, $_9F$, and $_{10}Ne$—belonging to Groups I through VIII, respectively. The third period also consists of eight elements. The fourth period starts with $_{19}K$ and $_{20}Ca$, belonging to Groups I and II, respectively. But the next ten elements—$_{21}Sc$, $_{22}Ti$, $_{23}V$, $_{24}Cr$, $_{25}Mn$, $_{26}Fe$, $_{27}Co$, $_{28}Ni$, $_{29}Cu$, and $_{30}Zn$—do not resemble the elements of the second and third periods under Groups III through VIII. Those ten elements appearing between Groups II and III are called the *transition elements* forming the *first transition group.* They are metals, and have similar properties. After these transition elements, the fourth period is completed by six more elements, belonging to Groups III through VIII, respectively.

The fifth period contains eight elements for eight groups, and a set of ten transition elements between Groups II and III.

The sixth period, in addition to eight elements in Groups I through VIII, contains 15 elements called *rare-earth elements* or *lanthanides*, and a third set of transition elements between Groups II and III. These rare-earth elements resemble one another so closely that it is hard to distinguish between them chemically.

The seventh period starts with $_{87}$Fr and $_{88}$Ra, followed by a big group of 15 elements called *actinides*.

Those elements with atomic numbers $Z = 93$ to 103 or higher have been produced only artificially, in the laboratory, and are mostly unstable. Actinides which have $Z > 92$ are called *transuranic elements*.

Thus we see that the elements are arranged in seven periods, eight groups, three transition series, the lanthanides, and the actinides. Figure 7.19 also shows the periodic atomic structure, and presents a plot of ionization potential versus atomic number Z. The high maximum ionization potentials for He, Ne, Ar, Kr, Xe, and Rn indicate the closing of shells for $Z = 2, 8, 18, \ldots$, that is, for $Z = 2n^2$, where $n = 1, 2, 3, 4, 5$, and 6. So they are inert. Note, however, that the alkali metals, on the other hand, have high chemical activity because each has one electron *outside* the closed shell, and hence they have very *low* ionization potentials. The properties of other elements are in between these two extremes.

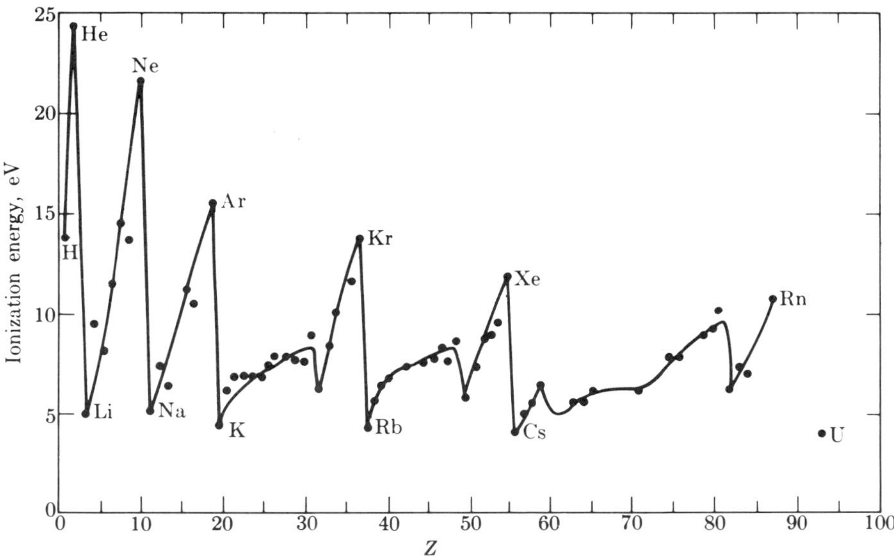

Fig. 7.19 The periodic structure of the elements is evident from this plot of ionization potential versus Z for different elements.

SUMMARY

Pauli's exclusion principle states that "No two electrons in an atom can be in the same quantum state" or "No two electrons in an atom can have identical values for a set of four quantum numbers (n, l, m_l, m_s) or (n, l, j, m_j)."

The total number of quantum states N_t for a given n are $N_t = 2n^2$, and the shells are filled according to the following order

$$1s^2 2s^2 2p^6 3s^2 3p^6 4s^2 3d^{10} 4p^6 5s^2 4d^{10} \cdots$$

The energy levels of sodium (Na) are

$$E_n = -Z_{\text{eff}} \frac{Rch}{n^2},$$

and the transitions of electrons from one level to another are classified into four series: sharp, principal, diffuse, and fundamental. The transitions are governed by the selection rules

$$\Delta l = \pm 1 \qquad \text{and} \qquad \Delta j = 0, \pm 1.$$

Magnetic moment is expressed in units of Bohr magnetons, $\mu_\beta = e\hbar/2m$. For an atom, the ratio of the magnetic dipole moment, in units of Bohr magnetons, to the angular momentum, in units of $\hbar$, is called the *gyromagnetic ratio*. That is,

$$\frac{\boldsymbol{\mu}_l/\mu_\beta}{\mathbf{L}/\hbar} = g_l,$$

with similar expressions for g_s and g_j. The ratio is also called Landé g factor.

The change in energy when an atom with a magnetic moment $\boldsymbol{\mu}$ is placed in a uniform external field $\mathbf{B}$ is given by

$$\Delta E = \boldsymbol{\mu} \cdot \mathbf{B}.$$

According to Larmor's theorem, the angular frequency of precession is given by

$$\omega_L = g_l \frac{e}{2m} B,$$

with similar expressions for ω_S and ω_J.

One can prove the space quantization of **L**, **S**, and **J** by means of the normal Zeeman effect, the Stern–Gerlach experiment, and the anomalous Zeeman effect, respectively.

In the normal Zeeman effect, the observed frequencies are

$$\nu = \nu_0, \qquad \nu = \nu_0 \pm \left(\frac{eB}{4\pi m}\right).$$

In the Stern–Gerlach experiment, the force acting on an atomic dipole is

$$F_z = -\frac{d}{dZ}(\Delta E_S), \qquad \text{where } \Delta E_S = g_s m_s \frac{eh}{4\pi m} B.$$

Characteristic x-rays are produced whenever electrons are removed from the innermost shells. The electrons from the outer shells jump to the inner shells to be nearer to the nucleus. When this happens, there is a simultaneous emission of x-rays.

According to Moseley's law

$$\sqrt{\nu_{K\alpha}} \propto Z,$$

or, in general,

$$\nu = C_n(Z - S)^2.$$

PROBLEMS

1. What are the configurations of the electrons of atoms of the following elements? (a) Si ($Z = 14$), (b) Cl ($Z = 17$), (c) Br ($Z = 35$), (d) Ca ($Z = 20$).
2. What are the ground-state configurations of electrons of atoms of the following elements? In each case, also draw the schematic representation of the electron configuration. (a) cobalt ($Z = 27$), (b) aluminum ($Z = 13$), (c) arsenic ($Z = 33$), (d) silver ($Z = 47$), (e) sulfur ($Z = 16$).
3. Consider an atom with an electron configuration of $1s^2 2s^2 2p^1$. What are the values of l, s, and j, and the corresponding values of the angular momenta L, S, and J, respectively?
4. Consider an atom with an electron configuration of $1s^2 2s^2 2p^6 3s^2 3p^6 4s^2 3d^1$. What are the values of l, s, and j, and the corresponding values of the angular momenta L, S, and J, respectively?
5. Suppose that atoms could be formed having electrons with principal quantum numbers up to $n = 8$. How many elements could be formed? Why are there no stable elements with $Z > 92$?
6. Identify the possible transitions which would take place if sodium atoms were left in the following excited states. (a) 4P, (b) 4D, (c) 4F, (d) 5P, (e) 5F.
7. The ionization energy of a sodium atom is 5.13 eV; the wavelengths of the sodium D lines are 5890 Å and 5896 Å. From this information, calculate the energies of the levels from which the sodium D lines originate. What is the separation, in eV, between these two levels?
8. The series limit of the principal series of a certain element is 43,486 cm^{-1}. Calculate the ionization potential for this element.
9. Write the electron configuration of Li ($Z = 3$). Assuming Li to be a hydrogenlike atom, calculate the ionization energy of the 2s valence electron. How does this calculated value compare with the experimental value of 5.39 eV? What are the values of the screening constant and the effective positive charge?

10. Calculate (a) the effective positive charge, and (b) the screening constant for the valence electron of sodium in the ground state. The ionization energy of sodium is 5.14 eV.

11. Consider the spectrum of a singly ionized Mg atom ($Z = 12$). How do the wavelengths of the typical transitions in this case differ from those in a Na atom?

12. Show that if a magnetic or electric dipole is placed in a uniform magnetic field, the net force acting on the dipole will be zero.

13. Suppose that the electron in a hydrogen atom is moving in one of the three Bohr orbits. For each orbit, calculate (a) the electric current, (b) the magnetic dipole moment. What do you conclude from the variation in those quantities for the three orbits?

14. Calculate the precessional frequencies of the s, p, and d electrons in an atom placed in a magnetic field of 2.0 webers/m^2.

15. Consider an atom, placed in a magnetic field of 1.0 weber/m^2, which has $l = 2$ and a magnetic moment of 2 Bohr magnetons. Calculate (a) the rate of precession, (b) the torque on the atom, given that the magnetic moment makes an angle of 45°.

16. Calculate the energy needed to reverse the direction of spin of an electron of an atom placed in a magnetic field **B**.

17. *Spin resonance* (*the spin flip experiment*). Spin flip is the transition of the spin of an electron from a parallel to an antiparallel orientation, and vice versa. Such transitions can be induced by a radiofrequency signal. Calculate the frequency of a signal which would cause such a transition in a magnetic field of 1.0 weber/m^2.

18. Draw an energy-level diagram for, and calculate the separation between the adjacent normal Zeeman components for the 3D state of an atom placed in a magnetic field of 1.5 webers/m^2. Assume that the resultant spin angular momentum is zero.

19. Draw an energy-level diagram for, and calculate the separation between the adjacent normal Zeeman components for the 4F state of an atom placed in a magnetic field of 1.5 webers/m^2. Assume that the resultant spin angular momentum is zero.

20. Suppose that the separation between two components of a normal Zeeman pattern is 4×10^{10} cycles/sec. What should the value of the magnetic field **B** be so that these lines may be seen resolved by a spectrometer capable of resolving lines separated by 0.05 Å?

21. By measuring the separation $\Delta\nu$ between different components of a normal Zeeman pattern in a known magnetic field B, one can calculate the value of e/m. Explain how.

$$\left[\Delta\nu = \frac{eB\,\Delta m_l}{4\pi m}, \quad \Delta m_l = \pm 1, 0.\right]$$

22. A certain transition in an atom of wavelength $\lambda = 4226$ Å, when observed in a magnetic field of 1 weber/m^2, exhibits a normal Zeeman pattern whose components are separated by 1.4×10^{10} cycles/sec. Calculate the value of e/m.

23. Show that the total angular momentum and the total magnetic moment of an atom with (a) a closed subshell, (b) a closed shell are zero.

★24. Calculate the Landé g factors and the total magnetic moments of atoms in the states $^2D_{5/2}$ and $^2D_{3/2}$.

★25. Calculate the Landé g factor and the total magnetic moment of an atom in the state $^2F_{7/2}$.

26. In the Stern–Gerlach experiment, silver atoms are heated to 1000°K. Then a narrow beam of these atoms is passed through an inhomogeneous magnetic field of gradient 0.75 weber/m²/cm over a distance of 10 cm. The beam travels another 12 cm in a fieldfree space before hitting a photographic plate. Calculate the separation between the two components of the beam on the plate. Use the average velocity obtained from the relation $\frac{1}{2}m\bar{v}^2 = \frac{3}{2}kT$.

27. A beam of hydrogen atoms with velocity 10^6 cm/sec passes through a magnetic field of gradient 0.5 weber/m²/cm for a distance of 50 cm. Calculate (a) the force exerted on the atoms, (b) the vertical displacements.

28. A beam of silver atoms with a velocity of 10^5 cm/sec passes through a magnetic field of gradient 0.50 weber/m²/cm for a distance of 10 cm. What is the separation between the two components of the beam as it comes out of the magnetic field?

29. Estimate the wavelength of the K_α lines of (a) light, (b) medium, (c) heavy elements. What do you conclude from this observation, and what is your explanation for your conclusions?

30. Before an atom emits an x-ray, it is at rest. To conserve momentum, when it emits an x-ray, it must recoil. Given that the mass of the atom is M and the energy of the x-ray is $h\nu$, what is the recoil energy of the atom?

31. What energy is needed to excite cadmium atoms so that L-series x-rays will be observed? What energy is needed so that all series will be observed?

32. Calculate the wavelength, the frequency, and the energy of the K_α line in uranium ($Z = 92$).

33. Consider the emission of the K_α line from silver ($Z = 47$). Calculate the ratio of the recoil energy of the silver atom to the energy of the emitted K_α photon.

34. The wavelength of the K x-ray of copper is 1.377 Å, while the critical voltage for production of these K x-rays is 9 kV. What is the value of h/e?

35. Using the following data, make a plot of $\sqrt{\nu}$ versus Z for the K_α x-rays emitted by these elements.

Element	Z	λ (Å)
Magnesium	12	9.87
Sulfur	16	5.36
Calcium	20	3.35
Chromium	24	2.29
Cobalt	27	1.79
Copper	29	1.59

From this plot,estimate A and S in the expression $\sqrt{\nu} = A(Z - S)$, where $A = \sqrt{C_n}$.

REFERENCES

1. Wolfgang Pauli, *Zeit. für Physik* **31,** 765 (1925)
2. P. A. M. Dirac, *Proc. Roy. Soc.* **A112,** 661 (1926)
3. Werner Heisenberg, *Zeit. für Physik* **38,** 411 (1926)
4. D. R. Hartree, *Proc. Camb. Phil. Soc.* **24,** 89, 111 (1928)
5. Linus Pauling, *Proc. Roy. Soc.* **A114,** 181 (1927)
6. R. A. Millikan and I. S. Brown, *Phys. Rev.* **23,** 764 (1924); A. Landé, *Zeit. für Physik* **25,** 46 (1924)
7. J. C. Slater and N. H. Frank, *Electromagnetism*, New York: McGraw-Hill, 1947
8. E. R. Cohen and J. W. M. DuMond, *Revs. Mod. Phys.* **37,** 537 (1965)
9. Atam P. Arya, *Fundamentals of Atomic Physics*, Boston: Allyn and Bacon, 1971
10. H. Goldstein, *Classical Mechanics*, Reading, Mass.: Addison-Wesley, 1959
11. Pieter Zeeman, *Phil. Mag.* **5,** 43, 226 (1897)
12. O. Stern and W. Gerlach, *Zeit. für Physik* **8,** 110 (1921); **9,** 349, 353 (1922); *Ann. Physik* **74,** 673 (1924)
13. H. G. J. Moseley, *Phil. Mag.* **26,** 1024 (1913); **27,** 703 (1914)

Suggested Readings

1. H. E. White, *Introduction to Atomic Spectra*, Chapters V, VII to X; New York: McGraw-Hill, 1934
2. Atam P. Arya, *Fundamentals of Atomic Physics*, Chapters X, XI and XII; Boston, Mass.: Allyn and Bacon, 1971
3. R. M. Eisberg, *Fundamentals of Modern Physics*, Chapter 11; New York: John Wiley, 1961
4. G. Herzberg, *Atomic Spectra and Atomic Structure*, Chapter II; New York: Dover Publications, 1944
5. Linus Pauling and S. Goudsmit, *The Structure of Line Spectra*, New York: McGraw-Hill, 1930
6. F. K. Richtmyer, E. H. Kennard, and John N. Cooper, *Introduction to Modern Physics*, sixth edition, New York: McGraw-Hill, 1969

CHAPTER 8
STRUCTURE AND SPECTRA OF MOLECULES

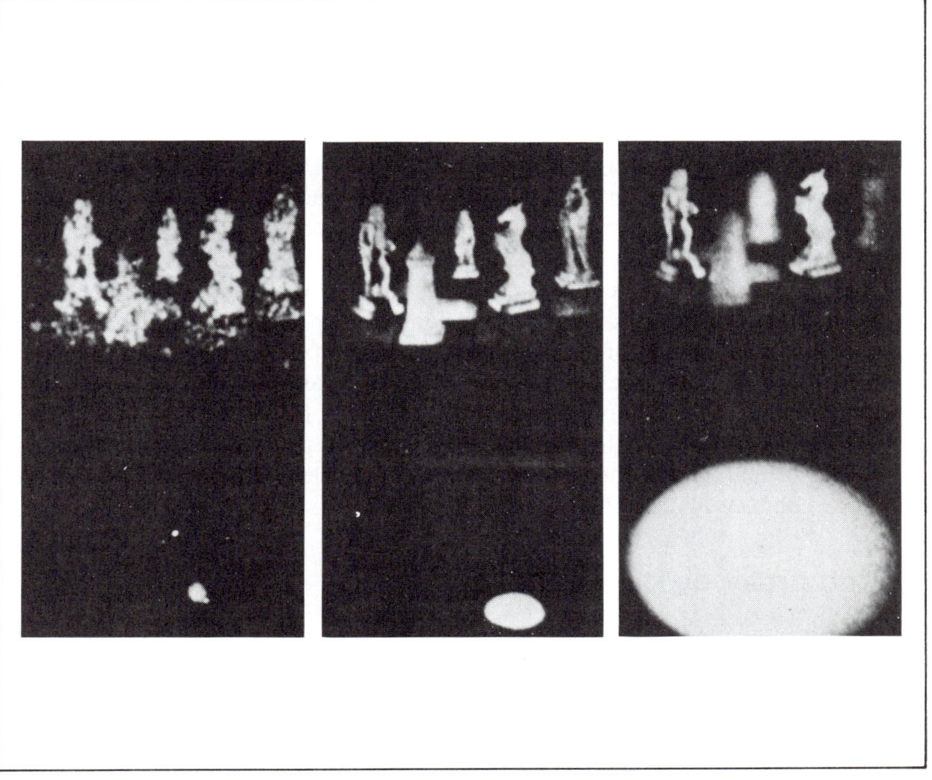

Reconstruction of an image by means of a hologram. Quasimonochromatic radiation from a laser at 6328 Å is reflected from some chessmen to form the hologram. These reconstructions, which use the same source, show less speckle and better resolution, but less depth of focus, as greater portions of the hologram (at lower center) are illuminated. The laser beam used is almost a half-millimeter in diameter (as in first picture, with successive pictures using increasing beam size). (From E. N. Leith and J. Upatniehs, J. Sci. Am. **212**, *6, 24, 1965)*

8.1 INTRODUCTION

We have seen how the development of quantum mechanics led to the qualitative and quantitative explanation of many aspects of atomic structure and spectra. When we are dealing with multi-electron atoms, solving the Schrödinger wave equation becomes complicated and hence only qualitative results are obtained. As a matter of fact, for any system containing more than two bodies, solving the Schrödinger equation—though it is possible in principle—is so involved even with modern computers that one concentrates on obtaining qualitative results only.

We would expect that, in the study of molecules and solids (which obviously consist of a very large number of particles), quantum mechanics would be of no use at all. But this is not true. By the qualitative application of quantum mechanics, scientists have been able to explain many aspects of molecules and solids which remained a mystery until the last two decades. In this chapter we shall investigate the structure and spectra of molecules, while Chapters 9 and 10 will be devoted to the structure and properties of solids.

We know that as atoms are brought together, molecules and solids are formed. It is safe to assume that, in this combination process, the nuclei of the atoms never come very close to each other. Thus we may neglect any nuclear interaction between them (although there is coulomb interaction between the nuclei). Actually the formation of molecules and solids takes place through the interaction of the electron waves of different electrons in the atoms, as we shall see.

Many scientists today are conducting complex experiments in hopes of finding out the type of bonding that holds different atoms together. We can only hope that, by studying molecules and solids, mankind can solve many problems of biology, such as cell division, heredity, and various diseases (cancer, of course, being at the top of the list), in addition to problems of physics, such as the working of lasers and masers, the mechanical, thermal, and electrical properties of solids, and superconductivity, to name but a few.

8.2 STRUCTURE OF MOLECULES

When two atoms combine to form a stable system, it is a *diatomic molecule*; if three atoms combine, it is a *triatomic molecule*. A stable system of more than three atoms is called a *polyatomic molecule*. The simplest are the diatomic molecules, so we shall limit our discussion to these. To form a stable molecule, the atoms must be bound together by an attractive force. Hence the energy of the atoms when they are close together must be less than the sum of the energies of the atoms when separated by large distances.

The four common types of bonding by which atoms in a molecule or solid may be bound together are: *ionic bonding*, *covalent bonding*, *van der Waals bonding*, and *metallic bonding*. There are a very few examples of pure bonding, because in most molecules the bonding is a mixture of different types. The types of bonding that occur most commonly are ionic (also called *heteropolar*) and covalent (also called *homopolar*) bonding.

a) Ionic bonding.[1,2] Examples of ionic (or heteropolar) bonding are the bonds in NaCl, KCl, and other salts. Take NaCl as an example. A NaCl molecule forms when an electron shifts from the sodium to the chlorine atom, thus forming Na^+ and Cl^- ions, which are then attracted to each other electrostatically, forming a stable NaCl molecule.

The sodium atom, $_{11}Na$, consists of 11 electrons, with an electron configuration of $1s^2 2s^2 2p^6 3s^1$. It belongs to the alkali-metal group, the first group of the periodic table. The 3s electron in the sodium atom is very weakly bound and can be removed by adding an energy of 5.1 eV:

$$Na + 5.1\ eV = Na^+ + e^-. \tag{8.1}$$

Sodium (like other alkali metals in general) is called *electropositive* because it easily loses an electron to form a positive ion. This results in a configuration consisting of closed electron shells, like that of the inert gases.

The chlorine atom, $_{17}Cl$, consists of 17 electrons, with an electron configuration of $1s^2 2s^2 2p^6 3s^2 3p^5$. Like the other halogens, chlorine belongs to the seventh group of the periodic table. The chlorine atom lacks one electron to complete the 3p subshell and form a tightly bound system. Its outermost shell, in other words, offers a ready home to any stray electron. When the chlorine atom does capture an electron, the result is the formation of a Cl^- ion and a release of 3.8 eV of energy:

$$Cl + e^- = Cl^- + 3.8\ eV. \tag{8.2}$$

The Cl^- ion, and other halogen ions in general, is said to be *electronegative*, and is said to have an *electron affinity* of 3.8 eV.

Thus, in order to form Na^+ and Cl^- ions (separated by an infinite distance) from Na and Cl atoms, one must add 1.3 eV of energy to the system. In other words, by adding Eqs. (8.1) and (8.2), we get

$$Na + Cl + 1.3\ eV = Na^+ + Cl^-. \tag{8.3}$$

Figure 8.1 shows the transfer of an electron and the formation of ions, and Fig. 8.2 shows the differences in energy between the two particles. Of course, the Na^+ and Cl^- ions are separated by an infinite distance, and have spherically symmetric charge distributions (positive for Na^+ and negative for Cl^-), because each have outermost electron shells that are closed (or filled).

When Na^+ and Cl^- are brought together, a molecule is formed. Let us say that these ions come together until they are separated by a distance of only 4 Å (which is a typical equilibrium distance). At this distance, electrostatic attraction takes place between these ions. The amount of energy emitted is equal to the coulomb potential energy between $+e$ and $-e$ separated by a distance of 4 Å:

$$\begin{aligned} V &= \frac{1}{4\pi\epsilon_0}\left(\frac{e^2}{r}\right) \\ &= 9 \times 10^9\ \frac{\text{N-m}^2}{\text{coulomb}^2}\left(-\frac{(1.6 \times 10^{-19}\ \text{coulomb})^2}{4 \times 10^{-10}\ \text{m}}\right)\frac{1}{1.6 \times 10^{-19}\ \text{J/eV}} \\ &= -3.6\ \text{eV}. \end{aligned} \tag{8.4}$$

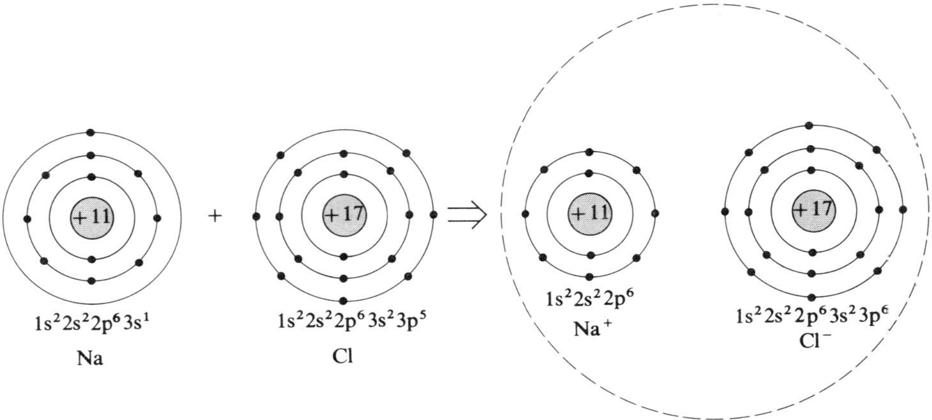

Fig. 8.1 When sodium and chlorine atoms are brought together, ionic bonding results in the formation of a sodium chloride molecule.

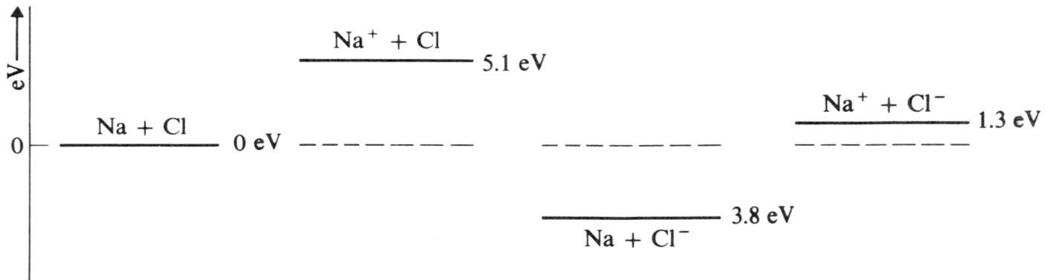

Fig. 8.2 Schematic representation of the differences in energy of sodium and chlorine atoms and ions in different states.

This means that, at a distance of 4 Å, Eq. (8.3) takes the form

$$\mathrm{Na} + \mathrm{Cl} + 1.3\ \mathrm{eV} = \overset{\leftarrow 4\ \text{Å} \rightarrow}{(\mathrm{Na}^+ + \mathrm{Cl}^-)} + 3.6\ \mathrm{eV}$$

or

$$\mathrm{Na} + \mathrm{Cl} = \overset{\leftarrow 4\ \text{Å} \rightarrow}{(\mathrm{Na}^+ + \mathrm{Cl}^-)} + 2.3\ \mathrm{eV}. \tag{8.5}$$

Thus, when Na^+ and Cl^- ions are at a distance of 4 Å, the net result of their attraction is that 2.3 eV of energy is emitted, and a stable NaCl molecule is formed. As these ions are brought still closer (Fig. 8.3), the attraction increases. However, when the separation between the ions becomes very small indeed, and equals the distance between the nuclei of these ions, a strong *repulsive* force starts to act between them. This repulsion, which is due to Pauli's exclusion principle, causes ions that are at short distances from each other to be repulsed regardless of the

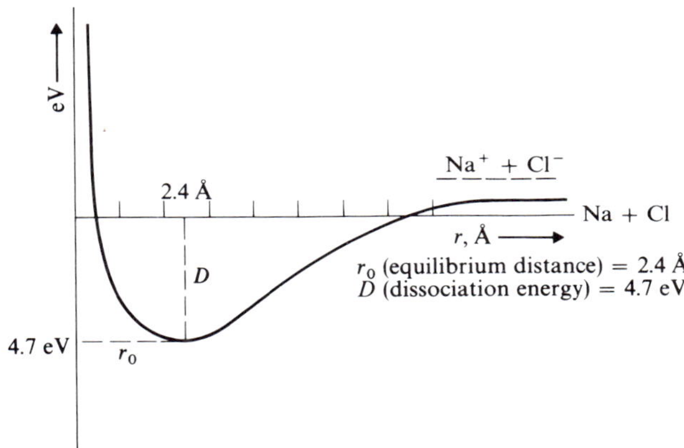

Fig. 8.3 Potential energy between Na^+ and Cl^- ions versus the distance between them. Note that the energy needed to form Na^+ and Cl^- ions at infinite separation is taken to be 1.3 eV. A stable NaCl molecule is formed when the ions are 2.4 Å apart.

type of bonding mechanism. [When two ions are very close, the electron clouds of the two start overlapping; or, in terms of quantum mechanics, the electron wave functions of the core electrons start overlapping. But the shells of electrons near the cores of the atoms are already completely filled, and according to the Pauli exclusion principle *some* of these electrons must jump to higher energy states. As the ions are brought closer, more and more electrons must go to higher energy states so as to prevent overlapping of the wave functions; this is the only way to satisfy Pauli's exclusion principle.]

Figure 8.3 shows the plot of the potential energy versus distance r between the two ions. Since the ions attract one another at large distances and repel one another at very short distances, there must be some *equilibrium distance* r_0 at which the energy will be a minimum, and hence at which the molecule will exist as a stable entity. This minimum energy is called the *dissociation energy* of a molecule. It is equal to the energy needed to separate a molecule in its lowest energy state into its components.

As we see from Fig. 8.3, the NaCl molecule has an equilibrium distance of $r_0 = 2.4$ Å and a dissociation energy of 4.7 eV. The portion of the molecule containing the Na nucleus is positive, while the portion containing the Cl nucleus is negative. Thus an ionic molecule is a *polar* molecule and the bonding is called *heteropolar* (i.e., dissimilar poles). Also, since it is a polar molecule, it has a permanent *electric dipole moment*. (In simple form, electric dipole moment = either charge × separation.)

b) Covalent bonding.[1-3] A completely different type of bonding, called covalent bonding (also called *homopolar bonding*), is responsible for the formation of stable diatomic molecules, such as H_2, N_2, etc. The attractive force between the two atoms in such a diatomic molecule is due to the sharing of one or two pairs of electrons by both atoms. These shared electrons spend more time in between the atoms than anywhere else, thereby producing an attractive force.

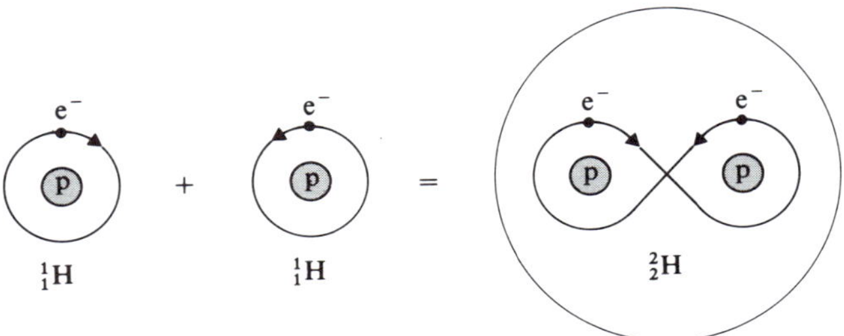

Fig. 8.4 When two hydrogen atoms are brought together to form a molecule, the two electrons are shared by both protons, as shown. It is not possible to associate either electron with a particular nucleus.

Look at Fig. 8.4, for example. It shows that when two hydrogen atoms (each with one electron) are brought together, the two electrons are shared by both atoms. Suppose we considered the formation of the H_2 molecule to be due to the transfer of an electron from one atom to the other, producing ions $_1H^+$ and $_1H^-$. In that case, the electrostatic energy between these ions would be positive for all separation distances, and hence no bonding would be possible! The bonding of two identical, neutral atoms cannot be explained in terms of classical theories. The formation of covalent bonds is a process that is entirely quantum mechanical. Only by solving the Schrödinger wave equation and making use of the Pauli exclusion principle can we explain the bonding in molecules such as H_2.

Now look at Fig. 8.5: For the H_2 molecule, the total electrical potential energy is given by

$$V_T = \frac{e^2}{4\pi\epsilon_0}\left(-\frac{1}{r_{11}} - \frac{1}{r_{12}} - \frac{1}{r_{21}} - \frac{1}{r_{22}} + \frac{1}{r_e} + \frac{1}{r_p}\right), \tag{8.6}$$

where the first two terms in the parentheses correspond to the interaction of proton p_1 with the two electrons, the third and fourth terms are interactions of proton p_2 with the two electrons, the fifth term is the interaction between the two electrons, and the sixth term is the interaction between the two protons. Ideally we should solve the Schrödinger wave equation using this potential energy, then calculate the minimum energy that would correspond to the equilibrium distance. This problem has been analyzed in great detail, but we shall state the salient features without going into any mathematical details.[4,5,6]

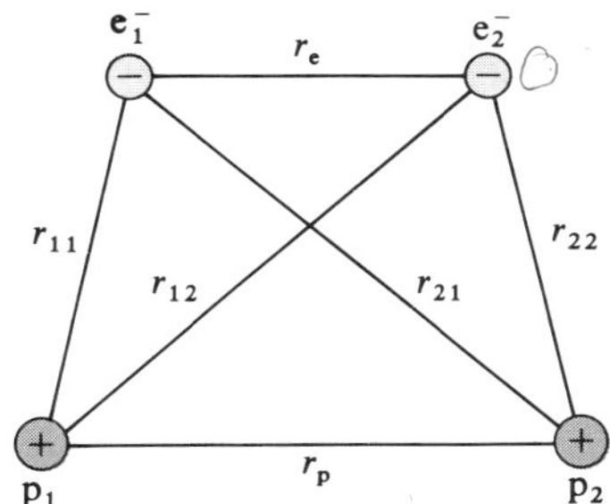

Fig. 8.5 The configuration of electrons and protons in a H_2 molecule for the purpose of calculating total coulomb potential energy.

Suppose our two hydrogen atoms are separated by a distance greater than 1.06 Å, twice the radius of the first Bohr orbit. In this case each electron is associated with the nucleus of *its own* atom. The wave function of each electron is $e^{-r/a}$. In the one-dimensional model, we write this as $e^{-|x|/a}$.

These wave functions are as shown in Fig. 8.6(a). If we bring these atoms closer, so that the distance between them is less than 1.06 Å, the wave functions of the two electrons start overlapping (see Fig. 8.6b). There is no way of distinguishing between the two electrons, so we can't associate one electron with one particular nucleus (a proton in this case).

However, there are two distinct ways in which the two hydrogen atoms can be brought together [see Fig. 8.6b(i) and (ii)]. In (i), the spins of the two electrons are antiparallel, while in (ii), the spins of the two electrons are parallel. The wave functions resulting from the addition of ψ_1 and ψ_2 are the *symmetric wave function* $\psi_S \approx \psi_1 + \psi_2$ [see Fig. 8.6a(i)]; and the *antisymmetric wave function* $\psi_A \approx \psi_1 - \psi_2$ [see Fig. 8.6b(ii)].

The probability of locating the electron is given by the square of the wave function. Figure 8.6c(i) shows this distribution for $|\psi_S|^2$, while Fig. 8.6c(ii) shows it for $|\psi_A|^2$. Obviously, for antiparallel spins, the probability of finding the electron between the two nuclei is much larger than the probability of finding the electron on the other two sides of the nuclei. For parallel spins, the probability of finding the electron between the two nuclei is small (zero at the exact midway point).

The density of the points in Fig. 8.6d(i) and (ii) also shows these probabilities. A greater density of points indicates a higher probability. Whether the overlapping of the two electron wave functions leads to the symmetric wave function ψ_S or the antisymmetric wave function ψ_A is of great significance. According to quantum mechanics, when two hydrogen atoms are brought together, the total energy of the system increases when the spins of the electrons are parallel, and decreases when they are antiparallel. Of course, when the two atoms are very close, the repulsion between the two nuclei overcomes everything else.

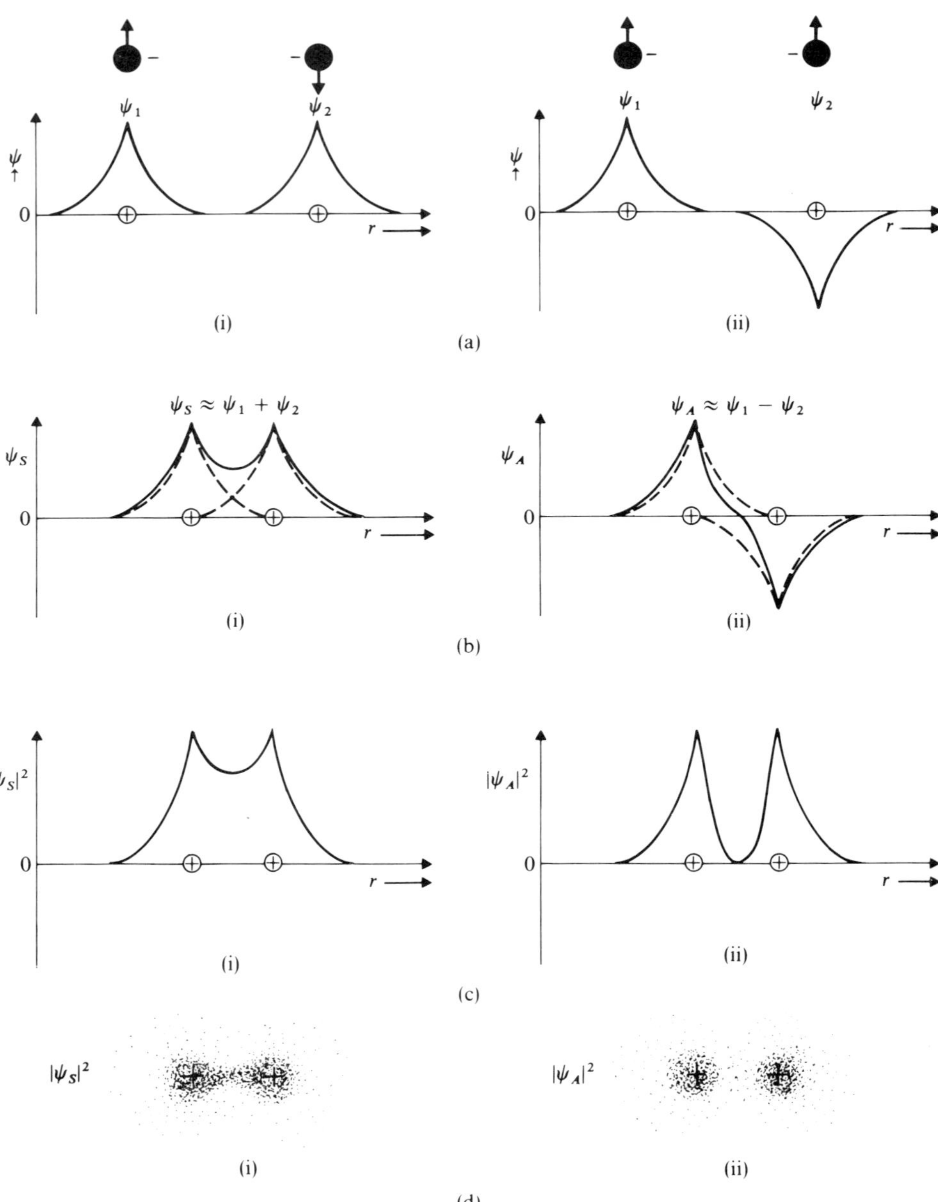

Fig. 8.6 (a) Plots of wave functions ψ_1 and ψ_2 of two electrons when they are far apart: (i) spins are antiparallel, (ii) spins are parallel. (b) Linear combinations of wave functions ψ_1 and ψ_2 (c) Probability distributions for wave functions ψ_S and ψ_A. (d) Density of points in a given volume is proportional to the probability of finding an electron in that volume.

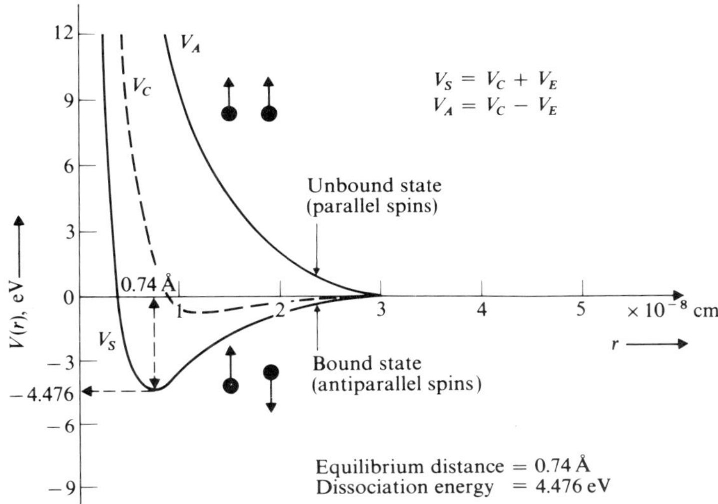

Fig. 8.7 Total potential energy versus r for two hydrogen atoms. Curve V_S is for the symmetric wave function and has a minimum, thereby resulting in the formation of a bonded H_2 molecule. Curve V_A is for the antisymmetric wave function, and does not result in a bonded molecule. If there were no exchange term, there would be no difference in the two curves, as shown by the dashed curve V_C, and no molecules would form. $V_S = V_C + V_E$ and $V_A = V_C - V_E$, where V_E is the contribution from the exchange term.

Figure 8.7 shows the plots of the energies $V(r)$ versus the separation distance r between the nuclei for the two cases.[7] It is quite obvious that only if the spins of the electrons are antiparallel will minimum energy and a bonded system be possible. Parallel spins would render such bonding impossible. For the case of two hydrogen molecules for which the spins of the two electrons are antiparallel, the bonding energy is −4.476 eV and the equilibrium distance $r_0 = 0.74$ Å (see Fig. 8.7).

The force binding the two hydrogen atoms in a molecule arises from energy that is called *exchange energy*.[5,6] When we wish to find the difference between the energy for parallel and for antiparallel spins, we have to once again apply the Pauli exclusion principle.

★ Suppose we denote the two electrons by α and β, and their positions by x_1 and x_2. For simplicity, we can say that the two electrons are at positions 1 and 2 with wave functions $\psi_\alpha(1)$ and $\psi_\beta(2)$; the wave function of the system is $\psi_\alpha(1)\psi_\beta(2)$. If α and β exchange positions, the wave function of the system is $\psi_\alpha(2)\psi_\beta(1)$. According to the Pauli exclusion principle, the two possible wave functions are the *symmetric wave function* ψ_S of the system, given by

$$\psi_S = \frac{1}{\sqrt{2}}[\psi_\alpha(1)\psi_\beta(2) + \psi_\alpha(2)\psi_\beta(1)] \tag{8.7}$$

and the *antisymmetric wave function* ψ_A given by

$$\psi_A = \frac{1}{\sqrt{2}}[\psi_\alpha(1)\psi_\beta(2) - \psi_\alpha(2)\psi_\beta(1)]. \tag{8.8}$$

The square of these wave functions gives the probability distribution for the antiparallel and parallel spins cases, respectively:

$$|\psi_S|^2 = C + E \tag{8.9}$$

$$|\psi_A|^2 = C - E \tag{8.10}$$

where

$$C = \tfrac{1}{2}[|\psi_\alpha(1)\psi_\beta(2)|^2 + |\psi_\alpha(2)\psi_\beta(1)|^2] \tag{8.11}$$

and

$$E = \psi_\alpha(1)\psi_\beta(1)\psi_\alpha(2)\psi_\beta(2). \tag{8.12}$$

The term E is called the *exchange term* or *interference term*, and has no classical analogy. (Note that the exchange implied here is that of the spins of the two indistinguishable electrons, say from ↓↑ to ↑↓.) As the name indicates, the exchange term E will be zero if there is no overlapping of the wave functions. If we are looking in the region of one electron, while the other electron is far away, so that there is no overlapping of the wave functions, either $\psi_\alpha(1) = 0$ or $\psi_\beta(2) = 0$; hence $E = 0$ and $|\psi_S|^2 = |\psi_A|^2$. But if ψ_α and ψ_β overlap, $E \neq 0$, and from Eqs. (8.9) and (8.10), $|\psi_S|^2 \neq |\psi_A|^2$. Thus it is due to this exchange term that we can see the differences for parallel and antiparallel spins in Figs. 8.6(d) and 8.7. For the case of antiparallel spins, we can find both electrons simultaneously in between the two protons. This naturally results in attractive forces between each electron and both protons.

Note that covalent molecules such as H_2 and O_2, unlike ionic molecules, do not have permanent electric dipole moments. Molecules which are not formed by ionic or covalent bonding may be formed by still another type: the *van der Waals bond* (named after its discoverer), a very weak type of bonding. We'll discuss it in Chapter 9 in connection with the structure of solids. For now, let's just say that an example of van der Waals bonding is the H_2O molecule. That will give us some idea of how important this type of bond is.

8.3 SPECTRA OF DIATOMIC MOLECULES

As we are all aware, we can get a good idea of the structure and energy levels of molecules of a substance by studying their emission and absorption spectra; that is, by examining the interaction of electromagnetic radiation (photons) with the molecules.[8] The energy levels of molecules, which are of course much more complicated than those of atoms, may be conveniently divided into three groups.

a) Energy levels resulting from rotational motion of molecules; these are in the energy range of $\sim 10^{-4}$ eV.

b) Energy levels resulting from vibrational motion of the nuclei of molecules; these are in the energy range of $\sim 10^{-1}$ eV.

c) Energy levels due to excitation of the electrons in molecules; these are the same as those due to the excitation of the electrons of atoms, and are in the range of $\sim$1–10 eV.

a) Rotational spectra of diatomic molecules

Figure 8.8 shows the simplest motion of a diatomic molecule as a whole, which is pure rotation about its center of mass, CM. Let the masses m_1 and m_2 of the two nuclei be located at distances r_1 and r_2, respectively, from the center of mass, and r_0 be the distance between the two nuclei. Neglecting the masses of the electrons as compared to those of the nuclei, and using the definition of center of mass, we may write

$$m_1 r_1 = m_2 r_2. \tag{8.13}$$

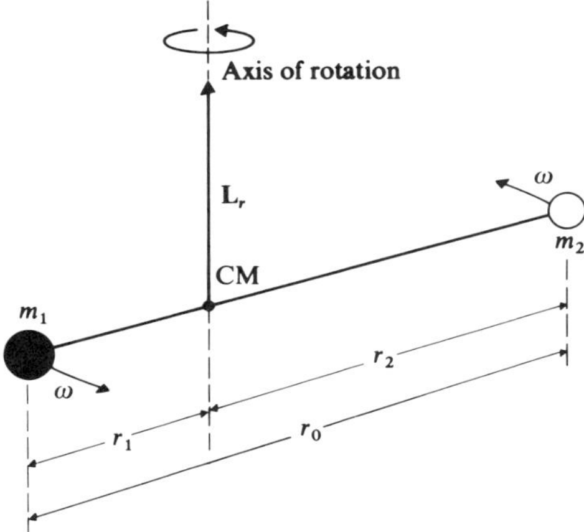

Fig. 8.8 Rotational motion of a diatomic molecule about an axis through the center of mass.

The moment of inertia I of this molecule about the axis passing through the center of mass and perpendicular to the line joining the masses m_1 and m_2 is given by

$$I = m_1 r_1^2 + m_2 r_2^2. \tag{8.14}$$

Using Eq. (8.13), we obtain

$$I = \left(\frac{m_1 m_2}{m_1 + m_2}\right)(r_1 + r_2)^2 \tag{8.15}$$

or

$$I = \mu r_0^2, \tag{8.16}$$

where μ is the reduced mass, $\mu = m_1 m_2/(m_1 + m_2)$, and $r_0 = r_1 + r_2$. Equation (8.16) implies that the rotation of two masses m_1 and m_2 may be replaced by that of a single mass μ located a distance r_0 from the axis of rotation.

Given that ω is the rotational frequency, the angular momentum L_r of the molecule is given by

$$L_r = I\omega. \tag{8.17}$$

In analogy with the rotational motion of the electron in the hydrogen atom, in which the angular momentum L is quantized, and is given by

$$L = \sqrt{l(l+1)}\,\hbar, \qquad \text{where } l = 0, 1, 2, 3, \ldots,$$

the angular momentum of the molecule is also quantized, and may be written as

$$\boxed{L_r = \sqrt{K(K+1)}\,\hbar, \qquad K = 0, 1, 2, 3, \ldots,} \tag{8.18}$$

where K is the *rotational quantum number*. Thus the rotational kinetic energy E_r of a rotating diatomic molecule is given by

$$E_r = \tfrac{1}{2}I\omega^2 = \frac{(I\omega)^2}{2I} = \frac{L_r^2}{2I}. \tag{8.19}$$

Or, substituting for L_r from Eq. (8.18), we get

$$\boxed{E_r = K(K+1)\frac{\hbar^2}{2I}.} \tag{8.20}$$

Thus the possible rotational energy levels of $K = 0, 1, 2, 3, \ldots,$ are $E_r = 0$, $2(\hbar^2/2I)$, $6(\hbar^2/2I)$, $12(\hbar^2/2I), \ldots,$ respectively, as shown in Fig. 8.9.

To measure these energy levels experimentally, one observes the absorption spectra of these diatomic molecules. Let us consider a transition in which a diatomic molecule in a lower (initial) rotational-energy state E_i absorbs a photon of energy $h\nu$ and is left in the upper (final) rotational-energy state E_f, such that

$$h\nu = E_f - E_i = [K_f(K_f + 1) - K_i(K_i + 1)]\frac{\hbar^2}{2I}. \tag{8.21}$$

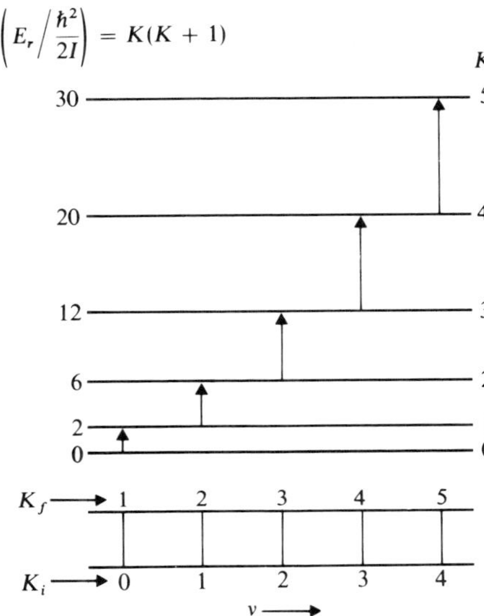

Fig. 8.9 Pure rotational energy levels of a diatomic molecule, plus the transitions observed in the absorption rotational spectrum of diatomic molecules.

When we are dealing with the spectra of molecules, just as in the case of the spectra of atoms, there are certain selection rules that we must observe. Only those rotational transitions are allowed which obey the selection rule

$$\Delta K = \pm 1. \tag{8.22}$$

That is, only transitions between adjacent levels are allowed; and $+1$ corresponds to absorption, while -1 corresponds to emission. Thus in Eq. (8.21), if $K_i = K$, then $K_f = K + 1$. Hence Eq. (8.21) takes the form

$$h\nu = 2(K+1)\frac{\hbar^2}{2I} \tag{8.23}$$

or

$$\boxed{\nu = (K+1)\frac{\hbar}{2\pi I}.} \tag{8.24}$$

Thus, for $K = 0, 1, 2, 3, \ldots,$

$$\nu = \frac{\hbar}{2\pi I}, \quad 2\frac{\hbar}{2\pi I}, \quad 3\frac{\hbar}{2\pi I}, \quad 4\frac{\hbar}{2\pi I}, \ldots,$$

respectively, which means that the spectral lines resulting from rotational motion are equally spaced, as shown in Fig. 8.9. Because of the low energies of the rotational levels ($E_r \sim 10^{-4}$ eV), the frequencies of these lines fall in the far-infrared and microwave regions, corresponding to the wavelength range from 10^{-3} to 10 cm.

Now suppose that initially all the molecules were in the ground state. What then? According to the selection rules, only the molecules in the ground (lowest) level or state would be excited, and we would expect only one spectral line, corresponding to this transition. But this is not true, because even at room temperature the thermal energies of the molecules, $\sim 4 \times 10^{-2}$ eV (see Example 8.1), are much larger than the rotational energies, which are $\sim 10^{-4}$ eV. Hence not only are some of the molecules in the ground state, but also many are in the first few excited states as well. Thus, in the absorption spectra, we see that molecules in levels other than the ground state are available for excitation to higher states.

Example 8.1. Show that—even at room temperature—a substantial number of molecules of a gas are in excited rotational states.

The kinetic energy E of the molecules of a gas at room temperature is $\frac{3}{2}kT$, where k is the Boltzmann constant and T is the absolute temperature. That is,

$$E = \tfrac{3}{2}kT = \tfrac{3}{2} \times 1.38 \times 10^{-23}\ \text{J/°K} \times 300\text{°K}\ \frac{1}{1.6 \times 10^{-19}\ \text{J/eV}}$$
$$= 3.9 \times 10^{-2}\ \text{eV}.$$

Let us say that the wavelength of a typical rotational transition is 1 cm (= 0.01 m), or the energy separation ΔE between the two rotational levels is given by

$$\Delta E = h\nu = \frac{hc}{\lambda} = \frac{6.626 \times 10^{-34}\ \text{J-sec} \times 3 \times 10^{8}\ \text{m/sec}}{0.01\ \text{m} \times 1.6 \times 10^{-19}\ \text{J/eV}}$$
$$\simeq 10^{-4}\ \text{eV},$$

which is much smaller than the thermal energies available. Hence, even at room temperature, a good fraction of the total number of molecules are in excited rotational states.

It is essential to note that interaction between incident electromagnetic radiation and molecules is possible only if the molecules have permanent electric dipole moments. We can observe the pure rotational spectrum only for molecules that are ionically bonded; i.e., for heteropolar and not for homopolar molecules.

Figure 8.10 shows a pure rotational spectrum of HCl in the gaseous phase.[9] By measuring the frequency interval between the absorption lines, we may use Eq. (8.24) to calculate I. From the relation $I = \mu r_0^2$ (since μ is known if the atomic masses are known), we can calculate the equilibrium distance or bond length r_0. Thus, for the HCl molecule, $r_0 = 1.3$ Å; while for the CO molecule, $r_0 = 1.1$ Å.

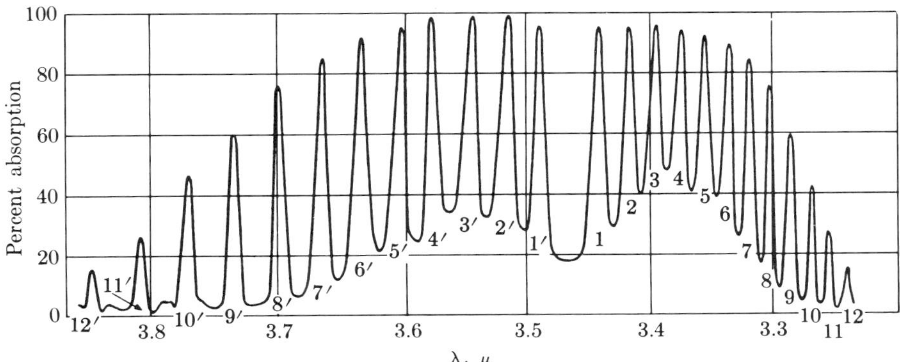

Fig. 8.10 Pure rotational absorption spectrum of diatomic molecules of HCl in the gaseous phase. [From G. Herzberg, *Molecular Spectra and Molecular Structure*, New York: Van Nostrand, 1939]

b) Molecular vibrational spectra

The two atoms in a diatomic molecule not only have rotational motion, but also vibrate about their equilibrium position r_0. The displacement measured from r_0 is given by $(r - r_0)$. The variation of potential energy $V(r)$ with r, explained in Section 8.2, is reproduced in Fig. 8.11. If we limit our discussion to small displacements, the actual potential curve may be approximated by a parabola (the dashed curve in Fig. 8.11). This parabolic potential may be written as

$$V(r) = \tfrac{1}{2}k(r - r_0)^2, \tag{8.25}$$

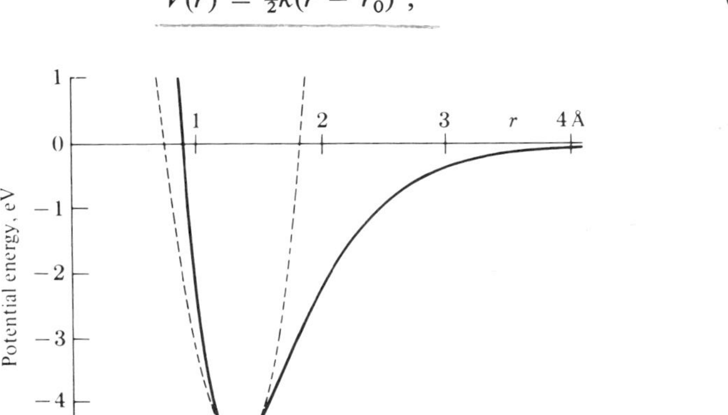

Fig. 8.11 The continuous curve is a plot of the actual potential $V(r)$ between two atoms of a diatomic molecule versus the separation distance r between the atoms. The dashed curve is a parabolic approximation of the actual one, and represents the potential of a simple harmonic oscillator.

where k is a constant. We may replace $(r - r_0)$ by x, and write

$$V(x) = \tfrac{1}{2}kx^2. \tag{8.26}$$

This is the familiar form of the potential of a simple harmonic oscillator.

In order to obtain the vibrational energy levels, let us solve the Schrödinger wave equation,

$$-\frac{\hbar^2}{2\mu}\frac{d^2\psi(x)}{dx^2} + \tfrac{1}{2}kx^2\psi(x) = E\psi(x). \tag{8.27}$$

We discussed the solution of this problem of the simple harmonic oscillator in Section 3.7. There we found that the energy of the oscillator is quantized, and that the total vibrational energies E_v of the allowed levels of the oscillator are given by

$$\boxed{E_v = (v + \tfrac{1}{2})\hbar\omega_0, \qquad v = 0, 1, 2, 3, 4, \ldots,} \tag{8.28}$$

where v is the *vibrational quantum number* and ω_0 is the classical angular frequency $\omega_0 = 2\pi\nu_0 = \sqrt{k/\mu}$.

Thus, for $v = 0, 1, 2, 3, 4, \ldots$, the vibrational levels have energies $\frac{1}{2}\hbar\omega_0$, $\frac{3}{2}\hbar\omega_0$, $\frac{5}{2}\hbar\omega_0$, $\frac{7}{2}\hbar\omega_0, \ldots$, as shown in Fig. 8.12. Note that, according to classical theory, the oscillator may be at rest and hence have zero energy, but according to quantum mechanics the minimum energy of vibration is $\frac{1}{2}\hbar\omega_0$, not zero.

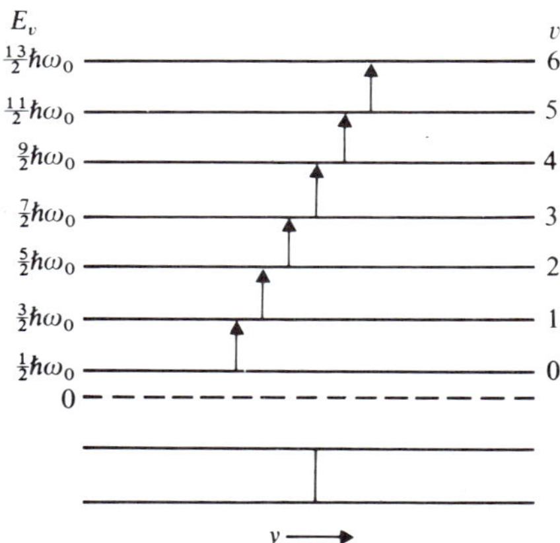

Fig. 8.12 Pure vibrational energy levels of a diatomic molecule. The absorption spectrum will result in only one vibrational frequency.

A transition between an initial level of energy E_i and a final level of energy E_f gives a photon of energy:

$$\begin{aligned} h\nu &= E_f - E_i = (v_f + \tfrac{1}{2})\hbar\omega_0 - (v_i + \tfrac{1}{2})\hbar\omega_0 \\ &= (v_f - v_i)\hbar\omega_0. \end{aligned} \tag{8.29}$$

According to the selection rules, only those transitions are allowed for which

$$\Delta v = (v_f - v_i) = \pm 1, \tag{8.30}$$

where $\Delta v = +1$ corresponds to absorption of energy and $\Delta v = -1$ corresponds to emission of energy. This means that Eq. (8.29) takes the form

$$h\nu = E_f - E_i = \hbar\omega_0. \tag{8.31}$$

That is, all transitions have a single frequency, as shown in Fig. 8.12. Vibrational transitions, like rotational transitions, occur only in those diatomic molecules which have permanent electric dipole moments, such as HCl, but homopolar molecules like N_2 and H_2 do not exhibit pure vibrational spectra.

Example 8.2. The molecules of HCl show a strong absorption line of wavelength $\lambda = 3.465$ microns. Assuming that this line is due to vibrational motion, calculate the force constant for the HCl bond.

The vibrational motion of a molecule is similar to that of a simple harmonic oscillator for which $F = -kx$ and $V(x) = -\int_0^x F(x)\,dx = \frac{1}{2}kx^2$. Hence

$$\nu_0 = \frac{1}{2\pi}\sqrt{\frac{k}{\mu}}$$

or

$$\begin{aligned} k &= 4\pi^2\nu_0^2\mu \\ &= 4\pi^2\left(\frac{c}{\lambda_0}\right)^2\mu, \end{aligned}$$

where $c = 3 \times 10^8$ m/sec, $\lambda_0 = 3.465 \times 10^{-6}$ m, and the reduced mass μ is

$$\begin{aligned} \mu &= \frac{m_1 m_2}{m_1 + m_2} = \frac{(1.0087\text{ u})(35.453\text{ u})}{(1.0087\text{ u}) + (35.453\text{ u})} \\ &= \frac{35.74}{36.46}\text{ u} = 0.98\text{ u} = 0.98 \times 1.67 \times 10^{-27}\text{ kg} \\ &= 1.63 \times 10^{-27}\text{ kg}. \end{aligned}$$

Therefore

$$\begin{aligned} k &= 4\pi^2\left(\frac{3 \times 10^8\text{ m/sec}}{3.\ 65 \times 10^{-6}\text{ m}}\right)^2 \times 1.63 \times 10^{-27}\text{ kg} \\ &= 480\text{ newtons/meter}. \end{aligned}$$

Vibrational transitions in diatomic molecules are observed in the near-infrared region. The energies of vibrational transitions are about 100 times larger than those of rotational transitions. This means that, when we are exciting vibrational levels, we cannot avoid exciting rotational levels as well, and there are many rotational levels associated with each vibrational level, as shown in Fig. 8.13. The total molecular energy E_{vr} of these levels, which is due to both vibration and rotation of the molecules, is

$$E_{vr} = E_v + E_r = (v + \tfrac{1}{2})\hbar\omega_0 + K(K + 1)\frac{\hbar^2}{2I}, \tag{8.32}$$

where $\hbar^2/2I \sim 10^{-4}$ eV, while $\hbar\omega_0 \sim 10^{-1}$ eV. Hence each vibrational level has a fine structure, as shown in Fig. 8.13. The transitions between these vibrational

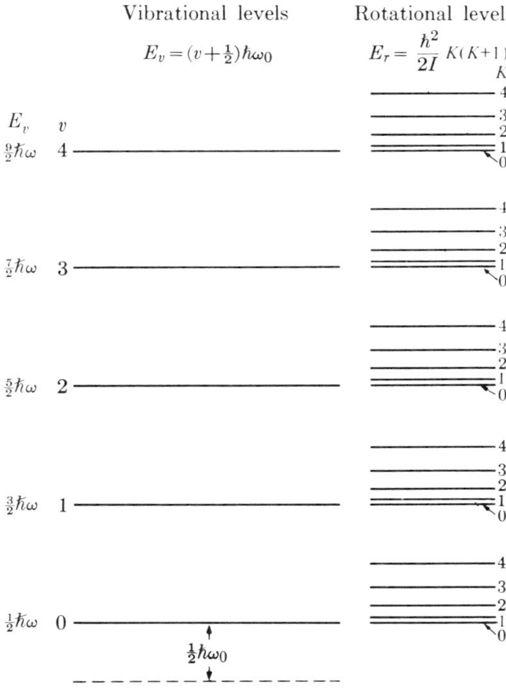

Fig. 8.13 Vibrational excitation of diatomic molecules leads to the excitation of many rotational levels corresponding to each vibrational level, thus resulting in the fine structure of the vibrational levels.

levels lead to a vibrational-rotational spectrum, the frequencies of which are given by

$$\begin{aligned} h\nu &= (E_{vr})_f - (E_{vr})_i \\ &= \left[(v_f + \tfrac{1}{2})\hbar\omega_0 + K_f(K_f + 1)\frac{\hbar^2}{2I}\right. \\ &\qquad\left. - (v_i + \tfrac{1}{2})\hbar\omega_0 - K_i(K_i + 1)\frac{\hbar^2}{2I}\right]. \end{aligned} \tag{8.33}$$

Imposing the selection rules $\Delta K = \pm 1$ and $\Delta v = \pm 1$, we obtain

$$\nu = \nu_0 + (K + 1)\frac{\hbar}{2\pi I}. \tag{8.34}$$

Figure 8.14 shows the resulting vibrational-rotational spectrum. Note that the frequencies corresponding to different transitions are equally spaced (spacing = $\hbar/2\pi I$) on both sides of the central frequency ν_0, with $\Delta K = +1$ transitions on

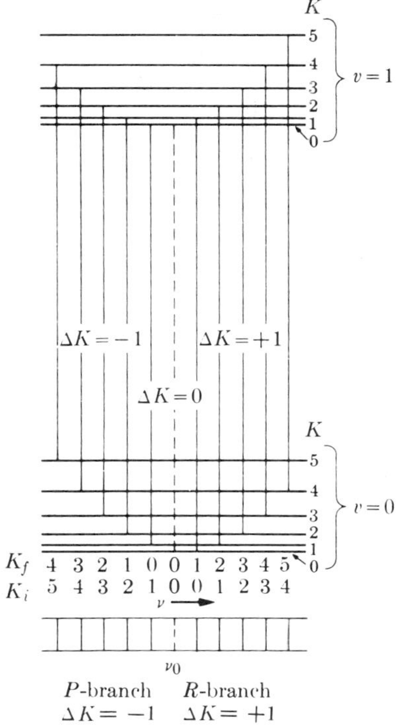

Fig. 8.14 Vibrational-rotational transitions of a diatomic molecule for a vibrational transition between $v = 0$ and $v = 1$. The central missing line (shown dashed) of frequency ν_0 is not allowed by selection rules.

the right (called the R-branch) and $\Delta K = -1$ transitions on the left (called the P-branch). The central line of frequency ν_0 is missing because it is not allowed by the selection rule; that is $\Delta K = 0$ is forbidden.

Figure 8.15 shows the vibrational-rotational absorption spectrum of HCl molecules. The chlorine has two isotopes, of mass numbers 35 and 37, and hence gives two slightly different values of ν_0 and I. This leads to slightly displaced spectra. Each transition has two close lines, as shown for the two isotopes ^{35}Cl and ^{37}Cl in $^1H^{35}Cl$ and $^1H^{37}Cl$, respectively, as shown in Fig. 8.15.

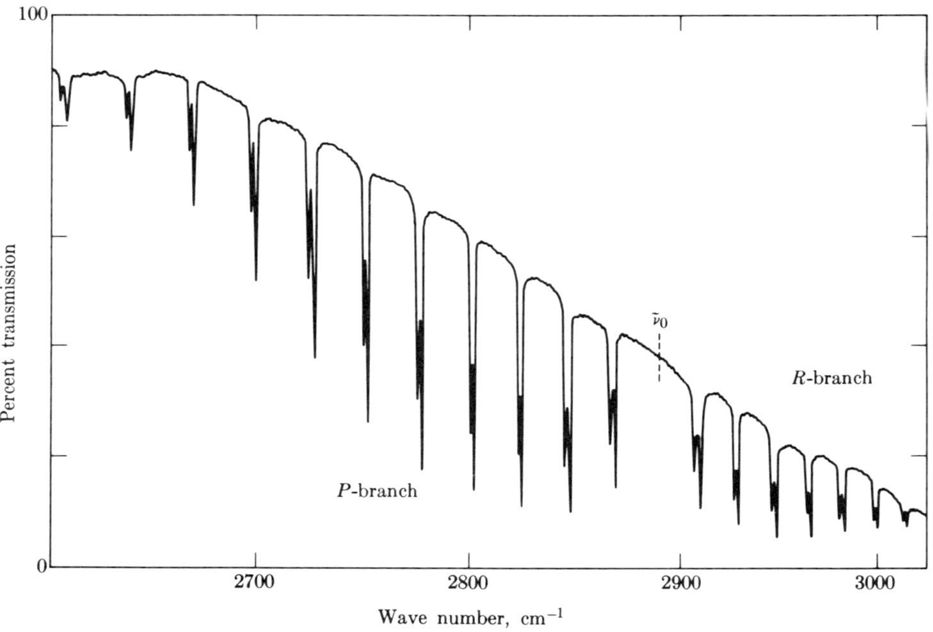

Fig. 8.15 The vibrational-rotational absorption spectrum of HCl molecules in the gaseous phase. The big dips are due to $^1H^{35}Cl$ molecules and the smaller dips to $^1H^{37}Cl$ molecules. Note the missing central line of frequency ν_0. [From Alonso and Finn, *Fundamental University Physics*, Volume III. Reading, Mass.: Addison-Wesley, 1968]

c) Electronic transitions in molecules

If the excitation energy is high, say from 1 to 10 eV, the situation becomes quite complicated. High excitation energy causes the electronic states to become excited, in addition to causing excitation of the vibrational and rotational states. When the electron is in an excited state, the molecular configuration is changed, and this leads to a change in the potential energy. Hence the vibrational and rotational levels, the moment of inertia, and ω_0 all change.

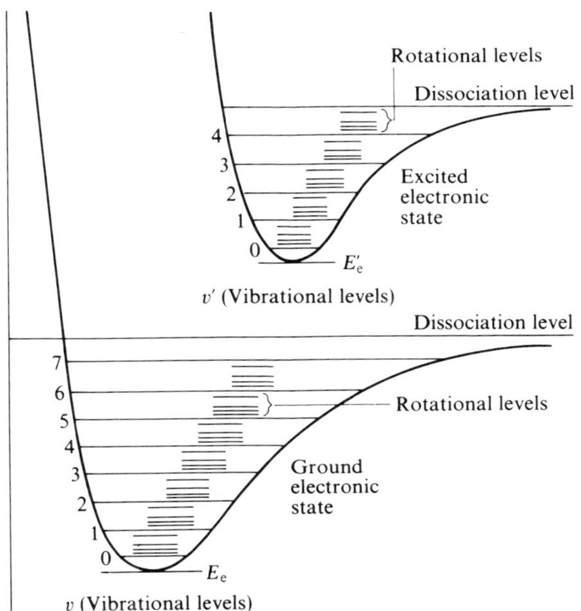

Fig. 8.16 Energy-level diagram showing the rotational and vibrational levels associated with the first two excited electronic states of a diatomic molecule.

Figure 8.16 shows two different excited electronic states of a diatomic molecule (upper and lower electronic states); corresponding to each there are vibrational and rotational levels. The electronic transitions are in the visible and the ultraviolet region.

The approximate energy of a molecule when it experiences an electronic transition is given by

$$E = E_e + E_v + E_r = E_e + (v + \tfrac{1}{2})\hbar\omega_0 + K(K + 1)\frac{\hbar^2}{2I}, \tag{8.35}$$

where E_e is the energy of the electron at the minimum of the potential, while the change in energy that takes place during an electronic transition between the initial and final state is

$$\begin{aligned} \Delta E_e &= (E_e)_f - (E_e)_i, \\ \Delta E_v &= (E_v)_f - (E_v)_i, \\ \Delta E_r &= (E_r)_f - (E_r)_i. \end{aligned} \tag{8.36}$$

Or we may write the frequency of the transition as

$$\nu = \frac{\Delta E}{h} = \frac{\Delta E_e}{h} + \frac{\Delta E_v}{h} + \frac{\Delta E_r}{h}$$
$$= \nu_e + [(v_f + \tfrac{1}{2})\nu_{0f} - (v_i + \tfrac{1}{2})\nu_{0i}]$$
$$+ \left[K_f(K_f + 1)\frac{\hbar}{2\pi I_f} - K_i(K_i + 1)\frac{\hbar}{2\pi I_i}\right]. \tag{8.37}$$

Only those transitions are allowed which obey certain selection rules. Thus $\Delta K = 0, \pm 1$, but 0 to 0 is forbidden. There is no restriction on v, but the transitions take place according to the Franck–Condon principle,[10] which we shan't discuss here. Electronic transitions give rise to closely packed lines, converging to a "head" on one side of the band. In a typical case, for a given electronic state, there are many vibrational levels, and for each vibrational level there are many rotational levels. As an example, Fig. 8.17 shows bands (groups of lines corresponding to different wavelengths) in a portion of the molecular spectrum of cyanogen (CN).

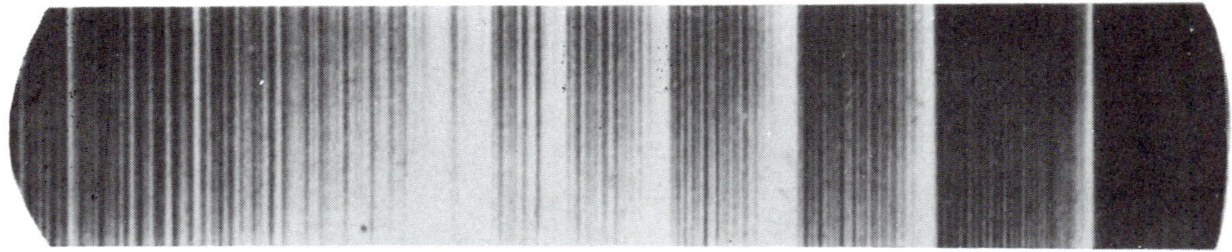

Fig. 8.17 Typical diatomic molecular spectrum (rotational, vibrational, and electronic) showing bands in a portion of the molecular spectrum of cyanogen (CN) in the ultraviolet region.

8.4 THE THREE STATISTICS

We are usually not interested in the behavior of an individual particle in a system or assembly of particles, but would like to know the overall behavior of the whole system. That is, we are interested in macroscopic and not microscopic properties. Statistical mechanics is a branch of physics which deals with the macroscopic properties of a system of particles. As the name indicates, statistical mechanics enables us to predict the *most probable behavior*, which in most cases is sufficient, and usually agrees with experimental results. We may state the problem of statistical mechanics as follows.

Suppose we are dealing with a system consisting of a large number of molecules N. According to the kinetic theory of gases, this system has a well-defined mean energy $\bar{E}$, which is a function of temperature T only. The problem of statistical mechanics is to determine how the total energy E of the system is distributed among the N molecules, keeping in mind that $\bar{E}$ must be equal to E/N. The distribution functions, or more accurately the *probability distribution functions*, are temperature-dependent functions. Their nature is determined by the type of system, i.e., whether the system consists of molecules, electrons, protons, neutrons, atoms, or photons, etc. According to statistical mechanics, if the particles in a system are completely distinguishable, we can calculate facts about the system by classical statistics, which results in the *Maxwell–Boltzmann distribution function*. If, on the other hand, the particles are indistinguishable, we use *quantum statistics* to analyze the system. Furthermore, if the indistinguishable particles have half-integral spins, we use the *Fermi–Dirac distribution function*; while if the particles have integral spin, they are governed by the *Bose–Einstein distribution function*. Let us now discuss these statistics briefly.

a) Maxwell–Boltzmann statistics[11]

Maxwell–Boltzmann statistics (or classical statistics) apply to systems consisting of completely distinguishable particles. They hold in the case of particles which are separated by large distances, such as molecules in a gas. The spins of the particles do not play any part in classical statistics.

Let us consider a system containing a large number of molecules N, with a total energy E distributed among i energy states $E_1, E_2, \ldots, E_i$ in the system. Since all the molecules are distinguishable, the energy state E_1 may be occupied by any one of the N molecules; i.e., there are N different ways of filling the E_1 state. After the first state is filled, there are $(N - 1)$ ways of filling the E_2 state. Similarly, state E_3 can be filled $(N - 2)$ different ways, and so on, until energy state E_i has only one choice. Thus the total number of ways of filling the i energy states is

$$N(N - 1)(N - 2) \cdots 3 \cdot 2 \cdot 1 = N! \tag{8.38}$$

Let us remove the restriction that there is only one molecule in each state. Let there be N_1 molecules in energy state E_1, N_2 molecules in energy state $E_2, \ldots, N_i$ molecules in energy state E_i, such that

$$N_1 + N_2 + \cdots + N_i = \sum_i N_i = N \tag{8.39}$$

and

$$N_1E_1 + N_2E_2 + \cdots + N_iE_i = \sum_i N_iE_i = E. \tag{8.40}$$

The number of possible arrangements that result from such a distribution of molecules among these energy states is called the *thermodynamic probability* P, and is given by

$$P = \frac{N!}{N_1!\, N_2! \cdots N_i!} = \frac{N!}{\Pi N_i!}. \tag{8.41}$$

(Note that rearrangement of N_i molecules in energy state E_i among themselves does not give a new arrangement.) The maximum value of P gives the most probable distribution, and may be obtained by causing a variation in P, that is, $\delta P = 0$, and combining this with the conditions of Eqs. (8.39) and (8.40). Without going into further mathematical details, we simply give the results. The average number of particles, $N(E_i)$, in a state with energy E_i according to Maxwell–Boltzmann (M–B) statistics is given by (dropping the subscript i)

$$N(E) = Ae^{-E/kT}, \tag{8.42}$$

where A is a constant, T is the absolute temperature at which the particles in a system are in equilibrium, and k is the Boltzmann constant, which is equal to 1.38054×10^{-23} joule/°K. Often it happens that there is more than one state with the same energy, i.e., the levels are degenerate, in which case Eq. (8.42) in a more general form is written as

$$N(E) = g(E)Ae^{-E/kT}, \tag{8.43}$$

where $g(E)$ is the *statistical weight factor* if we are dealing with discrete energy levels, and *density of the states* if we are dealing with a continuum of levels. For discrete excited states,

$$g(E) = (2 \times \text{quantum number of the state} + 1). \tag{8.44}$$

Figure 8.18 shows the plot of $N(E)/g(E)$ versus E at two different temperatures, according to Eq. (8.43). Note that these plots predict the *average* behavior of the molecules, not the exact behavior. Note also that this average behavior is not similar to the average behavior used in quantum mechanics.

The M–B statistics have been successfully applied to ideal gases; they have also been used to calculate specific heats of diatomic molecules.

b) Fermi–Dirac statistics[12]

Fermi–Dirac statistics apply to systems containing identical, indistinguishable particles of spin $\frac{1}{2}$ (or half-integer spins). These particles must obey the Pauli exclusion principle. Such particles are called *fermions*, after the Italian physicist Enrico Fermi. Examples are electrons in metals and systems of neutrons or protons.

In Maxwell–Boltzmann statistics, we assumed that there is no restriction on the number of particles that can be in a given level. But this is not true of fermions, which must obey Pauli's exclusion principle. This means that there can be only

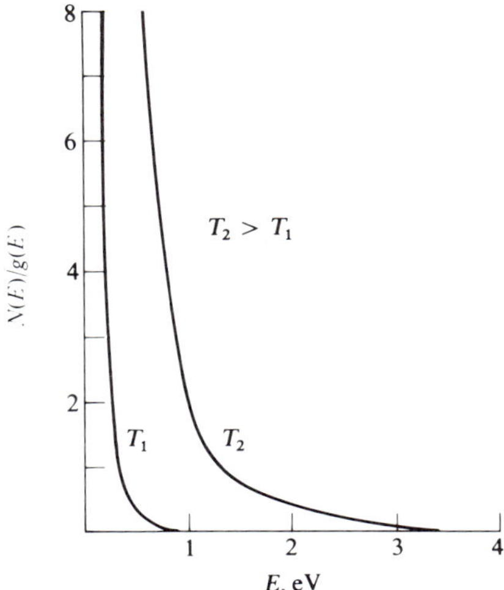

Fig. 8.18 Plot of the distribution function $N(E)/g(E)$ versus E, according to Maxwell–Boltzmann statistics, for systems at two different temperatures.

one particle in each state, and hence the total number of particles must be less than or equal to the total number of states available. Such considerations applied to fermions led to the following distribution, called the Fermi–Dirac (F–D) distribution law:

$$N(E) = \frac{g(E)}{e^{(E-E_F)/kT} + 1}, \tag{8.45}$$

where $N(E)$ is the number of particles in an energy state E, $g(E)$ is the statistical weight factor, and E_F is the *Fermi energy*. E_F is independent of temperature and is constant in many physical problems.

Figure 8.19 shows the plots of $N(E)/g(E)$ versus E for different temperatures. We shall talk about a typical application of F–D statistics to the electrons in metals in Chapter 9.

c) Bose–Einstein statistics[13]

Bose–Einstein (B–E) statistics apply to systems containing identical, indistinguishable particles of zero or integral spins. Hence Pauli's exclusion principle does not apply. Such particles are called *bosons*; an example is a system containing photons or phonons (acoustical photons are called *phonons*). The Bose–Einstein distribution

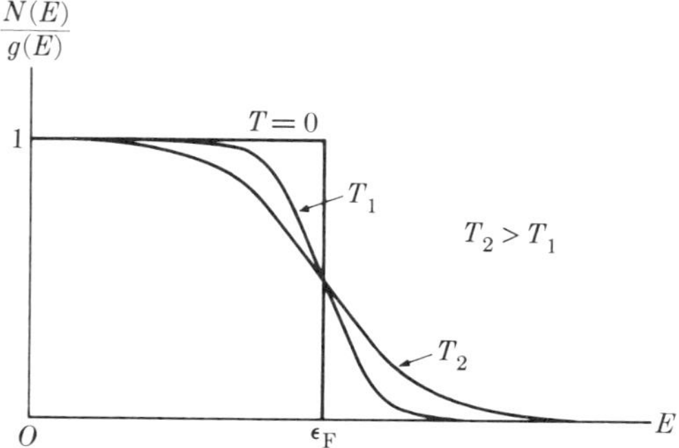

Fig. 8.19 Plot of the distribution function $N(E)/g(E)$ versus E, according to Fermi–Dirac statistics, for systems at two different temperatures.

law is given by.

$$N(E) = \frac{g(E)}{e^{\alpha}e^{E/kT} - 1}. \tag{8.46}$$

A typical case in which this law applies is that of a gas composed of photons for which $\alpha = 0$ and $E = h\nu$. This reduces Eq. (8.46) to

$$N(E) = \frac{g(E)}{e^{E/kT} - 1}. \tag{8.47}$$

Figure 8.20 shows the plot of $N(E)/g(E)$ versus E. Note that for $E \gg kT$, the exponential term in Eq. (8.47) is very large and -1 may be dropped, reducing Eq. (8.47) to Eq. (8.42). That is, B–E statistics reduces to M–B statistics. On the other hand, at low energies, $E \ll kT$ and -1 predominates. This makes $N(E)/g(E)$ much larger for B–E statistics than for M–B statistics at low energies.

8.5 THE LASER[14–16]

Few discoveries have produced such a great impact on the field of optics as the discovery of lasers. The laser is a device for producing very intense, almost unidirectional (or highly directional), monochromatic and coherent visible light beams. The name *laser* stands for *Light Amplification by the Stimulated Emission of Radiation.* Similarly, a device for microwave amplification is called *maser.* This name stands for *Microwave Amplification by the Stimulated Emission of Radiation.* Masers were developed in 1954. Lasers began to be developed in

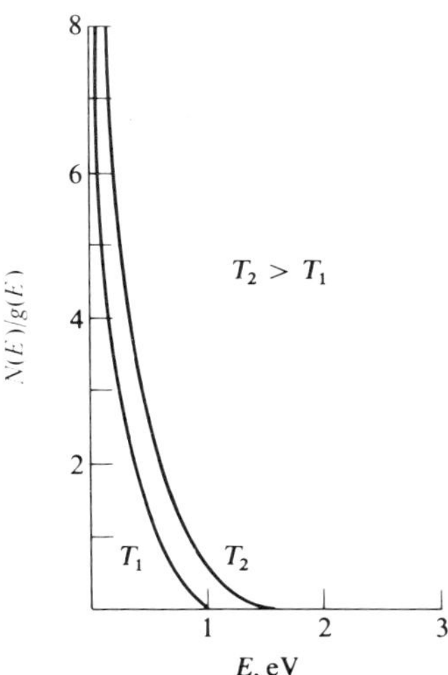

Fig. 8.20 Plot of the distribution function $N(E)/g(E)$ versus E, according to Bose–Einstein statistics, for systems at two different temperatures.

1958. Today different types of lasers—solid-state and gaseous-state lasers—are used for producing light at frequencies from the far infrared to the ultraviolet regions. Let's look at the basic principle of lasers and see how they work.

A beam of photons interacts with free atoms in the ground state or the excited state in many ways. Of these, three processes which are of particular interest are shown in Fig. 8.21. Suppose that a beam of photons of energy $h\nu = E_2 - E_1$ is incident on a sample in which some of the atoms are in the ground state, of energy E_1, while others are in an excited state, of energy E_2. If the photon interacts with an atom in the ground state E_1, the atom absorbs the photon and is left in the excited state E_2, as shown in Fig. 8.21(a). This process is called *induced or stimulated absorption.* Once the atom is in the excited state E_2, it can decay (or drop back to a lower energy state) by two different processes: *spontaneous emission,* as shown in Fig. 8.21(b), or *induced or stimulated emission,* as shown in Fig. 8.21(c). In spontaneous decay, the radiation is emitted in all directions and has energy $h\nu = E_2 - E_1$. In induced emission, the presence of photons of energy $h\nu = E_2 - E_1$ induces the atoms in state E_2 to decay by emitting photons traveling in the direction of the incident photon. This results in a unidirectional coherent beam, as shown in Fig. 8.21(c). Note that for every incident photon, we have *two* outgoing photons. That is why light amplification takes place. Now we can see that the idea is to have a device in which there will be more induced or stimulated emission than spontaneous emission, so as to have a unidirectional, coherent beam.

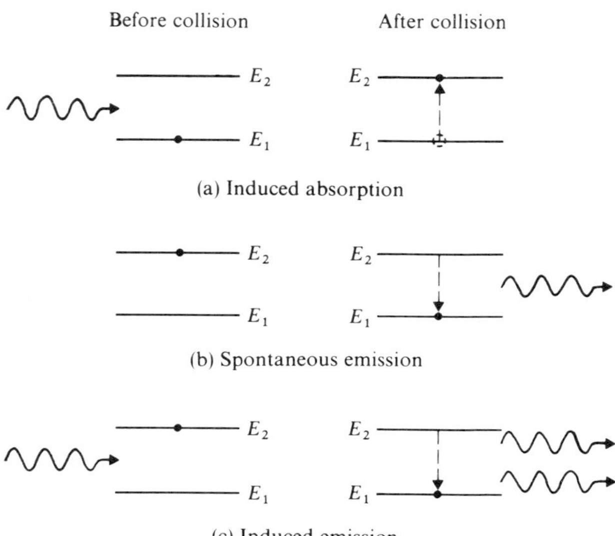

Fig. 8.21 Interaction of photons with atoms in energy states E_1 and E_2. (a) Induced absorption, (b) spontaneous emission, and (c) stimulated or induced emission.

According to Boltzmann's statistics, if a sample contains a large number of atoms, N_0, at temperature T, then in thermal equilibrium the numbers of atoms in energy states E_1 and E_2 are

$$N_1 = N_0 e^{-E_1/kT} \tag{8.48}$$

and

$$N_2 = N_0 e^{-E_2/kT}, \tag{8.49}$$

where k is Boltzmann's constant. For an ordinary optical source of light for which $T \sim 10^3$°K, $E_1 < E_2$ and $N_1 > N_2$, as shown in Fig. 8.22(a). Also, in these circumstances, the probability of induced emission is much less than the probability of spontaneous emission. Since spontaneous transitions occur in a random fashion, a visible source of light emits incoherent radiation. The basic requirement of a laser is to have predominantly *induced* transitions, so that radiation is emitted in the same direction as the incident radiation, and also has a definite phase relationship. This gives the radiation a much greater intensity as compared to the light from an ordinary source. One can achieve this by causing *population inversion*, so that there are more atoms in the higher energy state E_2 than in the lower energy state E_1. That is,

$$E_1 < E_2 \quad \text{but} \quad N_1 < N_2,$$

as shown in Fig. 8.22(b). Let's see how this works.

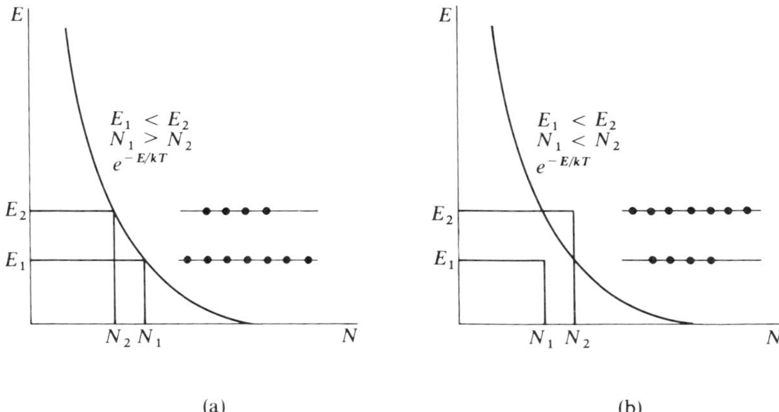

Fig. 8.22 (a) In a normal Boltzmann distribution, when $E_1 < E_2$, $N_1 > N_2$. (b) When the population is inverted, $E_1 < E_2$ and $N_1 < N_2$.

Consider a material whose atoms—when raised from the ground state E_1 to the excited state E_3—do *not* decay back to state E_1 because such transitions are forbidden by the selection rules (Fig. 8.23). On the other hand, atoms in the excited state E_3 (which has a mean life of $\sim 10^{-8}$ sec) decay spontaneously to state E_2. If it so happens that the lifetime of state E_2 is much longer than 10^{-8} sec—i.e., if it is a metastable state—the atoms reach state E_2 much faster than they leave state E_2. This leads to an increased number of atoms in state E_2 as compared to

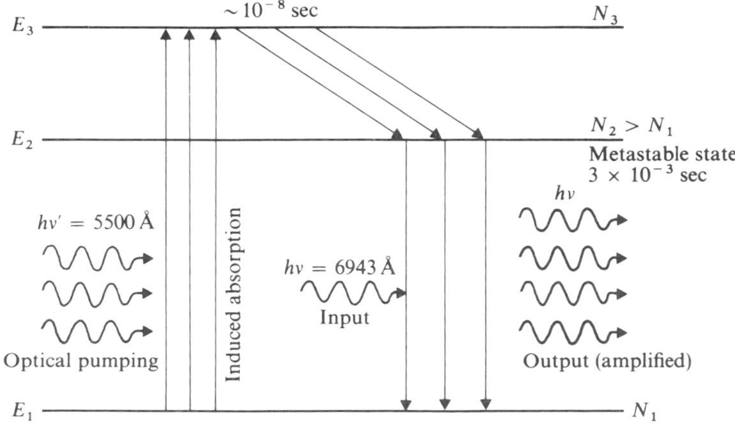

Fig. 8.23 Schematic representation of the working of a ruby laser, showing optical pumping, population inversion, and amplification.

the number in E_1, which means that population inversion has been achieved. This process, illustrated in Fig. 8.23, is called *optical pumping.*

After population inversion is achieved, state E_2 is exposed to a beam of photons of energy $h\nu = E_2 - E_1$, which causes induced or stimulated emission (Fig. 8.23). This gives rise to a coherent, highly intense beam of photons of energy $h\nu$, traveling in the direction of the incident beam.

The materials used in lasers are usually gases and solids; a few are liquids. Most common are the *helium–neon laser* and the *ruby laser.* The ruby laser consists of aluminum oxide (Al_2O_3) doped with $\sim$0.05% chromium (Cr) atoms replacing Al atoms. The laser action is due to the Cr atoms, while Al_2O_3 is the host material. Figure 8.23 shows the energy-level diagram of Cr ions in the ruby laser. The state E_2 is a metastable state with a lifetime of 3×10^{-3} sec. When light of wavelength 5500 Å is incident on the ruby laser, the Cr ions are raised from E_1 to E_3. The decay of E_3 to E_1 is forbidden. Hence ions in state E_1 decay spontaneously to state E_2. The state E_2 is metastable, so population inversion takes place. Thus, if a few photons of wavelength 6943 Å (corresponding to the red light of ruby) are incident on the metastable state, the stimulated emission is more than the spontaneous emission from state E_2.

The ruby laser, in the form of a cylinder (Fig. 8.24), is placed in a supporting material such as glass. The ends of the glass are optically ground and polished so as to form a cavity. Or the ruby crystal itself may be polished, and the supporting material may be dispensed with. The polished surfaces (of the glass or the ruby) reflect the beam many times before emerging.

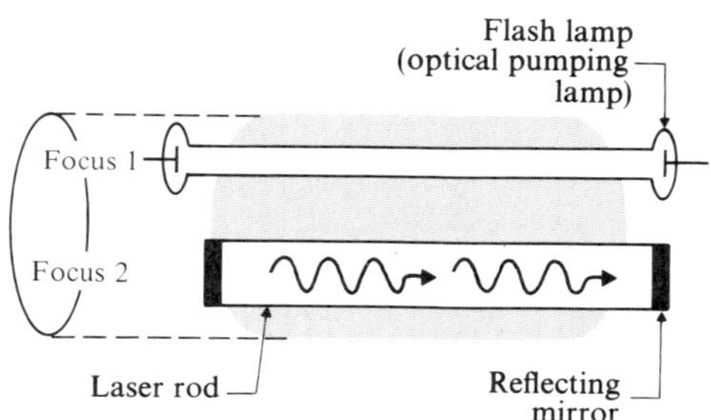

Fig. 8.24 A typical solid-state laser, with an optical pumping lamp.

The optical pumping is done by an external source, as shown in Fig. 8.24. The amplifying action is initiated by the few spontaneous transitions of ruby red light from state E_2. Figure 8.25 is a photograph showing a high-intensity laser beam at work.

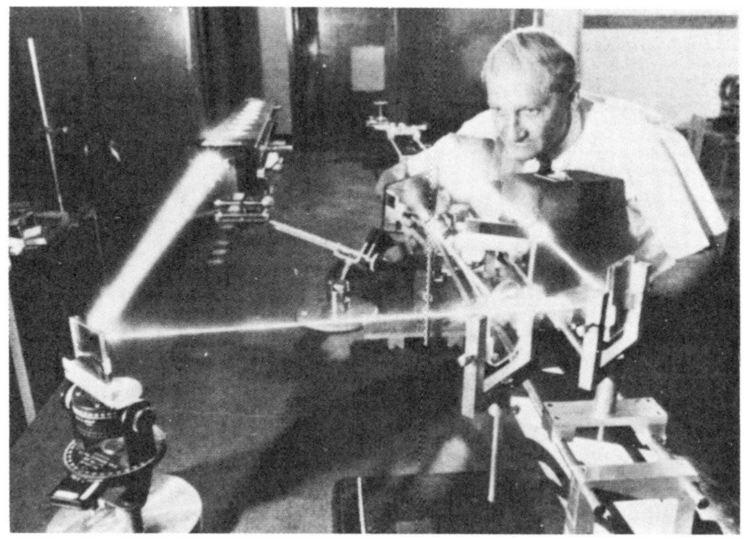

Fig. 8.25 A high-intensity laser beam is used by Stanford research engineer Matt Lehmann in making the first three-dimensional movie without a lens. (Photo by Associated Press)

SUMMARY

There are four common types of bonding by which atoms in molecules and solids may be held together: ionic, covalent, van der Waals, and metallic. Of these, the first two are the most common in molecules.

Ionic (or heteropolar) bonding in a molecule is the result of the shifting of an electron from one atom to the other, thus forming positive and negative ions, which then are attracted to one another electrostatically, forming a stable molecule.

Covalent (or homopolar) bonding in a molecule results from the attractive force due to the sharing of one or two pairs of electrons by both atoms in the molecule.

Pure rotational motion of the atoms in a diatomic molecule about their center of mass results in rotational energy levels given by

$$E_r = K(K+1)\frac{\hbar^2}{2I}, \qquad K = 0, 1, 2, 3, \ldots,$$

and the frequencies of rotational transitions are given by

$$\nu = (K+1)\frac{\hbar}{2\pi I}.$$

Vibrations of the atoms in diatomic molecules about their equilibrium position r_0 result in vibrational levels given by

$$E_v = (v + \tfrac{1}{2})\hbar\omega, \qquad v = 0, 1, 2, 3, 4, \ldots,$$

and the frequency of vibrational transitions is given by

$$h\nu_0 \; (= \hbar\omega_0), \qquad \text{where } \nu_0 = \frac{1}{2\pi}\sqrt{\frac{k}{\mu}}$$

In addition to the rotational and vibrational transitions, transitions can also be caused by electronic excitation of the atoms in the molecules.

There are three statistics which predict the most probable behavior of a system of particles: *Maxwell–Boltzmann statistics*, or classical statistics, applies to systems consisting of completely distinguishable particles. *Fermi–Dirac statistics* applies to systems containing identical, indistinguishable particles of half-integer spins which obey the Pauli exclusion principle. *Bose–Einstein statistics* applies to systems containing identical, indistinguishable particles of zero or integral spins.

Laser stands for *L*ight *A*mplification by the *S*timulated *E*mission of *R*adiation. A laser is a device in which there are more induced or stimulated emissions than spontaneous emissions. This results in a very intense, highly directional, monochromatic, and coherent visible light beam. The He-Ne laser and ruby laser are commonly used. The lasing action is achieved by causing population inversion in the material.

Maser stands for *M*icrowave *A*mplification by *S*timulated *E*mission of *R*adiation; its working is similar to that of the laser.

PROBLEMS

1. The dissociation energy of a NaCl molecule is -3.58 eV. Express this energy in units of kcal/mole.
2. For a HCl molecule, the equilibrium distance is 1.27 Å; the ionization potential is 13.8 eV. Calculate (a) the coulomb energy between the two ions at the equilibrium distance, (b) the dissociation energy.
3. The ionization energy of potassium is 4.34 eV, and the electron affinity of fluorine is 4.07 eV/electron. The equilibrium distance is 2.67 Å. Calculate (a) the coulomb energy between the two ions at the equilibrium distance, (b) the dissociation energy.
4. For a N_2 molecule, the equilibrium distance is 1.09 Å. The ionization energy of N_2 is 15.51 eV. Calculate (a) the coulomb energy of nitrogen ions at the equilibrium distance, (b) the dissociation energy.
5. The Bohr correspondence principle discussed in Section 5.6 also applies to the rotational motion of molecules. Prove that, for large rotational quantum numbers, the frequency of a photon resulting from a transition between adjacent energy levels

of a diatomic molecule is the same as the classical frequency of rotation of the molecule about its center of mass.

6. The equilibrium distance between the two nuclei of HI molecules is 1.61 Å. Calculate the energies of the rotational levels of this molecule and the wavelengths of the first two allowed transitions.

7. Calculate (a) the moment of inertia, (b) the equilibrium distance for a HCl molecule from the following data. The rotational spectrum of HCl molecules yields the following wave numbers (number of waves/cm) for the first three rotational lines: 81, 105, and 125.

8. HCl molecules consist of two molecules, $^1H^{35}Cl$ and $^1H^{37}Cl$, which have slightly different weights. Assume that the equilibrium distance for both of them is the same: 1.27 Å. (a) Calculate the first three rotational levels for each. (b) Calculate the wavelengths of the first two transitions for each. (c) Calculate the difference in the wavelengths between the first two transitions for each.

9. How does the increase in the vibrational energy of a molecule affect its moment of inertia?

10. Consider the vibrational motion of two masses M and m tied by a spring of elastic constant k. Show that these masses vibrate with a frequency ν given by

$$\nu = \frac{1}{2\pi}\sqrt{\frac{k}{\mu}},$$

where μ is the reduced mass and k is the force constant for the bond.

11. Given that the spacing between the vibrational levels of a CO molecule is 8.45×10^{-2} eV, calculate the value of the force constant k of the bond in a CO molecule.

12. Are the thermal energies of a molecule at room temperature sufficient to cause excitation of the vibrational levels? Is it likely that an F_2 molecule will be vibrating in its first excited vibrational state at room temperature? The force constant for the bond is 966 newtons/meter.

13. The H_2 molecule shows a strong absorption line at 2.3×10^{-6} m. Calculate (a) the force constant for the H_2 bond, (b) the zero-point energy, (c) the energies of the first three excited levels.

14. Assume that the force constant of the H_2 molecule is the same as that of D_2 (deuterium, an isotope of hydrogen). The H_2 molecule shows an absorption line at 2.3×10^{-6} m. What will be the vibrational frequency observed from D_2?

15. Show that, at high energy as compared to thermal temperatures—that is, $E \gg kT$—the Bose–Einstein as well as Fermi–Dirac distribution functions approach the classical Maxwell–Boltzmann distribution function.

16. Using the Maxwell–Boltzmann distribution, calculate the relative number of hydrogen molecules in the first three rotational levels at room temperature.

17. Using the Maxwell–Boltzmann distribution function, calculate the relative number of hydrogen molecules in the first three vibrational levels at 5000°K. How are the rotational levels affected at this temperature?

18. Calculate the temperature at which 5% of the H_2 molecules will be in the first excited rotational level. The bond length of the H_2 molecule is 0.74 Å.

19. Calculate the temperature at which 5% of the CO molecules will be in the first excited vibrational level (0.127 eV).

20. Starting from the Maxwell–Boltzmann distribution and the relation

$$N(v)\,dv = N(E)\,dE,$$

where v is the velocity of the molecules, show that

$$N(v) = A'v^2 e^{-(1/2)mv^2/kT},$$

where A' is a constant dependent on T, m, and k. Make a plot of $N(v)$ versus v.

REFERENCES

1. J. Spic, *Chemical Binding and Structure*, New York: Macmillan, 1964
2. Linus Pauling, *The Nature of the Chemical Bond*, Ithaca, N.Y.: Cornell University Press, 1960
3. H. Gray, *Electrons and Covalent Bonds*, New York: W. A. Benjamin, 1964
4. Max Born and J. R. Oppenheimer, *Ann. Physik* **84,** 457 (1927)
5. F. Hund, *Zeit fur Physik* **40,** 742, **42,** 93 (1927); also R. S. Milliken, *Phys. Rev.* **32,** 186 (1928)
6. Linus Pauling and D. M. Yost, *Proc. Nat. Acad. Sci.* **18,** 414 (1932)
7. Linus Pauling, *Chem. Revs.* **5,** 173 (1928)
8. G. C. King, *Spectroscopy and Molecular Structure*, New York: Holt, Rinehart and Winston, 1964
9. G. Herzberg, *Spectra of Diatomic Molecules*, New York: D. Van Nostrand, 1950
10. R. N. Dixon, *Spectroscopy and Structure*, Chapter 6, London: Methuen, 1965
11. J. C. Maxwell, *Trans. Roy. Soc.* **157,** 49 (1867); *Phil. Mag.* **35,** 129, 185 (1868); L. Boltzmann, *Wien. Ber.* **58,** 517 (1868)
12. Enrico Fermi, *Z. Physik* **36,** 902 (1926); P. A. M. Dirac, *Proc. Roy. Soc.* **A112,** 661 (1926)
13. Robert D. Reed and R. R. Roy, *Statistical Physics*, Chapter 12, New York: Intext Educational Publishers, 1971
14. J. P. Gordon, H. Z. Ziegler, and C. H. Townes, *Phys. Rev.* **95,** 282 (1954)
15. R. C. Jensen and G. R. Fowles, *Proc. IEEE* **52,** 1350 (1964)
16. B. A. Lengyel, *Introduction to Laser Physics*, New York: John Wiley, 1966

Suggested Readings

1. G. Herzberg, *Molecular Spectra and Molecular Structure*, Chapters 2, 3, and 4, New York: D. Van Nostrand, 1950
2. Atam P. Arya, *Fundamentals of Atomic Physics*, Chapter 14. Boston: Allyn and Bacon, 1971
3. R. C. Johnson, *An Introduction to Molecular Spectra*, London: Methuen, 1959
4. H. C. Allen, Jr., and P. C. Cross, *Molecular Vib-Rotors*, New York: John Wiley, 1963
5. E. U. Condon, "Electronic Structure of Molecules," Chapter 7 in *Handbook of Physics*, New York: McGraw-Hill, 1958
6. D. K. C. MacDonald, *Introductory Statistical Mechanics*, New York: John Wiley, 1963
7. G. J. F. Troup, *Masers and Lasers*, second edition, London: Methuen, 1963
8. A. L. Schalow, "Optical Masers," *Sci. Am.*, June 1961; "Advances in Optical Lasers," *Sci. Am.*, July 1963
9. B. Lengyel, *Introduction to Laser Physics*, New York: John Wiley, 1966
10. F. W. Sears, *Thermodynamics, The Kinetic Theory of Gases, and Statistical Mechanics*, Reading, Mass.: Addison-Wesley, 1964.

CHAPTER 9
THE STRUCTURAL PROPERTIES OF SOLIDS

Snow crystals enlarged 50 times (photomicrographs by Vincent J. Schaefer of the Munitalp Foundation in Schenectady, New York). Hexagonal symmetry of crystals is due to their molecular structure. (From A. M. Buswell and W. H. Rodebush, Sci. Am. ***194****, 4, 508, 1950)*

9.1 PHYSICAL STATES OF MATTER

When a very large number of atoms and molecules are brought together, it is the nature and strength of the interactions between them that determines which of the three states—gas, liquid, or solid—will be formed. In gases, the average distance between molecules is very large compared to the size of the molecule. The forces between the molecules are therefore much weaker than the forces that hold atoms together in the molecules. At the other extreme, in solids, the forces holding atoms or molecules together are of the order of molecular binding forces, and the distances between the atoms or molecules are almost fixed. Hence solids have rigid structure, which gives rise to elastic behavior. Liquids have characteristics that fall between those of the gaseous and the solid states.

Solid materials are of two kinds: (a) a *crystalline solid* consists of a regular arrangement of atoms in a repeated three-dimensional pattern. When solids exhibit such regularity or periodicity, they are said to form a crystal lattice. (b) An *amorphous solid* (one without form) consists of atoms or molecules tightly bound to one another, but having little or no geometric regularity or periodicity in their arrangement. Figure 9.1 illustrates the difference between a crystalline and an amorphous solid. Examples of amorphous solids are glass, wood, paper, and plastic.

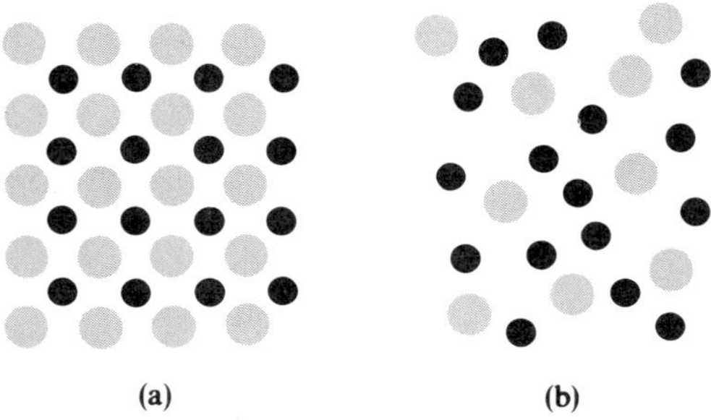

Fig. 9.1 (a) A crystalline solid, (b) an amorphous solid.

Some solids can be found in both crystalline and amorphous forms. Figure 9.2(a) shows crystalline B_2O_3 (boron trioxide), with oxygen atoms in a hexagonal array. Figure 9.2(b) shows amorphous B_2O_3, with only a short-range ordering. This form is a vitreous (glassy) substance.

We shall devote this chapter and the next one to the physical properties and structures of simple crystalline solids only. Organic solids, high polymers, and ceramics are specialized fields, so we shall not discuss them here.

Let us ask ourselves three questions about crystal structures: (a) When many atoms are brought together, why do they form a periodic structure in the lowest energy state? (b) What is the nature and the origin of the forces that hold these atoms together? (c) What are the allowed energy levels of the electrons in atoms of crystals?

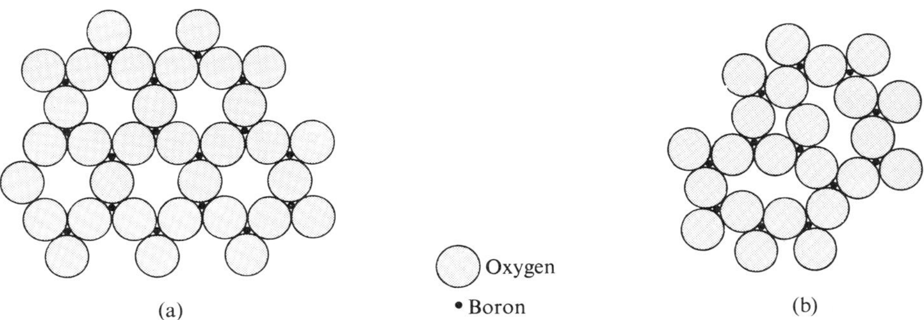

Fig. 9.2 (a) Crystalline boron trioxide (B_2O_3), in which oxygen atoms form a hexagonal array in a long-range ordering. (b) Amorphous B_2O_3 has only a short-range ordering.

The answer to (a) is not yet known; of necessity, we have to accept periodicity in the structure as a fact.

The answer to (b) is: Short-range attractive electrical forces between atoms cause stable configurations to be formed in crystals. However, these forces have to be greater than the forces generated by thermal agitation of the atoms. Bonding forces in solids are like bonding forces in molecules.

The answer to (c) comes from analogy with molecules. When many atoms are brought together in a solid, each level splits into many closely spaced levels, resulting in the formation of *energy bands*. We need to understand these bands in order to understand many properties of solids; we shall discuss them in detail in Chapter 10.

The *cohesive* or *binding energy* of a crystal is a measure of the stability of the crystal. The greater a crystal's binding energy, the greater the amount of energy needed to separate it into its constituent atoms. The binding energy equals the energy of the crystal in the lowest energy state (at absolute zero) minus the sum of the ground-state energies of the isolated atoms. Binding energy is thus a *negative* quantity. It is proportional to the volume of the crystal, and is usually expressed in electron volts per atom or per molecule (1 eV/atom = 23.052 kilocalories per mole).

9.2 BONDING IN SOLIDS

According to the nature and strength of their bonding, the crystals that compose solids may be classified into five categories, although many fall into intermediate categories.

a) *Ionic.* Binding energies about 5–10 eV per molecule; example: sodium chloride.
b) *Covalent.* Binding energies about 10 eV per molecule; example: diamond.
c) *Hydrogen-bonded.* Binding energies about 0.5 eV per molecule; example: ice.

d) *Molecular.* Binding energies up to 0.1 eV; example: methane, CH_4.

e) *Metallic-bonded.* Binding energies $\sim$1–5 eV per molecule; example: sodium.

Ionic and covalent bondings in crystals are the same as ionic and covalent bondings in molecules, discussed in Chapter 8. Therefore, in the following discussion, we shall assume that we already understand the nature of these two kinds of bonding.

a) Ionic crystals. An ionic crystal consists of positive and negative ions in a regular array. As explained in Chapter 8, ionic bonding takes place when atoms that lose electrons easily (elements of group I) combine with atoms that have tightly bound electrons (elements of group VII). The former give up electrons to the latter, so that the former become positive and the latter become negative. In a stable equilibrium, the attractive forces between oppositely charged ions exceed the repulsive forces between similarly charged ones.

Two types of structures are commonly found in ionic crystals: (a) *face-centered-cubic* (fcc) structures, such as sodium chloride (see Fig. 9.3), (b) *body-centered-cubic* (bcc) structures, such as cesium chloride (see Fig. 9.4). The separation between like ions in NaCl is 5.62 Å and in CsCl is 4.11 Å. The *coordination number*, or the number each ion has of nearest neighbors of the opposite kind, is 6 for NaCl and 8 for CsCl.

The cohesive or the internal potential energy of an ionic crystal is the sum of (a) the net attractive electrostatic forces between ions and (b) the short-range repulsive forces due to nuclear repulsion and repulsion of electrons in closed shells

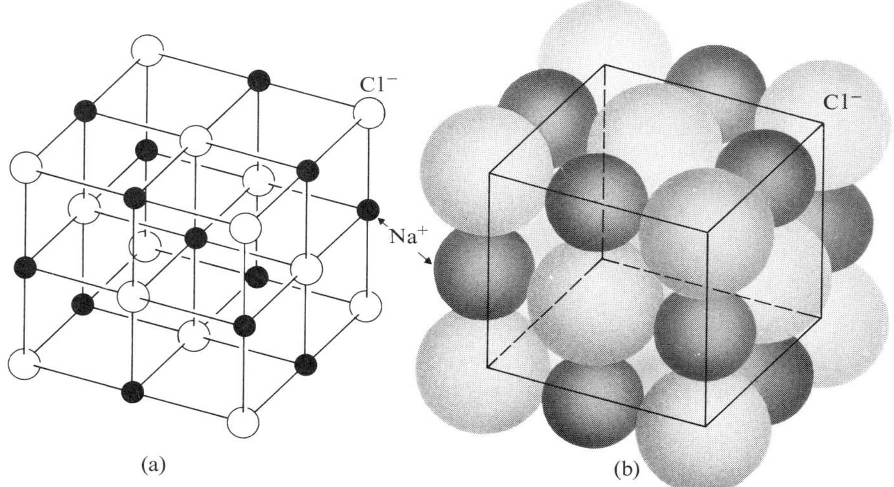

Fig. 9.3 Structure of a faced-centered-cubic crystal, NaCl. (a) Geometrical arrangement, (b) scale model.

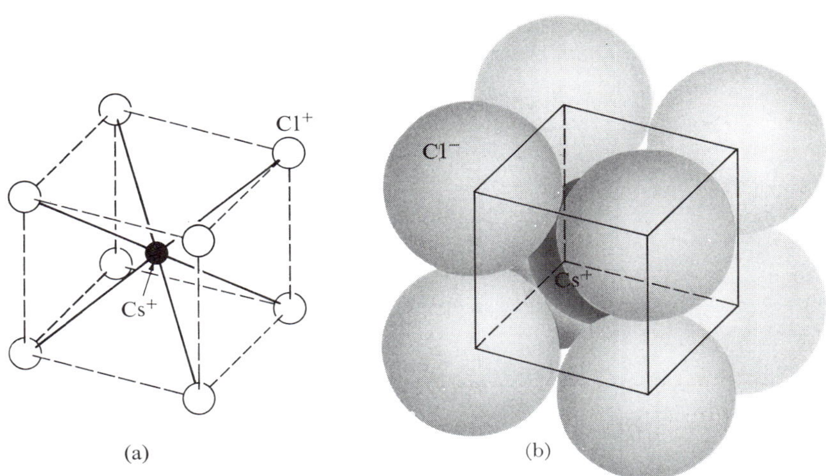

Fig. 9.4 Structure of a body-centered-cubic crystal, CsCl. (a) Geometrical arrangement, (b) scale model.

of different ions. (If only net attractive forces, without any repulsive forces, were present, the crystal would coalesce.) The net attractive electrostatic (or coulomb) potential energy, V_{coul} ($F = -\partial V/\partial r$), of an ionic crystal is given by

$$V_{\text{coul}} = -\alpha \frac{e^2}{(4\pi\epsilon_0)r}, \tag{9.1}$$

where r is the distance between two nearest ions and α is *Madelung's constant*, which depends on the geometry of the crystal,[1] and whose value lies between 1.6 and 1.8. For face-centered-cubic crystals such as NaCl, $\alpha = 1.7476$, and for a body-centered-cubic crystal such as CsCl, α is 1.7627.

Example 9.1. Show that for a face-centered-cubic crystal such as NaCl the Madelung constant α is equal to 1.7476.

To find the Madelung constant, we calculate the net coulomb potential energy V_{coul} acting on any ion. Consider a Na^+ ion in a NaCl crystal (Fig. 9.3). There are six neighboring Cl^- ions near to this Na^+ ion, each at a distance r. Therefore the potential energy of this Na^+ ion and the six Cl^- ions is

$$V_6 = -6 \frac{e^2}{4\pi\epsilon_0 r}.$$

The next-nearest neighbors of this Na^+ ion are another 12 Na^+ ions, each at a distance of $\sqrt{2}\, r$. Thus the potential energy between the first Na^+ ion and the 12 other Na^+ ions is

$$V_{12} = +12 \frac{e^2}{4\pi\epsilon_0 \sqrt{2}\, r}.$$

Next in the series is the group of 8 Cl^- ions, each at a distance of $\sqrt{3}\,r$; and so on. If we add all these and other terms, we get

$$\begin{aligned} V_{\text{coul}} &= V_6 + V_{12} + V_8 + \cdots \\ &= -\frac{1}{4\pi\epsilon_0}\frac{e^2}{r}\left(6 - \frac{12}{\sqrt{2}} + \frac{8}{\sqrt{3}} - 3 + \cdots\right) \\ &= -1.7476\,\frac{1}{4\pi\epsilon_0}\frac{e^2}{r}. \end{aligned}$$

When we compare this with Eq. (9.1), we get $\alpha = 1.7476$.

The German physicist Max Born[2] approximated the repulsive force by the potential

$$V_{\text{repul}} = A\,\frac{e^2}{4\pi\epsilon_0 r^n}, \tag{9.2}$$

where A and n are constants, n being greater than 1, making it a short-range force.

Thus the total effective potential, E_p, is given by

$$\begin{aligned} E_p &= V_{\text{coul}} + V_{\text{repul}} \\ &= -\frac{e^2}{4\pi\epsilon_0}\left(\frac{\alpha}{r} - \frac{A}{r^n}\right). \end{aligned} \tag{9.3}$$

Since, at the equilibrium distance, $r = r_0$, the potential energy is minimum, that is, $(dE_p/dr)_{r=r_0} = 0$, and Eq. (9.3) gives $A = \alpha(r_0^{n-1})/n$. The potential energy at the equilibrium distance is thus

$$\boxed{E_{p,r_0} = -\frac{\alpha}{4\pi\epsilon_0}\frac{e^2}{r_0}\left(1 - \frac{1}{n}\right).} \tag{9.4}$$

Example 9.2. For a NaCl crystal, $r_0 = 2.81$ Å, $\alpha = 1.7476$, and $n = 9$. Calculate the net cohesive energy per atom.

The *lattice energy*—i.e., the energy needed to form a crystal from its ions (rather than from atoms)—according to Eq. (9.4) is

$$\begin{aligned} E_{p,r_0} &= -\alpha\,\frac{1}{4\pi\epsilon_0}\frac{e^2}{r_0}\left(1 - \frac{1}{n}\right) \\ &= -1.7476 \times 9 \times 10^9\ \text{N-m}^2/\text{coul}^2\,\frac{(1.60 \times 10^{-19}\ \text{coul})^2}{2.81 \times 10^{-10}\ \text{m}}\left(1 - \tfrac{1}{9}\right) \\ &= -1.27 \times 10^{-18}\ \text{J}\,\frac{1}{1.6 \times 10^{-19}\ \text{J/eV}} \\ &= -7.94\ \text{eV/ion pair}. \end{aligned}$$

In other words, the lattice energy per ion is $\frac{1}{2}(-7.94) = -3.97$ eV. From this we must subtract the energy needed to form Na^+ and Cl^- ions from Na and Cl atoms, which is equal to $(+5.14 - 3.61) = +1.53$ eV/ion pair, or 0.77 eV per ion. Hence the net cohesive energy per atom is

$$E_{\text{cohesive}} = (-3.97 + 0.77) \text{ eV/atom}$$
$$= -3.20 \text{ eV/atom}.$$

This agrees with the experimentally observed value of -3.28 eV/atom.

Due to the strength of their ionic bonds, most ionic crystals are extremely hard. They are usually brittle, and have high melting points. Also ionic crystals, because they have very few free electrons, are poor conductors of heat and electricity. The electrons in ionic crystals are all paired off, so that their ions have completely filled electron shells. This means that the ions have no magnetic moments, and thus most ionic crystals are diamagnetic.

b) Covalent crystals. The bonding in covalent (or valence) crystals is similar to the bonding in covalent molecules, such as H_2, discussed in Chapter 8. Each atom in a covalent bond contributes an electron, and such electrons are shared equally by both the atoms. These sharing electrons spend more time in between the atoms than anywhere else.

Examples of covalent-bonding crystals are diamond, silicon, silicon carbide, and germanium. Figure 9.5 shows a crystal of diamond, which has tetrahedral

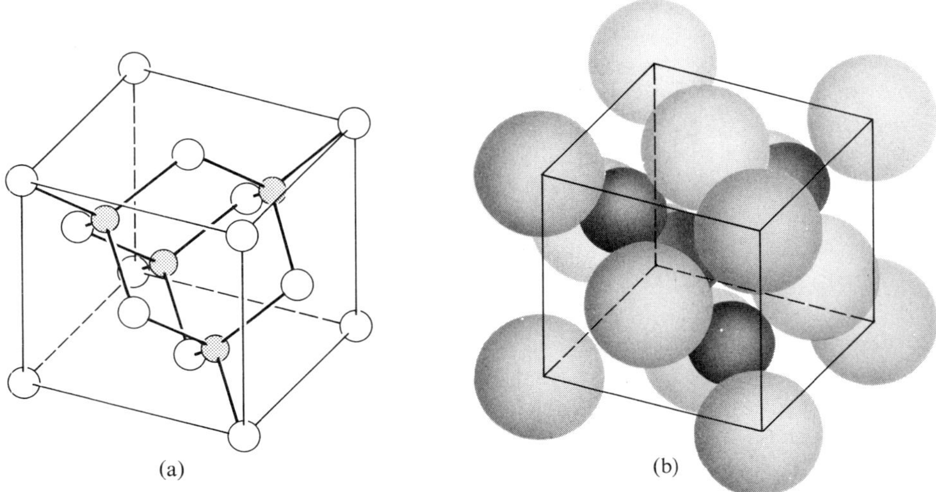

Fig. 9.5 Covalent bonding crystal with tetrahedral structure, diamond. (a) Geometrical arrangement, (b) scale model.

structure, resulting from the fact that each carbon atom forms covalent bonds with four other atoms. The length of each bond—i.e., the separation between two carbon atoms—in a diamond crystal is 1.54 Å. The coordination number of carbon is 4.

Covalent crystals have cohesive energies of ~ 6 to 12 eV per atom, and hence covalent bonds are very strong. For example, the cohesive energy of carbon in a diamond crystal is 7.4 eV per atom, and that of SiC is 12.3 eV per atom. Such a rigid electronic structure explains why crystals that bond by covalent bonding are extremely hard and difficult to deform. Diamond is the hardest substance known, and SiC is the industrial abrasive Carborundum. As we would expect, covalent crystals have very high melting points and are insoluble in most liquids. Again—as in the case of ionic crystals—there are very few free electrons. Hence the covalent crystals are poor conductors of heat and electricity. The energies of the first excited levels in covalent crystals are high (~ 6 eV in diamond), compared to visible photons, which are between 0.8 and 3.1 eV. Therefore no absorption of light can take place, which is the reason why many covalent crystals are transparent.

c) Hydrogen-bonded crystals. A neutral hydrogen atom, which has one electron, is expected to form a covalent bond with one other hydrogen atom. But this is not always true. Under certain conditions, hydrogen atoms appear to bond with other electronegative atoms—such as oxygen, fluorine, or nitrogen—thus forming a *hydrogen bond*: Some examples are H_2O, HF, and NH_3. Hydrogen-bonded crystals are much like ionic crystals, except in two respects: (i) their bonds are much weaker, (ii) they have permanent electric dipole moments (i.e., strongly polar molecules).

The melting and boiling points of hydrogen-bonded crystals are much higher than those of compounds that are similar to them. For example, the boiling point of H_2O is 100°C. Compare this with the boiling points of the nonmetallic hydrides H_2Te, H_2Se, and H_2S, which are -2°C, -42°C, and -60°C, respectively.

The hydrogen bond is formed as follows: In nonmetallic hydrides, the parent atoms have such high affinity for electrons that they pull the electron away from the hydrogen atom completely, leaving behind a proton with a positive charge. This proton attracts the negative charge of an adjacent molecule, forming chains of molecules such as H_2F_2, H_3F_3, H_4F_4, $H_5F_5, \ldots$, which may also be denoted by $HF \cdot HF$, $HF \cdot HF \cdot HF$, $HF \cdot HF \cdot HF \cdot HF, \ldots$. An example of such a chain formation[3] is ice, as shown in Fig. 9.6. Water molecules are tetrahedral. Such an open structure explains why ice occupies a larger volume than water. In other words, ice has a low density. When ice melts big clusters of chains are broken into smaller clusters, thereby occupying less volume. This means that the density of water at 0°C is much less than it is when the water is at 4°C, at which water's density achieves its maximum value.

d) Molecular crystals. Molecular crystals are formed from *non*polar molecules,

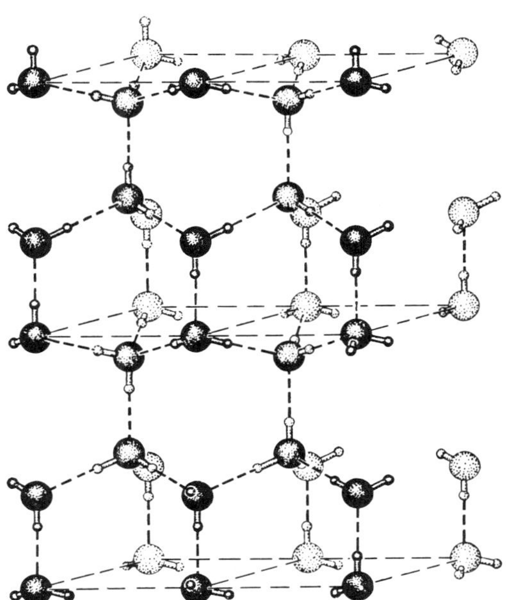

Fig. 9.6 Chain structure of water molecules in ice. [From Linus Pauling, *The Nature of the Chemical Bond*, Ithaca, N.Y.: Cornell University Press, 1960]

that is, from molecules which do not have permanent electric dipole moments. The electrons in these molecules are completely paired off, so there can be no covalent bonding between atoms of two of them. For example, the inert gases—He, Ne, Ar, Kr, Xe, and Rn—whose outer shells are completely filled with electrons, solidify as molecular crystals. Other examples are Cl_2, I_2, CO_2, and CH_4, when in the solid phase. These molecules retain their individual ties even in the solid phase.

The molecules in molecular crystals are bound by intermolecular forces called van der Waals forces.[4] These forces vary as $1/r^7$, and hence are very weak. However, when there are no ionic, covalent, or metallic bondings, van der Waals bonds are responsible for the condensation of gases into liquids and the freezing of liquids into solids. Even though the molecules of such crystals do not have permanent electric dipole moments, they do have instantaneous electric dipole moments, which over a period of time average out to zero. It is the interaction between these instantaneous electric dipoles that produces van der Waals forces.

Because of their very low bonding energies (~ 0.1 eV), molecular crystals have very low melting and boiling points. They don't have much mechanical strength, are easily deformable, and are poor conductors of heat and electricity.

Note: Some of the important properties of matter in bulk—such as surface tension, friction, and viscosity—are due to van der Waals forces.

e) Metallic crystals. If atoms which have weakly bound valence electrons are brought together to form a solid, each valence electron is closer to a given nucleus than it would be if it belonged to a single isolated atom. Thus the potential energy of an electron in a crystal is less than that of an electron in an isolated atom. This decrease in potential energy leads to the formation of metallic bonds. The energy available due to the decrease in potential energy is not all used for bonding; part of it is used for increasing the kinetic energies of the valence electrons. Thus the outermost electrons are almost free (or are very loosely bound to the ions). These "free" electrons in the crystal belong to no particular ion. Thus they move more or less freely throughout the crystal lattice (while the positive ions remain fixed), and are said to form an *electron gas.* Metallic bonding is the result of attraction between the positive ions and the electron gas. These free electrons in the metal may have energies between zero and a certain maximum E_F, called the *Fermi energy.* For example, in lithium, the Fermi energy is 4.72 eV.

It is the presence of the free electrons (or the electron gas) that accounts for the excellent thermal and electrical conductivities of the metals. Also, since these electrons can absorb any amount of energy, they easily absorb visible photons, and this explains the opacity of metals. Metallic bonds are weaker than ionic or covalent bonds, but stronger than van der Waals bonds. For example, the cohesive energies of copper and lead are 3.5 eV/atom and 2.0 eV/atom, respectively.

★9.3 CLASSIFICATION OF CRYSTAL STRUCTURES[5]

In Section 9.2 we encountered some examples of crystal structure such as the face-centered-cubic structure of NaCl, the body-centered-cubic structure of CsCl, the covalent tetrahedral structure of diamond, etc. So we ask: Are there an infinite number of different crystal structures? No. Crystal systems can be classified into a definite number of groups. Let us now get acquainted with the terminology necessary to describe them.

The smallest grouping in a crystal which is representative of the crystal structure is called a *unit cell* or *motif.* We can define a crystal as a regular array of units, i.e., the unit is repeated at regular intervals along any and all directions of the crystal. In a crystal, the environment at any one location is identical in all respects to the environment at a corresponding point anywhere else in the crystal.

Figure 9.7 shows a three-dimensional crystal in which the parallelepiped *ABCDEFGH* is a unit cell of the lattice, and is determined by the basis vectors **a**, **b**, and **c**. All translational displacements of *ABCDEFGH* by integral multiples of the vectors **a**, **b**, and **c** along these three directions translate it to some other region of the crystal identical to its original environment. Thus the whole crystal may be constructed by repeating this process for all possible combinations of the integral multiples of the basis vectors or crystal axes, **a**, **b**, and **c**. Any point at a distance $\mathbf{r}'$ from the origin may be reached from any other point at a distance $\mathbf{r}$ by

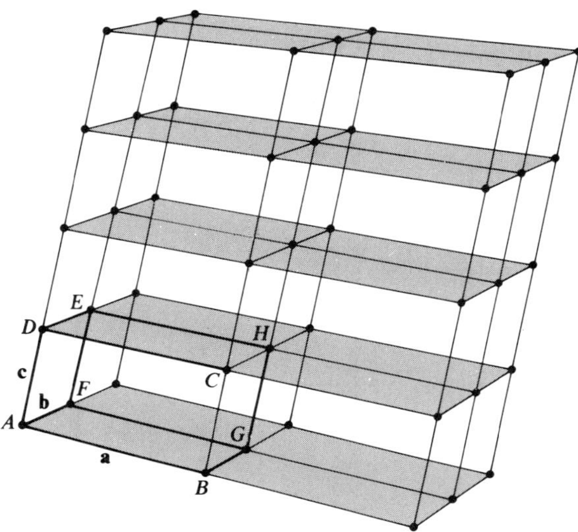

Fig. 9.7 Lattice of a three-dimensional crystal in which the parallelepiped $ABCDEFGH$ is a unit cell of the lattice.

the relation

$$\mathbf{r}' = \mathbf{r} + \mathbf{T}, \tag{9.5}$$

where

$$\mathbf{T} = n_1\mathbf{a} + n_2\mathbf{b} + n_3\mathbf{c}. \tag{9.6}$$

Here $\mathbf{T}$ is called the *translation operator* and n_1, n_2, n_3 are arbitrary integers.

Let us now give the following definition. *A unit cell is any polyhedron by means of which we can construct a crystal by repeated application of the translation operation.* The array of points generated by the translation operation is the *lattice*; each point in this lattice is a *lattice point.*

We choose the lengths of the vectors **a**, **b**, **c**, and the angles α, β, γ between these vectors arbitrarily, which might lead us to think that there is an infinite number of different types of lattices. But this is not true. By using certain symmetry properties, we can divide lattices into a finite number of groups. First let us say that a *symmetry operation* is one that, after it has been performed, leaves the crystal environment invariant. There are four main types of symmetry operations: (i) *translation,* (ii) *rotation,* (iii) *reflection,* and (iv) *inversion.*

We have already discussed the translation operation. An object has a *rotation symmetry* about an axis if, after it has been rotated through an angle θ, it has the same environment as it did before rotation. An object is said to have a *reflection symmetry* if, after it has been reflected along a line (in two dimensions) or a plane (in three dimensions), it remains unchanged. An object has *inversion symmetry*

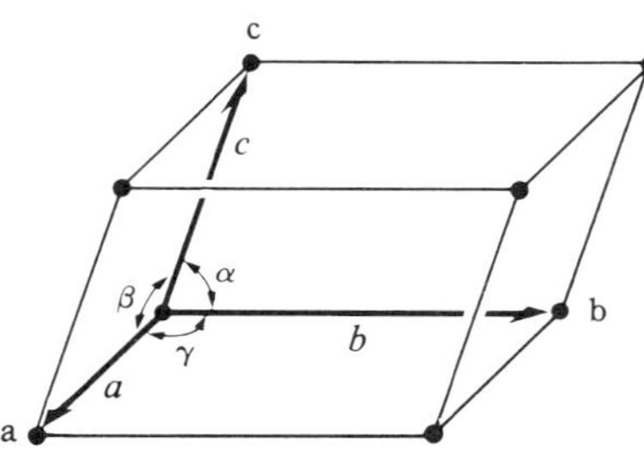

Definition of a, b, c, and α, β, γ

1. Cubic

$a = b = c$
$\alpha = \beta = \gamma = 90°$

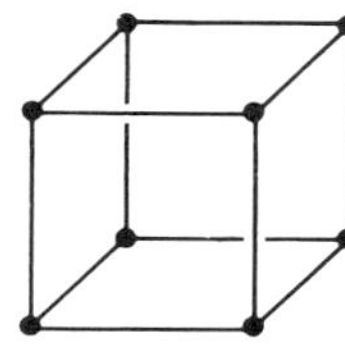
Simple cubic

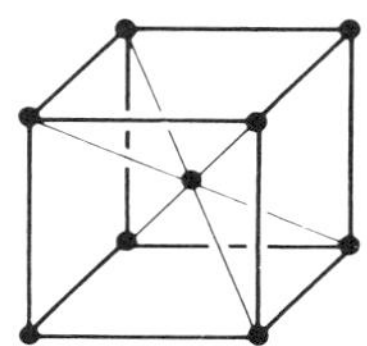
Body-centered cubic

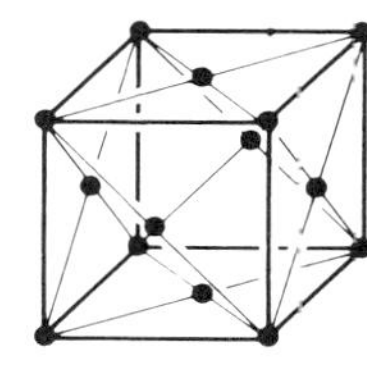
Face-centered cubic

2. Tetragonal

$a = b \neq c$
$\alpha = \beta = \gamma = 90°$

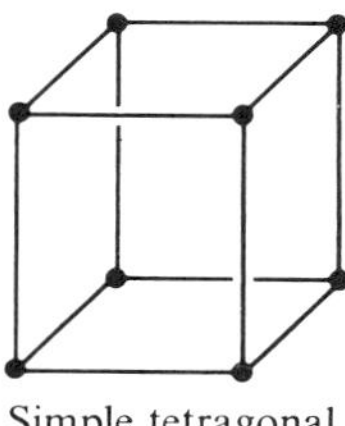
Simple tetragonal

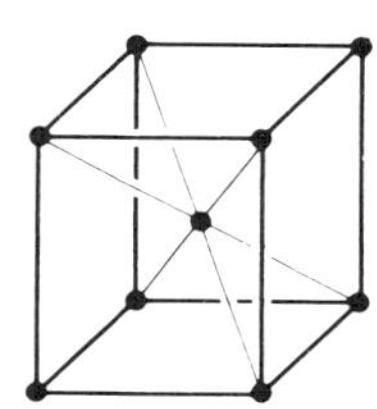
Body-centered tetragonal

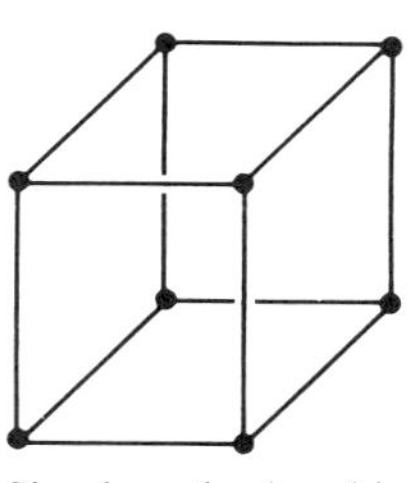
Simple orthorhombic

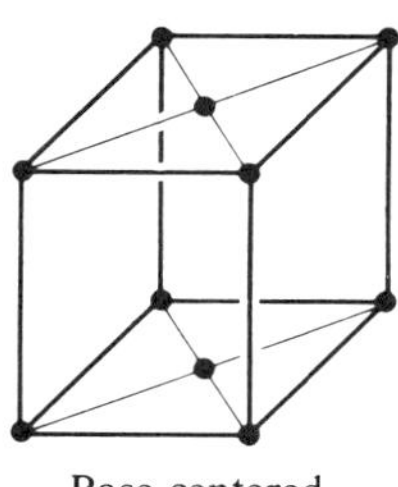
Base-centered orthorhombic

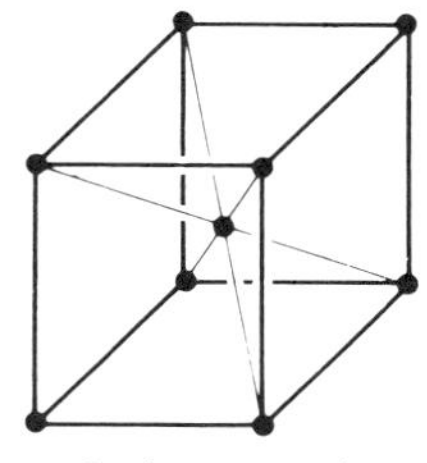
Body-centered orthorhombic

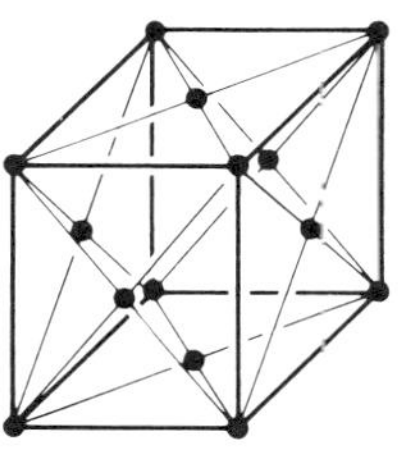
Face-centered orthorhombic

3. Orthorhombic $a \neq b \neq c$
$\alpha = \beta = \gamma = 90°$

if, after it has been inverted through a point, it changes a left-handed system to a right-handed system, and vice versa. Operations (ii), (iii), and (iv), or any combination of these, are said to form a *point group*.

Combination of the point groups with the translation group led the French scientist Auguste Bravais[6] to conclude, in 1866, that there are only 14 kinds of (three-dimensional) space lattices that can occur in crystals. These 14 space lattices, known as *Bravais lattices*, are shown in Fig. 9.8, which also classifies them into seven crystal systems. Table 9.1 (on page 314) gives this information in tabular form.

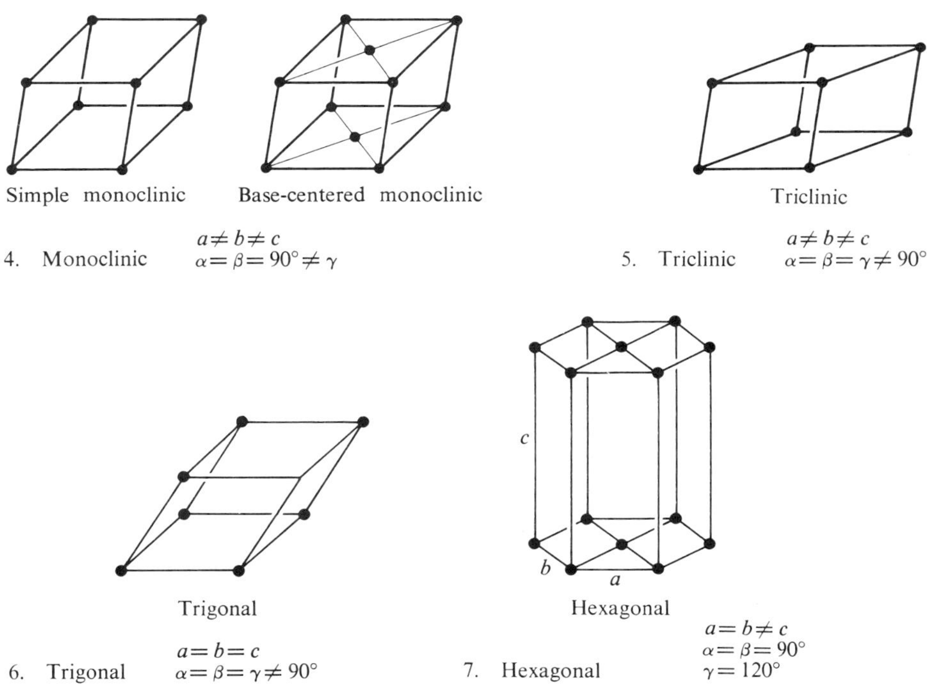

Fig. 9.8 Fourteen possible space lattices, known as *Bravais lattices*, that result from the combination of a point group with a translation group. These are classified into seven crystal systems.

Table 9.1

The Seven Crystal Systems and the Fourteen Bravais Lattices

Crystal system	Number of Bravais lattices in a system	Type of Bravais lattice	Characteristics of unit cell	Lengths and angles that must be specified
1. Cubic	3	simple body-centered face-centered	$a = b = c$ $\alpha = \beta = \gamma = 90°$	a
2. Tetragonal	2	simple body-centered	$a = b \neq c$ $\alpha = \beta = \gamma = 90°$	a, c
3. Orthorhombic	4	simple base-centered body-centered face-centered	$a \neq b \neq c$ $\alpha = \beta = \gamma = 90°$	a, b, c
4. Monoclinic	2	simple base-centered	$a \neq b \neq c$ $\alpha = \beta = 90° \neq \gamma$	a, b, c γ
5. Triclinic	1	simple	$a \neq b \neq c$ $\alpha = \beta = \gamma \neq 90°$	a, b, c α
6. Trigonal (rhombohedral)	1	simple	$a = b = c$ $\alpha = \beta = \gamma \neq 90°$	a α
7. Hexagonal	1	simple	$a = b \neq c$ $\alpha = \beta = 90°$ $\gamma = 120°$	a, c

9.4 POINT DEFECTS AND DISLOCATIONS

One seldom finds crystals with perfect lattices. All real crystals, no matter how formed, have some imperfections. The study of the properties and effects of imperfections in crystal structures is part of the study of solid state physics. Certain properties of solids—such as mechanical strength, electrical behavior, and optical quality—depend on the presence of impurities. On the other hand, the presence of imperfections does not practically alter properties such as paramagnetism and electrical conductivity in metals. Imperfections of crystal lattices fall into two groups: (a) point defects, and (b) dislocations.

a) Point defects. Point defects, which are always localized, may be divided into three categories: (i) *vacancies*, (ii) *substitutional impurities*, and (iii) *interstitial impurities*. (See Fig. 9.9.)

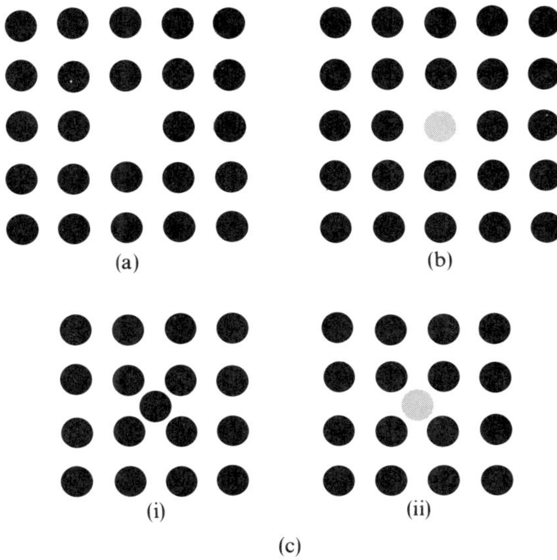

Fig. 9.9 (a) Point defect due to a vacancy. (b) Substitutional impurity. (c) Point defect due to an interstitial atom.

Figure 9.9(a) shows a point defect due to a vacancy at some lattice point. An atom is missing from a given site. To create such a vacancy, an atom must have enough energy to break its bonds with its neighbors and leave the site. It may get this energy—about 1 to 2 eV for crystals—from thermal vibrations at high temperatures. Thus this type of defect increases with increasing temperature.

Figure 9.9(b) shows a substitutional impurity: An atom in a regular array has been replaced by some foreign atom. Figure 9.9(c) shows point defects due to an interstitial atom: An extra atom squeezes in between the regular constituents of the lattice. This atom may be of the same type as the rest of the lattice atoms [Fig. 9.9c(i)] or it may be an atom of some impurity [Fig. 9.9c(ii)]. A simultaneous positive-ion vacancy and negative-ion vacancy is called a *Schottky defect*. A simultaneous positive-ion interstitial substitution and positive-ion vacancy is called a *Frenkel defect*. These defects also occur in crystals other than ionic crystals.

The presence of any one of the point defects in a crystal leads to a rearrangement of atoms in the neighborhood of the site of the defect. Scientists have learned a great deal from the study of ionic conductivity in imperfect ionic crystals that are otherwise nonconductors.

Point defects can be produced by heating crystals to very high temperatures, or by bombarding them with high-energy particles, which can easily knock out or dislocate lattice atoms. Radiation damage, however, changes many characteristics of the crystals.

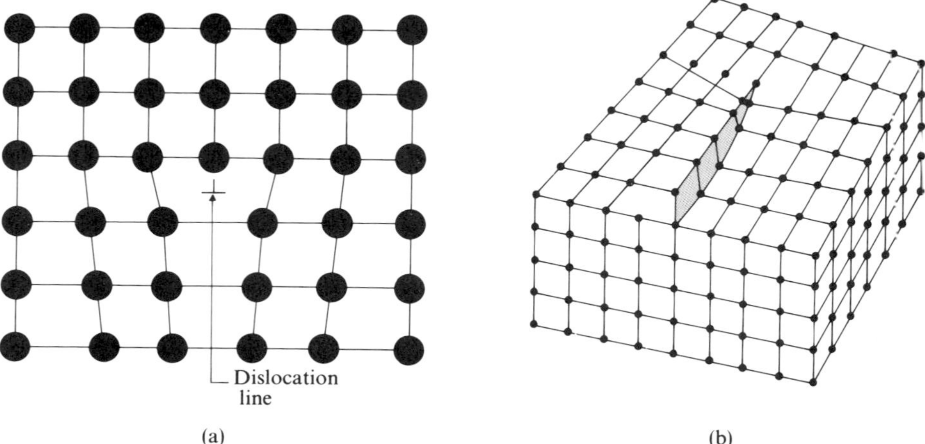

Fig. 9.10 Line defects or dislocations: (a) edge dislocation, (b) screw dislocation.

b) Dislocations. Line defects in crystals are called *dislocations.* Suppose that there is a line of atoms that do not have the same coordinate as the other atoms in the crystal. Then the crystal is said to have a dislocation (see Fig. 9.10). Dislocations not only change electric and optical properties, but they help us to understand the plastic behavior of solids and the ductility of metals. Dislocations in a crystal can reduce the strength of the bonding to as much as 1/100th of the value the crystal had before. The energy needed to produce such dislocations varies from 3 to 10 eV.

There are two main types of dislocations: *edge dislocations* (Fig. 9.10a) and *screw dislocations* (Fig. 9.10b). Visualize an edge dislocation as being an extra layer of atoms that has been pushed halfway into the crystal, or a layer of atoms removed halfway. In the process of rearranging, the atoms create a serious distortion in the structure. Think of a screw dislocation this way: Suppose that a cut is made into a perfect crystal and then one side of the cut is displaced (or slipped) with respect to the other. The dislocation itself is the boundary line between slipped and unslipped portions of the crystal. Usually both types of dislocations occur simultaneously.

9.5 FREE-ELECTRON THEORIES OF METALS

The thermal, electrical, and magnetic properties of solids are determined by the intrinsic charge, the magnetic moment, and the distribution of the electrons surrounding the atoms. Depending on the arrangement of its electrons, a solid may be classified as (a) an insulator, or (b) a metal. (a) In an *insulator*, most of the electrons are tightly bound to the atom; very few are free to move around in the crystal. (b) In a *metal*, there are always some electrons which are loosely bound to their respective atoms and are free to move throughout the crystal. Let us now consider the properties of metals in the light of the free-electron theory.

a) Classical free-electron theory. Paul Karl Ludwig Drude, in 1900, suggested that, in metals, positive ions are fixed while valence electrons are free to move anywhere they wish inside the crystal. These free electrons constitute a free-electron gas (a gas like any other, which follows the rules of the kinetic theory of gases). He called them *conduction electrons*, and the electrons in the closed shell *ion-core electrons*. In 1909, H. A. Lorentz[7] incorporated the following assumptions, leading to the formulation of the Drude–Lorentz *classical free-electron theory*: (i) Classical Maxwell–Boltzmann statistics applies to the free-electron gas. (ii) Repulsion between free electrons is negligible. (iii) The positive ions produce a constant potential field.

This theory successfully predicts many qualitative aspects of thermal and electrical conductivities. For example, it predicts Ohm's law and the relation between thermal and electrical conductivities. However, it fails to predict correct temperature dependence of electrical conductivity. It also fails to predict the magnitudes of the electronic specific heats of metals, and such magnetic properties as diamagnetic and paramagnetic susceptibility. The theory is valid enough, however, to warrant further discussion of its features.

a) *Electrical and thermal conductivities.* According to Ohm's law, the current i in metals is proportional to the applied voltage V. That is,

$$i = \frac{V}{R}, \tag{9.7}$$

where R is the resistance. Let us define current density $j = i/A$, where A is the cross-sectional area of the conductor, and $V = \mathscr{E}l$, where $\mathscr{E}$ is the electric field and l is the length of the conductor. Then we may write Ohm's law, Eq. (9.7), as

$$jA = \frac{\mathscr{E}l}{R} \quad \text{or} \quad j = \frac{l}{AR}\,\mathscr{E}.$$

That is,

$$\boxed{j = \frac{1}{\rho}\,\mathscr{E} = \sigma\mathscr{E},} \tag{9.8}$$

where

$$\sigma = \frac{1}{\rho} = \frac{l}{A}\frac{1}{R}.$$

Here ρ is the *resistivity* and σ is the *conductivity* of the metal.

According to the classical free-electron theory, we obtain the expression

$$\sigma_{\text{theo}} \simeq \frac{10^6}{\sqrt{T}}, \tag{9.9}$$

where T is the absolute temperature. However, experimentally we find that, except at low temperatures, the following relation is true:

$$\sigma_{\text{expt}} \propto \frac{1}{T}. \tag{9.10}$$

On the contrary, the classical theory correctly predicts that the ratio of the electrical conductivity σ to the thermal conductivity K is proportional to the absolute temperature T. That is,

$$\boxed{\frac{1}{T}\frac{\sigma}{K} = \text{constant} \simeq 2.5.} \tag{9.11}$$

The ratio σ/K is called the *Wiedemann–Franz ratio.*

b) *Specific heat.* According to the kinetic theory of gases, if the electron gas of a metal behaves like an ideal gas, its specific heat should be $\frac{3}{2}R$, where R is the universal gas constant. Experimentally it is found that the specific heat of a free-electron gas is temperature dependent, according to the relation

$$\boxed{C_{v\,\text{elec}} = 10^{-4}\,RT.} \tag{9.12}$$

These and many other shortcomings of the classical free-electron theory were overcome by the introduction of a new model by Sommerfeld. Let us now discuss that model.

b) Quantum-mechanical free-electron theory (Sommerfeld model).[8,9] Sommerfeld, in 1928, suggested two modifications of the classical theory. First, he said that the free electrons in a metal must be treated quantum mechanically, thus leading to the idea that these electrons have energies that are discrete. Second, he said that, for the electron gas, we must use Fermi–Dirac statistics and not Maxwell–Boltzmann statistics. So let us divide our discussion into two parts.

i) Quantum mechanics and Fermi–Dirac statistics applied to free electrons in metals

ii) Application of the theory to the properties of metals

i) *Quantum mechanics and Fermi–Dirac statistics applied to free electrons in metals.* According to Sommerfeld, the potential inside a metal crystal may be assumed to be constant so that there is no net force acting on the electrons inside the crystal. The motion of the electrons is restricted to the crystal by the presence of high potential barriers at the boundaries of the crystal. Let us visualize the crystal as a cubical box of length L containing free electrons. Chapter 3 showed that the energy of an electron in a one-dimensional infinite well of width L is given by

$$E_n = n_x^2 \frac{h^2}{8mL^2}, \qquad n_x = 1, 2, 3, \ldots, \tag{9.13}$$

where m is the mass of the electron. When we extend this to a three-dimensional case, i.e., to a cubical box of length L, we may rewrite Eq. (9.13) as

$$E = (n_x^2 + n_y^2 + n_z^2)\frac{h^2}{8mL^2} = n^2 \frac{h^2}{8mL^2}, \tag{9.14}$$

where

$$n^2 = n_x^2 + n_y^2 + n_z^2, \qquad n_x = 1, 2, 3, \ldots; \quad n_y = 1, 2, 3, \ldots; \quad n_z = 1, 2, 3, \ldots .$$

Figure 9.11(a) shows the plot of E versus n. We can show that, even for values as small as $L = 0.1$ cm, the separation between the energy levels is $\sim 10^{-14}$ eV, which is so small that the levels may be assumed to be closely packed, forming a *quasi-continuum* (seeming continuum) of levels. Each energy level is designated by four quantum numbers, n_x, n_y, n_z, and m_s, where $m_s = \pm\frac{1}{2}$. Since m_s can take only two values, each level can be occupied by two electrons, as shown in Fig.

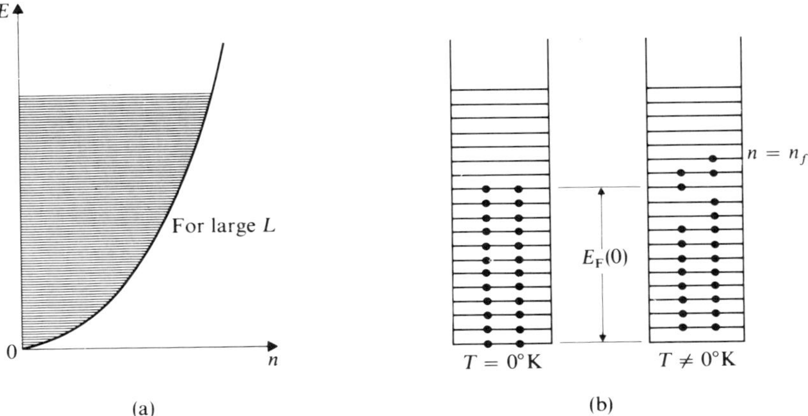

Fig. 9.11 (a) The plot of E versus n, showing quasi-continuum levels for a one-dimensional box of length L. (b) Occupation of energy levels by electrons.

9.11(b). The ground-state energy, the energy of the last electron added to the system —i.e., the energy of the electron in the uppermost level, with $n = n_f$, as shown in Fig. 9.11(b)—is a function of n_f and L. This maximum electron energy, corresponding to $n = n_f$ at zero absolute temperature, is called the *zero-point Fermi energy*, $E_F(0)$.

Usually one likes to know the ground-state or Fermi energy in terms of the total number of electrons and the size of the box. Suppose there are N electrons per unit volume, so that the total number of electrons in the system is NL^3. Designating each lattice point occupied by an electron by the quantum numbers n_x, n_y, n_z, and m_s, we get the following expression[9] for the zero-point Fermi energy:

$$\boxed{E_F(0) = \frac{\hbar^2}{2m}(3\pi^2 N)^{2/3}.} \tag{9.15}$$

Obviously, the Fermi energy is a function of the number of electrons N. Equation (9.15) yields the following values for the Fermi energies at zero-degree absolute temperature: Cs, 1.53 eV; Na, 3.12 eV; Ag, 5.51 eV.

Example 9.3. We can calculate the Fermi energy of sodium by making use of Eq. (9.15). Thus

$$E_F(0) = \frac{\hbar^2}{2m}(3\pi^2 N)^{2/3},$$

where $\hbar = 1.054 \times 10^{-34}$ J-sec, $m = 9.11 \times 10^{-31}$ kg, and N is the number of electrons per unit volume in sodium, calculated from

$$N = \frac{\rho N_A}{A} = \frac{0.9712\ \text{g/cm}^3 \times 6.02 \times 10^{23}/\text{g-mole}}{22.9898/\text{mole}}$$

$$= 2.53 \times 10^{22}/\text{cm}^3.$$

Therefore

$$E_F(0) = \frac{(1.054 \times 10^{-34}\ \text{J-sec})^2(3 \times \pi^2 \times 2.53 \times 10^{22}/\text{cm}^3)^{2/3}}{2 \times 9.1 \times 10^{-31}\ \text{kg}}$$

$$= 5.03 \times 10^{-23}\ \frac{\text{J}^2\text{-sec}^2}{\text{kg cm}^2} = \frac{5.03 \times 10^{-19}\text{J}}{1.6 \times 10^{-19}\ \text{J/eV}}$$

or

$$E_F(0) = 3.15\ \text{eV}.$$

To calculate the average kinetic energy $\bar{E}(0)$ of the electron in the metal at 0°K, we must know the *density of the states* $g(E)$, defined as the number of states per unit energy interval per unit volume. This is given by the relation

$$\int_0^{E_F(0)} g(E)\,dE = N \tag{9.16}$$

and

$$\bar{E}(0) = \frac{\int_0^{E_F(0)} g(E)\,E\,dE}{\int_0^{E_F(0)} g(E)\,dE}. \tag{9.17}$$

These equations, when combined with Eq. (9.15), yield

$$g(E) = \left[\frac{\pi}{2}\frac{8m}{h^2}\right]^{3/2} E^{1/2} \tag{9.18}$$

and

$$\boxed{\bar{E}(0) = \tfrac{3}{5}E_F(0).} \tag{9.19}$$

Note that, according to classical theory, $\bar{E}(0)$ should be zero at 0°K.

According to the above, at 0°K all the states up to energy $E_F(0)$ are completely filled, while states with $E > E_F(0)$ are completely empty. This leads to the following *zero-point distribution law*, according to which the probability $f(E)$ that the state is occupied is given by:

$$\begin{aligned} f(E) &= 1 \quad \text{for } E < E_F(0), \\ f(E) &= 0 \quad \text{for } E > E_F(0). \end{aligned} \tag{9.20}$$

Figure 9.12 shows a plot of $f(E)$ versus E.

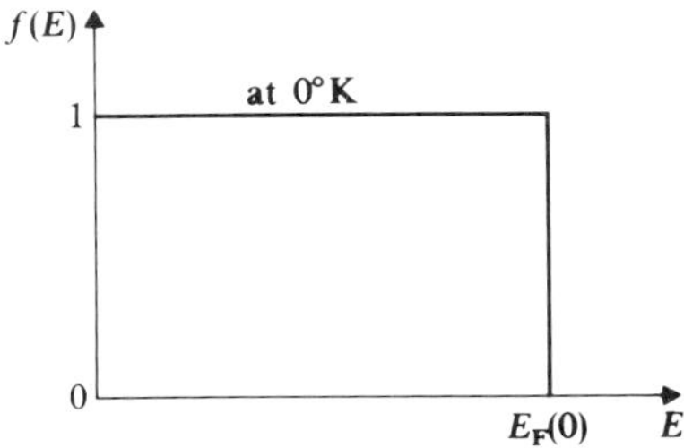

Fig. 9.12 A plot of the distribution function $f(E)$ versus E at absolute zero. Note that for $E < E_F(0), f(E) = 1$; that is, the states are completely filled. For $E > E_F(0), f(E) = 0$; that is, the states are completely empty.

Since electrons are fermions, the distribution of electrons in a metal in different energy states should be given by the Fermi–Dirac[10,11] statistics discussed in Chapter 8 and given by Eq. (8.46). That is,

$$N(E)\,dE = g(E)f(E)\,dE = \frac{g(E)\,dE}{e^{(E-E_F)/kT} + 1}. \tag{9.21}$$

That is,

$$f(E) = \frac{1}{e^{(E-E_F)/kT} + 1}, \tag{9.22}$$

where $N(E)$ is the number of electrons per unit energy in energy state E and E_F is the Fermi energy at temperature T°K, given by

$$\boxed{E_F \simeq E_F(0)\left\{1 - \frac{\pi^2}{12}\left[\frac{kT}{E_F(0)}\right]^2\right\}.} \tag{9.23}$$

Figure 9.13 shows a plot of $N(E)$ versus E for metals at different temperatures.

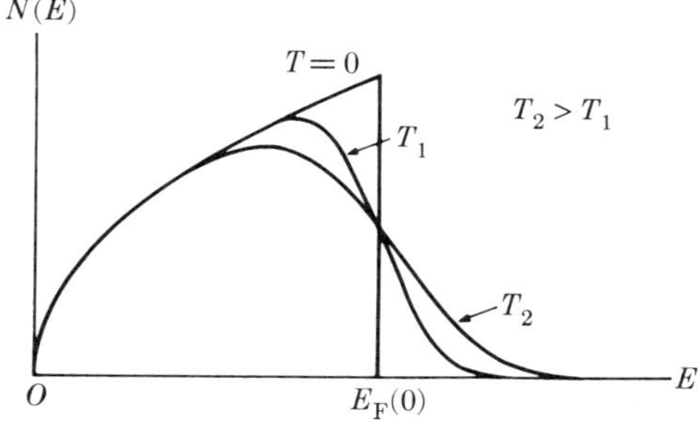

Fig. 9.13 The number of electrons per unit energy $N(E)$ versus E for a typical metal at different temperatures.

Raising the temperature above 0°K forces the electrons in the levels just below the energy E_F into excited levels just above E_F. At temperature T°K, only those electrons in levels with energies between $E_F - kT$ and E_F are excited into levels with energies between E_F and $E_F + kT$. (Other electrons, if they are excited, have to go into occupied states; this is not possible, according to the Pauli exclusion principle.) Thus the fraction of electrons in a metal that can be excited by means of external sources—such as thermal, electric, or magnetic sources—is equal to the ratio of the shaded area to the total area, as shown in Fig. 9.14. That is,

$$\frac{\text{shaded area}}{\text{total area}} = \frac{kT}{E_F} \simeq \frac{0.03}{3.00} = \frac{1}{100}. \tag{9.24}$$

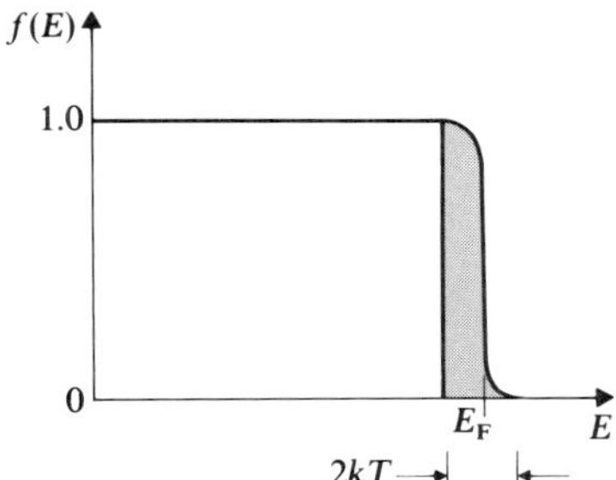

Fig. 9.14 The ratio of the shaded area to the total area is equal to the fraction of the electrons that undergo transitions when the temperature of a metal crystal is raised to T.

ii) *Application of theory to properties of metals.* Now we can explain some of the properties of metals using the quantum theory of free electrons.

a) Electrical conductivity of metals. According to quantum theory, free electrons in metals occupy definite energy states, according to Fermi–Dirac statistics. As shown in Fig. 9.15, in the absence of any external field, the electrons move randomly. That is, there are as many electrons moving with a given velocity in the $+X$ direction as in the $-X$ direction. Thus there is no net current. As soon as an electric field $\mathscr{E}$ is applied, the electrons are accelerated toward the positive terminal of the field. This acceleration means an increase in velocity and hence in energy. But only those electrons with energies near the Fermi energy E_F can acquire enough energy to be redistributed as shown in Fig. 9.16. The net excess of electrons along the $+X$ axis results in electrical current, and contributes to electrical conductivity. Calculations based on this theory show that conductivity σ is inversely proportional to the absolute temperature T, as we would expect from experimental evidence.

b) Thermal conductivity of metals. The high thermal conductivity of metals, and its relation to electrical conductivity, can be easily understood. In metals, conduction takes place on account of the vibrations of atoms as well as free electrons.

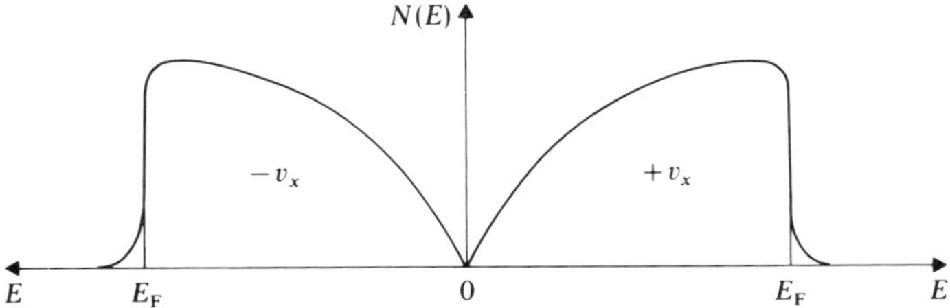

Fig. 9.15 In the absence of any external field, there are as many electrons with a given velocity moving in the $+X$ direction as in the $-X$ direction; i.e., electrons move randomly.

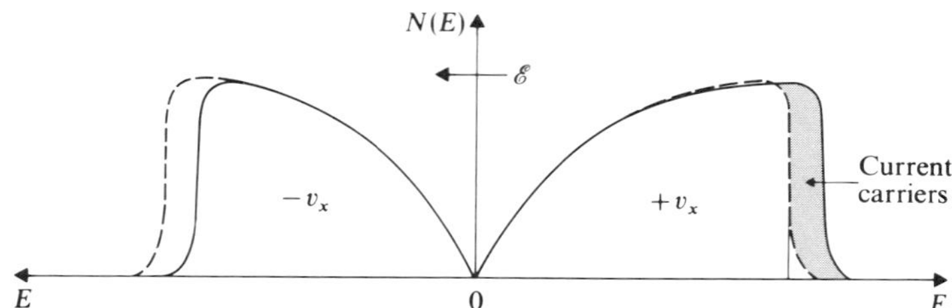

Fig. 9.16 The application of an electric field $\mathscr{E}$ in the direction shown results in the redistribution of electrons, leading to an excess in the positive X direction. This excess of electrons constitutes electrical current.

In nonmetallic materials, conduction takes place primarily on account of the vibrations of atoms. Instead of talking about vibrations of atoms, let us talk in terms of *phonons*. (In analogy with a photon, which is a quantum of radiational energy, a *phonon* is a quantum of vibrational energy.) The number of phonons increases with increasing temperature, but the mean free paths of phonons remain small (10–100 Å). Thus, for materials other than metals, the conductivities are very small. But even though the mean free paths of phonons in metals are small, the paths of free electrons are long. So it is the free electrons that cause the good thermal conductivities of metals. The same free electrons that are responsible for thermal conductivity are responsible for electrical conductivity in metals. So we should expect some relation between them.

The free-electron quantum theory predicts the Wiedemann–Franz law discussed earlier—that $(1/T)(\sigma/K) = \text{constant}$.

c) Electronic specific heats of metals. The specific heat of metal may be written as (remembering that electrons have three degrees of freedom)

$$C_v = C_v(\text{atomic}) + C_v(\text{electronic}) = 3R + \tfrac{3}{2}R = \tfrac{9}{2}R. \tag{9.25}$$

But it is well known that the specific heat of all solids above a certain temperature is equal to $3R$. Thus we must get rid of $\frac{3}{2}R$. According to Fermi–Dirac statistics, at any temperature T, only those electrons which are within $E_F \pm kT$ (recall Fig. 9.14) absorb heat. Accordingly the fraction of such electrons is $2kT/E_F$. Hence the electronic specific heat is given by

$$C_v(\text{electronic}) = \tfrac{3}{2}R \times \frac{2kT}{E_F} = \frac{3Rk}{E_F}T = \gamma T, \tag{9.26}$$

where $\gamma = 3Rk/E_F$ is a very small quantity; hence the contribution from C_v(electronic) is negligible, and

$$C_v \simeq C_v(\text{atomic}).$$

SUMMARY

Solids can be classified as *amorphous*, in which the atoms have little or no geometrical regularity, and *crystalline*, in which the atoms have regular arrangement, or in which a group of atoms fall in a repeated three-dimensional pattern.

The cohesive or binding energy of a crystal is a measure of the stability of the crystal. The greater a crystal's binding energy, the greater the amount of energy needed to separate it into its constituent atoms.

According to the nature and the strength of their bonding, solids may be classified into five categories: (a) ionic crystals, (b) covalent crystals, (c) hydrogen-bonded crystals, (d) molecular crystals, and (e) metallic-bonded crystals. The *coordination number* is the number each ion has of nearest neighbors of the opposite kind.

⋆ A *unit cell* or *motif* is the smallest grouping in a crystal which is representative of the structure of that crystal. Or it may be defined as any polyhedron that, when the operation of translation is applied repeatedly, will duplicate the original crystal. The array of points generated by the operation of translation is called the *lattice*.

⋆ The four principal types of symmetry operations are: (i) translation, (ii) rotation, (iii) reflection, and (iv) inversion. Use of these led Bravais to conclude that there are only 14 kinds of space lattices, known as *Bravais lattices*. These are classified into 7 crystal systems.

Imperfections of crystal lattices fall into two groups: (a) point defects and (b) dislocations. Point defects, which are always localized, may be divided into three categories: (i) vacancies, (ii) substitutional impurities, and (iii) interstitial impurities. Dislocations, which are line defects in crystals, may be classified as *edge dislocations* and *screw dislocations*.

In a material which is a good insulator, the electrons are tightly bound to the atoms, and are not free to move in the crystal. In metals, which are good conductors, there are always some electrons which are loosely bound to their respective atoms, and are free to move throughout the crystal.

The classical free-electron theory of metals, as formulated by Drude and Lorentz, correctly predicts Ohm's law and the relation between thermal and electrical conductivities. However, the theory fails to predict the dependence of electrical conductivity on temperature. It also fails to correctly predict the magnitudes of the electronic specific heats; and diamagnetic and paramagnetic susceptibilities.

These difficulties were overcome by the introduction of the quantum-mechanical free-electron theory by Sommerfeld. According to him, electrons in a metal must be treated quantum mechanically, and one must use Fermi–Dirac statistics instead of Maxwell–Boltzmann statistics.

The ratio of electrical conductivity σ to thermal conductivity K is called the Wiedemann–Franz ratio.

PROBLEMS

1. Show that, for a body-centered-cubic crystal such as CsCl, the Madelung constant is 1.7627.
2. Show that the Madelung constant for a linear monatomic lattice is 1.38.
3. Derive Eq. (9.4) from Eq. (9.3).
4. From Eq. (9.4), calculate the value of n for a NaCl crystal, given that $\alpha = 1.7476$, $r_0 = 2.81$ Å, and the lattice energy $E_{p,r_0} = -185.7$ kcal/ion-pair. Using this value of n, calculate V_{repul} from Eq. (9.2). How does this value compare with the value of V_{coul} given by Eq. (9.1)? What is the significance of the difference between these two values?
5. Calculate the potential energy of cesium chloride, given that the interatomic distance at equilibrium is 3.56 Å and n is 11.5.
6. For a LiCl crystal, the equilibrium distance between ions is 2.57 Å and the cohesive energy is 6.8 eV per ion pair. From these data, calculate the value of n.
7. For a KCl crystal, the Madelung constant is 1.7476 and the equilibrium distance between ions is 3.14 Å. The ionization energy of potassium is 4.34 eV and the electron affinity of chlorine is 3.61 eV. Calculate the cohesive energy of KCl. How does this value compare with the measured value of 6.42 eV? Explain.
8. Consider a crystal of size 1 cm^3. Using Eq. (9.14), show that the energy separation ($\Delta E = E_{n+1} - E_n$) is of the order of 10^{-14} eV.
9. By assuming that two electrons can be placed in each state, use Eq. (9.14) to calculate the energy of the last electron in a sodium crystal of size 1 cm^3. How does this value compare with the one calculated by using Eq. (9.15)? Explain the difference in the values.
10. Consider a crystal of size 0.1 cm^3. Represent a state of the system by (n_x, n_y, n_z) and calculate the energies corresponding to the following states: (a) E_1 for (3, 3, 3) and E_2 for (2, 3, 3), (b) E_3 for (10, 10, 10) and E_4 for (11, 10, 10). (c) Calculate $E_2 - E_1$ and $E_4 - E_3$. What do you conclude from these calculations?
11. Calculate the Fermi energy of aluminum at 0°K. Calculate the average electron energy at 0°K.
12. Using Eqs. (9.15) and (9.23), calculate the Fermi energy for copper at 0°K and 300°K.
13. Suppose that kinetic theory is applicable to a Fermi gas. Calculate the pressure of a Fermi gas at 0°K.
14. What fraction of the free electrons in sodium can be excited by a thermal source of temperature (a) 100°K, and (b) 1000°K? The Fermi energy of sodium is 3.2 eV.
15. Approximately what fraction of the free electrons in copper are in the excited states (a) at room temperature (b) at 1087°C? The Fermi energy of copper is 7.1 eV.
16. Show that, if $E_F(0) = 5$ eV and $\bar{E}(0) = \frac{3}{5}E_F(0)$, the classical temperature corresponding to this $\bar{E}(0)$ is $\sim 10^{4}$°K.

17. The Fermi energy of copper at 0°K is 7.1 eV. From the relation $\bar{E} = \frac{3}{5}E_F(0) = \frac{1}{2}m\bar{v}^2$, calculate $\bar{v}$ at 0°K. How does this value compare with $\bar{E} \simeq 0.025$ eV and $\bar{v} \simeq 10^6$ cm/sec for a gas at room temperature?
18. Suppose that we define a Fermi temperature $T_0 = E_F(0)/k$. Calculate T/T_0 for sodium at 300°K and 1000°K.
19. Suppose that we define a Fermi temperature $T_0 = E_F(0)/k$. Calculate T/T_0 for silver at room temperature. The Fermi energy of silver is 5.51 eV.
20. Given that the Fermi energy of aluminum is 11.7 eV, what is its electronic specific heat at 1000°K?
21. Calculate the Fermi energy of zinc, given that its electronic specific heat is $\sim 1.3 \times 10^{-4}\ T$ kcal/kg °K.

REFERENCES

1. H. M. Evjen, *Phys. Rev.* **39,** 680 (1932)
2. Max Born, *Atomic Physics*, seventh edition, New York: Hafner, 1962
3. Linus Pauling, *The Nature of the Chemical Bond*, Ithaca, N.Y.: Cornell University Press, 1960
4. H. Margenau, *Revs. Mod. Phys.* **11,** 1 (1939)
5. F. Seitz, *Z. Krist.* **88,** 433 (1934); **94,** 100 (1936)
6. A. Bravais, *Etudes crystallographiques*, Paris: Gautier–Villars, 1866; or German translation, *Klassiker der exakten Wissenschaften* **90,** 1897
7. H. A. Lorentz, *The Theory of Electrons*, second edition, New York: Dover Publications, 1952
8. A. Sommerfeld, *Z. Physik* **47,** 1 (1928)
9. A. Sommerfeld and N. H. Frank, *Revs. Mod. Phys.* **3,** 1 (1931)
10. Enrico Fermi, *Z. Physik* **36,** 902 (1926)
11. P. A. M. Dirac, *Proc. Roy. Soc.* **A112,** 661 (1926)

Suggested Readings

1. Robert A. Levy, *Principles of Solid State Physics*, New York: Academic Press, 1968
2. John S. Blakemore, *Solid State Physics*, Philadelphia: W. B. Saunders, 1969
3. C. Kittel, *Introduction to Solid State Physics*, third edition, New York: John Wiley, 1966
4. L. V. Azaroff and J. J. Brophy, *Electronic Processes in Materials*, New York: McGraw-Hill, 1963
5. A. J. Dekker, *Solid State Physics*, Englewood Cliffs, N.J.: Prentice-Hall, 1957
6. R. Sproull, *Modern Physics*, New York: John Wiley, New York, 1963

CHAPTER 10
BAND THEORY AND ITS APPLICATIONS

Scientists test a superconducting electromagnet; the small cylindrical object is removed from liquid helium bath (temperature, minus 450°F). A later experimental supermagnet made of niobium-tin wire produced a field strength of 78,000 gauss. (Photograph courtesy Bell Telephone Laboratories, Murray Hill, N.J.)

10.1 INTRODUCTION

The free-electron theory—together with the quantization condition and Fermi–Dirac statistics—was able to explain many different phenomena, such as the thermal and electrical properties of solids. However, there remained many important aspects of solids that were not yet accounted for. For example, we would like to know why some solids are insulators, some are conductors, and some are semiconductors. These and many other phenomena can be satisfactorily explained by the *quantum theory of solids*, also called the *band theory of solids*. There are, basically, two opposite approaches, but the results obtained are the same. One approach is to modify the free-electron model by considering the motion of an *electron in a periodic potential*, also called the *zone theory*. The second is the so-called *tight-binding approximation*, also called the *band theory*. We'll deal with both, and then, later on, we shall discuss some interesting applications of these theories, such as semiconductors and superconductivity.

10.2 ELECTRONS IN A PERIODIC POTENTIAL (ZONE THEORY)

In the free-electron theory it was assumed that the motion of electrons in a crystal is completely free, and that these electrons can roam anywhere within the crystal. It was further assumed that high potential barriers at the surfaces of the crystal prevent these electrons from leaving the crystal. But this is not really true. The actual potential is periodic, as shown in Fig. 10.1 for the case of a one-dimensional crystal. The periodicity of the potential is the result of coulomb interaction between the moving electron and the periodic charge distribution arising from the positive ions located at the lattice sites.

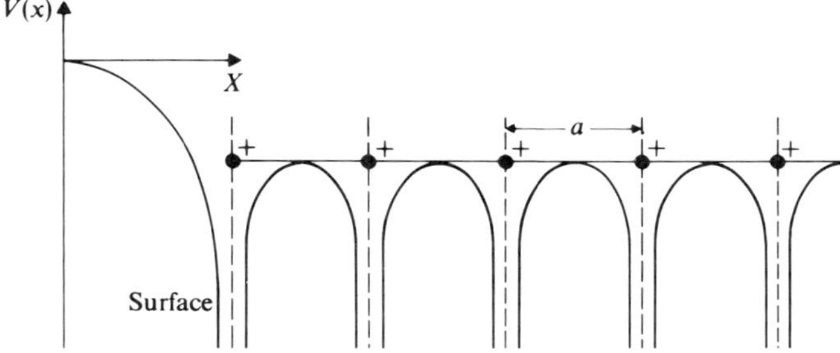

Fig. 10.1 The periodic potential due to a one-dimensional, perfectly periodic crystal lattice. Note that, at the extreme left, an abrupt change in the potential occurs at the surface of the crystal.

Finding the allowed energy levels of an electron involves solving the Schrödinger wave equation for an electron moving in such a periodic potential.[1,2] We can greatly simplify the problem, without affecting the final results, if we replace the potential of Fig. 10.1 by a periodic potential with rectangular barriers, as in Fig. 10.2. This is the so-called *Kronig–Penney Model*.[3] The Schrödinger wave

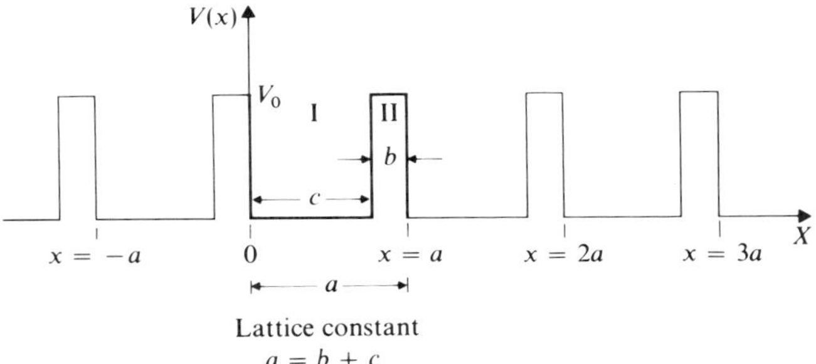

Fig. 10.2 A series of square-well potentials used in the Kronig–Penney model to investigate the behavior of electrons in a periodic lattice.

equation to be solved is

$$\frac{d^2\psi(x)}{dx^2} + \frac{2m}{\hbar^2}[E - V(x)]\psi(x) = 0, \tag{10.1}$$

where $V(x) = 0$ for regions of type I and $V(x) = +V_0$ for regions of type II. The effect of the periodic potential on the wave function is to change the wave function of the electron e^{ikx} to

$$\psi(x) = u(x)e^{ikx}, \qquad k = \frac{\sqrt{2m[E - V(x)]}}{\hbar} \tag{10.2}$$

and $u(x)$ is the modulating amplitude, which is almost unity at the center of the well and appreciably different from unity near the boundary of each well. Since $u(x)$ repeats from lattice to lattice (with a = lattice spacing), it must satisfy the condition

$$u(x + a) = u(x). \tag{10.3}$$

Equations (10.2) and (10.3) together constitute *Bloch's theorem.*[4] By using Bloch's theorem and the periodic boundary conditions for continuity of a wave function and its derivative, we obtain the relation

$$\cos ka = P\frac{\sin \alpha a}{\alpha a} + \cos \alpha a, \tag{10.4}$$

where P is a constant given by

$$P = \operatorname*{Limit}_{\substack{b\to 0\\ \beta\to\infty}} (\tfrac{1}{2}\beta^2 ab) = \frac{ma}{\hbar^2} V_0 b \tag{10.5}$$

and

$$\alpha = \frac{\sqrt{2mE}}{\hbar}, \qquad \beta = \frac{\sqrt{2m(V_0 - E)}}{\hbar}. \tag{10.6}$$

We can obtain the allowed energy levels of an electron in a periodic potential by the following two interpretations of Eq. (10.4).

a) Interpretation of the right side of the equation

Figure 10.3 shows the oscillatory curve which is the plot of the right-hand side of Eq. (10.4) versus αa. The vertical dashed lines are the limits of $+1$ and -1 imposed by the left side of Eq. (10.4), that is, by $\cos ka$. For real values of k, the physically meaningful solutions of Eq. (10.4) must lie within these two limits, which are shown by the heavy vertical lines along the αa axis. These energies—the energies which an electron moving in a periodic potential is allowed to take—are called *allowed bands* or *allowed regions* or *allowed zones.* They are shown shaded in Fig. 10.3. Between the allowed bands are the *forbidden bands* or *forbidden regions* or *forbidden zones* of energies, which the electron cannot take while it is moving through a periodic potential.

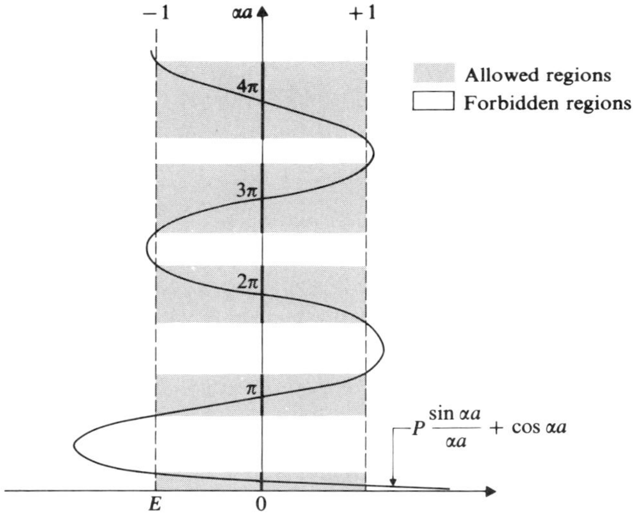

Fig. 10.3 Allowed and forbidden regions for electrons in a periodic potential of a periodic lattice, obtained by using the Kronig–Penney model. Heavy vertical lines obtained by imposing the condition $\cos k\alpha = \pm 1$ are the allowed energies of the electrons.

Note that if $V_0 b$ becomes very large, allowed zones become very narrow and the electron cannot move freely. This is what happens in the case of electrons that are tightly bound to the nucleus. On the other hand, if V_0 approaches zero, the allowed zones spread so much that the forbidden zones disappear. This is what happens in the case of valence electrons in an atom. For $V_0 = 0$, the situation reduces to the case of free electrons.

b) Interpretation of the left side of the equation

We can achieve a still more useful and direct interpretation of Eq. (10.4) by considering the left side of this equation. The function cos ka can assume only those values which correspond to allowed values of E. The heavy lines in Fig. 10.4 show the plot of these allowed values of E versus ka. If we project these allowed values to the right, we get the allowed bands or zones, and between these bands are the forbidden bands.

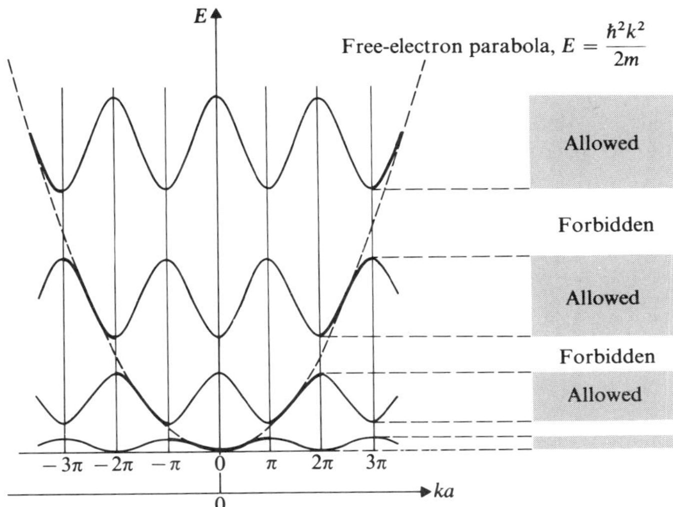

Fig. 10.4 The plot of E versus ka for an electron in a periodic potential resulting in allowed and forbidden energy regions. The allowed energies are shown by the heavy curves.

The dashed-line parabola in Fig. 10.4 corresponds to the case of a free electron for which $E = k^2\hbar^2/2m$. Note that the heavy lines depart slightly from the dashed-line parabola only in the neighborhood of $\pm n\pi$, where n is an integer. This means that the electron moving in a periodic potential behaves like a free electron for most values of k, except those near $\pm n\pi$. For large values of E, the two are almost similar.

Also note that the discontinuities in the allowed values of E occur at cos $ka = \pm 1$. That is,

$$ka = \pm n\pi, \qquad n = 0, 1, 2, 3, \ldots, \tag{10.7}$$

and if we substitute $k = 2\pi/\lambda$, we get

$$n\lambda = 2a, \tag{10.8}$$

which corresponds to the Bragg condition, provided that we substitute $\theta = 90°$ in $n\lambda = 2a \sin \theta$. This implies that the discontinuities represent locations at which the electron waves are reflected backward. The region that contains electrons with momentum such that $0 < k < \pi/a$ [that is, $n = 1$ in Eq. (10.7)] is called the *first Brillouin zone.* The *second Brillouin zone* is one that contains electrons with k values between π/a and $2\pi/a$. The third and fourth and higher Brillouin zones may be defined the same way.

10.3 BAND THEORY OF SOLIDS (TIGHT-BINDING APPROXIMATION)

Insulators, conductors, and semiconductors may be distinguished by (in addition to other factors) their electrical conductivities. *Insulators* are poor conductors of electricity; examples are diamond, quartz, and most covalent and ionic solids. *Conductors* are good conductors of electricity; examples are silver and other metals. The conductivity of *semiconductors* is in between that of insulators and conductors; examples are germanium (Ge) and silicon (Si). Let us now discuss the formation of these three types of materials according to the band theory.

The *tight-binding approximation theory*[5] assumes that each electron is tightly bound to its nucleus—an assumption completely opposite to that made in the zone theory. Yet the results obtained are almost the same. It is convenient to use zone theory when dealing with metals, but band theory for insulators and semiconductors.

To find the energy levels of electrons in solids, we need to know the energy levels of the individual atoms as they are brought together to form a solid. Imagine that we are bringing two hydrogen atoms together to form a molecule. The two 1s levels (one for each atom) combine to form two new molecular levels with differing energies. In the case of solids, there are many more atoms and the situation becomes complicated. For example, suppose that we bring together six sodium atoms, each with an electron in the 3s state. We will have six resulting wave functions, as shown in Fig. 10.5. The combination in Fig. 10.5(a) is the perfectly symmetric wave function, and has the least energy, while the one in Fig. 10.5(b) is the perfectly antisymmetric wave function, and has the most energy. The energies of the rest of the combined wave functions are between the two extremes, but they also depend on how close these atoms are brought together.

The result is the formation of energy bands, as shown in Fig. 10.6. Actually, because of spin degeneracy (one state resulting from spin pointing up and another state from spin pointing down), the total number of energy states will be doubled. Also, if there is a large number of atoms, the energy levels in the bands will be almost continuous, as shown in Fig. 10.6(c) for the 1s, 2s, 2p, and 3s levels of N isolated sodium atoms. This situation clearly engenders the formation of 1s, 2s, 2p, and 3s allowed energy bands and the forbidden energy bands. In general, in the formation of any solid from N individual atoms, there are $2N$ states in each

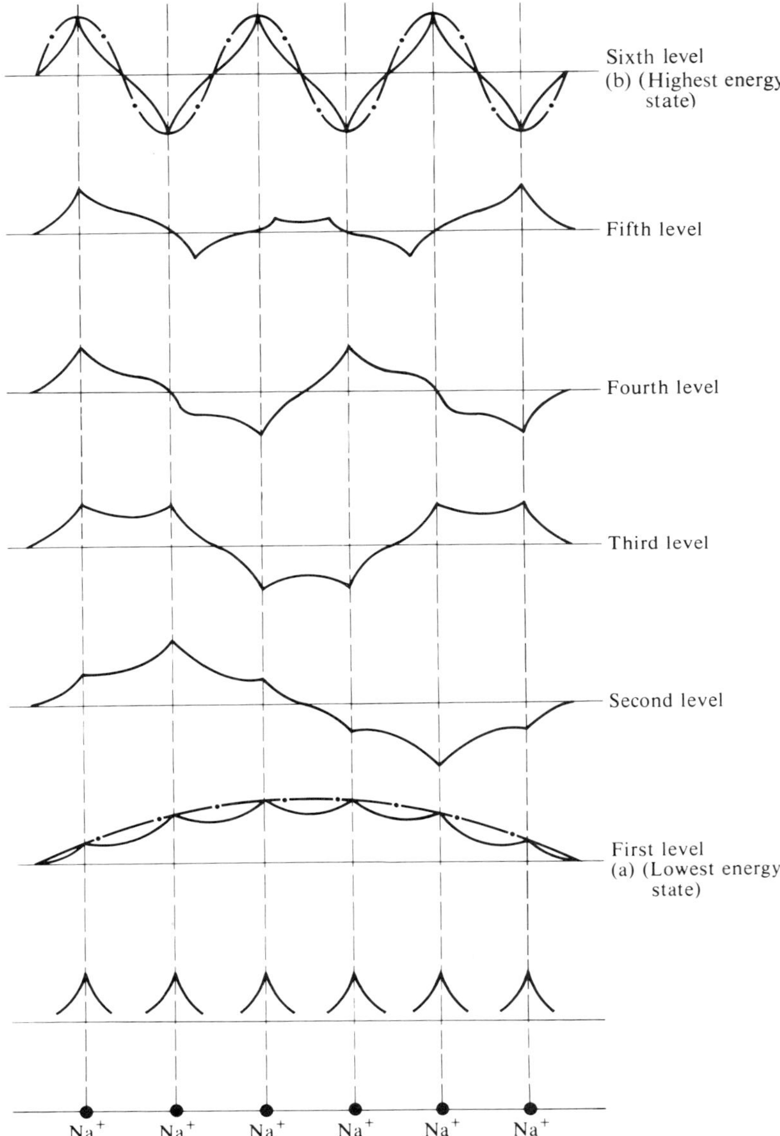

Fig. 10.5 (a) Perfectly symmetric wave function, with minimum energy, resulting from the combination of six sodium atoms, each with an electron in the 3s state. (b) Perfectly antisymmetric wave function, with maximum energy. The other four combinations have energies between these two extremes.

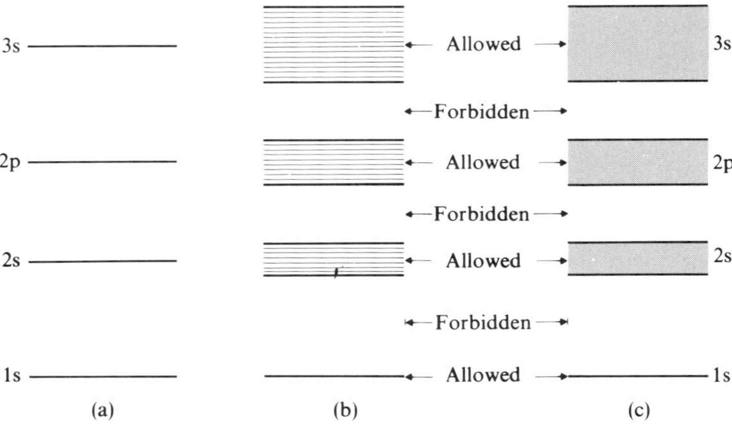

Fig. 10.6 (a) The energy levels of an isolated atom. (b) A solid containing a small number of atoms; note that the energy bands have discrete energy levels. (c) A solid containing a large number of atoms; note that the energy levels in the bands are almost continuous.

1s, 2s, 3s, . . . band, $6N$ states in each 2p, 3p, 4p, . . . bands, $10N$ states in each 3d, 4d, . . . band, and so forth. That is,

$$[\text{Number of states in each } l\text{-band}] = 2(2l + 1)N. \tag{10.9}$$

When solids are formed, four types of bands form, as shown in Fig. 10.7. This leads to the classification of solids into insulators, semiconductors, and conductors. Let's have a look at each.

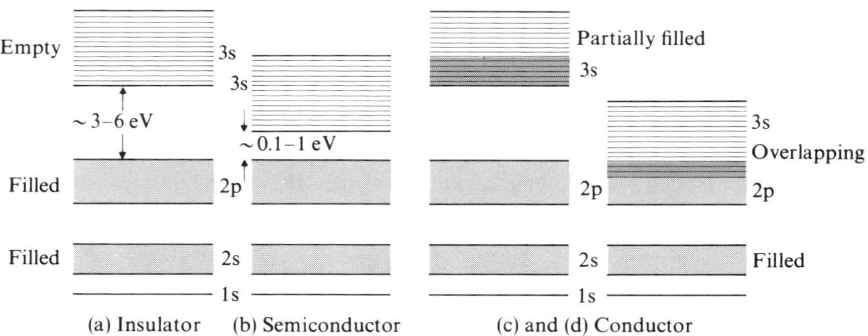

Fig. 10.7 Solids can be classified into three types, according to the way the energy bands are formed. (a) A solid is an insulator if the gap between the bands is very large. (b) A solid is a semiconductor if the gap between the bands is very small. (c) and (d) A solid is a conductor if the upper band is partially filled, as in (c), or if the filled and the empty band overlap, as in (d).

a) *Insulators.* Materials for which the band formation is like that in Fig. 10.7(a) are called insulators. The characteristic of this band formation is that the forbidden region between the highest filled band and the lowest empty band is very wide, $\sim$3 to 6 eV. Even if we thermally excite the electrons (at any physically achievable temperature) or apply an electric field to them, very few electrons in the filled band move to this empty band. However, because of the Pauli exclusion principle, the electrons cannot move about in the filled band. So there is no way to obtain a free-electron current. That's why solids of this type are poor conductors of electricity and are classified as insulators. To present some idea of this, Table 10.1 lists the widths of the forbidden bands between the highest filled and the lowest empty band for some insulators, and compares them with those of certain semiconductors. The uppermost filled or partially filled band is called the *valence band.*

Table 10.1

Some Forbidden Energy Gaps

Insulators	Gap, eV	Semiconductors	Gap, eV
Diamond (C)	5.33	Silicon (Si)	1.14
Zinc oxide (ZnO)	3.2	Germanium (Ge)	0.67
Silver chloride (AgCl)	3.2	Tellurium (Te)	0.33

b) *Semiconductors.* Materials for which the band formation is like that in Fig. 10.7(b) are called *semiconductors.* The characteristic of the formation of this band is that the forbidden region between the highest filled band and the lowest empty band is very narrow, $\sim$0.1 to 1 eV (Table 10.1). We can definitely move the electrons of a semiconducting material from the highest filled band to the empty band without too much trouble. We do this by means of thermal excitation or by applying an electric field. Then these few electrons in the empty band are available for electric current, because they can move about in the band. Figure 10.8(a) and (b) shows distributions of electrons in a semiconductor at 0°K and 300°K. (The distributions are governed by the laws of statistical mechanics discussed in Chapter 8.)

There is still another mechanism that causes electric current: The vacancies (or empty places left behind when an electron moves) that remain near the top of the uppermost filled band are called *holes.* These holes behave like positive electrons, and can contribute to the flow of electric current, as shown in Fig. 10.8(c). When an electric field is applied, the electron below the hole may gain enough energy to jump up and occupy the hole. So to all intents and purposes, the *hole* moves to a lower energy state. Such a motion of a hole also contributes to the electric current.

c) *Conductors.* Materials for which the band formation is like that in Fig. 10.7(c) or (d) are called *conductors.* The characteristic of the formation of this band is

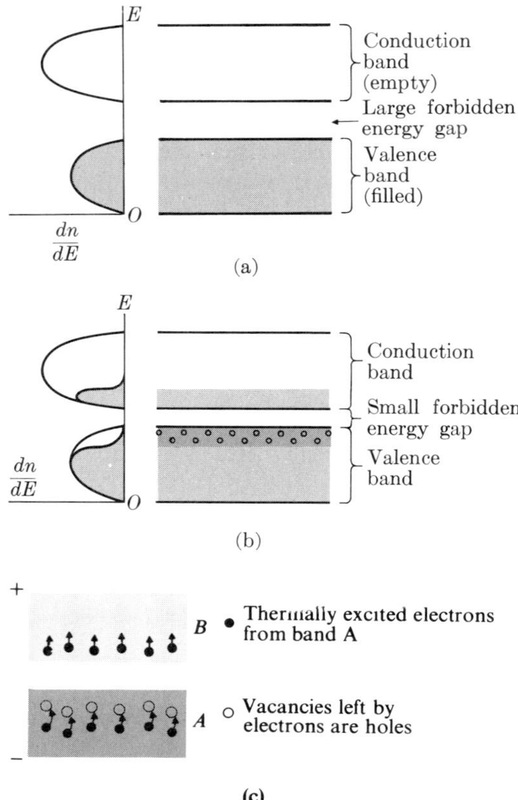

Fig. 10.8 Distribution of electrons in conduction and valence bands of a pure semiconductor at (a) 0°K, and (b) 300°K (room temperature). (c) The hole left behind by a departed electron is said to have moved when an electron from a neighboring atom moves into the hole.

that the valence band is either partially filled (as shown in Fig. 10.7c) or (if it *is* filled) the next allowed empty band overlaps with the filled valence band (as shown in Fig. 10.7d). In either case there are unoccupied states for electrons in the uppermost band. And these electrons are available to carry current. We call such solids conductors. Even when they are nonmetallic, they exhibit all the characteristics of metallic conductors.

Let's look at a few examples. Sodium ($Z = 11$) has an electronic configuration of $1s^2 2s^2 2p^6 3s^1$. Thus, in a solid containing N sodium atoms, there are $2N$ 1s electrons, $2N$ 2s electrons, $6N$ 2p electrons, and N 3s electrons. This implies that the 1s, 2s, and 2p bands are completely filled, while the 3s band is only half filled. Hence sodium and all other alkali metals should be conductors. And they *are*!

The situation, however, is slightly complicated. The equilibrium distance between the sodium atoms in the solid form is $r_0 = 3.67 \times 10^{-8}$ cm. At this distance, according to the tight-binding approximation theory, the 1s, 2s, and 2p bands remain undisturbed and separated, but the bands corresponding to the 3s and 3p states overlap. This means that to the N conduction electrons in the 3s states, there are N 1s and $6N$ 2p states, i.e., a total of $7N$ states available for conduction of electrons.

Next consider magnesium, Mg ($Z = 12$), whose electronic configuration is $1s^2 2s^2 2p^6 3s^2$. According to the tight-binding approximation theory, because the 1s, 2s, 2p, and 3s bands are completely filled, magnesium should be an insulator. But it isn't! As in the case of sodium, magnesium's 3s and 3p bands overlap, and hence, out of the $2N$ 3s states and $6N$ 3p states, only $2N$ states are filled. The rest are available for electron conduction. Thus all solids like magnesium—whose valence band is filled but which has states available for electron conduction because of overlapping of bands—are *conductors*. In fact, sometimes they are called *semimetals*. For example, in transition elements, the 3d, 4s, and 4p bands overlap; and in rare-earth elements, the 4f, 5d, 6s, and 6p bands overlap. Hence these elements are conductors.

Let us now consider the most important case of distinction between insulators and semiconductors according to the band theory. Take diamond, C (one of the covalent solids) whose electronic configuration is $1s^2 2s^2 2p^2$. As N atoms of carbon are brought together, three bands—1s, 2s, and 2p—should be formed. The 1s and 2s bands should be completely filled, but of the $6N$ states of the 2p band, only $2N$ states should be filled.

However, this is *not* what happens. When carbon atoms are being brought together to form a solid, the 2s and 2p levels start broadening into bands. As the interatomic spacing decreases further, these bands start overlapping, as shown in Fig. 10.9; as these distances get smaller and smaller, the continuum of 2s and 2p states that had a total of $2N + 6N = 8N$ states splits into two bands, each containing $4N$ quantum states. When the carbon is in a stable solid state, the four electrons of the carbon atom that used to be in the 2s and 2p states now fill the lower band with $4N$ electrons, while the upper band, containing $4N$ states, is empty. The equilibrium distance of diamond is $\sim 1.5 \times 10^{-8}$ cm; at this point, the separation between the lowest filled and the upper empty band is ~ 5 eV, which is a large forbidden energy gap. That's why diamond is a good insulator.

The band scheme for carbon shown in Fig. 10.9 also applies to semiconductors such as Si and Ge, except that the bands correspond to different atomic-energy levels. But there is one very fundamental and important difference: The equilibrium separation of the atoms in Si and Ge are larger than that in diamond, so that there are much narrower energy gaps in Si and Ge. The width of the forbidden band, which is ~ 5 eV in diamond, is 1.2 eV in silicon and 0.7 eV in germanium— (as shown in Fig. 10.9). The narrow energy gaps in Si and Ge imply that more electrons from the completely filled lower or valence bands can be thermally excited to the

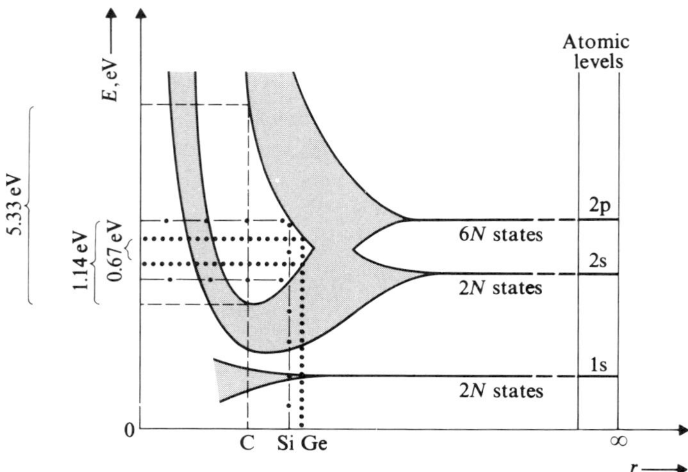

Fig. 10.9 Formation of energy bands in diamond (C), silicon (Si), and germanium (Ge). Note that gaps between the energy bands depend on the equilibrium distances.

empty conduction bands. And when this happens, these solids become conducting. Always bear in mind that both electrons and holes contribute to electric current.

10.4 INTRINSIC AND IMPURITY SEMICONDUCTORS

In Section 10.3 we talked about the kind of semiconductors in which the transformation of electrons to the conduction band and the creation of holes in the valence band is accomplished purely by means of thermal excitation. We call these *intrinsic semiconductors*. The electrons and holes are *intrinsic charge carriers*, and the resulting conductivity is *intrinsic conductivity*.

We can increase the conductivity of an intrinsic semiconductor tremendously, however, by adding certain impurities to the otherwise pure materials. By doing so, we get *impurity semiconductors*.[6,7] There are two different types of impurities that may be added: *n*-type and *p*-type. Each gives a different type of impurity semiconductor. We'll look at both types.

Let us introduce a small amount of phosphorus or arsenic (or any other element of Group V of the periodic table) as an impurity into an otherwise pure crystal of silicon or germanium. Adding an impurity means replacing an atom of silicon or germanium at a lattice site by an atom of the impurity. Silicon or germanium atoms each have four valence electrons, while atoms of Group V elements each have five valence electrons. Four of the electrons of the impurity atom form covalent pair bonds with the electrons of the neighboring atoms of the semiconductor, as shown in Fig. 10.10(a). But that fifth electron is only very weakly bound to the impurity atom by electrostatic forces.

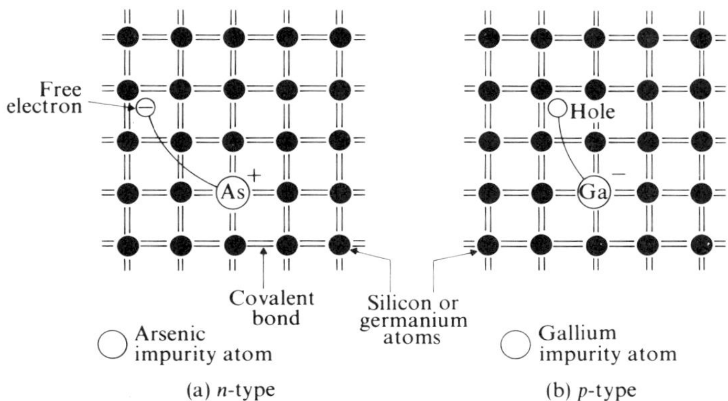

Fig. 10.10 (a) An *n*-type semiconductor; extra electrons are created by transfer of electrons from donor atoms. (b) A *p*-type semiconductor; extra holes are created by transfer of electrons to acceptor atoms.

Thus each impurity atom in the semiconductor has one extra electron. This electron can't be accommodated in the already filled original valence band. So it occupies a discrete energy level just below the conduction band (Fig. 10.11a). The separation between this new level and the conduction-band level may be only a few tenths of an electron volt. Thus Group V impurity atoms are very easily ionized (at temperatures above $\sim 20°K$), and the extra electrons jump into the conduction band, where they contribute to the electrical conductivity of the semiconductor. Note that these electrons are *in addition to* the electron–hole pairs produced by thermal excitation of the pure semiconductor. Thus overall there are more electrons than holes to serve as charge carriers. That's why these materials are called *n-type semiconductors*, or negative semiconductors. The Group V atoms are called *donor* atoms, because each impurity atom donates an additional free electron to the semiconductor. The part of the electrical conductivity that is caused by the impurity atom is *impurity conductivity*.[6,7]

Now, instead of adding Group V atoms to a silicon or germanium crystal, let

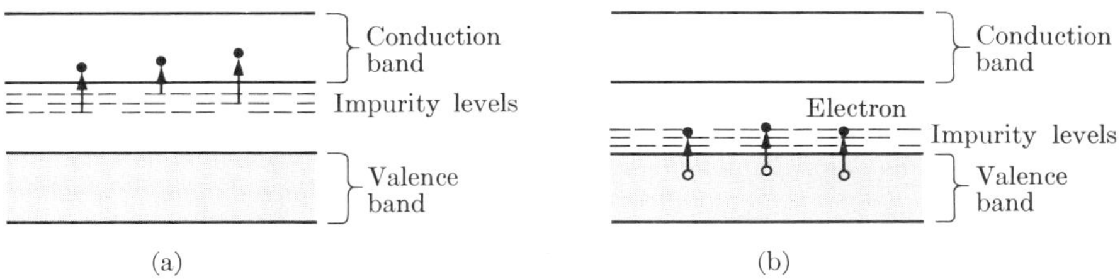

Fig. 10.11 Impurity levels in (a) *n*-type materials, and (b) *p*-type materials.

us add atoms of Group III—such as Al, Ga, or In—as impurities. The situation is quite different now. Group III atoms each have only three valence electrons. When a Group III atom is substituted into a semiconductor, such as a Si or Ge crystal, the three valence electrons of the Group III atom form covalent pair bonds with electrons of the neighboring atoms of the semiconductor. But that fourth available electron of the semiconductor lacks an electron with which it can form a bond. This is equivalent to saying that there is a hole created at the site of the impurity atom. An electron from the nearest semiconductor atom helpfully jumps in to fill this hole, thus imposing a negative charge on the impurity atom. And the hole is transferred to the atom from which the electron came (see Fig. 10.10b). In this case, the semiconductor (Ge or Si) is the host atom, while the Group III impurity atom is the *acceptor* atom, because it has accepted an electron from the covalent bond structure of the semiconductor, leaving a migratory hole behind.

Group III impurity atoms introduce vacant discrete energy levels very near the top of the host's completely filled valence bands (Fig. 10.11b). Again, it doesn't take very much energy to move the hole from the impurity atom, or to move an electron from the valence band to fill the hole. Except at very low temperatures, all holes are migratory. These holes behave like positive charge carriers, and are indistinguishable from holes created by thermal excitation. Because crystals of this type always have an excess of holes, or positive charge carriers, they are called *p-type semiconductors*, or positive semiconductors.

The conductivity of either *n*-type or *p*-type impurity semiconductors is greater than the conductivity of an intrinsic semiconductor. All we need is *one* impurity atom per *million* intrinsic semiconductor atoms to produce a significant change in the conductivity! The impurities are introduced by diffusion.

10.5 SEMICONDUCTOR DEVICES[8–12]

People these days are always talking about solid-state devices. But what they really should say is *semiconductor devices.* In recent years, semiconductor materials have been used in making many useful articles of great technological importance: rectifiers, transistors, photocells, switching circuits, voltage regulators, amplifiers, and countless other devices. They have often replaced electronic diode and triode vacuum tubes. Before we talk about these semiconductor devices, let's explain what is meant by *contact potential.*

Different metals require different amounts of energy to remove an electron from their surface. This energy, usually measured in electron volts, is the *work function* of the metal, denoted by ϕ. When two different metals or semiconductors are placed in contact, a potential difference is established between them. This contact potential, V_c, is equal to the difference between the work functions of the two materials in contact:

$$V_c = \phi_1 - \phi_2. \tag{10.10}$$

We can explain the existence of a contact potential as follows: Suppose we have two kinds of crystals, each with a different energy level. And suppose we place these two crystals in contact, so that electrons from one crystal can move into the other. Because of the difference between their Fermi energies, electrons in the material with the higher Fermi energy find unoccupied energy states in the material with the lower Fermi energy, and they start to fill these states. This process continues until the Fermi energy in both crystals is the same. This transformation of electrons makes one crystal positive with respect to the other, and hence establishes a potential difference. This is the *contact potential.*

We shall use this concept of contact potential—sometimes also called *diffused potential*—in discussing two semiconductor devices: (a) the semiconductor diode or *p-n* junction, and (b) the junction transistor (*n-p-n* or *p-n-p*).

a) *Semiconductor diode or p-n junction.* Suppose we have two crystals of the same semiconductor material, such as germanium. One is a *p*-type and the other is an *n*-type (see Fig. 10.12a). We place these two crystals in contact to form a *p-n* junction. Holes flow from left to right, and electrons from right to left. This double flow continues, producing positive layers on the right and negative layers on the left side of the junction, respectively. In the equilibrium state, the contact potential (Fig. 10.12b) is established. This prevents any further flow of holes and electrons across the junction. The width of the transition region at the junction is of the order of a few hundred angstroms.

How would the current flow across such a junction if we limited our discussion to the motion of holes only? (The motion of electrons is always equal and opposite to the motion of holes.) There would be free (or conduction) electrons in the *n*-type material. Therefore there would be a recombination of holes and electrons, and hence the number of holes in the *n*-type material would decrease. This would lead to a continuous flow of a hole current, I_1, from the *p*-side to the *n*-side. Due to thermal excitation, hole–electron pairs would be continuously produced in the *n*-type material. The flow of these holes to the *p*-side would produce the current I_2. When in equilibrium, the two hole currents would be equivalent, that is, $I_1 = I_2$.

Similar reasoning applies to the electrons. In an impurity semiconductor, there are always more current carriers of one sign than the other. Those that predominate are *majority carriers*; the others are *minority carriers.* In *n*-type materials, the electrons are the majority carriers and holes are the minority carriers. In *p*-type materials, the reverse holds true.

What happens if we apply a potential difference V (Fig. 10.12c) such that the positive terminal of the voltage source is connected to the *p*-type semiconductor and the negative terminal to the *n*-type? The height of the potential barrier decreases, which in turn increases the current I_1. There is a large supply of holes in the *p*-side, so the current I_1 rapidly increases with increasing voltage. The current $I = I_1 - I_2$ is as shown in Fig. 10.13. [Note that I_2 is not much affected by the application of V.] Such an arrangement is a *forward-biased p-n junction.*

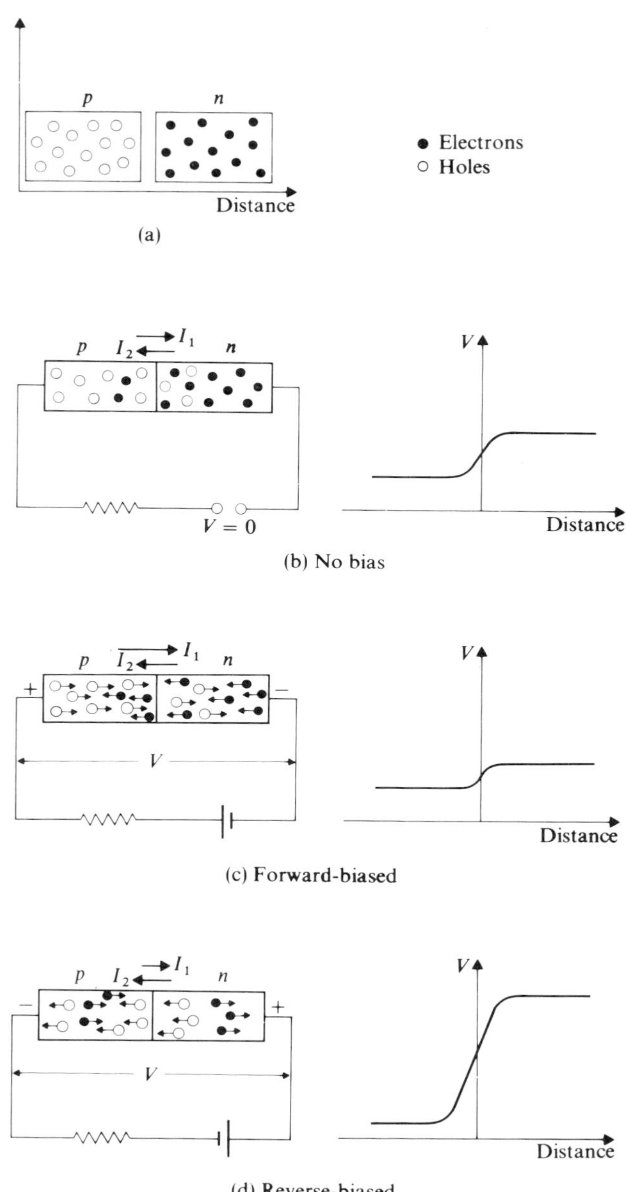

Fig. 10.12 (a) p- and n-type materials before contact. (b) p- and n-type materials after contact. (c) Potential difference V applied as shown causes p-n junction to be forward-biased. (d) Potential difference V applied as shown causes p-n junction to be reverse-biased.

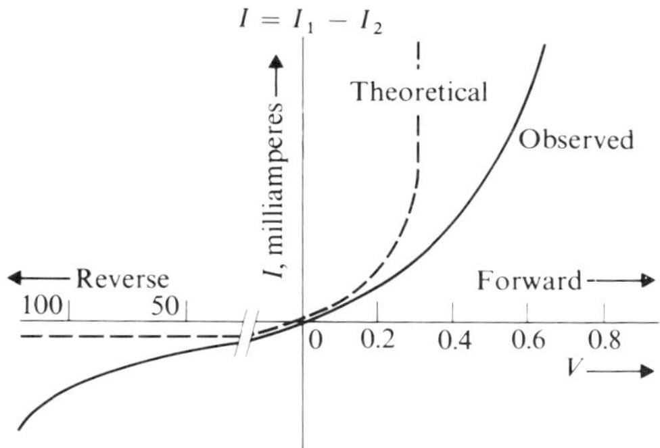

Fig. 10.13 Plot of current I resulting from application of voltage V across a *p-n* junction. Voltage applied in the direction from p to n is considered positive.

Consider the opposite case (Fig. 10.12d). That is, suppose we connected the negative terminal to the *p*-type and the positive terminal to the *n*-type. This would increase the potential difference across the junction, which would cause the reduction of current I_1 (because not many holes from the *p*-side could go to the *n*-side), while the current I_2 would not change. The *n-p* junction in these situations is called a *reverse-biased p-n junction.*

The plot of I $(= I_1 - I_2)$—that is, the net current across the *p-n* junction versus the applied voltage V (see Fig. 10.13)—indicates that the flow of current is favored in the direction from p to n. This means that we can use a *p-n* junction as a rectifier just like a vacuum-tube diode or triode. The *p-n* junction has the advantage in that it uses much less energy.

b) *Junction transistor.* The junction transistor is equivalent to a triode vacuum tube. It consists of a single slab of semiconductor in which, if the central portion has *p*-type conductivity, the other two parts have *n*-type, and vice versa. The central region is very thin (about a few thousandths of an inch), so that electrons crossing one junction find themselves in the other. The resulting junction transistor is either *n-p-n* or *p-n-p* type. We can also think of a junction transistor as being a combination of two *p-n* junctions.

Figure 10.14(a) shows an *n-p-n* transistor. There are many ways in which it may be used in an electronic circuit. Figure 10.14(b) shows the *n-p-n* transistor used as an amplifier. It is obvious that the junction on the left is forward-biased and the one on the right is reverse-biased. The central portion is the *base*, the left portion is the *emitter*, and the right portion is the *collector*. A circuit of this type is called a *common-base operation.*

When both regions in a *p-n* junction are highly doped—i.e., if they have very high amounts of impurities—the device is called a *tunnel diode.* When a junction transistor has no minority carriers, and its operation depends only on majority carriers, the device is called a *field-effect transistor* (FET).

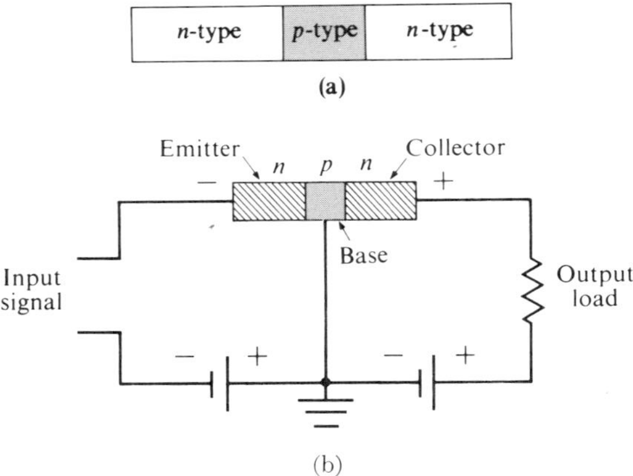

Fig. 10.14 (a) *n-p-n* transistor. (b) *n-p-n* transistor as an amplifier.

10.6 SUPERCONDUCTIVITY[13–17]

We know from experiment that the resistivity of a metal decreases with decreasing temperature.[13] The reason is that, as the temperature decreases, the amplitudes of the vibrating atoms also decrease, which increases the mean free path of the electrons. This decreases the resistivity or increases the conductivity of the metal. It has been found,[13] however, that in the neighborhood of 4°K many metals exhibit abnormally high conductivity. This is the *superconducting state*, and this property is called *superconductivity*.

The phenomenon was first discovered by Kamerlingh Onnes, in 1911, when he measured the resistivity of mercury at low temperature. It is obvious from the results (Fig. 10.15) that the resistance of mercury drops very abruptly to almost zero below 4.2°K. Later experiments proved that not only metals, but poor conductors like lead and indium exhibit superconductivity at low temperatures. Table 10.2 lists the *critical* (or *transition*) *temperature*, T_c, for some elements and compounds. Below its critical temperature, a material becomes superconducting.

Table 10.2

Transition Temperatures of Some Superconducting Elements and Compounds

Element	T_c, °K	Compound	T_c, °K
Tungsten	0.01	$ZrAl_2$	0.30
Cadmium	0.56	AuBe	2.64
Aluminum	1.19	NiBi	4.25
Mercury	4.15	Nb_3Al	17.5
Niobium	9.46	Nb_3	18.05
Technetium	7.92–8.22	$Nb_3Al_{0.8}Ge_{0.2}$	20.05

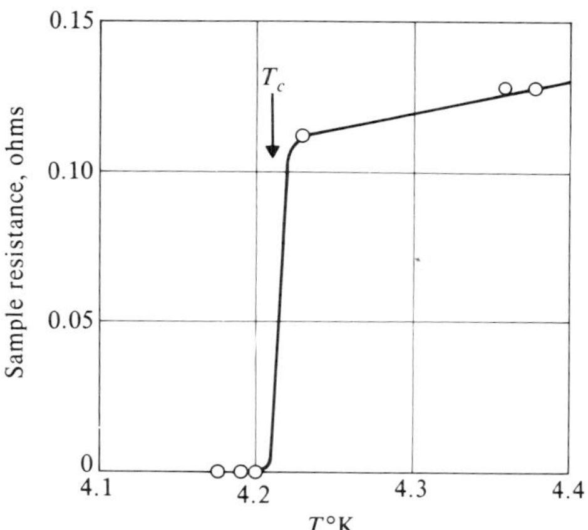

Fig. 10.15 Variation of resistance with temperature for mercury, showing a transition from normal to superconducting state at 4.15°K. [From H. K. Onnes, *Leiden Comm.* **124C** (1911)]

In addition to this property of superconductivity—which may also be called the disappearance of resistivity below the critical temperature T_c—there is another important property of the superconductive state: the *exclusion of flux*, or the *Meissner effect.*

Suppose that we cool a superconductor until it is below the critical temperature and then place it in a magnetic field. The magnetic field cannot exist inside the conductor and hence the magnetic flux is excluded from the superconductor (Fig. 10.16).

When a material is in a superconductive state, it is also perfectly diamagnetic.

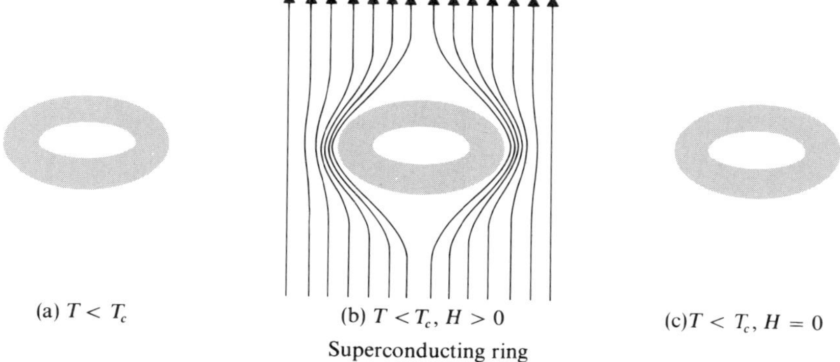

Fig. 10.16 (a) Superconductor at temperature $T < T_c$. (b) Superconductor at $T < T_c$ in a magnetic field H. Note the exclusion of flux from the interior of the ring. (c) Magnetic flux reduces to zero as H goes to zero.

According to Lenz's law, the current induced in the superconductor is of such a magnitude that all the flux is completely excluded from the interior of the superconductor. That is, inside the superconductor, at $T < T_c$, $B = 0$.

This property has a very interesting application to the case of a *superconducting ring*. Figure 10.17(a) shows an ordinary conductor at low temperature and Fig. 10.17(b) shows a superconductor at low temperature, both in magnetic fields. When a ring made of a superconductive material is placed in a magnetic field at $T > T_c$, the lines of force can pass through the ring. Now suppose that we lower the temperature of the ring so that $T < T_c$. Then the lines of force cannot pass through the cross section of the ring, *but* there are still lines of force passing through the hole, as shown.

Now suppose we switch off the field. The flux outside the ring disappears. What happens to the flux inside the hole of the ring? It is trapped, because it cannot pass through the walls of the superconductor. A large current is induced in the conductor by the outside collapsing field. The collapse of the magnetic field outside the superconducting ring induces a large current inside the ring itself, and

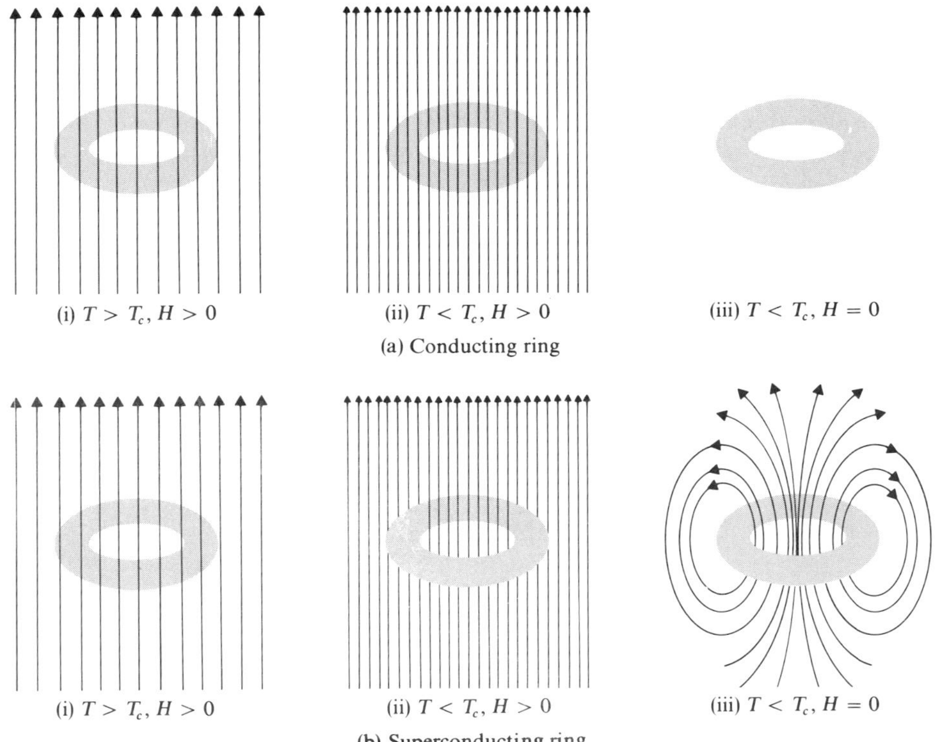

Fig. 10.17 Behavior of ordinary conductor and superconductor under different conditions.

this maintains the trapped flux, as shown in Fig. 10.17(b). Thus if the current is started in the ring-shaped superconductor, even when we switch off the field, the current continues to flow indefinitely inside the ring, because the resistivity in a superconductor is almost zero. Therefore such a device shows no losses. Investigators who have performed such experiments have been unable to detect a decrease in the current, even over periods of years!

The first successful explanation of superconductivity, given by J. Bardeen, L. N. Cooper, and J. R. Schrieffer,[16] is called the *BCS theory*. There is evidently strong electron–lattice interaction in superconducting material at low temperatures. It is this interaction that leads to the astonishing characteristics of superconductors which we have been discussing.

SUMMARY

Classification of solids into insulators, conductors, and semiconductors is explained by the *quantum theory*. There are two basic approaches to this problem: (1) the *zone theory*, or the modification of the free-electron model by considering the motion of an electron in the periodic potential. (2) The so-called *tight-binding approximation theory*, also called the *band theory*.

When we replace the periodic potential by the periodic potential of rectangular barriers, this is the *Kronig–Penney model*. We can solve the Schrödinger wave equation for an electron in this potential by using Bloch's theorem, and by this means we can find the allowed and forbidden energy bands for electrons in a given solid.

The region that contains electrons with momentum between 0 and π/a is called the *first Brillouin zone*; electrons with momentum between π/a and $2\pi/a$ are in the second Brillouin zone, and so forth.

Insulators are materials whose band formation is such that the forbidden region (or energy gap) between the highest filled band and the lowest empty band is very wide, ~ 3 to 6 eV. In *semiconductors*, the energy gap between the highest filled band and the lowest empty band is very narrow, ~ 0.1 to 1 eV. In *conductors*, the valence band is either partially filled or, if it is completely filled, the next empty band overlaps the valence band.

Intrinsic semiconductors are those in which the jumps of electrons to the conduction band and the creation of holes in the valence band are accomplished by means of thermal excitation. The conductivity of an intrinsic semiconductor can be increased tremendously by the addition of certain impurities to the otherwise pure semiconductor, to create *impurity semiconductors*. These are of two types: *p*-type and *n*-type.

When two different metals or semiconductors are placed in contact, a potential difference is brought about. This is equal to the difference between the work functions of the two and is called *contact potential*, $V_c = \phi_1 - \phi_2$. It is also called *diffused potential.* Semiconductor diodes (*p-n* junction) and junction transistors (*n-p-n* or *p-n-p*) may be used for all sorts of electronic devices, replacing vacuum tubes.

In the neighborhood of 4°K, many materials exhibit abnormally high conductivity. They are said to be in a *superconducting state*, and this property is called *superconductivity.* The *critical* (or *transition*) *temperature* is a temperature below which a given material is superconducting. Superconductors exhibit the *Meissner effect*, which means that, when they are below the critical temperature, the magnetic flux cannot cross the superconductor.

PROBLEMS

1. Using Eq. (10.4), (10.5), and (10.6), show that if V_0b becomes very large, the allowed energy zones in Fig. 10.3 become very narrow and hence the electrons cannot move freely.
2. Using Eqs. (10.4), (10.5), and (10.6), show that if $V_0 = 0$, the motion of the electron in the crystal is equivalent to that of a free electron.
3. Using Eq. (10.5), show that $P = maV_0b/\hbar^2$, provided that $b \to 0$ and $\beta \to \infty$. What is the justification for using these conditions?
4. Consider an electron in a crystal lattice described by the Bloch theory of the wave function. What is the probability of finding this electron in any unit cell of the crystal lattice?
5. Show that the wavelength of an electron given by the wave vector $k_x = n\pi/a$ satisfies the Bragg condition for reflection for a set of atomic planes that are perpendicular to the x axis and spaced a distance a apart.
6. Give two examples of materials which form *p*-type materials.
7. Give two examples of materials which form *n*-type materials.
8. Why are silicon and germanium frequently used in making solid-state (or semiconductor) devices, while tellurium is not?
9. The band structures of silicon and diamond are quite similar, but silicon has a metallic appearance, while diamond is transparent. Explain this difference, keeping in mind that the energy gap between the valence and the conduction bands in silicon is 1.14 eV and in diamond 5.33 eV.
10. Consider an arsenic atom replacing a silicon atom in a silicon crystal. The free (or unbound) electron "sees" a positive charge at the arsenic atom (see Fig. 10.10). The unbound electron moves in a Bohr-type orbit around the positive charge. How does the size of this orbit compare with the distance of 2.35 Å which lies between silicon atoms? The dielectric constant of silicon is 12.
11. Repeat Problem 10 for gallium instead of arsenic impurity atoms.
12. A semiconductor has a negative temperature coefficient of resistivity. Explain why.
13. List some outstanding uses of superconducting materials.

REFERENCES

1. F. Herman, *Proceedings of International Semiconductor Conference*, Paris, 1964
2. F. Herman *et al.*, *Proceedings of the Symposium on Energy Bands in Metals and Alloys*, Los Angeles, 1967
3. R. de L. Kronig and W. G. Penney, *Proc. Roy. Soc.* **A130,** 499 (1931)
4. F. Bloch, *Zeit. fur Physik* **23,** 555 (1928)
5. G. F. Koster, *Phys. Rev.* **98,** 901 (1955)
6. A. H. Wilson, *Proc. Roy. Soc.* **A133,** 458 (1931)
7. A. H. Wilson, *Proc. Roy. Soc.* **A134,** 277 (1931)
8. M. J. Morant, *Introduction to Semiconductor Devices*, Reading, Mass.: Addison-Wesley, 1964
9. D. E. Wooldridge, A. J. Ahearn, and J. A. Burton, *Phys. Rev.* **71,** 913 (1947)
10. K. G. McKay, *Phys. Rev.* **84,** 829 (1951)
11. J. W. Mayer and B. R. Gossick, *Rev. Sci. Instr.* **27,** 407 (1956)
12. S. S. Friedland, J. W. Mayer, and J. S. Wiggins, *Nucleons* **18,** 2, 54 (1960)
13. H. K. Onnes, *Leiden Comm.* **124C** (1911)
14. B. T. Matthias, T. H. Geballe, and V. B. Compton, *Revs. Mod. Phys.* **35,** 1 (1963)
15. N. Phillips, *Phys. Rev.* **134,** 385 (1964)
16. J. Bardeen, L. N. Cooper, and J. R. Schrieffer, *Phys. Rev.* **108,** 1175 (1957)
17. M. L. Cohen, *Phys. Rev.* **134,** 511 (1964)

Suggested Readings

1. L. V. Azaroff and J. J. Brophy, *Electronic Processes in Materials*, New York: McGraw-Hill, 1963
2. C. Kittel, *Introduction to Solid State Physics*, third edition, New York: John Wiley, 1966
3. Robert A. Levy, *Principles of Solid State Physics*, New York: Academic Press, 1968
4. John S. Blakemore, *Solid State Physics*, Philadelphia: W. B. Saunders, 1969
5. E. R. Jones, *Solid State Electronics*, Scranton, Pa.: Intext Educational Publishers, 1971
6. E. Spenk, *Electronic Semiconductors*, New York: McGraw-Hill, 1958
7. John P. McKelvey, *Solid State and Semiconductor Physics*, New York: Harper and Row, 1966
8. A. Leitner, *Introduction to Superconductivity*, a film produced in 1963, East Lansing, Mich.: Michigan State University

CHAPTER 11
STRUCTURE OF THE NUCLEUS

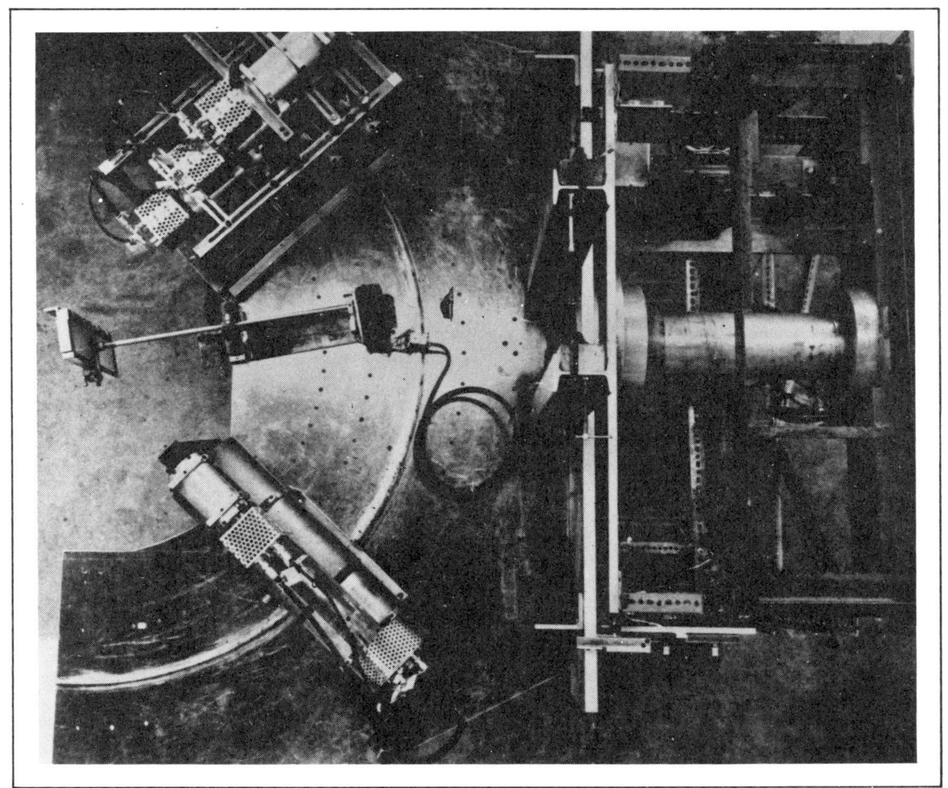

Double-scattering experiment at the University of Rochester to investigate nuclear forces. The first scattering target is located within the accelerator (not shown), while the second target is at left center. (Courtesy Dr. Robert E. Marshak)

11.1 INTRODUCTION

In Chapter 4 we saw how the Rutherford theory[1] of α scattering led to the establishment of the nuclear model of the atom. According to this model, most of the mass of the atom is concentrated in a small spherical volume, of radius $\sim 10^{-13}$ cm, called the *nucleus*. The electrons are distributed around the nucleus, also in a spherical volume, of radius $\sim 10^{-8}$ cm. The only constituents of *any* nucleus are neutrons and protons; electrons cannot reside inside it. Since the discovery of the neutron by Chadwick[2] in 1934, many more subnuclear particles, also referred to as elementary particles, have been discovered. We'll discuss them in the last chapter. Until then, let us talk about the nucleus and its properties.

Scientists who have worked on atomic, molecular, and solid-state physics have been able to develop theories that explain their experimental results, and in most cases make correct predictions of future discoveries. The situation in nuclear physics is somewhat different. There are still many questions which remain unanswered: (a) What is the exact form of the nuclear force, the force that holds neutrons and protons together in the nucleus? Nucleon is a generic name for neutron or proton. (b) There is no single model of the nucleus which can explain all its characteristics.

In spite of these difficulties, nuclear physics has made tremendous progress in the past 40 years. Furthermore, it presents a challenge which makes its study just that much more interesting.

We can break the study of nuclear physics down into the following subtopics.

a) Mass, size, and arrangement of constituents of the nucleus
b) Forces which hold the nucleus together
c) Laws governing the decay (or disintegration) of unstable nuclei
d) The investigation of different types of nuclear reactions
e) Instruments—detectors and accelerators—used in nuclear experiments

This division is of course an artificial one. And naturally there is no definite order in which these topics should be investigated.

11.2 BASIC PROPERTIES OF THE NUCLEUS

The following rundown is by no means a complete exposition of the basic properties of the nucleus, but it contains those essentials necessary for any logical discussion of nuclear physics.

a) *Constituents, charge, mass, and size of the nucleus.* A nucleus consists of Z protons and N neutrons. Z is the *atomic number*, while $N + Z = A$ is the *mass number*. Thus in any nucleus there are A *nucleons*. Since each proton has a charge $+1e$ and the neutron is neutral, each nucleus has a net charge of $+Ze$. The rest masses of proton and neutron are:

$$\text{Mass of proton} = m_\text{p} = 1.00727663\ \text{u} = 938.256\ \text{MeV}$$

$$\text{Mass of neutron} = m_\text{n} = 1.0086654\ \text{u} = 939.550\ \text{MeV}$$

When different constituents are brought together to form a nucleus, the final mass of the nucleus formed is less than the sum of the masses of the individual constituents, e.g., the mass of a nucleus is less than $Zm_p + Nm_n$. The energy equivalent of this mass difference is called the *binding energy*; it is responsible for holding the nucleons together in the nucleus.

As mentioned earlier, the size of the nucleus is much smaller compared to the size of the atom (nuclear radius $= \sim 10^{-13}$ cm, compared to atomic radius $= \sim 10^{-8}$ cm). The size of any atom is almost constant, but the size of any nucleus is a function of the number of nucleons A in the nucleus. The radius R of the nucleus (assuming that it is spherical in shape) is given by

$$R = r_0 A^{1/3} \simeq 1.35 \times 10^{-13} A^{1/3} \text{ cm}. \tag{11.1}$$

The value of r_0 depends on the type of experiment, and varies between 1.2×10^{-13} cm and 1.48×10^{-13} cm. (See Section 4.3.)

b) *Angular momentum of the nucleus.* Like the electron, the neutron and the proton each have intrinsic spins, with spin quantum number $= \frac{1}{2}$ and spin angular momentum $= \sqrt{\frac{1}{2}(\frac{1}{2} + 1)}\, \hbar = \frac{\sqrt{3}}{2}\hbar$. Each nucleon in the nucleus (like an electron in an atom) is assumed to have both orbital and spin angular momenta. Vector addition of the angular momenta of all the nucleons leads to the net total angular momentum **I**, called the *angular momentum of the nucleus.* According to quantum mechanics, the magnitude of the nuclear angular momentum I is given by

$$I = \sqrt{i(i + 1)}\, \hbar, \tag{11.2}$$

where $\hbar = h/2\pi$ and i is the total nuclear quantum number, called the *nuclear spin.* The name nuclear spin, although it is used very frequently instead of "total angular momentum of the nucleus," is actually misleading. i is a quantum number, either an integer or a half-integer. Like other angular momenta, the nuclear angular momentum also exhibits space quantization. Thus a nucleus with a spin i, when placed in an external magnetic field, can take $(2i + 1)$ different orientations, so that the angular momentum vector projected in the direction of the magnetic field has one of the following values (in units of $\hbar$), that is,

$$m_i = i, i - 1, i - 2, \ldots, 1, 0, -1, \ldots, -(i - 1), -i. \tag{11.3}$$

Experiments have shown a very close relationship between the nuclear spin i and the mass number A of the nucleus:

Nuclei with even mass numbers have zero or integral spins; those with odd mass numbers have half-integral spins. That is,

$$\boxed{\begin{array}{ll} i = 0, 1, 2, 3, 4, \ldots & \text{for even } A \\ i = \frac{1}{2}, \frac{3}{2}, \frac{5}{2}, \frac{7}{2}, \ldots & \text{for odd } A \end{array}} \tag{11.4}$$

In particular, even-even nuclei (nuclei with both Z and N even) have zero spin.

c) *Magnetic moment of the nucleus.* In atomic spectra, the magnetic moments of atoms are measured in units of *Bohr magnetons*, μ_β, defined as the magnetic moment associated with an atomic electron in orbital motion with an angular momentum of 1 $\hbar$. The magnitude of μ_β is

$$\mu_\beta = \frac{e\hbar}{2m_e} = 0.927 \times 10^{-23} \text{ joule/weber/m}^2, \tag{11.5}$$

where e is the charge on the electron, m_e is its mass, and c is the speed of light.

All nuclei have a net nuclear charge. Unless the total (or spin) angular momentum of the nucleus is zero, the motion of the nucleons inside the nucleus (like the motion of the electrons in the atoms) should give rise to nuclear magnetic moments. Furthermore, if we assume that the distribution of the nuclear charge is spherically symmetric,[3] the nucleus will give rise to a magnetic dipole moment only. The nuclear magnetic moments μ_I are measured in units of *nuclear magnetons* μ_N, defined as

$$\mu_N = \frac{e\hbar}{2m_p} = \frac{\mu_\beta}{1836} = 5.05 \times 10^{-27} \text{ joule/weber/m}^2, \tag{11.6}$$

where m_p is the mass of the proton, which is 1836 times the mass of the electron. The actual measured values of the nuclear magnetic moments are between $-3\mu_N$ and $+10\mu_N$. The positive sign means that the magnetic moment of the nucleus is in the same direction as the nuclear spin, while the negative sign means that it is in the opposite direction. The most significant are the experimentally measured values of the magnetic moments of a proton and a neutron.[4] The magnetic moment of a proton is

$$\mu_{\text{proton}} = +2.79276 \ \mu_N, \tag{11.7}$$

which is not 1 nuclear magneton, as might be expected. This indicates that the proton has a nonuniform charge distribution. Similarly, even though the net charge of a neutron is zero, it has a magnetic moment given by

$$\mu_{\text{neutron}} = -1.191315 \ \mu_N. \tag{11.8}$$

This indicates that the neutron has a nonuniform charge distribution, and that the direction of the magnetic moment is opposite to that of its intrinsic spin angular momentum vector. The magnetic moments of proton and neutron indicate that their charge distributions are very complex.[5]

d) *Nuclear forces.* We shall examine nuclear forces in greater depth later on, but for now let us just touch the high spots.

From the facts we already know about the basic properties of nuclei, we can conclude two things about nuclear forces.

i) Nuclear forces are much more attractive and stronger than coulomb forces.

ii) Nuclear forces are short-range.

The forces between nucleons—i.e., between proton and proton, neutron and neutron, and neutron and proton—must be attractive; otherwise there would be no stable nuclei. Also these attractive forces must be much stronger than the coulomb repulsive forces, otherwise there would be no stable heavy nuclei, because the repulsion between the large number of protons would disrupt the heavy nuclei.

From our study of the structures of atoms and molecules, we conclude that nuclear forces do not contribute to the formation or the properties of atoms and molecules. Hence nuclear forces must be short-range, extending from the center of the nucleus to its surface, or only a little beyond its surface.

If a proton is brought from an infinite distance toward a nucleus at 0, the variation of the potential $V(r)$ versus the distance r between the nucleus and the proton is as shown in Fig. 11.1. For $r > R$, the only force is the coulomb force, which is repulsive, and varies as $1/r^2$, while the potential varies as $1/r$, and is as shown. For $r < R$, the proton starts feeling the attractive force of the nucleus. This attractive nuclear force is much stronger than the coulomb repulsive force for $r < R$, and hence is represented by a negative potential (Fig. 11.1). The exact form of the potential between 0 and R is not known, but a constant value of $V(r)$, that is, $V(r) = -V_0$ for $r < R$, is sufficient to explain most of the experimental results.

Now suppose that, instead of a proton, a neutron is approaching the nucleus. The plot of $V(r)$ versus r is as shown in Fig. 11.2. Note that there is no coulomb

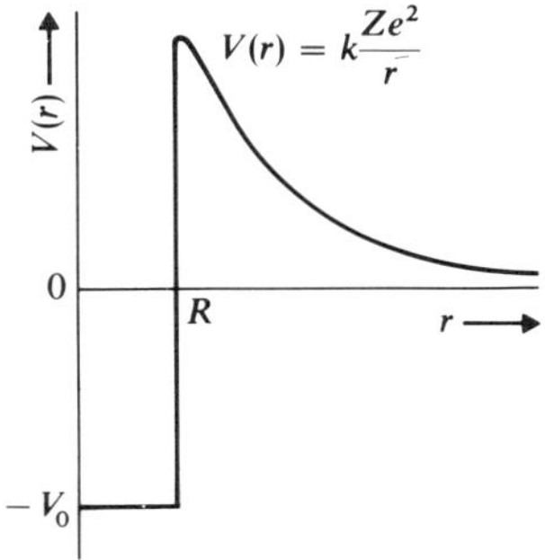

Fig. 11.1 Variation of potential $V(r)$ with r between nucleus and a proton as the proton is brought from an infinite distance toward the nucleus at 0.

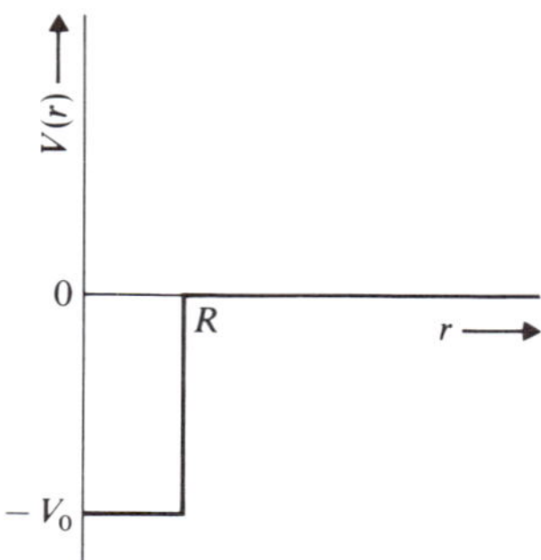

Fig. 11.2 Variation of potential $V(r)$ with r between nucleus and a neutron as the neutron is brought from an infinite distance toward the nucleus at 0.

force between the nucleus and the neutron, and hence $V(r) = 0$ for $r > R$. For $r < R$, the neutron starts feeling the attractive force, which is represented by a constant potential $-V_0$ for $r < R$. Typical values of V_0 and R which are in agreement with experiments are

$$V_0 \simeq -40 \text{ MeV},$$
$$R \simeq 1.4 \times 10^{-13} A^{1/3} \text{ cm}, \tag{11.9}$$

where A is the mass number of the nucleus. For a nucleus with $A = 125$, the radius of the nucleus is 7×10^{-13} cm.

e) *Nuclear models.* In atomic physics, the law of force is well established; hence all atomic models are constructed on the basis of this law. But for the nucleus, the exact form of the force law is not known. There is no single model which can explain satisfactorily all the characteristics of both stable and unstable nuclei. Many models are at present in use. (i) shell model, (ii) liquid-drop model, (iii) compound-nucleus model, (iv) optical model, (v) direct-interaction model, (vi) collective model, and (vii) Fermi-gas model.

11.3 BINDING ENERGY OF THE NUCLEUS

An atom denoted by ${}_Z^A X$ is formed by bringing together Z protons, $(A - Z)$ neutrons, and Z electrons, but its mass $M(A, Z)$ is less than the sum of the masses of its constituents (protons, neutrons, and electrons) in the free state. This decrease in mass ΔM is converted into energy ΔE, given by

$$\Delta E = \Delta M c^2. \tag{11.10}$$

This energy is released when the atom is formed. Once the atom has been formed, if we want to break the atom into its constituents, we have to supply energy equivalent to the energy released. This energy is called the *binding energy BE* ($= -\Delta E$), and is given by

$$BE = [Zm_p + (A - Z)m_n + Zm_e - M(A, Z)]c^2, \tag{11.11}$$

where m_p, m_n, and m_e are the masses of the proton, neutron, and electron, respectively. If we neglect the small binding energy of the hydrogen atom, we may replace $m_p + m_e$ by m_H, the mass of the hydrogen atom. Thus Eq. (11.11) may be written as

$$BE = [Zm_H + (A - Z)m_n - M(A, Z)]c^2. \tag{11.12}$$

The binding energy per nucleon, BE/A, may be written as

$$\boxed{BE/A = [Zm_H + (A - Z)m_n - M(A, Z)]c^2/A.} \tag{11.13}$$

Figure 11.3 shows the plot of the binding energy per nucleon, BE/A, for different nuclei, versus the mass number A. Almost all the nuclei lie on or close to a single smooth curve, with the exception of ^4He, ^{12}C, and ^{16}O. Some of the outstanding features of this curve are the following.

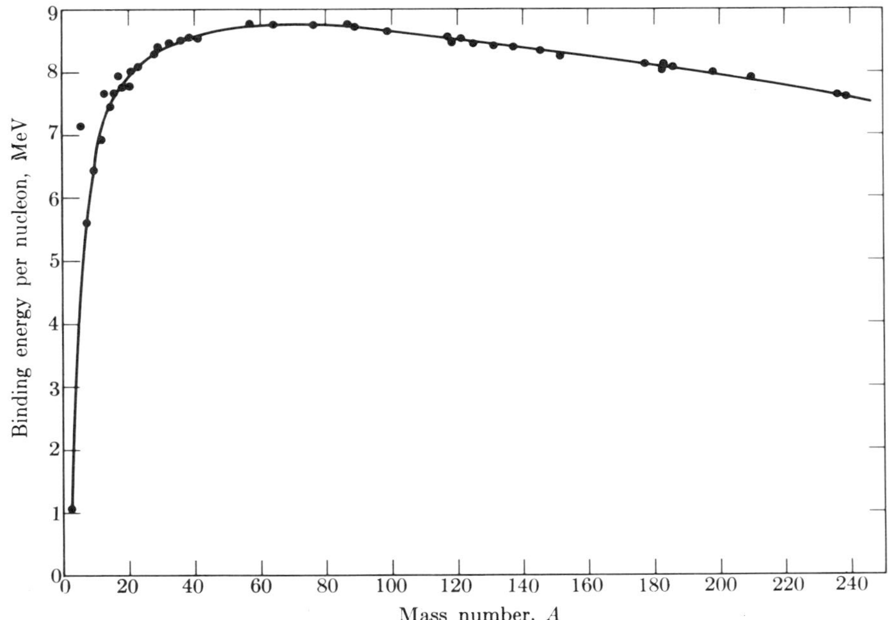

Fig. 11.3 The plot of the binding energy per nucleon, BE/A, of different nuclei, versus A, the mass number of the nucleus.

a) There is a flat maximum around $A = 50$, where the binding energy per nucleon, BE/A, is $\sim$8.8 MeV.

b) For low A, BE/A is low, but increases rapidly with increasing A, up to $A \simeq 20$.

c) There is not much variation between $A \sim 20$ and $A \sim 160$. The average value of BE/A in this region is $\sim$8.5 MeV,

d) For higher values of A, say $A > 140$, the BE/A decreases slowly and continuously with increasing A; it reaches a value of 7.6 MeV at $A = 238$ for ^{238}U.

We shall try to explain some of these features in the following section.

⋆11.4 NUCLEAR STABILITY[6–9]

Nuclei which have low BE/A for small A may be explained on the basis of the surface tension effect, derived from the *liquid-drop model*[6] Think of a nucleus as being a drop of liquid, in which the nucleons (protons and neutrons) take the place of the molecules in the drop. As shown in Fig. 11.4(a), the nucleons deep inside the nucleus are attracted from all sides by neighboring nucleons, while those on the surface are attracted from one side only. Thus the binding energy for the nucleons at the surface of the nucleus is smaller than the binding energy for the nucleons inside the nucleus. Since there is a larger fraction of surface nucleons in light nuclei than in medium and heavy nuclei (Fig. 11.4), the total binding energy (and hence the BE/A) is small for small A.

The low value of BE/A for nuclei with large A is due to the coulomb repulsive force. The protons inside the nucleus repel each other according to coulomb's

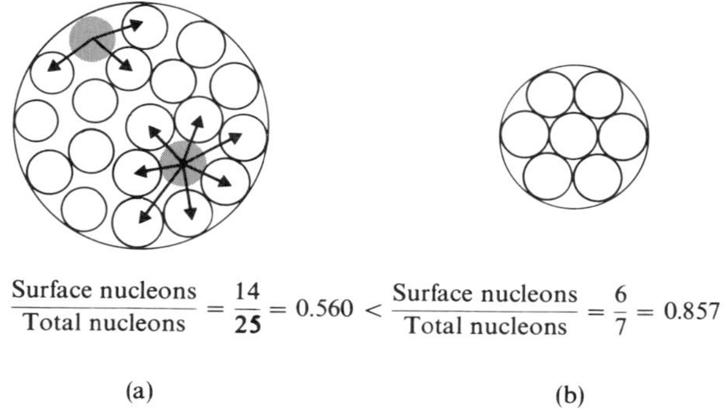

Fig. 11.4 Nucleons well inside the nucleus are attracted from all sides and hence are strongly bound, while those on the surface are not so strongly bound as shown in (a). In medium and heavy nuclei, the fraction of the nucleons on the surface as in (a) is much smaller than in light nuclei as in (b).

law, thereby decreasing the binding energy or increasing the mass of the nucleus. Coulomb forces are long-range, and hence every proton inside the nucleus repels every other proton, not just its neighbors. The repulsive force increases with increasing A, thereby resulting in smaller values of BE/A for large A (Fig. 11.3). We can calculate coulomb repulsion by assuming the nucleus to be a liquid drop in the form of a uniformly charged sphere of radius R and charge $+Ze$.

There is another important consequence of coulomb repulsion. Since coulomb repulsion has the effect of making nuclei unstable, heavy nuclei tend to be made up of more neutrons than protons. That such is the case is obvious from Fig. 11.5, which shows a plot of neutron number N versus proton number Z for all nuclei, stable as well as unstable (radioactive). Solid squares indicate stable nuclei; all others represent unstable radioactive nuclei. A curve drawn through the stable nuclei shows that its locus departs from the line $N/Z = 1$ for low Z toward the direction of a higher number of neutrons (heavier elements), reaching a value of $N/Z = 1.6$ for $A = 238$. This curve is called the *stability curve*. The isotopes on both sides of this curve are radioactive, and eventually decay in such a way that the final stable product lies on the stability curve.

A closer look at the stable nuclei in Fig. 11.5 reveals the following.

i) *Odd–even effect.* The total binding energy of a nucleus depends not only on the ratio of the number of neutrons to the number of protons, but on whether these numbers of neutrons and protons are odd or even. Table 11.1 shows all the stable nuclei classified into four possible groups: even–even, even–odd, odd–even, and odd–odd. Obviously even–even nuclei are the most abundant by far, while there are only five stable odd–odd nuclei. The stability of even–odd and odd–even lies between the two extremes.

Table 11.1

Classification of Stable Nuclei

A	Z	N	Number of stable nuclei
Even	Even	Even	156
Odd	Even	Odd	50
Odd	Odd	Even	48
Even	Odd	Odd	5
			259

ii) *Pairing of nucleons.* A survey of stable nuclei suggests that nucleons tend to form neutron–proton pairs. Thus nuclei that satisfy the condition $A = 2Z$ are more strongly bound together. Any deviation from $A = 2Z$ should decrease the binding energy. This effect is not to be confused with the coulomb repulsion effect.

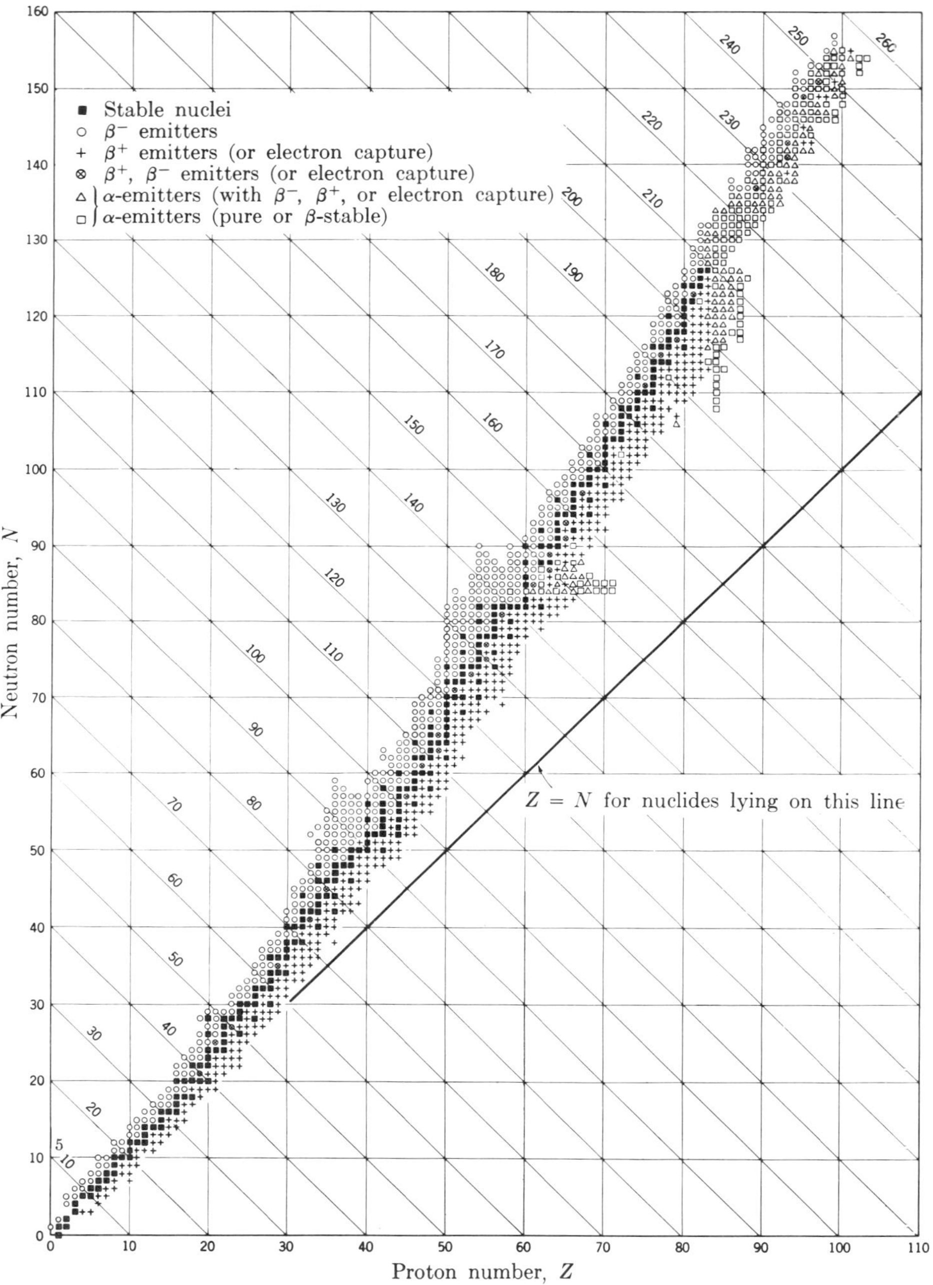

Fig. 11.5 A plot of neutron number N versus proton number Z for all known nuclei, stable and unstable. A curve through the stable nuclei starts with $N/Z = 1$ for low-A nuclei and reaches a value of $N/Z = 1.6$ for high-A nuclei.

11.5 THE NUCLEAR SHELL MODEL

In Chapter 6, we applied quantum mechanics to the structure of the atom and found that it correctly predicted the closure or filling of electronic shells at $2n^2$, where n is an integer. That is, electronic shells are completely closed (or filled up with electrons) for those elements for which $Z = 2, 10, 18, 36$, and 54, corresponding to elements He, Ne, Ar, Kr, and Xe, respectively. Therefore these elements should be chemically inactive. That such is the case is clear from Fig. 7.19, which shows that the ionization potentials for these elements are higher than those of their neighbors. We can refer to the numbers 2, 10, 18, 36, and 54—corresponding to extra-stable elements—as the *atomic magic numbers* of the periodic table.

Now what about the nucleus? Does it also exhibit some kind of shell structure? Although Fig. 11.3 indicates that the curve of BE/A versus A varies smoothly, detailed investigations of this curve and other properties of nuclei reveal that the periodicities in the properties of the nuclei might be due to a nuclear shell structure similar to the atomic shell structure. Evidence indicates that the numbers (N for number of neutrons and Z for number of protons),

$$\boxed{\begin{aligned} N &= 2,\ 8,\ 20,\ 28,\ 50,\ 82,\ \text{and } 126 \\ Z &= 2,\ 8,\ 20,\ 28,\ 50,\ \text{and } 82 \end{aligned}} \tag{11.14}$$

are the *nuclear magic numbers*. However, these magic numbers are not as strongly convincing as the atomic magic numbers. Let's look at some of the evidence for the nucleus having a shell structure, and hence for the existence of nuclear magic numbers.

a) *Evidence for the shell model.* Elements with the above magic numbers Z or N have many more isotopes or isotones, respectively, than their neighbors. For example, Sn ($Z = 50$) has 10 stable isotopes, while In ($Z = 49$) and Sb ($Z = 51$) each have only 2. Similarly, for $N = 20$, there are 5 stable isotones, while for $N = 19$, there is none, and for $N = 21$ there is only one. The same thing holds true for other magic numbers.

Even though Fig. 11.3 showed a smooth binding-energy curve through different points, a closer look at the binding energies of nuclei reveals a shell structure. Thus if the shells of the nuclei close at the magic numbers, the nuclides (or nuclei) corresponding to these magic numbers should be very stable. In other words, the binding energy of the last added neutron or proton to close the shell should be very large. This has indeed been found to be the case.[10]

We can find a still more convincing argument for the existence of magic numbers by looking at the plot of neutron-capture cross section versus neutron number[11]. Nuclei with magic numbers should have very low cross section, because

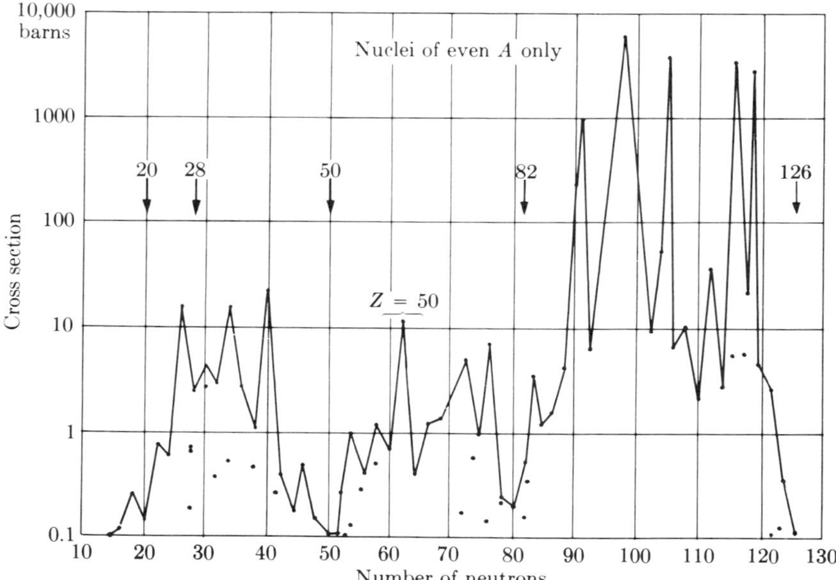

Fig. 11.6 Shell structure of the nucleus at N = 20, 28, 50, 82, and 126 is evident from this plot of neutron capture cross section versus neutron number. [From B. Flowers, *Prog. in Nuc. Physics* **2**, 235 (1952); London: Pergamon Press]

closed shells mean that there is no more vacancy. Figure 11.6, which shows neutron-absorption cross section versus number of neutrons, indicates that this is true. Note in particular the low neutron-capture cross section for doubly magic nuclei ${}^{40}_{20}\text{Ca}$ ($Z = 20$, $N = 20$) and ${}^{208}_{82}\text{Pb}$ ($Z = 82$, $N = 126$).

There is much more practical evidence for a system of shells in the nucleus, but we shall not go into the matter further. Let us instead consider the theoretical justification for the shell structure of the nucleus.

b) *Theory of the nuclear shell model.* In the case of the structure of the atom, we arrived at the system of electronic shells and hence the magic numbers theoretically by solving the Schrödinger wave equation. We could do this because we knew the form of the coulomb potential, which is proportional to $1/r$.

The nuclear shell structure or the nuclear magic numbers cannot be arrived at so easily, because we don't yet know the exact form of the nuclear force or the nuclear potential. Of course, we know definitely from our investigation of the nucleus that nuclear forces are strongly attractive and that they extend over a very short range from the center of the nucleus. Actually, the nuclear force goes to zero within $\sim 10^{-13}$ cm of the center. Usually it doesn't matter what the actual form of the nuclear force (and hence the nuclear potential) is, so long as it is strongly attractive and short-range. Whatever the form of the potential, the calculations of the shell model are based on these two assumptions.

i) Each nucleon moves freely in a force field described by the potential, which is a function of the radial distance from the center of the system.

ii) The Pauli exclusion principle applies; i.e., the energy levels or shells are filled according to the exclusion principle.

Let us assume that a nucleon of mass M with angular momentum $\sqrt{l(l+1)}\,\hbar$ is moving in a potential $V(r)$. Since we are assuming that $V(r)$ is independent of θ and ϕ, the Schrödinger wave equation to be solved (Section 6.2) is

$$\frac{d^2}{dr^2}(rR) + \frac{2M}{\hbar^2}\left[E - V(r) - \frac{l(l+1)\hbar^2}{2Mr^2}\right](rR) = 0, \tag{11.15}$$

where R is the radial wave function and E is the energy eigenvalue. Irrespective of the form of $V(r)$, so long as it is a function of r only, the same quantum numbers, n, l, j, m_j, result in both an atomic and a nuclear shell model. Also we use the notation s, p, d, f, g, . . . , standing for $l = 0, 1, 2, 3, 4, \ldots$, respectively. Since we want to find only the order of the levels corresponding to different states of the motion of the nucleon and not the exact binding energies, we can make use of the simpler potential instead of more accurate potentials for solving Eq. (11.15).

Figure 11.7 shows three simple potentials. Figure 11.7(a) is a square-well potential given by

$$\begin{aligned} V(r) &= -V_0 \qquad \text{for } r \leq R \\ &= 0 \qquad\quad \text{for } r > R. \end{aligned} \tag{11.16}$$

Figure 11.7(b) is a harmonic-oscillator potential given by

$$V(r) = -V_0 + \tfrac{1}{2}kr^2. \tag{11.17}$$

If we combine the two potentials, square-well and harmonic-oscillator, we have the potential shown in Fig. 11.7(c), given by

$$\begin{aligned} V(r) &= -V_0\left(1 - \frac{r^2}{R^2}\right) \qquad \text{for } r \leq R \\ &= 0 \qquad\qquad\qquad\qquad r > R \end{aligned} \tag{11.18}$$

Solving Eq. (11.15) for any one of the above potentials does not give all the magic numbers. However, using Eq. (11.18) gives all except magic number 28.

In 1949, a different and successful approach to the shell model was suggested independently by M. Mayer[12] and O. Haxel, J. Jensen and H. Suess.[13] According to them, in addition to the potential $V(r)$, there is a strong interaction

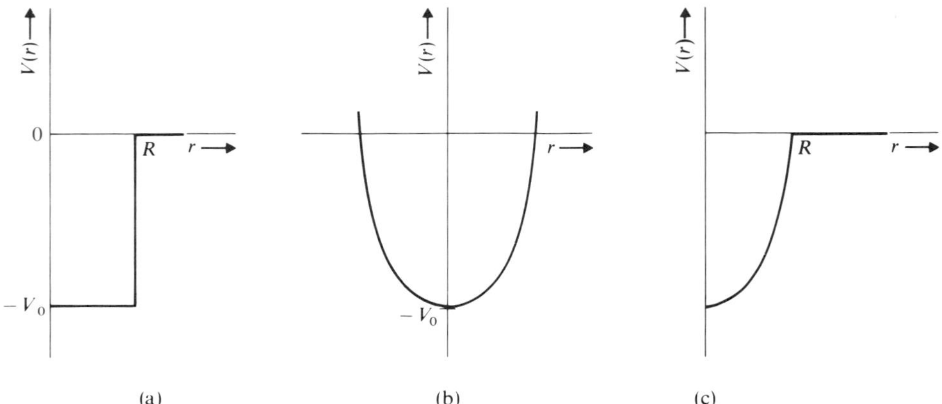

Fig. 11.7 Plots of the three simplest forms of potentials used for calculating shell models. (a) Square-well potential, (b) harmonic-oscillator potential, and (c) combined square-well and harmonic-oscillator potential.

between the orbital and spin angular momenta of each nucleon. The strength of this interaction depends on their relative orientation and on their magnitudes, and may be written as $V(r)\,\mathbf{S}\cdot\mathbf{L}$. The resulting energy levels obtained by solving the Schrödinger wave equation (Eq. 11.15) are

$$1s_{1/2},\ 1p_{3/2},\ 1p_{1/2},\ 1d_{5/2},\ 2s_{1/2},\ 1d_{3/2},\ 1f_{7/2},\ \ldots . \tag{11.19}$$

Figure 11.8 shows these levels for both protons and neutrons. Because of the coulomb repulsion between protons with high values of Z, the energy levels of the protons are slightly different from those of the neutrons. The number of protons or neutrons in any given level is $2j + 1$. Hence there are 2, 4, 2, 6, 2, 4, 8, ... nucleons in each of the states respectively given by Eq. (11.19). Figure 11.8 indicates clearly that closure of the shells takes place at nucleon numbers

2, 8, 20, 28, 50, 82, 126,

which are the desired magic numbers.

Note that Fig. 11.8 is not an energy-level diagram for any given nucleus; it is a diagram of the order of levels in the shell model. In any filled proton level there are $2j + 1$ protons, and in any filled neutron level there are $2j + 1$ neutrons.

c) *Nuclear spins and parities.* We can predict the spins and parities of nuclei in the ground state if we take the following into account. (1) In a filled energy level (of a shell or subshell), the orbital and spin angular momenta add in such a way that the total angular momentum is zero. (2) In levels that are not filled, the nucleons form proton pairs and neutron pairs, but no proton–neutron pairs. These assumptions lead to the following coupling rules.

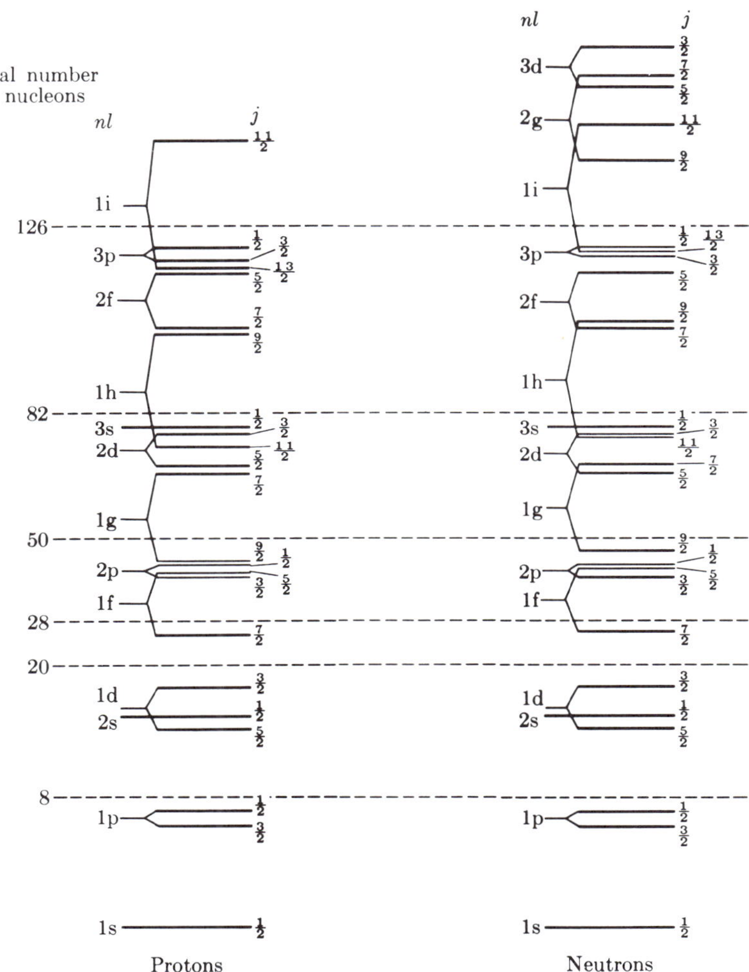

Fig. 11.8 Energy levels for neutrons and protons according to the shell model, obtained by solving the Schrödinger wave equation using a combination of square-well and oscillator potentials. This diagram also indicates the effect of spin–orbit coupling. The energy levels for protons have been corrected for the proton's coulomb repulsion energy. [From P. F. A. Klinkenberg, *Revs. Mod. Phys.* **24,** 63 (1952)]

Rule 1. The ground state of any even–even nucleus has zero total angular momentum and even parity, regardless of the number of protons and neutrons. That is,

$$\sum J_n = 0 \qquad \text{and} \qquad \sum J_p = 0,$$

where J_n and J_p are the total angular momenta of the neutrons and the protons, respectively.

Rule 2. (a) In a nucleus with N even and Z odd, $\sum J_n = 0$; hence the ground-state spin of the nucleus is determined by the last odd proton. (b) In a nucleus with Z even and N odd, $\sum J_p = 0$; hence the ground-state spin of the nucleus is determined by the last odd neutron. (c) In a nucleus with N odd and Z odd, the ground-state spin is determined by the last odd neutron and the last odd proton, and may have any value between $|J_n - J_p|$ and $|J_n + J_p|$. Only exact calculations give the correct spin.

Table 11.2

Spins of Some Odd Nuclei

Nucleus	Spin and parity (shell model)	Spin and parity (measured)
Neutron	$1s_{1/2}$	$\frac{1}{2}+$
Proton	$1s_{1/2}$	$\frac{1}{2}+$
${}^{3}_{1}H$	$1s_{1/2}$	$\frac{1}{2}+$
${}^{3}_{2}He$	$1s_{1/2}$	$\frac{1}{2}+$
${}^{7}_{3}Li$	$1p_{3/2}$	$\frac{3}{2}-$
${}^{11}_{5}B$	$1p_{3/2}$	$\frac{3}{2}$
${}^{13}_{6}C$	$1p_{1/2}$	$\frac{1}{2}$
${}^{17}_{8}O$	$1d_{5/2}$	$\frac{5}{2}$

Table 11.2 lists the spin configurations predicted by the shell model, as well as experimentally observed values. As an example, Fig. 11.9 shows the energy-level scheme of ${}^{43}_{20}Ca$. Since the last neutron is in the $1f_{7/2}$ state, the ground state of the

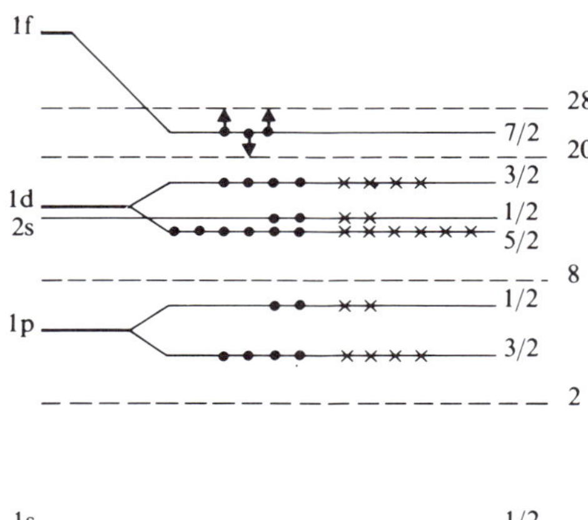

Fig. 11.9 Nucleon configuration of ${}^{43}_{20}Ca$ ($Z = 20$, $N = 23$). (x = proton and • = neutron). The ground state is $1f_{7/2}$.

$^{43}_{20}Ca$ nucleus is $1f_{7/2}$. Although the shell model does predict the nuclear spins of stable nuclei, there are many cases in which the predicted values do not agree with the measured ones.

The parity of a system is given by $(-1)^l$, where l is the orbital quantum number of the last odd nucleon. Thus, for a nucleon in the state s, d, g, . . . corresponding to $l = 0, 2, 4, \ldots$, the parity is even (+); while for a nucleon in the state p, f, h, . . . , corresponding to $l = 1, 3, 5, \ldots$, the parity is odd (−).

11.6 NUCLEAR FORCES

We are now quite familiar with two types of forces—*gravitational* and *electromagnetic*. We had a lot to say about electromagnetic (or coulomb) forces in the preceding chapters, when we were discussing the structure of atoms. In our discussion, we stressed that it is the coulomb forces that bind together the electrons in the atoms.

There are two other types of fundamental forces that are equally important: *nuclear force*, which is responsible for holding the constituents of a nucleus together, and *weak interactions*, which manifest themselves in certain decay processes (beta decay) of nuclei (see Chapter 12). The relative strengths of these four forces—nuclear, electromagnetic, weak, and gravitational—are 1, 10^{-2}, 10^{-13} and 10^{-39}, respectively.

In most of our study of physics and chemistry, the only information we had to have about the atomic nucleus was its charge (atomic number Z) and its mass or mass number A. Now we shall investigate less familiar but more important aspects of nuclei: the nuclear forces and their characteristics. We can best grasp the qualitative aspects of nuclear forces by observing the regularities in complex nuclei. We shall discuss these characteristics under the following headings: (a) strength and range of nuclear forces, (b) saturation of nuclear forces, (c) charge dependence of nuclear forces, and (d) the meson theory of nuclear forces.

a) *Strength and range of nuclear forces.* Stable nuclei owe their very existence to the fact that the average net nuclear force among the nucleons inside these nuclei is strongly attractive; otherwise the coulomb repulsion between the protons in the nuclei would disrupt them. We can see that this is true when we realize the magnitudes of the binding energies of the nucleons in the nucleus: ~8 MeV per nucleon. The force (or interaction) between any two nucleons is represented by a potential V such that the force F is

$$F = -\frac{\partial V}{\partial r} \tag{11.20}$$

which implies that nuclear forces are central forces. But unlike the electromagnetic forces, which extend to infinity—i.e., they have infinite range—the nuclear force has a very short range. The structures of atoms and molecules are never affected by nuclear forces. We may say that the range α of the nuclear force is either equal to the radius of the nucleus, or extends only slightly beyond the physical perimeter of it.

We can find the magnitude of the attractive potential responsible for nuclear force from the following considerations. Suppose that λ is the wavelength of a nucleon inside a nucleus of radius R. A necessary condition for the nucleon to be inside the nucleus is that its wavelength must fit into the nucleus. That is,

$$\lambda \leq 2R. \tag{11.21}$$

Also (using the deBroglie hypothesis) the wavelength λ, the kinetic energy K, the mass M, and the momentum p of the nucleon are:

$$\lambda = \frac{h}{p} = \frac{h}{\sqrt{2\,MK}} = \frac{1}{2\pi}\left(\frac{\hbar}{Mc}\right)\sqrt{\frac{Mc^2}{2K}}, \tag{11.22}$$

where $(\hbar/Mc)$ is the Compton wavelength of the nucleon, equal to 2.1×10^{-14} cm. Since the radius R of a typical medium-weight nucleus is $\sim 3 \times 10^{-13}$ cm, we combine this with the above two equations to obtain $K \simeq 25$ MeV (Problem 18). Thus, so that the nucleon is confined inside the nucleus, the potential energy $-V_0$ (the minus sign is due to the fact that the potential is attractive) should be equal to the binding energy + the kinetic energy. Since the binding energy is ~ 8 MeV per nucleon,

$$\begin{aligned} -V_0 &= BE + KE, \\ -V_0 &\simeq 8 \text{ MeV} + 25 \text{ MeV} \simeq 33 \text{ MeV}, \\ V_0 &\simeq -33 \text{ MeV}, \end{aligned} \tag{11.23}$$

while the range α of the nuclear force is given by

$$\alpha \simeq 2 \times 10^{-13} \text{ cm}. \tag{11.24}$$

Thus to describe the form of the nuclear potential requires two parameters: (i) the strength or depth of the potential well V_0, and (ii) the range α of the nuclear force, beyond which the potential falls very rapidly to zero. However, Eq. (11.23) doesn't tell us the exact variation of the potential between $r > 0$ and $r < \alpha$.

Some potentials that are commonly used are (1) the square well, (2) the harmonic oscillator, (3) the combination of square-well and harmonic oscillator, (4) the gaussian, (5) the exponential, and (6) the Yukawa (see Fig. 11.10). From experiments at low energies, we know that any one of these forms may be used without affecting the results, because, at low energies (< 100 MeV), the wavelength of an incident particle is very large compared to the range of the nuclear force. Therefore the incident particle cannot "detect" the shape of the potential. One

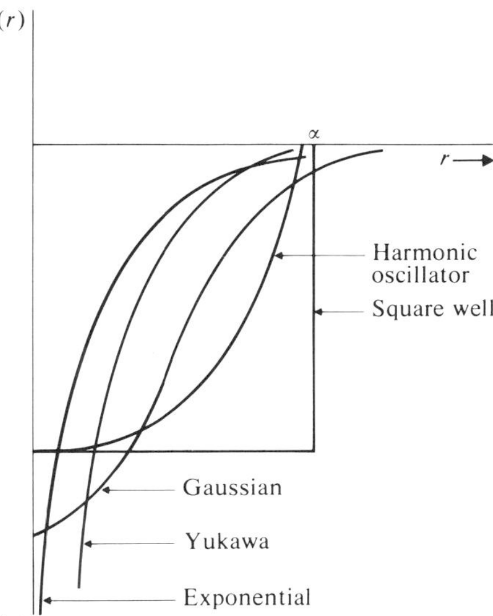

Fig. 11.10 Potentials commonly used to describe nuclear interactions (or forces).

can detect the shape of the potential only by using incident particles of very high energy. The square-well potential given by

$$\begin{aligned} V(r) &= -V_0 \qquad \text{for } r \leq \alpha \\ &= 0 \qquad \text{for } r > \alpha \end{aligned} \tag{11.25}$$

is the simplest of all. It is frequently used in the interpretation of experimental results. One can derive the form of the Yukawa potential theoretically.

Scattering experiments with particles having very high energies reveal that there is a repulsive core at the center of the nucleus, even though the overall effect of the nuclear force is attractive. If there were no repulsive force at the core, the nulceus would collapse to the size of the nuclear-force range α. That is, all nuclei would have the same size! Of course we realize that they do not. The size of the nucleus is a function of the mass number, $R = r_0 A^{1/3}$. As shown in Fig. 11.11, the range of the repulsive force at the core is much smaller than α. However, so long as we are dealing with low energies (which we shall be for the most part in this text), we shall ignore this repulsive force at the core, and for simplicity use the square-well attractive potential.

b) *Saturation of nuclear forces.* When we speak of the saturation of nuclear forces, we mean that a given nucleon can interact strongly with only a limited number of the nucleons surrounding it. To see this, recall the plot of BE/A versus A

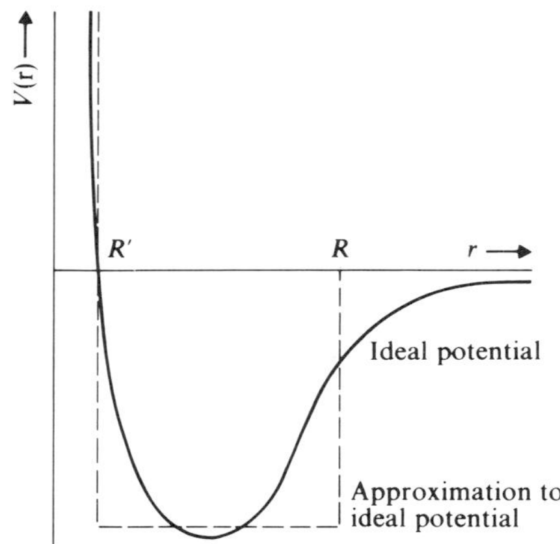

Fig. 11.11 The nuclear potential $V(r)$ has a repulsive force at the core for $r < R'$ and an attractive potential between $r = R'$ and $r = R\,(= \alpha)$. The net effect is an attractive potential. The dashed line is an approximation to the ideal potential well.

in Fig. 11.3. We said earlier that, except for low mass numbers, the binding energy per nucleon is almost constant ($BE/A \simeq 8$ MeV). That is,

$$\begin{aligned} BE/A &= \text{constant}, \\ BE &\propto A. \end{aligned} \tag{11.26}$$

If every nucleon interacted with every other nucleon, the forces would be unsaturated, and (as in the case of coulomb forces, which are long-range forces, and cause every charged particle to interact with every other particle) the result would be different. Suppose that there are A particles, and that every particle interacts with $(A - 1)$ other particles. This would mean that the binding energy would be proportional to $A(A - 1)/2$. (The factor $\frac{1}{2}$ is due to the fact that every nucleon is counted twice). That is,

$$BE \propto \frac{A(A-1)}{2},$$

or, for unsaturated forces, assuming $A \gg 1$,

$$BE \propto A^2. \tag{11.27}$$

Thus the nuclear forces are saturated, with $BE \propto A$; while the coulomb forces are unsaturated, with $BE \propto A^2$. A nucleon in a nucleus interacts *only with its neighbors*. Addition of more nucleons to the nucleus increases the total binding energy, but not the binding energy per nucleon. This characteristic of the nuclear forces is called *saturation*.

c) *Charge dependence of nuclear forces.* We can classify the forces between the nucleons as (i) the force between two protons, denoted by p-p, (ii) the force between two neutrons, denoted by n-n, and (iii) the force between a neutron and a proton, denoted by n-p. The *charge independence* of nuclear forces implies that p-p $\approx$ n-n $\approx$ n-p. If only p-p $\approx$ n-n, the forces are said to be *charge symmetric.*

For the light- and medium-weight stable nuclei (for which the coulomb repulsive force is negligible), $N \simeq Z \simeq A/2$. This implies that neutrons and protons have a tendency to go in pairs. The extraordinary stability of nuclei such as ${}^4_2\text{He}$, ${}^8_4\text{Be}$, ${}^{12}_{6}\text{C}$, ${}^{16}_{8}\text{O}$, etc., indicates that neutrons and protons add in pairs. From these statements we conclude that either

$$\text{n-p} > \text{p-p} \approx \text{n-n} \qquad \text{or} \qquad \text{n-p} \approx \text{p-p} \approx \text{n-n} \tag{11.28}$$

Further experimental evidence indicates that the nuclear forces are charge independent; that is, n-p $\approx$ p-p $\approx$ n-n.

From the discussion in this section, we may conclude the following.

i) *Nuclear forces are strongly attractive and have short range.*

ii) *Nuclear forces show saturation.*

iii) *Nuclear forces are charge independent.*

d) *The meson theory of nuclear forces.* Werner Heisenberg, in 1932, suggested that nuclear forces result from the constant exchange of heavy quanta (or massive particles) between two nucleons. The Japanese physicist Hideki Yukawa[14] established this fact theoretically in 1935. This phenomenon is like the exchange of photons resulting in coulomb force between two charged particles. According to the quantum-mechanical theory of fields, each electrical charge is surrounded by a cloud of "virtual" photons which are being constantly emitted and absorbed by the charged particle. When two charged particles are brought near each other, they exchange virtual photons; that is, one charged particle emits a photon and the other charged particle absorbs it, and vice versa. This constant interchange between the two results in the coulomb forces between them. The zero mass of the photons is what gives coulomb forces their long-range nature.

According to Yukawa, because of the short range of the nuclear forces, a nucleon is surrounded by a cloud of virtual *massive particles* (or heavy quanta), which the nucleon is constantly emitting and absorbing. If another nucleon is brought near to the first, a particle emitted by one may be absorbed by the other, and vice versa. This means that there is a constant transfer of momentum from one nucleon to the other, and hence a force is exerted between them. We can deduce the mass of these heavy quanta by using Heisenberg's uncertainty principle. An example will show how it's done.

Example 11.1. If a nucleon emits a virtual particle of rest mass m_π, it will lose an amount of energy ΔE, where

$$\Delta E = m_\pi c^2. \tag{11.29}$$

According to the uncertainty principle, we won't observe this loss of energy if this virtual particle (or another just like it) *returns* to the nucleus within time Δt, given by

$$\Delta t \sim \frac{\hbar}{\Delta E} = \frac{\hbar}{m_\pi c^2}. \tag{11.30}$$

Within this time, the maximum distance α the virtual particle can travel is

$$\alpha = c\,\Delta t, \tag{11.31}$$

where c is the maximum speed ($c = 3 \times 10^{10}$ cm/sec, the speed of light) with which the virtual particle can travel. Combining the above equations, we obtain

$$\alpha \sim \frac{\hbar}{m_\pi c}. \tag{11.32}$$

Thus, if we know the range of the nuclear force α, we can find m_π, and vice versa. So if $\alpha \simeq 1.2 \times 10^{-13}$ cm, we get

$$m_\pi \simeq \frac{\hbar}{\alpha c} \simeq 140 \text{ MeV}. \tag{11.33}$$

Experiment has shown that the heavy quanta involved in nuclear exchange exist in three different forms: neutral, negatively charged, and positively charged, called the *neutral pion* (or pi meson) π^0, the *negative pion* π^-, and the *positive pion* π^+, respectively. Their rest masses are 135 MeV, 139.6 MeV, and 139.6 MeV, respectively. Interactions between the nucleons which may result from the exchange of neutral or charged pions may be represented as follows.

i) Interaction due to exchange of π^0 (Fig. 11.12a) may be represented by

$$\text{n} \rightarrow \text{n} + \pi^0 \qquad \text{and} \qquad \pi^0 + \text{p} \rightarrow \text{p}.$$

ii) Interaction due to exchange of π^- (Fig. 11.12b) may be represented as

$$\text{n} \rightarrow \text{p} + \pi^- \qquad \text{and} \qquad \pi^- + \text{p} \rightarrow \text{n}.$$

That is, the neutron becomes a proton, and vice versa.

The other interactions may be represented similarly.

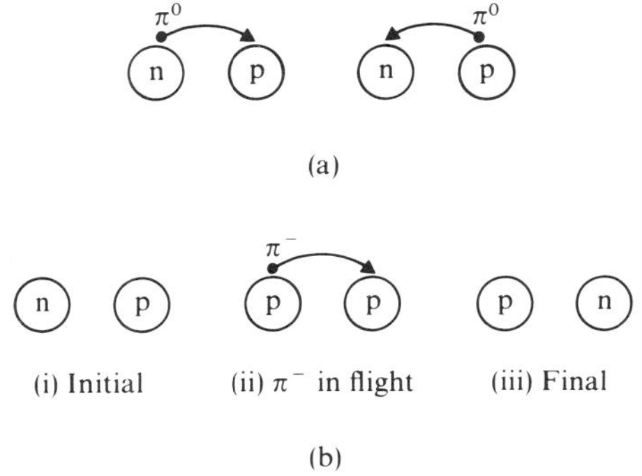

Fig. 11.12 (a) Interaction takes place between neutron and proton because of the exchange of π^0 mesons; this is interaction without exchange of charge. (b) The n-p interaction takes place because of the exchange of the π^- meson; in this interaction, charge is exchanged.

SUMMARY

Basic properties of the nucleus are as follows. A nucleus consists of Z protons and N neutrons, and hence has a mass number $A = Z + N$. Its charge is $+Ze$. If one assumes that it is spherical in shape, its radius $R = r_0 A^{1/3} = 1.35 \times 10^{-13} A^{1/3}$ cm. The angular momentum of the nucleus is $I = \sqrt{i(i + 1)}\,\hbar$, where i is the total angular momentum (or spin) quantum number of the nucleus; $i = 0, 1, 2, 3, 4, \ldots$ for even-A nuclei and $i = \frac{1}{2}, \frac{3}{2}, \frac{5}{2}, \ldots$ for odd-A nuclei. The magnetic moment of the nucleus is measured in units of nuclear magnetons, $\mu_N = e\hbar/2m_p c = \mu_\beta/1836$. The proton has a positive magnetic moment, while the neutron has a negative magnetic moment. This gives an idea of the complexity of the structure of nucleons. The forces holding nucleons together are both attractive and short-range.

The binding energy of the nucleus is the energy needed to break it into its constituents:

$$BE\,(= -\Delta E) = [Zm_{\text{H}} + (A - Z)m_{\text{n}} - M(A, Z)]c^2.$$

The average binding energy per nucleon, BE/A, is $\sim$8 MeV.

⋆ On a plot of N versus Z, a curve drawn through the stable nuclei shows that the locus of the curve departs from the line $N/Z = 1$ for low Z toward the direction of a higher number of neutrons (for heavier elements), reaching a value of $N/Z = 1.6$ for $A = 238$. According to the rule of the odd–even effect, even–even nuclei are most abundant, while odd–odd nuclei are the least abundant. A survey of stable nuclei suggests that nucleons tend to form neutron-proton pairs.

The nucleus, like the atom, exhibits a shell structure. Closure of the shells occurs when the shells have the following numbers of nucleons (these numbers are called *magic numbers*):

$$N = 2, 8, 20, 28, 50, 82, \text{ and } 126,$$

$$Z = 2, 8, 20, 28, 50, \text{ and } 82.$$

The order of the levels in the shell structure of the nucleus is

$$1s_{1/2},\ 1p_{3/2},\ 1p_{1/2},\ 1d_{5/2},\ 2s_{1/2},\ 1d_{3/2},\ 1f_{7/2}, \ldots$$

Nuclear forces are attractive and short range; in most cases they may be represented by a square-well potential of the form

$$\begin{aligned} V(r) &= -V_0 \qquad \text{for } r < \alpha, \\ &= 0 \qquad \text{for } r > \alpha, \end{aligned}$$

where $V_0 \simeq 40$ MeV and α (the range of the nuclear forces) $\simeq 2 \times 10^{-13}$ cm. Nuclear forces show saturation, which we know because of the fact that $BE/A =$ constant. Most probably nuclear forces are also charge independent.

Heisenberg suggested that the existence of nuclear forces may be explained by the exchange of pi-mesons (or pions) between nucleons; this theory was later developed by Yukawa. The mass of the particle exchanged by the nucleons is ~ 139 MeV.

PROBLEMS

1. Show that the density of nuclear matter is $\sim 10^5$ tons/mm^3. Assume that the mass of the proton is 1.67×10^{-27} kg, and that its spherical shape has a radius $R = r_0 A^{1/3}$, where $r_0 = 1.35 \times 10^{-13}$ cm.
2. The neutron has an intrinsic spin and no net charge, but it does have a negative magnetic moment, which we can account for by assuming that the neutron has some sort of charge distribution, say a positive core surrounded by a negative shell, or vice versa. What type of charge distribution would you assume (given corresponding sizes of the charge distributions) in order to account for the negative magnetic moment?
3. Assume that a spinning proton can be replaced by a positive point charge moving in a circular orbit of radius 1.35×10^{-13} cm. Calculate the current and magnetic moment. Compare the value of the magnetic moment with the one given in the text.
4. Like an electron, a free proton, when placed in an external magnetic field, exhibits space quantization. Calculate the energies of the two possible orientations of the proton in a magnetic field **B**. Given that $\mathbf{B} = 4$ webers/m^2: (a) What are the energies of the two states? (b) What is the difference in energy between the two states?

(c) What should be the energy of the incident photon so that when a proton in a lower energy state absorbs it, the proton undergoes transition to a higher energy state? (This phenomenon is known as nuclear magnetic resonance, NMR.)

5. The binding energy of the last neutron added to the nucleus (A, Z) is given by

$$BE(n) = [M(A-1, Z) + m_n - M(A, Z)]c^2.$$

Using this relation, calculate the binding energy of the last added neutron in the following three isotopes of lead $(Z = 82)$:

$$^{206}\text{Pb}, \quad ^{207}\text{Pb}, \quad \text{and} \quad ^{208}\text{Pb}.$$

Is there any connection between the $BE(n)$ values and the Z and N values of lead isotopes?

6. Calculate the binding energy of a neutron added to $^{235}_{92}\text{U}$. Compare this with the value of the binding energy per nucleon. Repeat this for $^{233}_{92}\text{U}$ and compare the two results.

7. Calculate the binding energy of the last added (a) neutron in $^{4}_{2}\text{He}$, or (b) proton in $^{16}_{8}\text{O}$ and $^{32}_{16}\text{S}$. Compare these values with the binding energy per nucleon.

8. Using the atomic masses and the masses of the proton, neutron, and electron, calculate:

a) The energy needed to remove a proton from ^{13}C
b) The energy needed to remove a neutron from ^{13}C
c) The energy needed to remove a proton from ^{56}Fe
d) The energy needed to remove a neutron from ^{56}Fe

Compare the results of (a) and (b) with those of (c) and (d), and explain the difference.

9. When a neutron and a proton combine, a nucleus deuteron is formed. A photon of 2.22 MeV energy can disintegrate the deuteron into its component neutron and proton. Given that the masses of the neutron and the proton are 1.008665 u and 1.007277 u, what is the mass of the deuteron?

10. According to the shell model, draw energy-level diagrams (both for neutrons and protons) for all nuclei in Table 11.2.

11. Draw energy-level diagrams for the following nuclei, showing the filling of the levels by neutrons and protons.

$$^{7}_{3}\text{Li}, \quad ^{17}_{8}\text{O}, \quad \text{and} \quad ^{41}_{20}\text{Ca}$$

What are the spins of these nuclei in their ground states?

12. According to the shell model, what are the spins and parities of the following nuclei in their ground states?

$$^{35}_{17}\text{Cl} \quad \text{and} \quad ^{95}_{42}\text{Mo}.$$

13. According to the shell model, what are the spins and parities of the following nuclei in (a) the ground state, and (b) the first excited state?

$$^{17}_{9}\text{F}, \quad ^{31}_{15}\text{P}, \quad \text{and} \quad ^{127}_{53}\text{I}.$$

14. The ground state of $^{137}_{56}$Ba is $\frac{3}{2}+$, while the first two excited states are $\frac{1}{2}+$ and $\frac{11}{2}-$. What different shell-model levels correspond to these states?

15. The spin of the $^{23}_{11}$Na nucleus has been measured to be $\frac{3}{2}$, while the shell model predicts a value of $d_{5/2}$. Can you account for this discrepancy by combining angular momentum states of more than one nucleon?

16. If the nucleus has a filled j shell minus one nucleon—i.e., if there is a hole in an otherwise-filled shell—the spin of the nucleus is given by the spin of the hole (or vacant) state. According to the shell model, what are the spins of the following nuclei in their ground states?

$$^{39}_{19}\text{K}, \quad ^{59}_{27}\text{Co}, \quad ^{49}_{22}\text{Ti}, \quad ^{115}_{49}\text{In}, \quad ^{87}_{38}\text{Sr}, \quad \text{and} \quad ^{137}_{56}\text{Ba}.$$

17. Using the shell model, explain why there are more stable isotopes corresponding to even Z than to odd Z.

18. By combining Eqs. (11.21) and (11.22), show that $K \simeq 25$ MeV.

REFERENCES

1. Ernest Rutherford, *Phil. Mag.* **21,** 669 (1911)
2. J. Chadwick, *Proc. Roy. Soc.* **A136,** 692 (1932)
3. Atam P. Arya, *Fundamentals of Nuclear Physics*, Boston: Allyn and Bacon, 1966, page 318
4. E. R. Cohen and J. W. M. DuMond, *Revs. Mod. Phys.* **37,** 537 (1965)
5. R. M. Littaner, H. F. Schopper, and R. R. Wilson, *Phys. Rev. Lett.* **7,** 141 (1961)
6. Niels Bohr and J. A. Wheeler, *Phys. Rev.* **56,** 426 (1939)
7. C. F. Von Weizsacker, *Z. Physik* **96,** 431 (1935); *Naturwiss* **24,** 813 (1936)
8. E. Feenberg, *Revs. Mod. Phys.* **19,** 239 (1947)
9. P. A. Seeger, *Nuclear Physics* **25,** 1 (1961)
10. J. A. Harvey, *Phys. Rev.* **81,** 353 (1951)
11. D. J. Hughes and D. Sherman, *Phys. Rev.* **78,** 632 (1950)
12. M. G. Mayer, *Phys. Rev.* **74,** 235 (1948); **75,** 1969 (1949); **78,** 16 (1950)
13. O. Haxel, J. H. D. Jensen, and H. E. Suess, *Phys. Rev.* **75,** 1766, L (1949)
14. Hideki Yukawa, *Proc. Phys., Math. Soc.*, Japan **17,** 48 (1935)

Suggested Readings

1. Atam P. Arya, *Fundamentals of Nuclear Physics*, Chapter I, V, X, XII. Boston: Allyn and Bacon, 1966
2. D. Halliday, *Introductory Nuclear Physics*, New York: John Wiley, 1962
3. H. A. Enge, *Introduction to Nuclear Physics*, Reading, Mass.: Addison-Wesley, 1966

4. *Proceedings of the Hamilton Conference on Nuclear Masses*, Toronto: University of Toronto Press, 1960

5. M. G. Mayer and J. H. D. Jensen, *Elementary Theory of Nuclear Shell Structure*, New York: John Wiley, 1955

6. L. R. B. Elton, *Introductory Nuclear Theory*, Chapters 3 and 4, Philadelphia: W. B. Saunders, 1959

7. J. M. Blatt and V. F. Weisskopf, *Theoretical Nuclear Physics*, Chapters 2, 3, and 4, New York: John Wiley, 1958

8. B. L. Cohen, *Concepts of Nuclear Physics*, New York: McGraw-Hill, 1971

CHAPTER 12
DECAY OF THE NUCLEUS

A typical nuclear spectroscopy laboratory for investigating α-, β-, and γ-radioactivity. (Courtesy Nuclear Physics Laboratory, West Virginia University)

12.1 DISCOVERY OF RADIOACTIVITY

Radioactivity was discovered by the French physicist Henri Becquerel[1] in 1896. He accidentally left a small quantity of a compound containing uranium lying on a photographic plate which was quite well wrapped. Imagine his surprise when he developed the plate and found images of the crystals of the uranium compound. Repeated experiments showed that some sort of radiation was coming from the compound. Intensive investigations by Becquerel, Madame Marie Curie,[2] Ernest Rutherford, Frederick Soddy, and many others led to the discovery of many isotopes of different elements (with $A > 210$), which emitted such radiation. This phenomenon of emission of radiation is called *radioactivity*; the elements which exhibit it are called *radioactive elements* (or radioactive nuclides).

Detailed investigations of different properties (such as penetrating power and specific ionization) of this radiation revealed that it consisted of three different types (or components): (a) alpha rays (α rays), (b) beta rays (β rays), and (c) gamma rays (γ rays).

Alpha rays have such a weak penetrating power that they are stopped by a thin sheet of paper. On the other hand, they cause very intense ionization in air (or gas). Beta rays cause much less ionization in air or gas, but are $\sim$100 times more penetrating than α rays. For example, β rays can penetrate a sheet of aluminum a few mm thick. They can be stopped only by a sheet of lead about one-tenth of an inch thick. Gamma rays cause much less ionization than β-rays, but are $\sim$100 times more penetrating. It takes lead sheets several inches thick to stop γ rays. All three types of radiation cause fluorescence (which on close examination is found to be a series of scintillations—small bright sparks) in certain substances.

The presence of the above three types of radiation can be demonstrated by means of a simple experiment (Fig. 12.1). A small amount of radioactive sample, such as radium, is placed at the bottom of a deep hole drilled in a lead block. This produces a well-collimated beam of radiation. A strong magnetic field is applied at right angles to the plane of this paper (Fig. 12.1). If the magnetic field points into the paper, the positively charged particles should be deflected to the left, the negatively charged particles to the right, and the uncharged particles should not be deflected at all. Of course, the amount of deflection depends on the mass, charge, and velocity of the particles. A photographic plate, when it is later developed, shows various exposure points. The α particles produce an image at the left only at one spot, indicating thereby that they are positively charged and monoenergetic. The β rays produce images at all spots to the right of the center of the photographic plate, indicating that they are negatively charged and have energies varying from zero to a certain maximum. And there is a central spot on the photographic plate corresponding to the γ rays, which are uncharged. Similar conclusions can be arrived at by applying an electric field.

Most α particles are emitted with velocities between $\sim 1.5 \times 10^7$ m/sec and $\sim 2.2 \times 10^7$ m/sec. Any group of α particles emitted from the same type of nuclei always has a definite velocity, and hence a definite energy. The values of e and e/m show that α particles have a mass of 4 u and a positive charge of 2 units. Thus α particles are fast-moving doubly ionized helium atoms, He^{2+}.

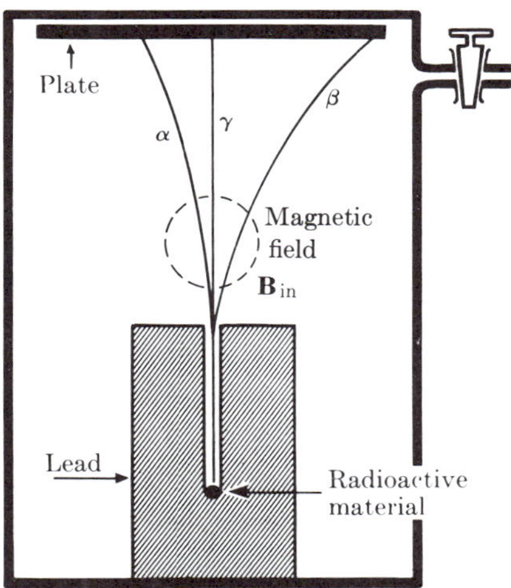

Fig. 12.1 Experimental arrangement showing the deflection of α, β, and γ rays by a magnetic field.

The velocities of β particles emitted from various nuclei range up to $0.99c$, where c is the speed of light ($c = 3 \times 10^8$ m/sec). A radioactive element emits β particles with energies varying between zero and a certain maximum, called the *endpoint energy*. Measurements of e and e/m reveal that β particles are fast-moving electrons.

The uncharged γ rays exhibit all the characteristics of electromagnetic waves. Actually, γ rays may be said to be x rays or photons of very short wavelengths, between $\sim 1.7 \times 10^{-8}$ m and $\sim 4.1 \times 10^{-6}$ m; they travel at the speed of light.

We may point out that α, β, and γ rays, though they are the types of radiation most frequently emitted, are not the only types emitted. Some nuclei disintegrate by emission of protons, neutrons, positrons (a *positron* is a particle with the same mass as the electron, but with a unit positive charge), or some other particle.

12.2 THE LAW OF RADIOACTIVE DECAY

When any nucleus disintegrates by emitting a particle or a gamma ray, or by capturing an electron from the atomic shell, the process is called *radioactive decay*, or simply *decay*. In any radioactive sample the disintegration of a nucleus is spontaneous. The overall activity is a prolonged process, varying from a few seconds to millions of years, depending on the type of nuclei in the sample. Radioactive decay is also a statistical process: This means that each undecayed nucleus has equal probability λ of decaying in the next second. According to the statistical theory, in time dt the probability of decay of each nucleus is $\lambda\, dt$. Given that there are N undecayed nuclei (or atoms) at time t, the number dN that will decay in the short time interval between t and $t + dt$ is given by

$$dN = -\lambda \, dtN. \tag{12.1}$$

The negative sign means that N decreases as t increases. The probability constant λ is called the *disintegration constant* or *decay constant*.

Assuming the initial condition that there are N_0 radioactive atoms at time $t = 0$, Eq. (12.1) yields, on integration,

$$\int_{N_0}^{N} \frac{dN}{N} = -\int_0^t \lambda \, dt$$

or

$$\boxed{N = N_0 e^{-\lambda t},} \tag{12.2}$$

N being the number of radioactive atoms present at any time t.

One can detect the presence of a radioactive sample not by the radioactive atoms present but by the radiation emitted by these atoms when they disintegrate. Hence the quantity of interest is the *activity* A of a sample, defined as the number of disintegrations per second. Thus, from Eq. (12.2),

$$\boxed{A = \left|\frac{dN}{dt}\right| = \lambda N_0 e^{-\lambda t} = \lambda N.} \tag{12.3}$$

The *half-life*, $t_{1/2}$, of any sample is defined as the time interval in which the number of undecayed atoms decreases by half (or the initial activity A_0 decreases to $A_0/2$). Thus substituting $N = N_0/2$ and $t = t_{1/2}$ in Eq. (12.2) yields

$$\frac{N_0}{2} = N_0 e^{-\lambda t_{1/2}}$$

or

$$\boxed{t_{1/2} = \frac{\ln 2}{\lambda} = \frac{0.693}{\lambda}.} \tag{12.4}$$

Since the unit of $t_{1/2}$ is seconds, the unit of λ is 1/sec.

The exponential nature of decay given by Eq. (12.2) implies that it takes an infinite time for a given radioactive sample to completely disintegrate. Individual radioactive atoms may have lifetimes of between zero and infinity. Hence it is meaningful to talk about the *average* or *mean life* τ, defined as

$$\tau = \frac{t_1 dN_1 + t_2 dN_2 + t_3 dN_3 + \cdots}{dN_1 + dN_2 + dN_3 + \cdots}. \tag{12.5}$$

This equation says that dN_1 nuclei have lifetime t_1, dN_2 have t_2, etc. Equation (12.5) in integral form (noting that $dN_1 + dN_2 + dN_3 + \cdots = N_0$) is

$$\tau = \frac{\int_0^{N_0} t\, dN}{\int_0^{N_0} dN} = \frac{\int_0^{N_0} t\, dN}{N_0}. \tag{12.6}$$

Substituting for dN from Eq. (12.2) and integrating, we obtain

$$\tau = \frac{\int_\infty^0 \lambda t N_0 e^{-\lambda t}\, dt}{N_0} = \int_0^\infty \lambda t e^{-\lambda t}\, dt = \frac{1}{\lambda},$$

$$\boxed{\tau = 1/\lambda.} \tag{12.7}$$

To illustrate the above equations, let us consider the radioactive isotope gold-198, $^{198}_{79}\mathrm{Au}$, which decays by emitting an electron with a half-life of 64.8 hr. The decay constant according to Eq. (12.4) is $\lambda = 2.96 \times 10^{-6}$ sec^{-1}, and the mean lifetime, according to Eq. (12.7), is $\tau = 93.5$ hr. We can find the number of radioactive atoms N and the activity λN at any time by using Eqs. (12.2) and (12.3), respectively. Figure 12.2(a) is a plot of N and λN versus time in units of half-life on a linear scale, while Fig. 12.2(b) shows the same plots on a semilogarithmic scale.

Example 12.1. Consider 1 g of radioactive radium-226 which has a half-life of 1620 years. Its decay constant is

$$\lambda = \frac{0.693}{t_{1/2}} = \frac{0.693}{1.62 \times 10^3 \text{ yr}} = 1.38 \times 10^{-11} \text{ sec}^{-1}.$$

There are 6.02×10^{23} atoms (Avogadro's number) in 1 g-atom of radium. Therefore 1 g of radium contains N radioactive atoms:

$$N = \frac{N_A}{A} = \frac{6.02 \times 10^{23}}{226} = 2.66 \times 10^{21} \text{ atoms}.$$

Hence the disintegration rate of radium (at time $t = 0$) is

$$|\lambda N| = 1.38 \times 10^{-11} \text{ sec}^{-1} \times 2.66 \times 10^{21} \text{ atoms}$$

$$\simeq 3.7 \times 10^{10} \text{ disintegrations/sec}.$$

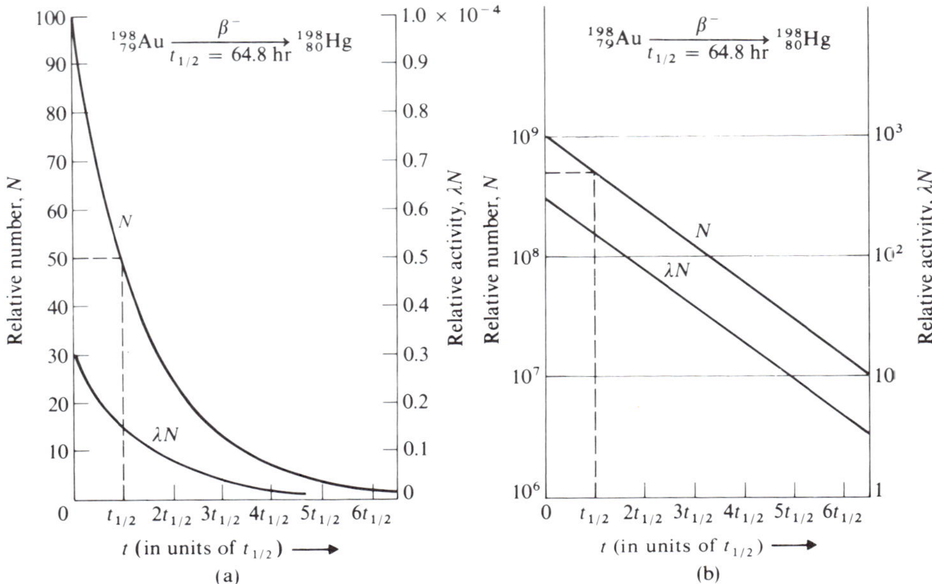

Fig. 12.2 (a) Linear plot of relative number of radioactive atoms N and activity λN versus time t for $^{198}_{79}Au$, which has a half-life of 64.8 hr. (b) Semilogarithmic plot of N and λN versus t for $^{198}_{79}Au$ in arbitrary units.

In the past, scientists have said that any radioactive sample which has 3.7×10^{10} disintegrations/sec has an activity of *one curie* (1 Ci) (1 millicurie $= 10^{-3}$ Ci and 1 microcurie $= 10^{-6}$ Ci). Recently, another unit of activity, called the *Rutherford*, has been introduced. A sample is said to have an activity of 1 rutherford (1 rd) if it has 10^6 disintegrations per second. *One millirutherford* (mrd) $= 10^3$ disintegrations/sec and *one microrutherford* (μrd) $=$ 1 disintegration/sec.

12.3 NATURAL RADIOACTIVITY AND RADIOACTIVE DATING

Radioactive nuclei can be divided into two categories: Those which are found in nature are said to exhibit *natural radioactivity*. Those which are produced in the laboratory are said to exhibit *artificial radioactivity*. About 1000 radioactive isotopes have been produced in the laboratory. There is only a small number ($\sim$65) of naturally occurring radioactive nuclei. Both artificially produced and naturally occurring radioactive isotopes often decay by successive disintegrations. The initial radioactive isotope is called the *parent*; its decay product is called the *daughter*. The daughter itself may be radioactive; *its* decay product is called the *granddaughter* and so on.

When scientists began to search for naturally occurring radioactive isotopes, they found that there are many radioisotopes among the elements with atomic numbers between $Z = 81$ and $Z = 92$, because at large Z, the coulomb repulsion between the protons makes the isotopes less stable. To decrease the coulomb repulsive force, the unstable nucleus decays by emitting α rays which causes it to lose two protons and two neutrons. This decay leaves the nucleus with an excess of neutrons. So *this* unstable nucleus decays by emitting β rays. (So that charge is conserved in beta decay, when an electron is emitted, a neutron is converted into a proton.) The resulting nucleus may still be unstable and may again decay by emitting α rays. This process of the chain of alternate alpha and beta decays continues until the nucleus reaches stability. Note that in alpha decay, the nucleus decreases its mass number by four units, that is, $\Delta A = 4$, and the charge on the nucleus decreases by two units, that is, $\Delta Z = 2$. On the other hand, if the nucleus decays by β or γ emission, there is no change in A; that is, $\Delta A = 0$. The radioactive isotopes between $Z = 81$ and $Z = 92$ may be divided into four series (n being an integer).

$A = 4n$	Thorium series
$A = 4n + 1$	Neptunium series
$A = 4n + 2$	Uranium series
$A = 4n + 3$	Actinium series

The name of the series is the same as that of the parent nucleus from which the series starts. A nucleus belonging to one of these series remains in that series, even after any number of decays.

Figure 12.3 shows the chain of decays in these two of the four series. All four series follow patterns of decay similar to that of the uranium series. The parent isotope in each series has a very long lifetime (1.39×10^{10} years, 4.5×10^9 years, and 7.15×10^8 years for the thorium, uranium, and actinium series, respectively), except for the parent of the neptunium series ($t_{1/2} = 2.2 \times 10^6$ years). Because of their short lifetimes, the radioactive isotopes in the neptunium series no longer occur in nature; they have decayed entirely since the formation of the universe, which is thought to have taken place about 10 billion years ago. Another characteristic common to all these series is that the stable end product of three of the four (except for Np) series is some isotope of lead ($^{208}_{82}Pb$, $^{206}_{82}Pb$, and $^{207}_{82}Pb$).

At the time the earth was formed, all elements—stable and unstable—were formed in varying amounts. Since most of the unstable nuclei had half-lives much shorter than the age of the earth, these have long since decayed away. Only the parents of the three series mentioned above, and 18 other radioactive isotopes, are still found in nature. Some of these are $^{40}_{19}K$ ($t_{1/2} = 1.3 \times 10^9$ years) $^{115}_{49}In$ ($t_{1/2} = 5 \times 10^{14}$ years), and $^{204}_{82}Pb$ ($t_{1/2} = 1.3 \times 10^{17}$ years).

Two important aspects of radioactive decay are (a) radioactive equilibrium, and (b) radioactive dating.

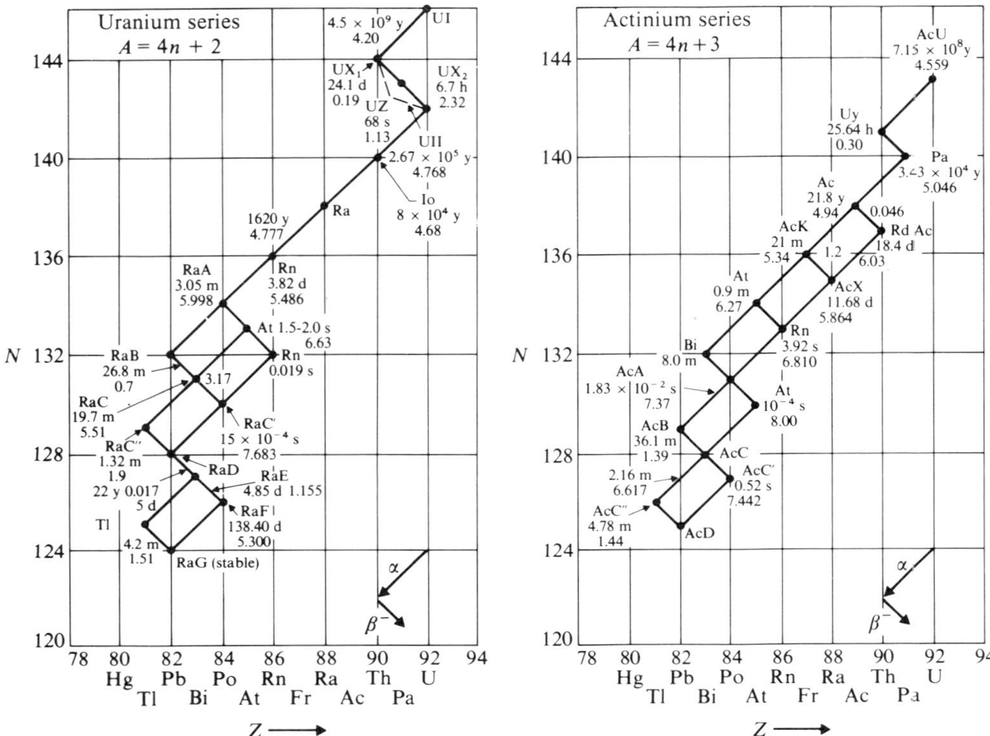

Fig. 12.3 Plot of number of neutrons N ($= A - Z$) versus number of protons Z for two of the four radioactive series represented by $A = 4n + 2$ and $4n + 3$ (uranium and actinium series), where n is an integer. Also shown are the corresponding half-lives and decay energies (in MeV). [From A. P. Arya, *Fundamentals of Nuclear Physics*, Boston: Allyn and Bacon, 1966]

a) *Radioactive equilibrium.* Observe from Fig. 12.3 that the half-life of the parent nucleus is very long compared to that of its daughter, granddaughter, or any other member in the series. After the parent nucleus has been decaying for a sufficiently long time, it reaches a state in which the number of radioactive atoms of any member of the series remains constant subsequently. In other words, the members of the series are in *radioactive equilibrium.* Under these circumstances the activities of all the members of the series are equal. We can derive this result mathematically as follows.

Suppose that there are $N_1, N_2, N_3, \ldots$ radioactive atoms of the type parent, daughter, granddaughter, ..., with decay constants $\lambda_1, \lambda_2, \lambda_3, \ldots$ and mean lives $\tau_1, \tau_2, \tau_3, \ldots$, respectively. Using the definition of activity, we may write the following equations

$$\frac{dN_1}{dt} = -\lambda_1 N_1, \tag{12.8}$$

$$\frac{dN_2}{dt} = \lambda_1 N_1 - \lambda_2 N_2, \tag{12.9}$$

$$\frac{dN_3}{dt} = \lambda_2 N_2 - \lambda_3 N_3, \tag{12.10}$$

$$\vdots \qquad \vdots$$

Equation (12.8) gives the rate of decay of N_1. Equation (12.9) means that radioactive atoms of type N_2 are produced at the rate $\lambda_1 N_1$ and decay at the rate of $\lambda_2 N_2$. Subsequent equations have a similar meaning. In the equilibrium state, $dN_2/dt = 0$, $dN_3/dt = 0, \ldots$, which leads to the following equations:

$$\lambda_1 N_1 = \lambda_2 N_2 = \lambda_3 N_3, \tag{12.11}$$

$$\frac{N_1}{\tau_1} = \frac{N_2}{\tau_2} = \frac{N_3}{\tau_3}. \tag{12.12}$$

[Of course, from Eq. (12.8), $dN_1/dt = -\lambda_1 N_1 = 0$, which is approximately correct because, for the parent, λ_1 is very small (τ_1 being very large) and the product $\lambda_1 N_1 \simeq 0$.] Equations (12.11) and (12.12) state the condition for *permanent* or *secular radioactive equilibrium.*

Suppose that the parent does *not* have a very long lifetime. To find $N_1, N_2, N_3, \ldots$, and the corresponding equations for the activities, we must solve Eqs. (12.8), (12.9), and (12.10) $\cdots$ by the usual methods (see Problem 7).

b) *Radioactive dating.* The ingenious method of radioactive dating, devised by W. F. Libby[3] in 1952, is used to estimate the age of organic relics which may be many thousands of years old.

Our atmosphere is constantly being bombarded by cosmic rays, which are very highly penetrating radiation of unknown origin bombarding the earth's atmosphere. Cosmic rays consist mostly of high-energy particles like neutrons and protons. Atmospheric nitrogen-14 absorbs fast-moving neutrons and changes into carbon-14, with the emission of a proton. That is,

$${}^{14}_{7}\mathrm{N} + {}^{1}_{0}\mathrm{n} \rightarrow {}^{14}_{6}\mathrm{C} + {}^{1}_{1}\mathrm{p}. \tag{12.13}$$

Carbon-14 is radioactive and decays by beta emission to nitrogen-14, with a half-life of 5730 years.

Thus the carbon dioxide molecules in the air contain a small fraction of radioactive carbon-14 atoms in addition to stable carbon-12 atoms. Living organisms consume molecules of carbon dioxide which contain *both* types of carbon. The intake of carbon-14 stops only when the organism dies. Subsequently the amount of carbon-14 decreases according to the radioactive-decay law, i.e., in 5730 years, there are only half as many atoms of carbon-14 left. So if we know the relative number of the two isotopes of carbon in a given sample, we can estimate the number of years the organism has been dead. The amount of carbon-14 is determined by measuring its activity, i.e., λN = activity.

Example 12.2. An archeologist finds a piece of wood in an excavated house which he knows to be of great antiquity. He brings the wood to you to examine. It weighs 50 grams and shows C-14 activity of 320 disintegrations per minute. Estimate the length of time which has elapsed since this wood was part of a living tree, assuming that living plants show a C-14 activity of 12 disintegrations per minute per gram. The half-life of C-14 is 5730 years.

Assume that the living tree, just before it died, had N_0 radioactive atoms. Hence its activity A_0 was

$$A_0 = \lambda N_0. \tag{i}$$

After the tree died, its radioactivity decreased exponentially with time. That is,

$$A = \left|\frac{dN}{dt}\right| = \lambda N = \lambda N_0 e^{-\lambda t}. \tag{ii}$$

Dividing Eq. (ii) by Eq. (i), we get

$$\frac{A}{A_0} = e^{-\lambda t}, \tag{iii}$$

where A_0 = 12 disintegration/min/gram. Also we have

$$A = \tfrac{320}{50} \text{ disintegrations/min/gram} \qquad \text{and} \qquad \lambda = \frac{0.693}{t_{1/2}} = \frac{0.693}{5730 \text{ years}}.$$

Therefore, from Eq. (iii), we obtain

$$t = \frac{\ln (A_0/A)}{\lambda} = \frac{\ln (50 \times 12/320)}{0.693} \times 5730 \text{ years}$$

$$= \frac{0.626}{0.693} \times 5730 \text{ years} = 5170 \text{ years}.$$

This means that the wood has been dead for 5170 years.

12.4 ALPHA DECAY

So far we have been discussing radioactivity in general. We have said that α,β, and γ emissions are the three most frequent modes by which nuclei decay. Now let us discuss these three different types of decay in some detail, starting with α decay. We shall divide our discussion into three parts: (a) spontaneous decay, (b) general characteristics, and (c) theory of alpha decay.

a) *Spontaneous decay.* If a parent nucleus ${}^{A}_{Z}\mathrm{X}$ disintegrates into daughter nucleus ${}^{A-4}_{Z-2}\mathrm{Y}$ and an alpha particle, ${}^{4}_{2}\mathrm{He}$, the process is called *alpha decay*. That is,

$$ {}^{A}_{Z}\mathrm{X} \rightarrow {}^{A-4}_{Z-2}\mathrm{Y} + {}^{4}_{2}\mathrm{He} \quad (\text{or } \alpha). \tag{12.14} $$

Note that, because of different Z, the chemical nature of the daughter is different from that of its parent. Figure 12.4 represents such a decay process schematically, showing a plot of $N\ (= A - Z)$, the neutron number, versus Z.

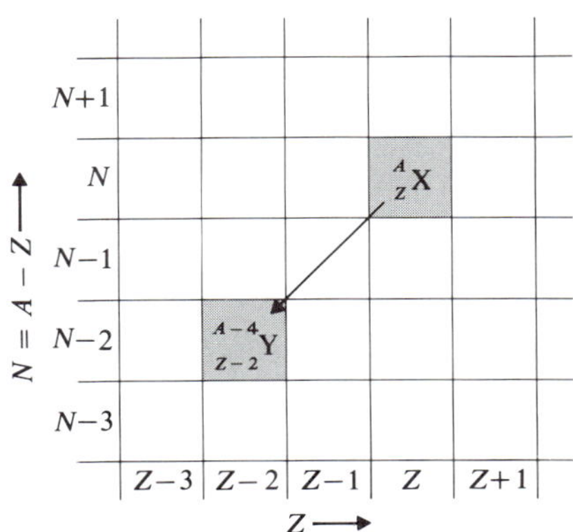

Fig. 12.4 Positions of the parent and the daughter on the N versus Z diagram when the parent decays by alpha emission.

We shall now use the principles of conservation of momentum and energy to find the condition under which a nucleus will decay by α emission. Let M_{p}, M_{d}, and m_α be the rest masses of the parent, daughter, and α-particle, respectively. The parent nucleus is at rest before decay, and its linear momentum is zero. Therefore the α particle and the daughter must leave in opposite directions (as shown in Fig. 12.5) in order to conserve momentum. Let E_i and E_f be the total energies of the system before and after decay. Applying the principle of conservation of energy, we obtain

$$ E_i = E_f \tag{12.15} $$

or

$$ M_{\mathrm{p}}c^2 = M_{\mathrm{d}}c^2 + K_{\mathrm{d}} + m_\alpha c^2 + K_\alpha, \tag{12.16} $$

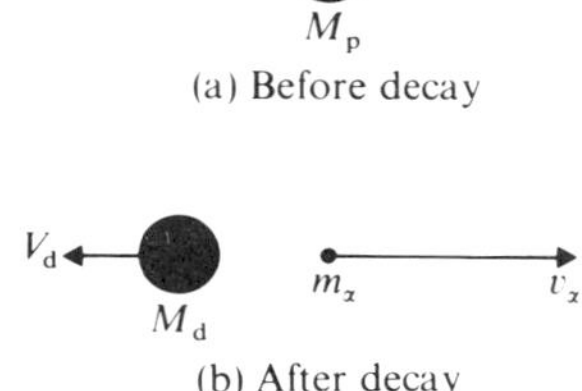

Fig. 12.5 (a) Parent nucleus at rest before decay. (b) After decay, α particle and daughter nucleus are emitted in opposite direction to conserve linear momentum.

where (using nonrelativistic expressions at low energies) $K_d = \frac{1}{2}M_d V_d^2$ and $K_\alpha = \frac{1}{2}m_\alpha v_\alpha^2$ are the kinetic energies of the daughter nucleus and the α particle, respectively. We may rewrite Eq. (12.16) as

$$K_d + K_\alpha = Q = (M_p - M_d - m_\alpha)c^2, \tag{12.17}$$

where Q is the total disintegration energy (equal to the net kinetic energy in the decay), which must be positive for spontaneous decay. Thus the condition for spontaneous α decay is that the rest mass of the parent nucleus must be greater than the sum of the rest masses of the daughter nucleus and the alpha particle. Usually, it is desirable to express the Q value in terms of atomic masses. We can do this by adding and subtracting Zm_e (where m_e is the mass of the electron) from the right-hand side of Eq. (12.17). The result is

$$\boxed{Q = [M(A, Z) - M(A - 4, Z - 2) - m(4, 2)]c^2} \quad \text{for } \alpha \text{ decay}, \tag{12.18}$$

where $M(A, Z)$ $[= M_p + Zm_e]$, $M(A - 4, Z - 2)$ $[= M_d + (Z - 2)m_e]$, and $m(4, 2)$ $[= m_\alpha + 2m_e]$ are the atomic masses of the parent atom, daughter atom, and helium atom, respectively.

Once we have established that a nucleus decays by α emission, we can calculate the kinetic energy of the α particle by using the principle of conservation of momentum and energy. That is,

$$M_d V_d = m_\alpha v_\alpha, \tag{12.19}$$

$$Q = K_d + K_\alpha = \tfrac{1}{2}M_d V_d^2 + \tfrac{1}{2}m_\alpha v_\alpha^2. \tag{12.20}$$

Substituting for V_d from Eq. (12.19) into Eq. (12.20), we have

$$Q = \tfrac{1}{2}M_d\left(\frac{m_\alpha v_\alpha}{M_d}\right)^2 + \tfrac{1}{2}m_\alpha v_\alpha^2, \tag{12.21}$$

$$Q = K_\alpha\left(\frac{m_\alpha}{M_d} + 1\right), \tag{12.22}$$

or

$$K_\alpha = \frac{Q}{1 + (m_\alpha/M_d)}. \tag{12.23}$$

It is reasonable to say that $m_\alpha/M_d \cong 4/(A - 4)$, and Eq. (12.23) takes the form

$$\boxed{K_\alpha \cong \frac{A - 4}{A} |Q|.} \tag{12.24}$$

Since A is usually very large, $A - 4/A \cong 1$, and hence the alpha particle carries most of the disintegration energy.

As an example, Fig. 12.6 shows schematically the decay of ${}^{228}_{90}$Th to ${}^{224}_{88}$Ra by α emission. Note that there are five α-groups of different energies. Except for the 5.421 MeV alpha decay (which is a ground-state to ground-state), the other four groups leave the daughter nuclei in the excited states. The excited nuclei reach the ground state by emitting gamma rays. Note that the existence of these different α energy groups and gamma rays proves the existence of discrete nuclear energy states.

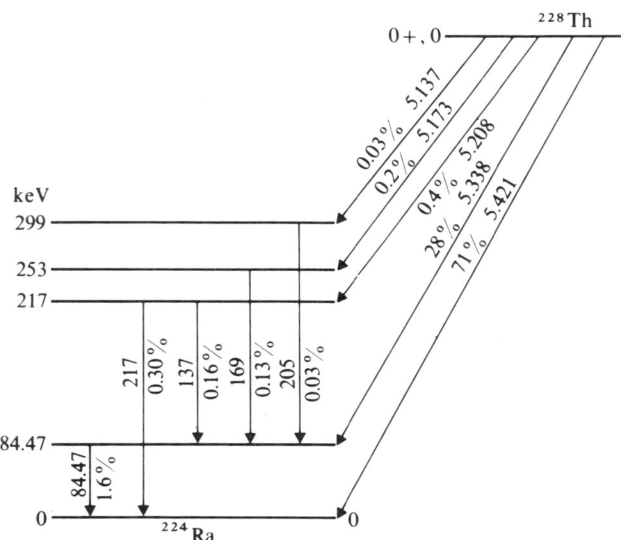

Fig. 12.6 Decay of ${}^{228}_{90}$Th → ${}^{224}_{88}$Ra + α. Slanting lines represent α groups, vertical lines represent γ rays. [From F. S. Stevens, F. Asaro, and I. Perlman, *Phys. Rev.* **107,** 1091 (1957)] Note that a small decrease in the energies of the α particles greatly reduces the intensities. Percentages are the intensities of γ rays and α rays. Energies of the γ rays are in keV, while those of α rays are in MeV.

Example 12.3. $^{240}_{94}Pu$ decays with a half-life of 6760 years by emitting two groups of alpha particles, with energy 5.17 MeV and 5.12 MeV. What are the disintegration energies?

According to Eq. (12.24),

$$K_\alpha \simeq \frac{A-4}{A}|Q| \quad \text{or} \quad |Q| = \frac{A}{A-4}K_\alpha.$$

Thus

$$Q_1 \simeq \tfrac{240}{236}\,5.17\text{ MeV} = 5.25\text{ MeV} \quad \text{and} \quad Q_2 \simeq \tfrac{240}{236}\,5.12\text{ MeV} = 5.20\text{ MeV}.$$

When Pu decays with Q_2 disintegration energy, the daughter nucleus is left in the excited state, and decays further with an emission of a gamma ray of energy $h\nu = Q_1 - Q_2 = 5.25 - 5.20 = 0.05$ MeV, as illustrated in the figure. This calculated value of 0.05 MeV for the energy of the gamma ray agrees well with the measured value of 0.045 MeV.

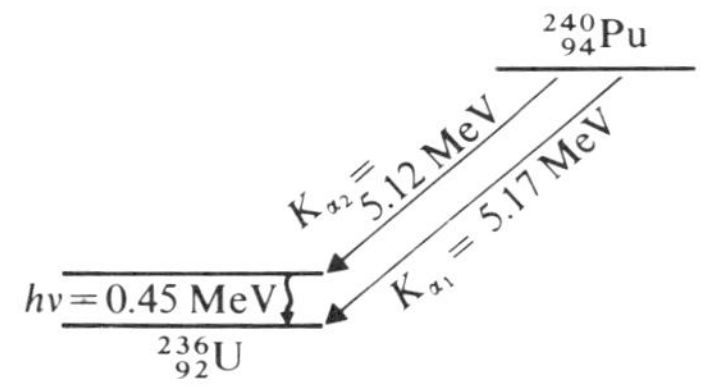

---

b) *General characteristics.* Experimental investigations of about 160 α emitters identified so far reveal several systematic trends in alpha decay. Empirical relations between the total disintegration energy E_0 ($E_0 = E_\alpha + E_d$) and the atomic number, mass number, and half-life (or decay constant) obtained from experimental data are useful in formulating theories of alpha decay.

The most striking of these trends is the relation between α energy and half-life of a nucleus. Most α particles are emitted with energies between 4 and 9 MeV. Longer-lived nuclei emit the least energetic alpha particles, while short-lived nuclei emit the most energetic alpha particles. The intensity of α groups of different energies in a given nucleus decreases rapidly with energy (see Fig. 12.6). Table 12.1 shows an example of this.

Table 12.1

Energy Versus Half-Life in Alpha Decay

Alpha emitter	E_α, MeV	$t_{1/2}$, sec	λ, sec^{-1}
^{238}Th	4.05	0.67×10^{18}	1.5×10^{-18}
^{218}Em	7.25	1.9×10^{-2}	36.4

Therefore observations reveal that *a change in the energy of a particle by a factor of 2 or 3 corresponds to a factor of* $\sim 10^{20}$ *in the half-life or decay constant.* The aim of any α-decay theory is its ability to explain these trends.

c) *Theory of alpha decay.* Classical theories do not explain the mechanism by which nuclei decay by α emission. We can best understand the problem by considering the following example. $^{214}_{84}$Po emits α particles which have energies of 7.68 MeV. When these α particles are scattered from a thin foil of $^{238}_{92}$U, they obey the Rutherford scattering law. There seems to be no absorption of these α particles by $^{238}_{92}$U nuclei, because 7.68-MeV α particles of $^{214}_{84}$Po do not have enough energy to cross the coulomb repulsive barrier, as shown in Fig. 12.7.

On the other hand, $^{238}_{92}$U itself emits α particles of kinetic energy 4.20 MeV, even though this is not enough energy to enable the particle to cross its own potential barrier (see Fig. 12.7). The paradox is that Po-α particles appear unable to cross the potential barrier, while the lower-energy U-α particles appear to do so.

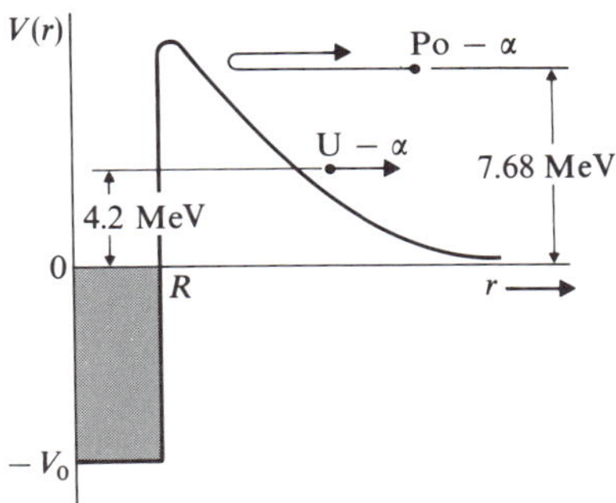

Fig. 12.7 Plot of $V(r)$ versus r for the potential well (shown shaded and extending up to R with $V_0 \approx 40$ MeV) and the coulomb potential barrier of a nucleus.

The paradox was resolved with the development of quantum mechanics. One of the first successes of quantum mechanics was its application to the theory of α decay given in 1928 simultaneously by George Gamow[4] and by R. Gurney and Edward Condon.[5] Presumably before decay, the parent nucleus consists of the daughter nucleus and the α particle sticking together. In the language of quantum mechanics, an α particle exists in one of the discrete energy states, say E_0 in Fig. 12.8, of the daughter nucleus. The motion of the α particle is restricted to a spherical region created by the potential barrier of the daughter. As it goes back and forth, the α particle presents itself again and again to the potential barrier, until the conditions are right for penetrating it (see Fig. 12.8). Classically, the α particle does not have enough energy to climb the potential barrier. But according to quantum mechanics, there is a probability that the *wave* associated with the

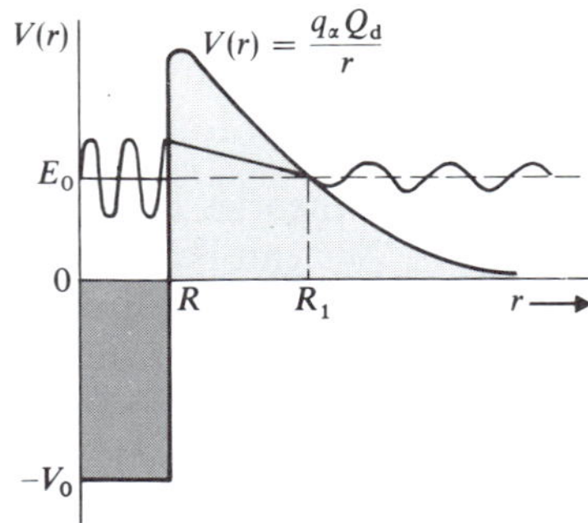

Fig. 12.8 Representation of α decay, showing how the wave associated with the α particle leaks through the coulomb barrier.

α particle can leak through the barrier as if there were a hole in it. This effect is called *tunneling*.[6] (Barrier penetration was explained in Chapter 2.)

This theory also explains why alpha-emitting nuclei do not decay immediately if the energies of the α particles are less than the height of the potential barrier. Note the relative amplitude of the wave in different regions, which indicates that the probability (equal to the square of the amplitude) of finding the α particle to the right of the barrier is much less than the probability of finding it to the left of the barrier; inside the barrier, the probability decreases exponentially. This explains why, for low E_0, α-particle emitters have long half-lives.

12.5 BETA DECAY

There are three mechanisms by which nuclei decay by beta emission.

a) *Negatron emission* (or β^- decay), in which a nucleus decays by emission of an electron

b) *Positron emission* (or β^+ decay), in which a nucleus decays by emission of a positron

c) *Electron capture* (EC decay), in which a nucleus decays by capturing an extra nuclear atomic electron.

In all three, there is no change in the mass numbers, that is, $\Delta A = 0$. Hence these processes are called *isobaric transformations.* In order to conserve charge in β^- decay, when an electron is emitted, a neutron is simultaneously converted into a proton. Thus the charge on the nucleus increases by one unit, that is, Z changes to $Z + 1$. Note: An electron or a positron *cannot* exist inside the nucleus. Hence we assume that the electron or positron is created at the time when the nucleus disintegrates. While the electron is in the process of being captured (in EC process) by the nucleus, it disappears, and its mass is converted into energy.

There are more than 1000 artificially produced radioactive isotopes which disintegrate by beta decay. These isotopes lie above and below the stability line as was shown in Fig. 11.5. The stability line passes through the stable nuclei on an N-versus-Z plot. Those beta emitters which lie *above* the curve decay by β^- emission to reach the stability line, while those lying *below* the curve decay by β^+ or EC processes to reach the stability line.

When an electron is in the process of being captured, it is the K electron which is captured most often (K-capture decay) because it lies closest to the nucleus. The capture of the M electron, N electron, etc., is less probable.

Suppose that a nucleus has captured a K electron. This leaves a vacancy or hole in the K shell. The electron from the outer shell, say the L shell, jumps to fill this hole. In the process, a K x-ray (characteristic of the daughter nucleus) is emitted. It is only by observing these x-rays that scientists have been able to detect decay by electron capture.

a) *Conditions for spontaneous decay.* By using the principle of conservation of energy, we can show whether a given unstable beta nucleus will decay by β^- emission, β^+ emission, or electron capture. The three processes discussed above may be written (representing X and Y as parent and daughter nuclei, respectively),

$$ {}^{A}_{Z}\mathrm{X} \rightarrow {}_{Z+1}^{\quad A}\mathrm{Y} + {}_{-1}^{\;\;0}\mathrm{e}\ (\beta^-), \tag{12.25} $$

$$ {}^{A}_{Z}\mathrm{X} \rightarrow {}_{Z-1}^{\quad A}\mathrm{Y} + {}_{+1}^{\;\;0}\mathrm{e}\ (\beta^+), \tag{12.26} $$

$$ {}^{A}_{Z}\mathrm{X} + {}_{-1}^{\;\;0}\mathrm{e} \rightarrow {}_{Z-1}^{\quad A}\mathrm{Y}. \tag{12.27} $$

[As we shall see below, these equations are not correct, because there is another particle which is emitted in these decays. But the rest mass of this particle is almost zero. So, if we take the kinetic energy of the beta particle emitted to be maximum, the conditions for beta decay given below are still correct.]

Consider Eq. (12.25) and let M_p, M_d, and m_e be the masses of X, Y, and e, respectively. Let K_d and $K_e^{\max}$ be the kinetic energies of the daughter and the electron. From the principle of conservation of energy, ($K_p = 0$, since X is at rest before decay),

$$ M_p c^2 = M_d c^2 + K_d + m_e c^2 + K_e^{\max}, \tag{12.28} $$

or the disintegration energy Q of this decay is defined as

$$ Q = K_d + K_e^{\max} = (M_p - M_d - m_e)c^2. \tag{12.29} $$

We replace the nuclear masses M_p and M_d by the atomic masses $M(Z)$ and $M(Z + 1)$ by making use of the relations

$$ M(Z) = M_p + Zm_e, $$

$$ M(Z + 1) = M_d + (Z + 1)m_e. $$

Then we may write Eq. (12.29) as

$$\boxed{Q = [M(Z) - M(Z + 1)]c^2} \qquad \text{for } \beta^- \text{ decay.} \tag{12.30}$$

Since for spontaneous decay Q must be positive, Eq. (12.30) states that β^- decay will occur only if the mass of the parent atom is greater than that of the daughter atom, i.e., *an unstable atom decays by electron emission if it is heavier than another atom whose Z is greater by* 1.

Similarly, the principle of conservation of energy applied to Eqs. (12.26) and (12.27) yields

$$\boxed{Q = [M(Z) - M(Z - 1) - 2m_e]c^2} \qquad \text{for } \beta^+ \text{ decay,} \tag{12.31}$$

and

$$\boxed{Q = [M(Z) - M(Z - 1)]c^2} \qquad \text{for EC decay,} \tag{12.32}$$

These equations have meanings similar to those of Eq. (12.30). Note that β^+ decay and EC decay are two competing processes. Figure 12.9 illustrates these three decays on an N-versus-Z plot.

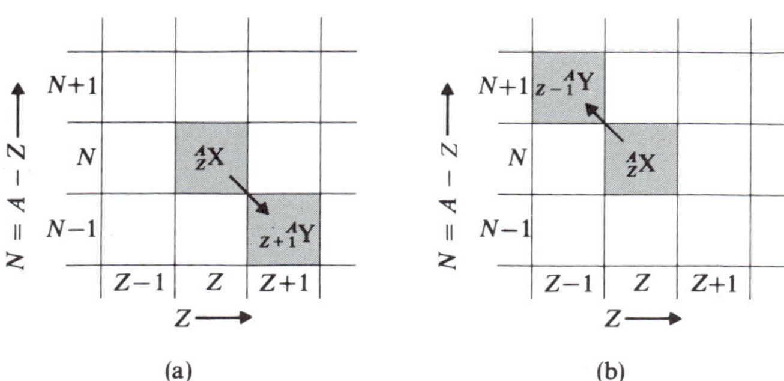

Fig. 12.9 Shifting of the radioisotopes on the N–Z diagram, (a) for β^- decay, (b) β^+ or EC decay.

Note that it is possible for an unstable atom to satisfy all three decay conditions given above. ^{64}Cu is such an atom. That is ^{64}Cu decays by β^-, β^+, and EC processes, with a half-life of 12.8 hours.

Example 12.4. Show that a radioactive $^{64}_{29}\text{Cu}$ isotope satisfies the conditions for decaying by β^-, β^+, and EC processes.

The three possible decays may be represented (as we see on page 402) by

$$^{64}_{29}\text{Cu} \rightarrow {}^{64}_{30}\text{Zn} + \beta^- + \bar{\nu}$$
$$^{64}_{29}\text{Cu} \rightarrow {}^{64}_{28}\text{Ni} + \beta^+ + \nu$$
$$^{64}_{29}\text{Cu} + {}_{-1}\text{e} \rightarrow {}^{64}_{28}\text{Ni} + \nu$$

The atomic masses of the three isotopes are

$$^{64}_{29}\text{Cu (63.9297 u)}, \qquad ^{64}_{28}\text{Ni (63.9280 u)}, \qquad ^{64}_{30}\text{Zn (63.9291 u)}$$

Assuming that the Q-values given by Eqs. (12.30), (12.31), and (12.32) are positive, all three decays will occur. Thus, according to Eq. (12.30), for β^- decay,

$$\begin{aligned} Q &= [M(Z) - M(Z + 1)]c^2 \\ &= [63.9297 \text{ u} - 63.9291 \text{ u}]c^2 \\ &= 0.0006 \text{ uc}^2 \times 931 \text{ MeV/uc}^2 = 0.56 \text{ MeV} > 0. \end{aligned}$$

From Eq. (12.31), for β^+ decay,

$$\begin{aligned} Q &= [M(Z) - M(Z - 1) - 2m_e]c^2 \\ &= [M(Z) - M(Z - 1)]c^2 - 2m_ec^2 \\ &= [63.9297 \text{ u} - 63.9280 \text{ u}]c^2 - 2m_ec^2 \\ &= 0.0017 \times 931 \text{ MeV} - 2(0.51) \text{ MeV} \\ &= 1.58 \text{ MeV} - 1.02 \text{ MeV} = 0.56 \text{ MeV} > 0. \end{aligned}$$

From Eq. (12.32), for EC decay,

$$\begin{aligned} Q &= [M(Z) - M(Z - 1)]c^2 \\ &= 0.0017 \times 931 \text{ MeV} = 1.58 \text{ MeV} > 0. \end{aligned}$$

Experimentally it is found that $^{64}_{29}\text{Cu}$ decays by three processes: (a) 40% of it disintegrates by β^- decay, to stable Zn, (b) $\sim$20% by β^+ decay, to stable Ni, and (c) $\sim$40% by EC, to stable Ni. The measured Q values of the three are 0.57 MeV, 0.66 MeV, and 1.66 MeV, respectively. The small differences are due to the inaccuracy in the mass of Cu.

---

b) *Characteristics of beta-ray spectra.* Figure 12.10 shows a plot of $N(K)$, the number of beta particles with kinetic energy K, versus K for the negative beta particles emitted from a radioactive sample of RaE (bismuth-210). This is a beta spectrum of RaE, and is typical of all beta (β^- or β^+) emitters.

The spectra of both β^- and β^+ emitters show the following characteristics.

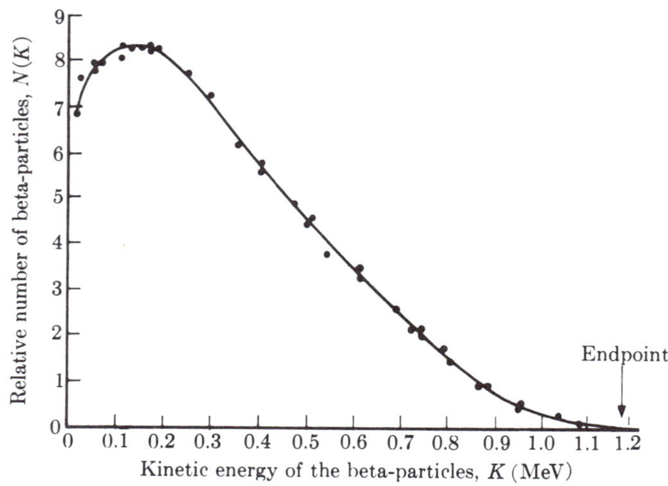

Fig. 12.10 Beta spectrum of RaE (Bi-210). [From G. J. Neary, *Proc. Roy. Soc.* **A175,** 71 (1940)]

a) Both β^- and β^+ emitters exhibit a continuous spectrum; this holds true for naturally occurring as well as artificially made beta emitters.

b) There are definite maxima for $N(K)$ and $N(p)$ [the number of electrons with momentum p] in the continuous spectrum (or distribution). The corresponding energy of the maxima depends on the type of nucleus undergoing beta decay.

c) The maximum or endpoint energy K_{max} is different for different nuclei, but is always approximately equal to the maximum disintegration energy.

Let us apply to beta decay the laws of conservation of (i) energy, (ii) linear momentum, and (iii) angular momentum.

If the daughter and the beta particle (β^- or β^+) really are the only two particles resulting from beta decay, the situation is very similar to that of α decay; i.e., it is a two-body problem. So that linear momentum is conserved, since initially the parent nucleus is at rest, the daughter and the electron are always emitted in opposite directions, both with the same amount of linear momentum. For a definite momentum, the beta particles emitted have a definite energy, i.e., they ought to be monoenergetic. However, we can see from the plots of $N(K)$ versus K or $N(p)$ versus p that this is not so, because of the continuous distribution.

Also, since the mass of the daughter nucleus is very large compared to that of the beta particle which has been emitted, the beta particles *should* take away almost all the disintegration energy. Again this does not agree with experimental results! Almost all the beta particles emitted have energies less than the maximum (or endpoint) energy; i.e., between 0 and K_e^{max}.

We run into another difficulty when we apply conservation of angular momentum. Before decay, there are A nucleons. After decay, there are still A nucleons, plus one beta particle. The spin of each nucleon is $\frac{1}{2}$, and the spin of a beta particle is also $\frac{1}{2}$. Thus if the system before decay has an integer (or half-integer) spin, after decay it has a half-integer (or integer) spin. Such changes from integer to half-integer or vice versa do not conserve angular momentum.

All these difficulties were overcome by the introduction of the *neutrino hypothesis.*

c) *The neutrino hypothesis and the theory of beta decay.* Wolfgang Pauli, in 1930, postulated the existence of a neutrino.[7] According to him, an additional particle, called a *neutrino* and denoted by ν, is emitted in the process of beta decay. In order to satisfy all the requirements of beta decay, the neutrino must be assigned the following properties.

i) It must have *zero or almost zero mass,* otherwise the maximum kinetic energy of the beta particle would not be equal to the disintegration energy.

ii) It must have *zero charge,* because the charge is already conserved without the neutrino.

iii) It must have a *spin* of $\frac{1}{2}$, so that if the spin of the system is an integer (or half-integer) before decay, it remains the same after decay.

The neutrino, like the electron and positron, cannot exist inside the nucleus. It is created at the time of decay. Just as a positron is the counterpart of an electron (i.e., an electron is a particle and a positron is its antiparticle), so a neutrino ν has its counterpart, called an *antineutrino,* $\bar{\nu}$. An antineutrino also has zero or almost zero mass, no charge, and a spin of $\frac{1}{2}$. Since both have the same properties, how do we distinguish between a ν and a $\bar{\nu}$? A neutrino ν is a *left-handed* particle (in the sense of a left-handed screw), with its spin angular momentum vector $\mathbf{S}_\nu$ antiparallel to its linear momentum $\mathbf{p}_\nu$ (Fig. 12.11a). An antineutrino $\bar{\nu}$ is a right-handed particle, with its spin angular momentum vector $\mathbf{S}_{\bar{\nu}}$ parallel to its linear momentum, as in Fig. 12.11(b). We know that ν and $\bar{\nu}$ are really two different particles because this has been verified experimentally.[8] This discovery leads to interesting and important consequences, which we'll talk about presently.

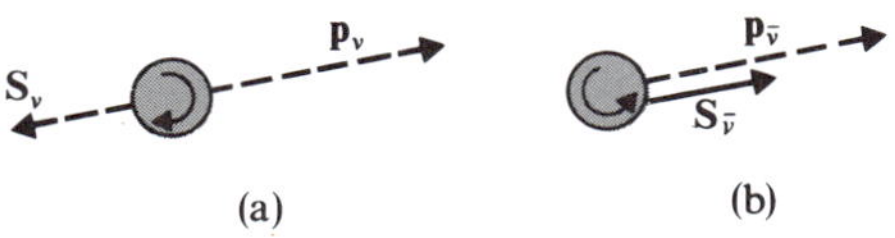

Fig. 12.11 (a) A neutrino, whose $\mathbf{S}_\nu$ is antiparallel to its $\mathbf{p}_\nu$. (b) An antineutrino, whose $\mathbf{S}_{\bar{\nu}}$ is parallel to its $\mathbf{p}_{\bar{\nu}}$.

In the light of the neutrino hypothesis, we may write the equations for beta decay as

$$ {}^{A}_{Z}\mathrm{X} \rightarrow {}^{\;\;A}_{Z+1}\mathrm{Y} + \beta^{-} + \bar{\nu}, \tag{12.33}$$

$$ {}^{A}_{Z}\mathrm{X} \rightarrow {}^{\;\;A}_{Z-1}\mathrm{Y} + \beta^{+} + \nu, \tag{12.34}$$

$$ {}^{A}_{Z}\mathrm{X} + {}^{\;\;0}_{-1}\mathrm{e} \rightarrow {}^{\;\;A}_{Z-1}\mathrm{Y} + \nu, \tag{12.35}$$

or we may write

$$\boxed{\begin{aligned} &\mathrm{n} \rightarrow \mathrm{p} + \beta^{-} + \bar{\nu}, \\ &\mathrm{p} \rightarrow \mathrm{n} + \beta^{+} + \nu, \\ &\mathrm{p} + {}^{\;\;0}_{-1}\mathrm{e} \rightarrow \mathrm{n} + \nu. \end{aligned}} \tag{12.36}$$

Note that the creation of a particle must be accompanied by the simultaneous creation of an antiparticle, like β^- and $\bar{\nu}$ in β^- decay and β^+ and ν in β^+ decay.

Thus, according to Eq. (12.36), β-decay is a three-body problem, not a two-body one (α decay, on the other hand, is a two-body problem). The linear momenta associated with the three bodies can be combined vectorially in an infinite number of ways to give a zero resultant linear momentum [because initially (before decay), the nucleus is at rest]. Figure 12.12 shows how. Beta particles therefore have energies between 0 and K_e^{max}. The recoiling daughter takes a small amount of energy, while the neutrino takes the rest.

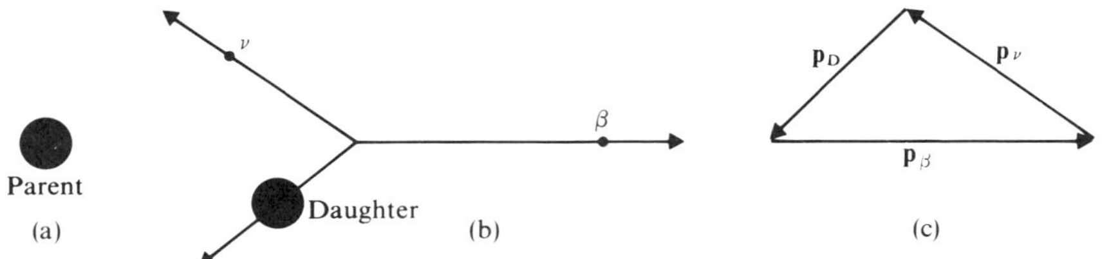

Fig. 12.12 (a) Parent nucleus before decay. (b) Beta decay of the parent nucleus yields daughter nucleus, beta, and neutrino (a three-body decay). (c) Since the linear momentum of the parent before decay is zero, the linear momenta of the three decay products must form a closed triangle, with the beta and neutrino having linear momentum between zero and a maximum value.

Free neutrons have been observed to decay according to Eq. (12.36), with a half-life of 932 ± 14 sec. As for protons, it is energetically not possible for them to decay.

What *type* of force (or interaction) causes β decay? According to Enrico Fermi,[9] there exists an interaction between the neutron, electron, and antineutrino (or the proton, positron, and neutrino) that causes a neutron to be transformed into a proton (or a proton into a neutron) with the simultaneous emission of an electron and an antineutrino (or a positron and a neutrino). This interaction is very weak and has a very short range; or it may even be a point interaction. This type of interaction which causes β-decay is called, in fact, *weak interaction.* The relative strengths of four different types of well-known interactions —nuclear, coulomb, weak, and gravitational—are 1, 10^{-2}, 10^{-13}, and 10^{-39}, respectively. (Notice how weak the gravitational interaction is!)

★12.6 NONCONSERVATION OF PARITY IN BETA DECAY

Parity is that property of a wave function that describes its behavior under reflection (or inversion) of the coordinate system, i.e., when the signs of the coordinates are changed. Let $\psi(x, y, z)$ be the wave function of a system. If this wave function changes sign on reflection, it is said to have *odd* parity, that is, $P = -1$. If the wave function does not change sign on reflection, it is said to have *even* parity, that is, $P = +1$. For the odd-parity wave function,

$$\psi(-x, -y, -z) = -\psi(x, y, z), \tag{12.37}$$

and for the even-parity wave function,

$$\psi(-x, -y, -z) = +\psi(x, y, z). \tag{12.38}$$

The parity of a system is given by the product of the parities of the individual particles or parts of the system. Until 1956, physicists assumed that parity was conserved in all types of nuclear interactions. That is, they thought that physical laws permitted the existence of a mirror image of a system (or motion). They thought, in other words, that the mirror image of an object is indistinguishable from the object itself, i.e., that left–right symmetry is conserved. The conservation of parity had been experimentally verified in strong interactions, but in 1956 its validity in other interactions was questioned by Tsung-Dao Lee and Chen Ning Yang.[10] According to them, it is possible that parity is not conserved in weak interactions such as beta decay. And they were right, as we shall soon see.

Consider a neutrino traveling toward a mirror with momentum $\mathbf{p}_\nu$ (Fig. 12.13), with its spin angular momentum vector in a direction opposite to that of $\mathbf{p}_\nu$. The mirror image of this neutrino has its spin angular momentum $\mathbf{S}_\nu$ going in the same directions as the linear momentum $\mathbf{p}_\nu$. (Note that the mirror does not change the direction of the angular momentum.) The motion depicted by the mirror image should be a motion possible to the neutrino. But this is not so, because, by our definition of the neutrino, the mirror image of a neutrino is a different particle: an antineutrino. We call this a *parity-nonconserving property* of the neutrino.

On these bases, Lee and Yang postulated that mirror effects do not occur in beta decay, unless light particles (such as ${}_{-1}e$, ν) change into antiparticles (${}_{+1}e$, $\bar{\nu}$).

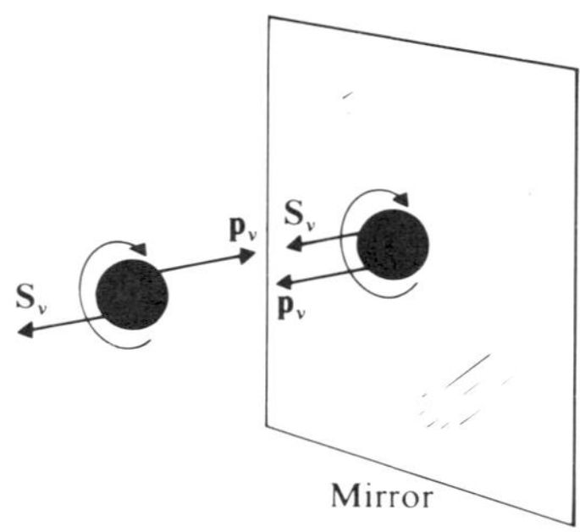

Fig. 12.13 The mirror image of a neutrino is not a neutrino but an antineutrino. This means that the neutrino is not conserving parity.

The experimental verification of this hypothesis was carried out by C. S. Wu *et al.*[11] They measured the angular distribution of beta particles emitted by ^{60}Co. They did this by putting a sample of ^{60}Co at 0.01°K in a magnetic field so as to align the nuclear spins in one direction. Then they switched off the magnetic field, and found that angular distribution of the beta particles emitted from ^{60}Co was *anisotropic*. That is, more beta particles were emitted in the direction of the alignment than in the opposite direction. This means that the mirror image of this situation would have to have more beta particles emitted in the direction opposite to the alignment. Figure 12.14 diagrams this situation. We can see that the system formed by the mirror is not identical to the original system. This demonstrates the nonconservation of parity in beta decay, and in fact in weak interactions in general.

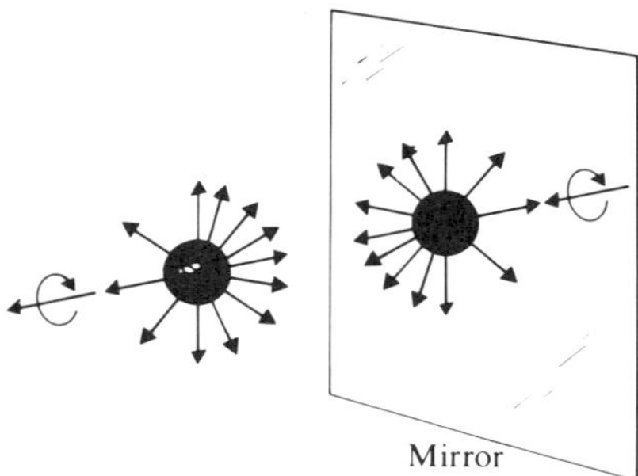

Fig. 12.14 A system of nuclei which results in emission of anisotropic radiation does not conserve parity. On the left, more particles are emitted in a direction opposite to the angular momentum vector. In its image (right) more particles are emitted along the direction of the angular momentum vector.

12.7 GAMMA DECAY

When a nucleus disintegrates by emitting alpha, beta, or any other kind of particle, it is usually left in an excited state. In fact, in most nuclear reactions, the recoiling nucleus is left in an excited state also. If the excited nucleus doesn't have enough energy to emit another particle, or if the decay by emission of another particle is slow, the nucleus decays by electromagnetic interaction, like this: A nucleus in the higher excited state with energy E_i makes a transition to a lower excited state (or ground state) of energy E_f. The excess energy,

$$\Delta E = E_i - E_f \tag{12.39}$$

is emitted by one of the following processes: (a) gamma-ray emission, (b) internal conversion, or (c) the creation of internal pairs.

Decays by gamma emission and internal conversion are much more frequent than decays by creation of internal pairs. Remember that all three processes are caused by electromagnetic interactions, and that internal conversion and the creation of internal pairs are alternatives to gamma emission. Internal conversion is more frequent in heavy excited nuclei. The creation of internal pairs (or the creation of an electron–positron pair from energy ΔE) is more frequent in light nuclei, and is possible only if ΔE is greater than 1.02 MeV. Let's talk about gamma-ray emission and internal conversion for a moment.

a) *Gamma-ray emission.* The gamma-ray spectra of excited nuclei consist of discrete energies (or sharp lines). This indicates that the nucleus (like the atom) has discrete energy levels. Figure 12.15 shows how nuclei, after decaying by (a) α emission and (b) β^- emission, are left in discrete excited energy states. If we

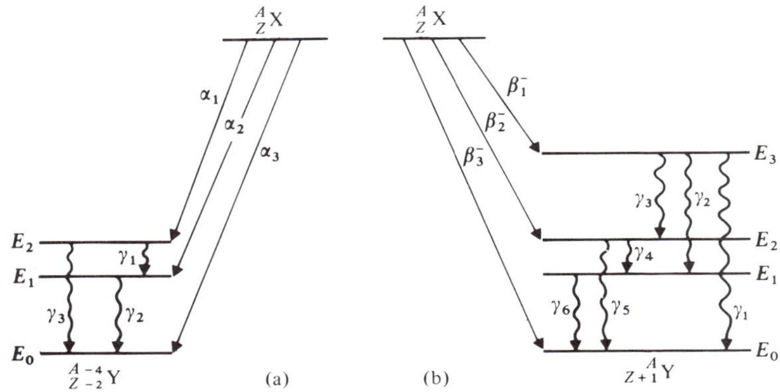

Fig. 12.15 Gamma rays are emitted whenever the nuclei are left in the excited states and the excited nucleus does not have enough energy to emit another particle. (a) α decay followed by gamma emission. (b) β^- decay followed by gamma emission.

neglect the small energy taken away by the daughter, the energy of the gamma ray (or photon) emitted is given by

$$h\nu = \Delta E = E_i - E_f, \tag{12.40}$$

where h is Planck's constant and ν is the frequency of the photon emitted. We can grasp the qualitative aspects of gamma emission by using classical electromagnetic theory, i.e., the oscillating charges in the nucleus give rise to radiation. But for finer details, one must use a quantum-mechanical treatment.[12]

General experimental observations indicate that, unlike the situation which results when atoms decay from their excited states, the nucleus, in its excited state, can have a mean half-life varying from $\sim 10^{-16}$ sec to many years. As for the selection rules, (i) the photon emitted must carry away at least one unit of angular momentum, and (ii) in gamma decay, parity is always conserved. For example, given that the spins of the initial and final states of the nucleus are $\mathbf{I}_i$ and $\mathbf{I}_f$, and that $\mathbf{L}$ is the angular momentum carried away by the photon, we have

$$\boxed{\mathbf{I}_i - \mathbf{I}_f = \mathbf{L}.} \tag{12.41}$$

That is,

$$(I_i + I_f) \geq L \geq |I_i - I_f|. \tag{12.42}$$

b) *Internal conversion.* Recall that internal conversion is an alternative to gamma decay. Let us say that a nucleus is in an excited state, and that one of the electrons in the K shell of the atom happens to be very close to it. Actually, according to quantum mechanics, a K electron spends a part of its time *inside the nucleus.* In such a case, the nucleus, instead of giving out a gamma ray, may de-excite (or decay) by giving its extra energy to that near-by electron in the closest atomic shell. Such electrons are thereby knocked out of the atoms entirely! We call these electrons *conversion electrons* and the process *internal conversion.* The kinetic energy K_e of the conversion electron is

$$K_e = \Delta E - I_B, \tag{12.43}$$

where $\Delta E = E_\gamma = E_i - E_f$ is the energy with which the gamma ray *would* have been emitted (except that it wasn't), and I_B is the atomic binding energy of the electron. Note that these conversion electrons appear as discrete lines superimposed on the continuous spectrum of beta particles, as shown in Fig. 12.16.

Suppose that a given radioactive sample emits N_γ gamma rays and N_e conversion electrons in the same time span. Then the ratio N_e/N_γ is called the *conversion coefficient* α. That is,

$$\boxed{\alpha = N_e/N_\gamma.} \tag{12.44}$$

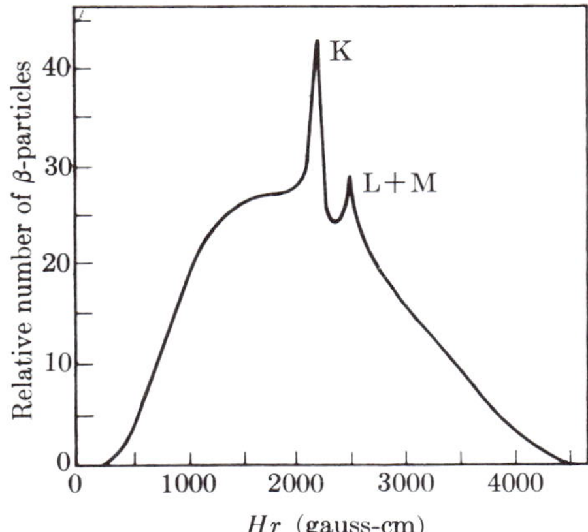

Fig. 12.16 The line spectrum superimposed on the continuous β^- spectrum of radioisotope Au-198 indicates conversion lines due to the emission of K and L electrons, which are emitted instead of gamma rays. Note that momentum is proportional to Hr, H being the magnetic field and r the radius of the electron path. [From C. Y. Fan, *Phys. Rev.* **87**, 258 (1952)] Note that the L and M conversion electron lines are not resolved.

Given that the N_e are K electrons, α is the K conversion coefficient, denoted by α_K. Similarly, α_L stands for the L conversion coefficient, and so forth.

If the electron from the K shell is removed by internal conversion or by electron capture (discussed in Section 12.5), there will be a vacancy in the K shell. This vacancy in the K shell is filled by an electron which jumps in from the L shell. During this process, a characteristic x-ray is emitted, which has energy $h\nu_K = (I_K - I_L)$, where I_K and I_L are the binding energies of the K and L electrons.

★12.8 THE MÖSSBAUER EFFECT[13,14]

The Mössbauer effect is the technique developed and theoretically explained by Rudolf L. Mössbauer in 1958 for producing recoilless emission and recoilless absorption of photons. The success of this method lies in the fact that, by using it, we can measure differences in energy of one part in 10^{13}, and it is the most precise method ever used in physics. Because of its accuracy, it is naturally used to observe the emission and absorption of resonance by nuclei, to study hyperfine structure, to investigate the energy-level structure in the solid state, to perform relativity experiments, and in many other ways.

Suppose that an atom in the excited state of energy E emits a photon of energy $h\nu$ and goes to the ground state of energy E_0 ($E > E_0$), and that a photon of the same energy $h\nu$ ($= E - E_0$) is incident on (or collides with) this same type of atom in the ground state. The atom absorbs the photon and is raised to the excited state of energy E. These two processes are called *resonance emission* and *resonance absorption*, respectively. In these considerations we have neglected the recoiling energy of the atom, which we can do only if the energy of the transition is small, as is the case in atomic transitions. However, in nuclear and other transitions, in which the transition energies are great, we cannot neglect the recoiling energy of the atom; then the resonance emission and absorption cannot occur, as we shall explain below.

Suppose that, during a transition from an excited state of energy E to the ground state of energy E_0, an atom or a nucleus emits a photon of energy $h\nu$. The conservation of linear momentum requires (nonrelativistically) that

$$\frac{h\nu}{c} = MV, \tag{12.45}$$

where M is the mass and V is the velocity of the atom or nucleus. The conservation of energy requires that

$$E - E_0 = h\nu + E_R = h\nu + \tfrac{1}{2}MV^2 \tag{12.46}$$

where $E_R = \frac{1}{2}MV^2$ is the recoil energy of the atom. Substituting for V from Eq. (12.45) into Eq. (12.46), we get

$$E - E_0 = h\nu + \frac{h^2\nu^2}{2Mc^2}. \tag{12.47}$$

Thus the energy $h\nu$ of the emitted photon is less than the transition energy $E - E_0$ by $h^2\nu^2/2Mc^2$. Similarly, in the process of resonance absorption, when an atom absorbs a photon of energy $h\nu$, the atom uses up, in the process of recoiling, an amount of energy $h^2\nu^2/2Mc^2$. These processes are illustrated in Fig. 12.17.

So we see that, if the system emits photons, and these photons are absorbed by atoms in the same system, the energy of the photons available for absorption is not $h\nu$, but $h\nu - 2E_R$, where

$$\boxed{\Delta E_R = 2E_R = \frac{h^2\nu^2}{2Mc^2} + \frac{h^2\nu^2}{2Mc^2} = \frac{h^2\nu^2}{Mc^2}.} \tag{12.48}$$

If the natural width Γ of the excited level is more than ΔE_R, resonance absorption is possible. On the other hand, if the width Γ is less than ΔE_R, resonance absorption is not possible.

According to the uncertainty principle, for an excited energy level of width $\Delta E = \Gamma$ and lifetime $\Delta t = \tau$, we get $\Gamma\tau \geq \hbar$. In the case of atomic transitions, the mean lifetime τ of an excited level is 10^{-8} sec. Therefore the width of the level is $\Gamma \sim \hbar/\tau \simeq 10^{-7}$ eV. The photons emitted have energies of the order of an

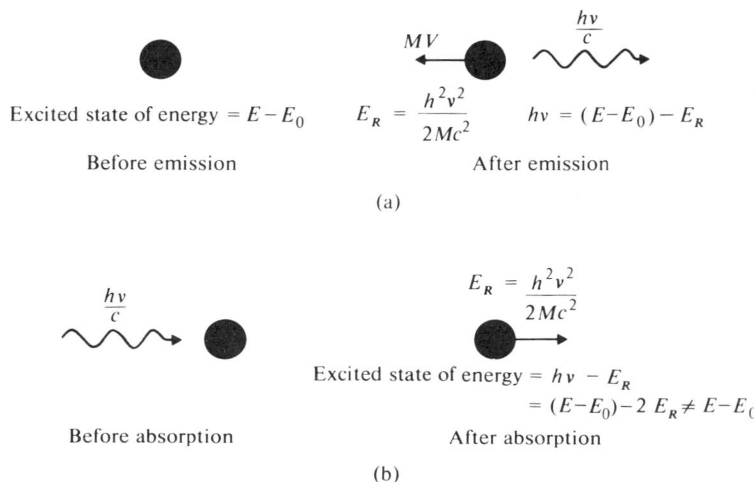

Fig. 12.17 (a) The atom emits a photon of energy $h\nu$ and recoils with energy E_R. (b) The photon is absorbed. For momentum to be conserved, the atom, after absorbing a photon, must recoil.

electron volt, say $h\nu = 1$ eV. For an atom of $A = 100$, $Mc^2 \simeq 10^{11}$ eV. Thus

$$\Delta E_R \simeq \frac{(1 \text{ eV})^2}{10^{11} \text{ eV}} = 10^{-11} \text{ eV}, \qquad \text{that is, } \Delta E_R \lll \Gamma.$$

Hence resonance absorption is always possible in atomic transitions.

On the other hand, in nuclear decay, $h\nu$ is approximately 100 keV $= 10^5$ eV, while Mc^2 is still 10^{11} eV (for $A = 100$). Thus $\Delta E_R = (10^5 \text{ eV})^2/10^{11} \text{ eV} = 10^{-1}$ eV, which is greater than the natural widths $\Gamma = 10^{-4}$ eV of the nuclear levels.

In conclusion, let us say that, in atomic systems, $\Gamma/\Delta E_R \approx 10^3$, while in nuclear systems, $\Gamma/\Delta E_R \approx 10^{-3}$. This means that resonance absorption is possible in atomic systems, but not always in nuclear ones.

There are two different ways of achieving resonance absorption for high-energy or nuclear transitions. One is to increase the width of the levels by some means so that the condition that $\Gamma > \Delta E_R$ is satisfied, which can be achieved by *Doppler broadening*. The other method is to reduce the loss of energy due to recoil so that ΔE_R is less than Γ. This is accomplished by the Mössbauer effect.

a) *The Doppler effect.* W. Davey and P. Moon[15] demonstrated in 1950 that by placing a source of photons on the rim of an ultracentrifuge rotor they could produce an external Doppler shift. For a rotor moving at a speed of 800 m/sec, this use of the Doppler effect increased the frequency of the emitted radiation enough to compensate for the loss of energy due to recoil. By varying the speed of the rotor, they were able to change the amount of resonance absorption, which made it possible for them to measure a mean lifetime of excited levels of $\sim 10^{-11}$ sec.

b) *The Mössbauer effect.* Rudolf L. Mössbauer devised a method which reduces the atom's energy loss due to recoil almost to zero. A high-energy photon-emitting system (such as a nucleus) is made a part of a crystal lattice. This is achieved by embedding the source in the crystal lattice while it is at a low temperature. Under these conditions, the whole mass of the crystal—not just the recoiling particle—takes part in the recoil. Of course the mass of the crystal is very large compared with the mass of the recoiling atom or nucleus, so, according to Eq. (12.48), the recoil energy ΔE_R is almost zero. Similarly, the target (which is of the same material as the source which is emitting photons) is made a part of the lattice at low temperature. Thus the condition that ΔE_R be less than Γ is satisfied and resonance absorption is possible.

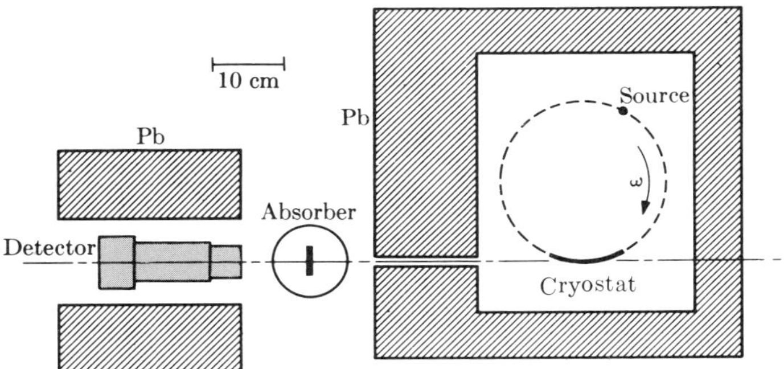

Fig. 12.18 R. L. Mössbauer's experimental arrangement to demonstrate recoilless resonance absorption. Both the source of photons and the absorber are placed inside the cryostat. The source is mounted on a wheel, which may be rotated in either direction with a variable velocity. [From R. L. Mössbauer, *Naturwiss* **45**, 538 (1958)]

Figure 12.18 shows the apparatus Mössbauer used. The first source he used was ^{191}Ir, which emits photons of energy 129 keV. By moving the source with a variable velocity, he was able to obtain a resonance curve, i.e., the rate of absorption of photons by Ir versus the velocity of the source, as shown in Fig. 12.19. Even if the source moves very slowly, the movement destroys the resonance. This verifies that ΔE_R is zero at zero velocity, and there is no need to produce any Doppler shift.

In 1959 Mössbauer obtained the resonance curve for ^{57}Fe, which gives photons of energy 14.4 keV, without even cooling the crystal. In addition to finding a peak of resonance absorption when the source of photons was at zero velocity, he found that there were other peaks which were due to the hyperfine splitting resulting from the nuclear spin of ^{57}Fe. This hyperfine splitting resulted

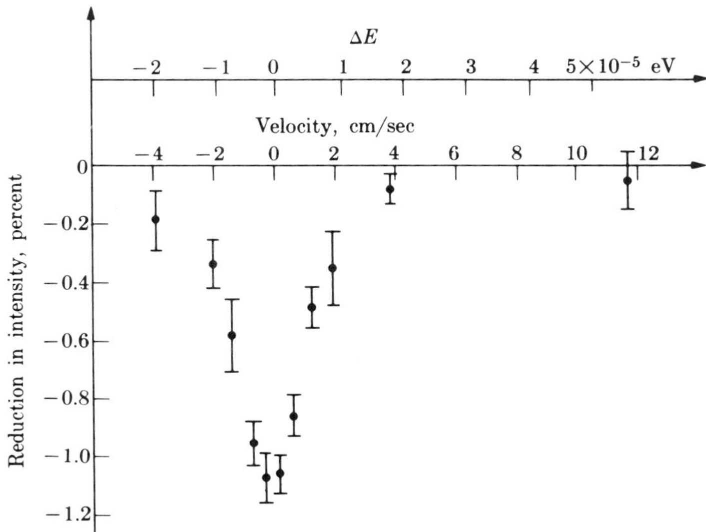

Fig. 12.19 Recoilless resonance absorption of 129-keV gamma radiation by atoms of ^{191}Ir. Curve shows the effect of the motion of the source of photons on the resonance curve at 88°K, resulting in a half-width Γ at half maximum absorption of

$$2\Gamma = (9.2 \pm 1.2)10^{-6} \text{ eV}.$$

[From R. L. Mössbauer, *Naturwiss* **45,** 538 (1958)]

from ^{57}Fe's nuclear spin of $i = \frac{3}{2}$ and ground-state spin of $j = \frac{1}{2}$. It is the Mössbauer technique's accuracy of one part in 10^{13} that resolves the hyperfine components.

SUMMARY

The phenomenon of emission of radiation by unstable nuclei is called radioactivity, and the elements which exhibit it are called radioactive nuclides (or elements). Alpha rays are fast-moving doubly ionized helium atoms, beta rays are fast-moving electrons or positrons, and gamma rays are electromagnetic waves.

The law of radioactive decay is $N = N_0 e^{-\lambda t}$. The activity of a sample is defined as the number of disintegrations per second, $A = \lambda N$. The half-life of a sample is the time interval in which the number of undecayed atoms in the sample decreases by half; $t_{1/2} = 0.693/\lambda$. The mean lifetime $\tau = 1/\lambda$.

The units of radioactivity are the *curie* (1 Ci), defined as 3.7×10^{10} disintegrations per second, and the *rutherford* (1 rd), defined as 10^6 disintegrations per second.

There are four natural radioactive series: thorium, neptunium, uranium and actinium. Elements in the neptunium series have decayed entirely because the parent atoms do not have as long a half-life as those of other series. The radioactive equilibrium reached is given by

$$\lambda_1 N_1 = \lambda_2 N_2 = \lambda_3 N_3 = \cdots;$$

this is the relation that is used for radioactive dating.

One condition which is necessary before a radioactive nucleus will decay by alpha emission is that the mass of the parent atom must be greater than the sum of the masses of the daughter and the helium atoms. According to the trend established for alpha decay, a change by a factor of 2 or 3 in the energy of the alpha particle corresponds to a factor of $\sim 10^{20}$ in the half-life or decay constant of the alpha emitter.

Alpha decay can be explained theoretically by means of quantum theory. The explanation involves calculation of the probability of penetration by an alpha particle through the coulomb potential barrier of the daughter nucleus.

There are three processes by which a nucleus disintegrates by beta decay: emission of an electron, emission of a positron, and capture of an atomic electron. For decay to happen, the Q value must be positive.

The general characteristics of the beta spectrum indicate that beta decay is a three-body problem. In beta decay, besides a beta particle and a daughter nucleus, the nucleus also emits a third particle, a *neutrino*. The neutrino (and also its opposite member, the *antineutrino*) has a rest mass of almost zero, no charge, and a spin of $\frac{1}{2}$. The theory of beta emission was formulated by Enrico Fermi. According to him, the force responsible for beta decay is called *weak interaction*. In beta decay, parity is not conserved.

The decay of an excited nucleus via the process of electromagnetic interactions takes place by one of three processes: gamma emission, internal conversion, and creation of an internal pair. The conversion coefficient is defined as the ratio $\alpha = N_e/N_\gamma$.

The angular momentum $\mathbf{L}$ of the gamma ray is given by $\mathbf{L} = \mathbf{I}_i - \mathbf{I}_f$.

★ Recoilless emission and recoilless absorption of photons were made possible by R. L. Mössbauer's method of reducing the atom's loss of energy due to recoil almost to zero. The end result is called the *Mössbauer effect*.

PROBLEMS

1. A sample of radioactive iodine-123 was observed to decay in the following manner.

Time elapsed, hr	Counts/min	Time elapsed, hr	Counts/min
0	99,968	25	26,448
5	77,105	30	20,789
10	58,860	35	15,567
15	44,932	40	12,246
20	34,646	45	9,163

Make the plot of the above data and calculate the following.

a) The half-life, decay constant, and mean life
b) Activity of the sample after 26 hours
c) Number of radioactive atoms after 26 hours

2. A sample of radioactive silver-105 was observed to decay in the following manner.

Time elapsed, days	Counts/min	Time elapsed, days	Counts/min
0	100,210	50	28,651
10	77,880	60	22,313
20	60,653	70	17,377
30	47,237	80	13,534
40	36,788	90	10,540

a) Calculate the half-life, decay constant, and mean life.
b) What will be the activity of this sample after three months?
c) Find the number of undecayed radioactive atoms in the sample after three months.

3. The half-life of radioactive cobalt-60 is 5.26 years.
a) Calculate the mean life and the disintegration constant.
b) What is the activity of a 1-g sample of ^{60}Co? Express this activity in curies and in rutherfords.
c) Each disintegration of ^{60}Co is accomplished by an emission of an electron, of energy 0.31 MeV, and two gamma rays, of energies 1.17 MeV and 1.33 MeV, as shown in the figure. Calculate, in joules per second and calories per second, the rate at which energy is given out from a 1-g radioactive sample.

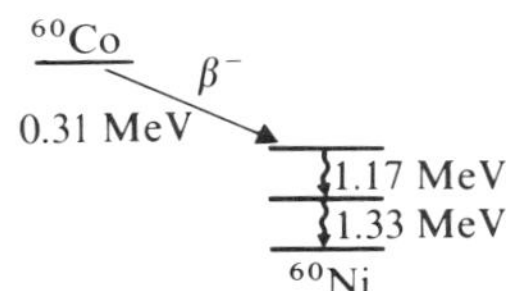

4. ^{144}Pm has a half-life of 365 days. How long will it take to reduce its activity from 1 millicurie to 10 microcuries?

5. What will be the mass of a 10-curie sample of cobalt-60, given that its half-life is 5.26 years?

6. The final decay product of ^{238}U, after it has emitted 8 α particles and 6 electrons, is ^{206}Pb. That is,

$$^{238}_{92}\text{U} \rightarrow {}^{206}_{82}\text{Pb} + 8\,{}^{4}_{2}\text{He} + 6\,{}^{0}_{-1}\text{e}.$$

If we start with 1 kg of ^{238}U, how much lead (^{206}Pb) will be formed in 10 million years, given that the half-life of ^{238}U is 4.5×10^9 years? How can this method be used to find the age of the earth?

7. Suppose that atoms of type 1 decay to type 2, and atoms of type 2 decay to stable atoms of type 3. The decay constants of 1 and 2 are λ_1 and λ_2, respectively. This

reduces Eqs. (12.8), (12.9), and (12.10) to the following

$$\frac{dN_1}{dt} = -\lambda_1 N_1, \qquad \frac{dN_2}{dt} = \lambda_1 N_1 - \lambda_2 N_2, \qquad \frac{dN_3}{dt} = \lambda_2 N_2.$$

Assume that at $t = 0$, $N_1 = N_{10}$, and $N_2 = N_3 = 0$. Show that the solution of the above equations yields

$$N_1 = N_{10} e^{-\lambda_1 t},$$

$$N_2 = \frac{\lambda_1}{\lambda_2 - \lambda_1} N_{10}(e^{-\lambda_1 t} - e^{-\lambda_2 t}),$$

$$N_3 = N_{10}\left(1 + \frac{\lambda_1}{\lambda_2 - \lambda_1} e^{-\lambda_2 t} - \frac{\lambda_2}{\lambda_2 - \lambda_1} e^{-\lambda_1 t}\right).$$

8. Consider the following radioactive decay series:

$$^{105}_{44}\text{Ru} \xrightarrow[t_{1/2}=4.5\text{ hr}]{\beta^-} {}^{105}_{45}\text{Rh} \xrightarrow[t_{1/2}=35\text{ hr}]{\beta^-} {}^{105}_{46}\text{Pd (stable)}.$$

Initially we have 10^6 radioactive atoms of $^{105}_{44}$Ru. Using the results obtained in Problem 7, plot the number of atoms of Ru, Rh, and Pd as functions of time.

9. The activity of a certain radioactive sample decreases by a factor of 8 in a time interval of 30 days. What is its half-life, mean life, and disintegration constant?

10. The half-life of $^{232}_{90}$Th is 1.41×10^{10} years ($= 4.45 \times 10^{17}$ sec). Assume that the universe was created 10 billion years ago. What fraction of Th atoms has survived?

11. ^{226}Ra decays by α emission, and has a half-life of 1620 years. Calculate the kinetic energies of the α particles, given that if the Q-values are 4.866 MeV and 4.673 MeV. What is the energy of the gamma ray emitted?

12. An α emitter decays by emission of two groups of α particles, with kinetic energies K_1 and K_2, where $K_1 > K_2$. The nucleus which emits the α particle with kinetic energy K_2 is left in the excited state, and hence decays to the ground state by emitting a γ ray. Calculate the energy of the γ ray in terms of K_1, K_2, and A (the mass number).

13. What is the velocity and kinetic energy of the daughter nucleus formed in the α decay (of energy 4.19 MeV) of the parent nucleus ^{238}U?

14. Show that a nucleus will decay by α emission only if

$$(B_1 A_1 - B_2 A_2) < 4(7A_2 - B_2),$$

where A_1 and A_2 are the mass numbers and B_1 and B_2 are the total binding energies of the parent and the daughter, respectively. Assume that the binding energy of the α particle is 28 MeV.

15. Assume that the radius of a nucleus is given by $R = 1.36 \times 10^{-13} A^{1/3}$ cm. From the purely classical point of view, what is the minimum energy needed by an α particle from this nucleus to penetrate the coulomb barrier of a ^{238}U nucleus?

16. Derive Eq. (12.31).

17. Derive Eq. (12.32).

18. Which of the following will decay by (i) β^- emission, (ii) β^+ emission, (iii) electron capture?

$$ {}^{14}_{6}\mathrm{C}, \quad {}^{22}_{11}\mathrm{Na}, \quad {}^{146}_{61}\mathrm{Pm}, \quad {}^{26}_{13}\mathrm{Al}, \quad {}^{60}_{27}\mathrm{Co} $$

19. A free neutron which is unstable decays into a proton by β^- emission with a half-life of 15 minutes. The mass of the neutron is 1.30 MeV greater than the mass of the proton. Calculate the maximum kinetic energies of (a) the β^- particles, (b) the recoil protons.

20. ${}^{64}_{29}\mathrm{Cu}$ decays with a half-life of 12.8 hr by β^-, β^+, and EC. What are the daughter nuclei produced in the three cases? Calculate the maximum kinetic energies of the β^- and β^+ particles and the maximum recoil energies of the three daughter nuclei.

21. A 100-mg sample of RaE is used in a calorimetric experiment. The half-life of RaE is 5 days, and it decays by emitting β particles of average energy 0.34 MeV. What is the rate of heat production? Assume that 100% of the energy absorbed is converted into heat.

22. ${}^{37}_{18}\mathrm{Ar}$ decays by capture of K electrons to ${}^{37}_{17}\mathrm{Cl}$, with a half-life of 35 days. Calculate the recoil energy and velocity of the Cl nucleus, assuming that the rest mass of the neutrino is zero. The total disintegration energy is 0.82 MeV.

23. ${}^{7}_{4}\mathrm{Be}$ decays by electron capture to ${}^{7}_{3}\mathrm{Li}$, with a Q-value of 0.863 MeV. The maximum recoil energy of Li is 55 eV. Calculate the kinetic energy of the neutrino, assuming that its rest mass is zero.

24. ${}^{198}_{79}\mathrm{Au}$ decays with a half-life of 2.7 days to ${}^{198}_{80}\mathrm{Hg}$ by emitting β^- particles of three different energies, as shown in the figure. That is, some nuclei are left in the ground state, while others are in one of the two excited states. The excited nuclei decay to the ground state by emitting gamma rays. What are the energies of the different gamma rays and beta rays emitted?

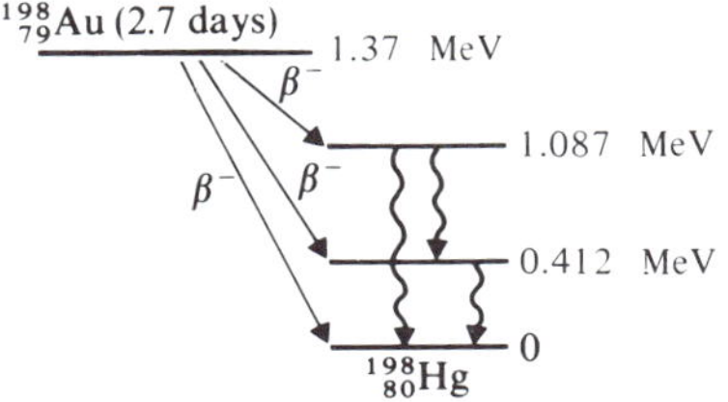

25. ${}^{137}_{55}\mathrm{Cs}$ decays by β^- emission, as shown in the figure. When the nucleus left in the excited state decays to the ground state, what are the energies of the β rays and of the photon emitted? Given that K- and L-shell binding energies of barium are 37.44 and 5.99 keV respectively, what are the energies of the K- and L-shell conversion lines?

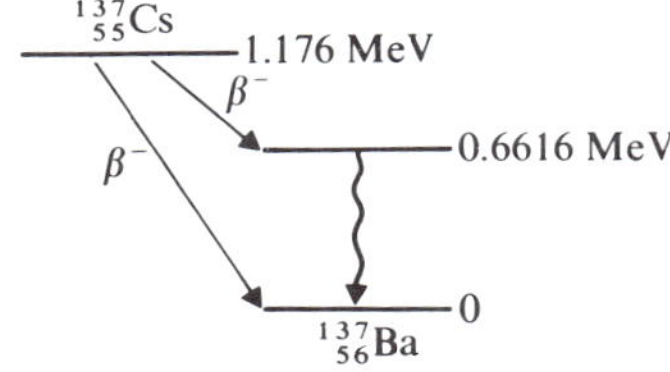

REFERENCES

1. Henri Becquerel, *Compt. Rend.* **122,** 420, 501, 689 (1896)
2. Marie Curie, *Compt. Rend.* **126,** 1101 (1898)
3. W. F. Libby, *Radioactive Dating*, second edition, Chicago: University of Chicago Press, 1955
4. George Gamow, *Z. Physik* **51,** 204 (1928)
5. R. Gurney and E. Condon, *Nature* **122,** 439 (1928)
6. Atam P. Arya, *Fundamentals of Nuclear Physics*, Boston: Allyn and Bacon, 1966
7. E. J. Konopinski, *Revs. Mod. Phys.* **15,** 209 (1943)
8. F. Reines and C. L. Cowan, Jr., *Phys. Rev.* **90,** 492 (1953); **113,** 273 (1959)
9. Enrico Fermi, *Z. Physik* **88,** 161 (1934)
10. T. Lee and C. Yang, *Phys. Rev.* **105,** 1671 (1957)
11. C. S. Wu, E. Ambler, R. W. Hayward, D. P. Hoppes, and R. P. Hudson, *Phys. Rev.* **105,** 1413 (1957)
12. J. M. Blatt and V. F. Weisskopf, *Theoretical Nuclear Physics*, New York: John Wiley 1952
13. R. L. Mössbauer, *Z. Physik* **151,** 124 (1958)
14. R. L. Mössbauer, *Naturwiss* **45,** 538 (1958); *Z. Naturforsch* **14a,** 211 (1959)
15. W. G. Davey and P. B. Moon, *Proc. Phys. Soc.* **A–66,** 956 (1953)

Suggested Readings

1. Atam P. Arya, *Fundamentals of Nuclear Physics*, Chapters II, VII, VIII, and IX, Boston, Mass.: Allyn and Bacon, 1966
2. E. Segré, editor, *Experimental Nuclear Physics*, Vol. IV, New York: John Wiley, 1959
3. R. D. Evan, *The Atomic Nucleus*, New York: McGraw-Hill, 1955
4. K. Siegbahn, editor, *Alpha-, Beta-, and Gamma-Ray Spectroscopy*, New York: Interscience, 1965
5. F. Ajzenberg, *Nuclear Spectroscopy*, edited by Selove, Parts A and B, New York: Academic Press, 1960
6. H. A. Enge, *Introduction to Nuclear Physics*, Reading, Mass.: Addison-Wesley 1966
7. Ernest Rutherford, J. Chadwick, and C. D. Ellis, *Radiation from Radioactive Substances*, New York: Macmillan, 1930
8. H. Frauenfelder, *The Mössbauer Effect*, New York: W. A. Benjamin, 1962
9. A. J. F. Boyle and H. E. Hall, *Rep. Prog. Phys.* **25,** 441 (1962)

CHAPTER 13
INTERACTION OF NUCLEAR RADIATION; DETECTORS AND ACCELERATORS

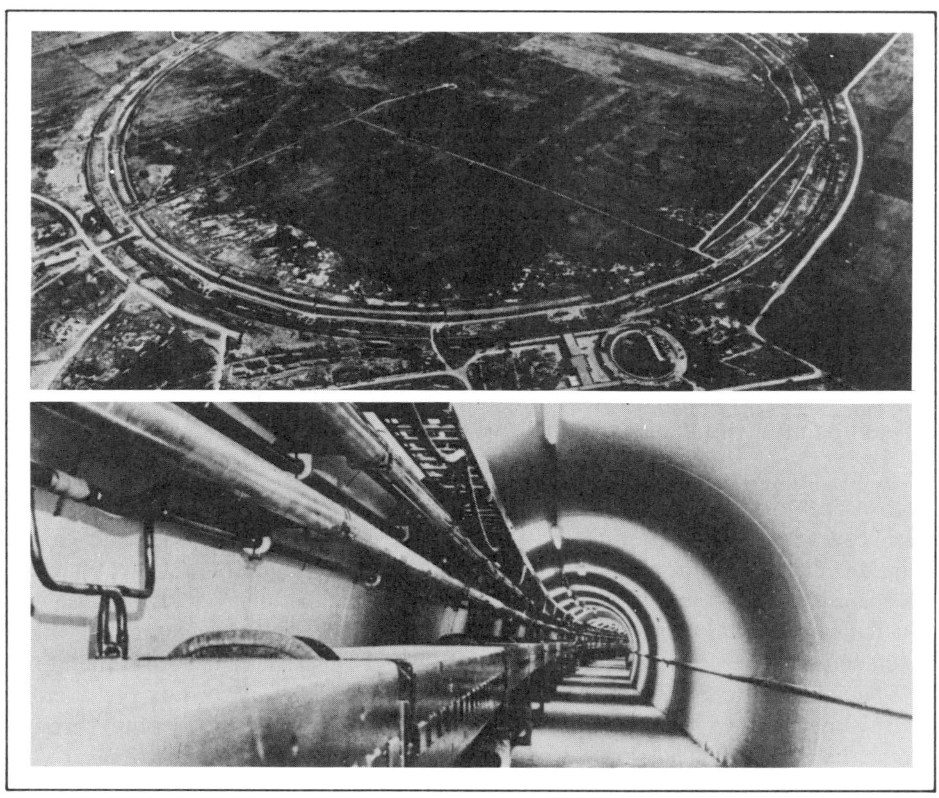

The national accelerator at Batavia, Illinois. (Above) Air view of main ring (4-mile circumference) of accelerator. Smaller ring of booster accelerator is in right foreground. (Below) Magnets encase beam tube along most of its 4-mile circumference. (Photograph courtesy National Accelerator Laboratory)

13.1 INTRODUCTION

Naturally occurring radioactive isotopes usually decay by emitting α, β, and γ rays, whose intensities and energies are very limited. To offset this limitation, scientists have developed many different types of accelerators and nuclear reactors. The result has been that we now have available beams of α, β, and γ rays, protons, deuterons, neutrons, and many other particles with a wide range of energies and high intensities, so that we can select what we need. Once the radiation has been produced, we need instruments to detect it. All radiation-detecting instruments, called *nuclear detectors*, or simply *detectors*, are based on the method by which the radiation interacts with matter. So, for convenience, we shall divide this chapter into three sections: (a) the interaction of radiation with matter, (b) different types of nuclear detectors, and (c) accelerators for producing fast-moving particles.

13.2 INTERACTION OF RADIATION WITH MATTER

Any type of radiation will travel indefinitely through a perfect vacuum (provided it does not decay). The only way the radiation can lose energy is by interacting with a gas, a liquid, or a solid while passing through it. And it is only during the process of the interaction of radiation with matter that we can identify the type of radiation and measure its intensity and energy. The kind of interaction by which energy is lost depends on the *type* of radiation: (a) the interaction of heavy charged particles with matter, (b) the interaction of electrons with matter, (c) the interaction of gamma rays with matter, (d) the interaction of neutrons with matter.

a) *The interaction of heavy charged particles with matter.*[1] As fast-moving particles such as protons, deuterons, or helium ions pass through a medium, because of their charge they interact with the atomic electrons of the medium. This causes them to collide with the electrons. In such collisions, although the incident particle loses a very small fraction of its energy, its path remains undeviated, because it is much heavier than the mass of the electron. After a collision, either the atom is left in an excited state or the electron completely detaches itself from the atom.

Now let us define a few terms. *Average ionization energy* $\bar{I}$ is the average energy needed to cause excitation or ionization of an atom of the medium through which the radiation is passing. When we know $\bar{I}$, we can calculate the range, or the *mean range*, which is the distance a particle travels through a medium from its source to the point at which its kinetic energy is zero.

The important quantity when we are dealing with the passage of charged particles through matter is the *stopping power* $S(E)$, which is the amount of energy lost per unit length (of the medium) by the incident particle in a given medium. That is,

$$S(E) = -\frac{dE}{dx} = n_{\text{ion}}\bar{I}, \tag{13.1}$$

where $S(E)$ is a function of the kinetic energy E of the particle (which is different in different materials), n_{ion} is the number of ion pairs produced per unit length, and $\bar{I}$ is the average excitation and ionization energy. If we know the $S(E)$ of a material, we can calculate the range R of a particle of energy E as follows:

$$R = \int_0^R dx = \int_0^E \left(-\frac{dE}{dx}\right)^{-1} dE = \int_0^E \frac{dE}{S(E)}. \tag{13.2}$$

The advantage of dealing with the stopping power lies in the fact that it is not necessary to measure it experimentally for different absorbers, because it can be calculated theoretically either from classical mechanics[2] or quantum mechanics.[3,4] The energy lost per unit length by a nonrelativistic particle is given by

$$\boxed{S(E) = -\frac{dE}{dx} = \frac{4\pi z^2 e^4 NZ}{m_0 v^2} \ln\left(\frac{2m_0 v^2}{\bar{I}}\right),} \tag{13.3}$$

where v is the velocity of the incident charged particle, ze is its charge, m_0 is the mass of the electron, N is the number of atoms per unit volume, and Z is the atomic number of the absorber.

More sophisticated calculations, taking into account relativistic effects, yield the following expression for the stopping power:

$$S(E) = -\frac{dE}{dx} = \frac{4\pi z^2 e^4 NZ}{m_0 v^2}\left[\ln\left(\frac{2m_0 v^2}{\bar{I}}\right) - \ln(1 - \beta^2) - \beta^2\right], \tag{13.4}$$

where $\beta = v/c$. Note that the above expressions do not contain the mass of the incident charged particle.

These expressions for stopping powers have been verified experimentally. When we know the stopping power, it is easy to calculate the range, using Eq. (13.1) for different charged particles in different materials.

Figure 13.1 shows the plots of dE/dx (in MeV/cm) versus E for different charged particles in air, and Fig. 13.2 shows the plots of range R versus energy E for different charged particles in aluminum.

In air, any charged particle loses, on the average, 30 eV of its energy in producing one electron–positive-ion pair. *Specific ionization* is the amount of ionization per unit length of the path of the incident beam. In most experimental work we want to know the thickness of an absorber that will be needed to stop the incident particles. Such a quantity is expressed in terms of equivalent thickness in units of mg/cm^2 (milligram/cm^2), defined as

$$\text{Equivalent thickness in mg/cm}^2 = \text{range} \times \text{density} \times 1000. \tag{13.5}$$

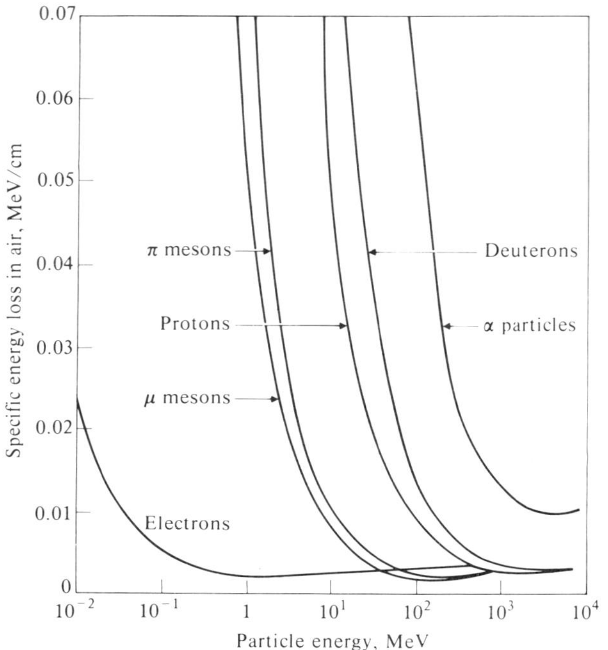

Fig. 13.1 Plots of dE/dx (in MeV/cm) versus E in air for different charged particles. [From Arthur Beiser, *Revs. Mod. Phys.* **24,** 273 (1952)]

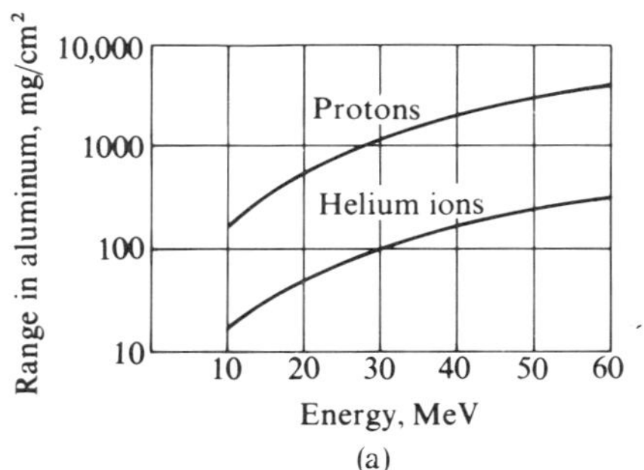

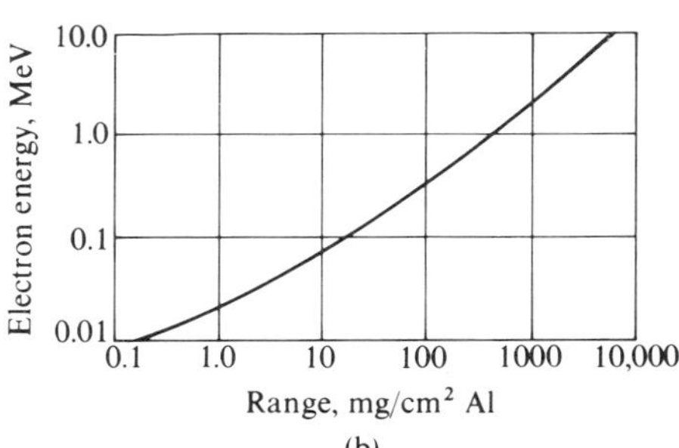

Fig. 13.2 Plots of range R versus energy E (a) for heavy charge particles and (b) for electrons in aluminum.

Example 13.1. What is the range of 5-MeV alpha particles in aluminum? (The relative stopping power of aluminum is 1600.) Assume that the range of alpha particles in air is given approximately by

$$R = 0.318E^{3/2}\ \text{cm}.$$

Since $E = 5$ MeV,

$$R = 0.318(5)^{3/2} = 3.56\ \text{cm}.$$

The relative stopping power of aluminum is given by

$$\text{RSP} = \frac{\text{Range in air}}{\text{Range in aluminum}}.$$

$$\text{Range in aluminum} = \frac{\text{Range in air}}{\text{RSP}} = \frac{3.56\ \text{cm}}{1600}$$

$$= 2.225 \times 10^{-3}\ \text{cm}.$$

The equivalent thickness according to Eq. (13.5) is:

$$\text{Equivalent thickness in mg/cm}^2 = \text{range} \times \text{density} \times 1000$$

$$= 2.225 \times 10^{-3}\ \text{cm} \times 2.7\ \text{g/cm}^3 \times 1000 = 6.1\ \text{mg/cm}^2.$$

---

b) *The interaction of electrons with matter.* The interactions of electrons with matter are much more complicated than the interactions of heavy charged particles with matter. In dealing with electrons, we have to take into account relativistic effects. And the fact that the incident electron *can* lose more than half its energy in a single collision with an atomic electron further complicates matters. We identify the electron which has more energy after the collision as the incident electron.

There are two main processes by which high-energy electrons lose their energy while passing through a medium.[5]

i) *Energy loss by inelastic collisions.* Taking into account relativistic correction, Eq. (13.4) for heavy charged particles takes the following form for fast-moving electrons:

$$\left(-\frac{dE}{dx}\right)_{\text{coll}} = \frac{2\pi e^4}{mv^2} NZ \left\{ \ln\left[\frac{mv^2 E}{2\bar{I}^2(1-\beta^2)}\right] - (2\sqrt{1-\beta^2} - 1 + \beta^2)\ln 2 + 1 - \beta^2 + \tfrac{1}{8}(1-\sqrt{1-\beta^2})^2 \right\}, \qquad (13.6)$$

where $\beta = v/c$ and E is the kinetic energy of the incident electron.

ii) *Energy loss by radiation* (*bremsstrahlung*). According to classical electromagnetic theory, an accelerated charged particle always gives out electromagnetic radiation. Thus, if a charged particle such as an electron or proton moves in the field of the nucleus, it is accelerated and radiates electromagnetic waves, called *Bremsstrahlung*.

The total energy loss by fast-moving electrons is given by

$$\left(-\frac{dE}{dx}\right)_{\text{total}} = \left(-\frac{dE}{dx}\right)_{\text{coll}} + \left(-\frac{dE}{dx}\right)_{\text{rad}}. \tag{13.7}$$

It is found that

$$\boxed{\frac{(dE/dx)_{\text{rad}}}{(dE/dx)_{\text{coll}}} \approx \frac{EZ}{1600\ mc^2},} \tag{13.8}$$

where $mc^2 = 0.51$ MeV is the rest-mass energy of the electron.

Figure 13.3 shows the rate at which electrons lose energy when they pass through lead. At high energies, the loss of energy due to radiation becomes much more important than the loss of energy due to collisions.

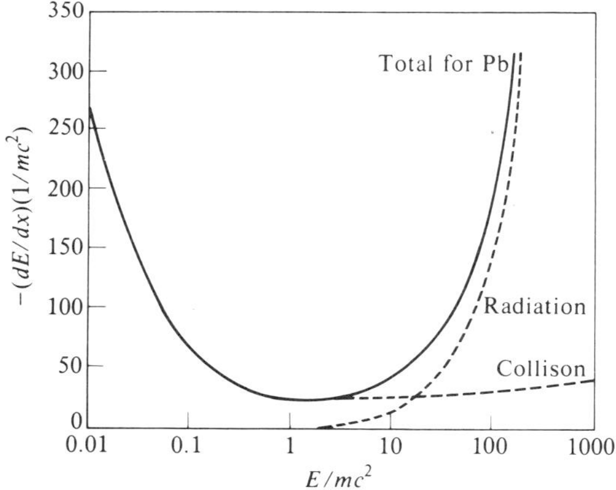

Fig. 13.3 Rates of energy loss by electrons versus E in lead. Note that energy is expressed in units of mc^2. [From W. Heitler, *The Quantum Theory of Radiation*, New York: Oxford University Press (1944)]

Compared to alpha particles or to heavy charged particles, electrons, given the same energy as a result of their collision, travel much longer distances in air or in any other medium, even though their paths are rather erratic. That is why thin metallic foils are commonly used to absorb electrons or beta particles. (Recall Fig. 13.2(b), with its plot of the range of electrons in aluminum.)

c) *The interaction of gamma rays with matter.*[6] The processes by which gamma rays lose energy when they interact with matter are different from those of heavy charged particles and electrons. The decrease in intensity ΔI in the beam of incident gamma rays is directly proportional to the incident intensity I and the thickness Δx of the material through which the gamma rays are traveling. That is,

$$\Delta I \propto -I\,\Delta x$$

or

$$\Delta I = -\mu I\,\Delta x, \tag{13.9}$$

where the constant of proportionality μ is the *linear absorption coefficient.* The negative sign means that I decreases as x increases. Integrating Eq. (13.9) with the condition that $I = I_0$ when $x = 0$ yields

$$\boxed{I = I_0 e^{-\mu x},} \tag{13.10}$$

where I is the intensity of the beam after it has passed through a thickness x of the material. Equation (13.10) means that the decrease in intensity of the beam varies exponentially with the thickness of the material; this implies that the intensity of the beam becomes zero only when x is infinite. Thus we speak of the *half-thickness* $x_{1/2}$, defined as that thickness of the absorber that will reduce the intensity of the incident beam to one-half its initial intensity. That is,

$$\frac{I}{I_0} = \frac{1}{2} = e^{-\mu x_{1/2}}$$

or

$$x_{1/2} = \frac{\ln 2}{\mu} = \frac{0.693}{\mu}. \tag{13.11}$$

Note that if one expresses the thickness x in centimeters, μ is given in cm^{-1}; while if one expresses x in g/cm^2, μ is expressed in cm^2/g.

Now let's look at the processes by which gamma rays lose energy to the absorber. Gamma rays interact with matter in many ways, but, fortunately, there are only three processes which cause significant decrease in intensity of the incident gamma-ray (or photon) beam. These three processes (which are effective in different energy ranges) are the following.

i) *The photoelectric effect* is dominant for photon energies between ~0.01 MeV and ~0.5 MeV. As we said in Section 2.4, in this process the incident photon is absorbed by one of the electrons of the atom. This electron is ejected from the atom with a kinetic energy K_e given by

$$K_e = h\nu - I_B, \tag{13.12}$$

where $h\nu$ is the incident photon energy and I_B is the binding energy of the orbital electron.

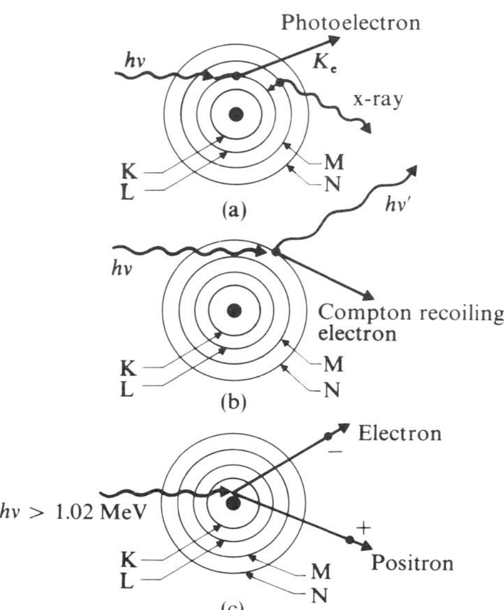

Fig. 13.4 The three processes by which gamma rays most often interact with matter. (a) Photoelectric process: The incident photon knocks out one of the inner electrons and the vacancy created is filled by an outer electron jumping in, with simultaneous emission of x-ray. (b) Compton process: The incident gamma ray interacts with one of the outer electrons and transfers a part of its energy to this electron. (c) Pair production: The incident photon converts into an electron–positron pair in the coulomb field of the nucleus.

Figure 13.4(a) illustrates this process. The vacancy created in the atom is filled by one of its outer electrons jumping in, with the simultaneous emission of a characteristic x-ray.

ii) *Compton scattering* is dominant for photon energies between ~0.1 MeV and ~10 MeV. As explained in Section 2.7, a photon of energy $h\nu$ collides with a free electron and loses part of its energy to this electron. This results in a photon being scattered at angle θ, with energy $h\nu'$ given by

$$h\nu' = \frac{h\nu}{1 + (h\nu/mc^2)(1 - \cos\theta)}. \tag{13.13}$$

Figure 13.4(b) illustrates this process schematically.

iii) *Pair production* starts with photons which have energy of 1.02 MeV, and increases with increasing photon energy. As we'll see in Section 13.3, in this process the incident photon disappears, and in its place an electron–positron pair is formed. That is, if $h\nu > 1.02$ MeV,

$$h\nu \doteq 2m_0c^2 + K_{e^-} + K_{e^+} \tag{13.14}$$

where K_{e^-} and K_{e^+} are the kinetic energies of the electron and positron, respectively. This process is illustrated schematically in Fig. 13.4(c).

Thus we may write the linear absorption coefficient μ as the sum of the three processes above:

$$\mu = \mu_\tau + \mu_\sigma + \mu_\kappa, \tag{13.15}$$

where μ_τ, μ_σ, and μ_κ are the absorption coefficients for the photoelectric, compton, and pair-production processes, respectively.

Figure 13.5 shows typical variations of these three, and their total μ versus the energy of the incident photons for different materials.

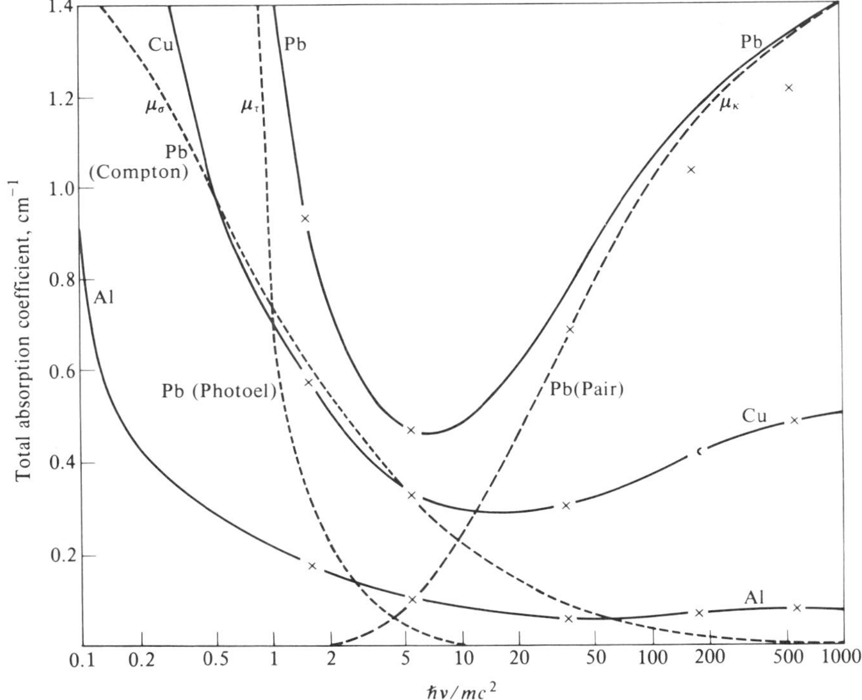

Fig. 13.5 Total absorption coefficient per centimeter versus energy of gamma rays for Al, Cu, and Pb. Also shown are plots of μ_τ, μ_σ, and μ_κ for Pb. [From W. Heitler, *The Quantum Theory of Radiation*, New York: Oxford University Press (1944)]

Example 13.2. The linear absorption coefficient μ of lead for 1-MeV gamma rays (from Fig. 13.5) is 0.74 cm^{-1}. Calculate (a) the half-thickness of lead for these gamma rays, and (b) the thickness of lead required to reduce the intensity of the gamma rays to $\frac{1}{1000}$ of its original value.

a) According to Eq. (13.11), the half-thickness is given by

$$x_{1/2} = \frac{\ln 2}{\mu} = \frac{0.693}{0.74\ \text{cm}^{-1}} = 0.94\ \text{cm}.$$

That is, lead which is 0.94 cm thick will reduce the intensity of the 2-MeV gamma rays to one-half its original value.

b) According to Eq. (13.10), $I = I_0 e^{-\mu x}$, and if $I/I_0 = \frac{1}{1000}$,

$$\tfrac{1}{1000} = e^{-\mu x}$$

or

$$x = \frac{\ln 1000}{\mu} = \frac{3 \ln 10}{0.74} = \frac{3(2.30)}{0.74},$$

$$x = 9.32\ \text{cm}.$$

That is, lead which is 9.32 cm thick is required to reduce the intensity of the 2-MeV gamma rays to $\frac{1}{1000}$ of its original value.

---

d) *Interaction of neutrons with matter.*[7] Even though neutrons are as heavy as protons, because they are neutral they do not interact with the atomic electrons. The magnetic moment of neutrons gives rise to only a very weak interaction. However, being neutral gives neutrons one advantage: They don't have to cross the coulomb barrier, so even slow neutrons can reach the nucleus without difficulty. The most important process by which neutrons are slowed down (and finally captured) is *scattering*, which may be elastic or inelastic. When neutrons experience inelastic collisions, the target nucleus is left in the excited state and eventually decays by emitting gamma rays. When neutrons experience elastic collisions with protons, it slows them down considerably. For example, a fast neutron, when it collides head-on with a proton in an elastic collision, may transfer all its energy to the proton. This is the reason for using paraffin to slow down beams of fast-moving neutrons. The slowed-down neutrons are easily captured by nuclei of the absorber; these nuclei de-excite by emitting gamma rays, which then interact as described above.

13.3 PAIR PRODUCTION AND PAIR ANNIHILATION

Pair production is an example of the conversion of electromagnetic energy into rest-mass energy, while pair annihilation is an example of the conversion of rest mass into electromagnetic energy. Furthermore, these processes confirm the relations of the special theory of relativity. So let's give these processes a closer look.

Pair production. When a photon of energy greater than 1.02 MeV strikes a heavy nucleus (or strikes a foil of high Z containing many heavy nuclei), the photon disappears, and in its place an electron–positron pair is formed. The positron is a particle having the same mass as the electron, but a positive charge of 1 unit. The conservation of energy requires that

$$h\nu = m_0^- c^2 + m_0^+ c^2 + K_- + K_+ + K_n,$$

where $h\nu$ is the energy of the incident photon, m_0^- and m_0^+ are the rest masses of the electron and the positron, K_- and K_+ are the kinetic energies of the electron and the positron, and K_n is the recoil energy of the nucleus. Since the mass of the electron is equal to that of the positron, $m_0^- = m_0^+ = m_0$, and K_n is almost negligible (because it is much heavier than either the electron or the positron). Hence we may write the above equation as

$$\boxed{h\nu \doteq 2m_0c^2 + K_- + K_+.} \tag{13.16}$$

Figure 13.6(a) shows pair production schematically. Note that if $K_- = K_+ = 0$, then $h\nu = 2m_0c^2 = 2(0.51 \text{ MeV}) = 1.02$ MeV. Thus, in order for a pair to be produced, a photon must have minimum energy equal to twice the rest-mass energy of the electron; that is, 1.02 MeV (or the wavelength of the photon should be 0.012 Å), which is the *threshold energy* for pair production. Figure 13.6(b) is a photograph of a cloud chamber, showing pair production.

We can show (Problem 15) that the presence of a heavy particle such as a nucleus is necessary for pair production to take place, because energy and momentum must be conserved. (Energy much higher than 1.02 MeV is needed to produce a pair if the recoiling particle is light, such as an electron; but there is no pair production in empty space.)

The existence of the positron was predicted theoretically by Paul A. M. Dirac[8] in 1930 and discovered experimentally by Carl D. Anderson[9] in 1932, when he examined a cloud-chamber photograph of cosmic radiation. [A note here about cosmic rays: High-energy radiation of unknown origin is constantly bombarding the earth's atmosphere. These cosmic rays contain photons of very high energies which produce electron–positron pairs. When observed under the influence of a magnetic field, the positrons exhibit paths that bend in a direction opposite to that of the electrons.]

Pair annihilation. What happens to positrons? Unlike electrons, they don't exist in free space. The process by which they are removed from circulation is called pair annihilation.

A positron keeps colliding with the surrounding atoms until its kinetic energy is almost zero. At this point it captures an electron from the medium and forms a *positronium atom* (see Fig. 13.7a). This atom is just like a hydrogen atom, except

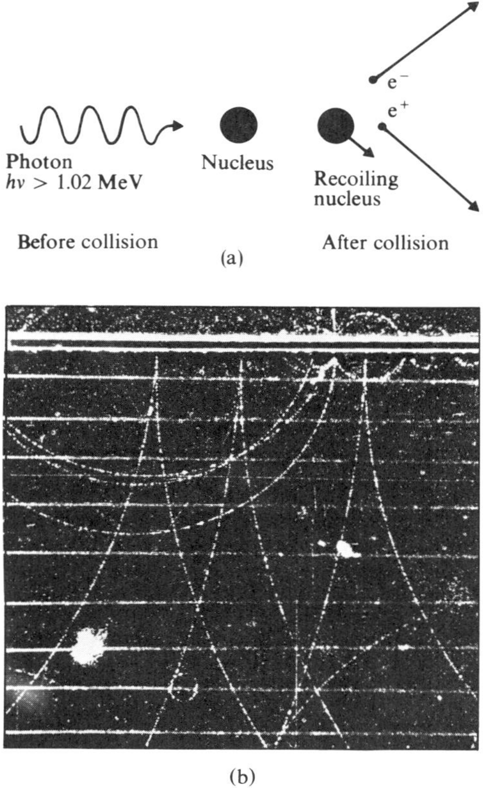

Fig. 13.6 (a) Schematic representation of pair production. (b) Cloud-chamber photograph showing production of electron–positron pairs. Gamma rays entering from top produce pairs while interacting with a lead sheet. [Courtesy of Radiation Laboratory, University of California]

that the proton (i.e., the hydrogen nucleus) has been replaced by a positron. The positronium atom, which is unstable, disappears in $\sim 10^{-10}$ sec, and in its place two photons appear (Fig. 13.7b). The conservation of linear momentum and energy requires that

$$\frac{h\nu_1}{c} = \frac{h\nu_2}{c} \tag{13.17}$$

and

$$2m_0c^2 = h\nu_1 + h\nu_2, \tag{13.18}$$

which implies that

$$h\nu_1 = h\nu_2 = m_0c^2 = 0.51 \text{ MeV}. \tag{13.19}$$

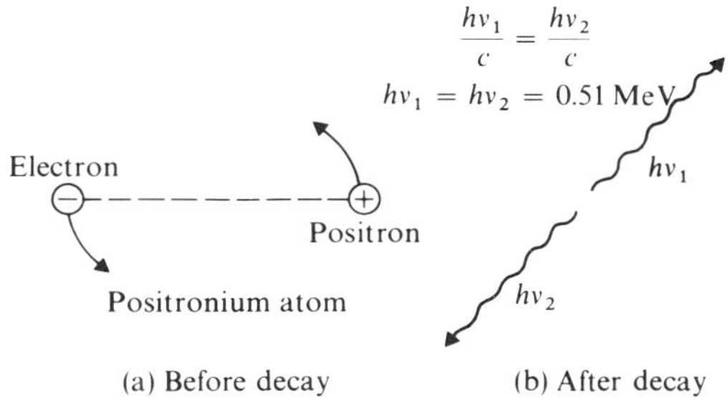

Fig. 13.7 (a) Formation of positronium atom, (b) annihilation of positronium atom.

Detecting 0.51-MeV photons emitted in opposite directions tells us that the electron–positron pair has annihilated into two photons.

The electron is called a *particle*, while the positron is called its *antiparticle*. Similarly, other particles, such as neutrons and protons, have their antiparticles—antineutrons and antiprotons—which have been produced in various laboratories (we'll get around to these in Chapter 15). Whenever a particle combines with its antiparticle, annihilation takes place, i.e., there is conversion of rest-mass energy into electromagnetic radiation. Our universe is made up of particles, and the antiparticles cannot exist for very long. Scientists are always postulating that there must be an antiuniverse somewhere, where antiparticles are stable, while particles are unstable!

We may explain pair production theoretically as follows. According to the theory of relativity, a particle of rest mass m_0 and momentum p has total energy E given by

$$E^2 = m_0^2c^4 + p^2c^2,$$

which means that

$$E = \pm\sqrt{m_0^2c^4 + p^2c^2}. \tag{13.20}$$

We assume that E is always $\geq m_0c^2$, and hence negative rest mass has no meaning. But according to Dirac,[8] either

$$E \geq m_0c^2 \qquad \text{or} \qquad E \leq -m_0c^2, \tag{13.21}$$

while E cannot have values between $-m_0c^2$ and $+m_0c^2$ (see Fig. 13.8). This span is called the *forbidden region*. The negative energy states ($E \leq -m_0c^2$) are assumed to be completely filled with electrons. But if any one of these electrons in a negative energy state is given an energy equal to or greater than $2m_0c^2 =$ 1.02 MeV, this raises the electron to a positive energy state $E \geq m_0c^2$, leaving behind an empty hole in the otherwise completely filled negative energy state. This hole is identified as a *positron* (Fig. 13.8).

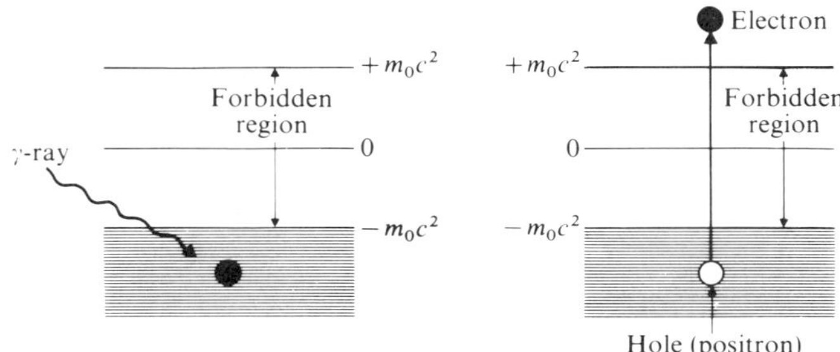

Fig. 13.8 Formation of an electron–positron pair according to hole theory given by Dirac.

13.4 DETECTORS OF NUCLEAR RADIATION

Nuclear radiation detectors, or devices for measuring nuclear radiation, have played an essential role in nuclear physics. Detectors used for detection and measurement of the energy of radiation are based on the interaction of radiation with matter (especially ionization and excitation). These detectors may be grouped as follows.

a) *Ionization chambers*
   - i) Simple ionization chamber
   - ii) Proportional counter
   - iii) Geiger–Müller counter

b) *Visual detectors* (for low energies)
   - i) Cloud chamber
   - ii) Diffusion chamber
   - iii) Nuclear emulsions (also used for high energies)

c) *Scintillation counters*

d) *Semiconductor detectors*

e) *High-energy detectors*
   - i) Bubble chamber
   - ii) Spark chamber
   - iii) Čerenkov detector

Which detector one chooses depends on the type of measurement. Modified forms of these detectors may often be used to fit a particular situation. Usually these detectors are used in combination with complicated electronic devices, which we shall not describe here. One cannot do justice to all these detectors in one text, so we shall give only a brief description of each with emphasis on those most recently developed and commonly used.

I. Ionization chambers[10,11]

The simplest ionization chamber (Fig. 13.9a) consists of a small volume of gas at atmospheric pressure contained in chamber *I*. Two electrodes *E* and *E′* placed inside the chamber are maintained at a high potential difference by means of a voltage source *V*. The incident radiation entering the chamber causes ionization, producing − and + ions, which are collected by plates *E* and *E′*. The voltage is kept high enough so that there is no appreciable amount of recombination. The charge collected results in an electrical current which may be measured by means of a current-measuring device, or it may be converted into an electrical pulse by the arrangement shown in Fig. 13.9(b). These pulses are amplified and then counted.

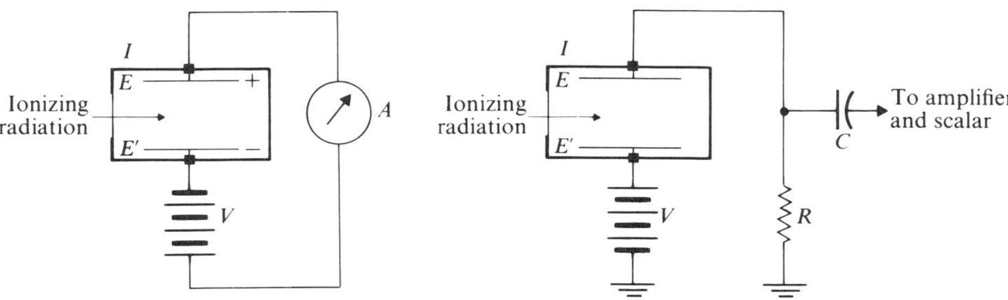

Fig. 13.9 (a) Ionization chamber containing a meter *A* measures ionization produced by a continuous beam of particles or x-rays. (b) Ionization chamber with a circuit arrangement such that the output is an electrical pulse; used for detecting individual particles.

Increasing the voltage *V* increases the relative number of ions collected, as shown in Fig. 13.10 for two different types of radiations—alpha particles and electrons. The detectors operating in region *C* are called proportional counters and those in region *D* are called Geiger counters. We can see that in the proportional-counter region the relative number of ions collected depends on voltage as well as type of radiation. The Geiger-counter region is almost flat; here the numbers of ions collected is independent both of the applied voltage and the type of radiation.

Physically the proportional counter is a modified form of an ionization chamber, as shown in Fig. 13.11. In the proportional counter, one electrode is a hollow cylinder and the other is a wire that runs inside the cylinder along its axis. The proportionality characteristic of this detector enables one to use it to differentiate between particles of varying ionization powers and energies. One big disadvantage is that the proportional counter needs a highly stabilized power supply to maintain the proportionality characteristic.

The Geiger–Müller counter (or GM counter, or simply Geiger counter[12]),

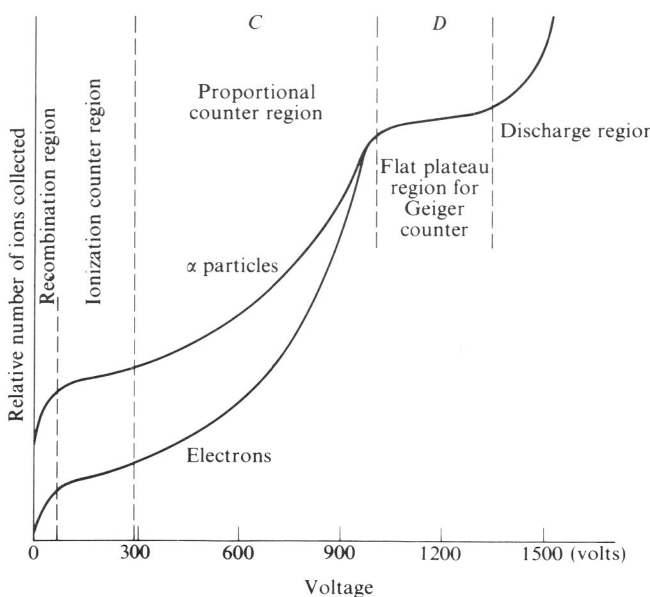

Fig. 13.10 Relative number of ions collected versus voltage V for a gas-filled detector of incident radiation consisting of α particles or electrons. Note the almost flat Geiger-counter region, which is independent of the type of radiation.

like the proportional counter, has a hollow metal-cylinder electrode and an electrode made of thin wire which operate as cathode and anode, respectively, enclosed in a thin glass tube (Fig. 13.11); but the voltage applied is such that the tube operates only in the Geiger region. The tube is filled to a pressure of about 10 cm of mercury with a mixture of 90% argon and 10% of some organic vapor (such as ethyl alcohol) or some halogen (such as Cl_2 and Br_2). The amplitude of

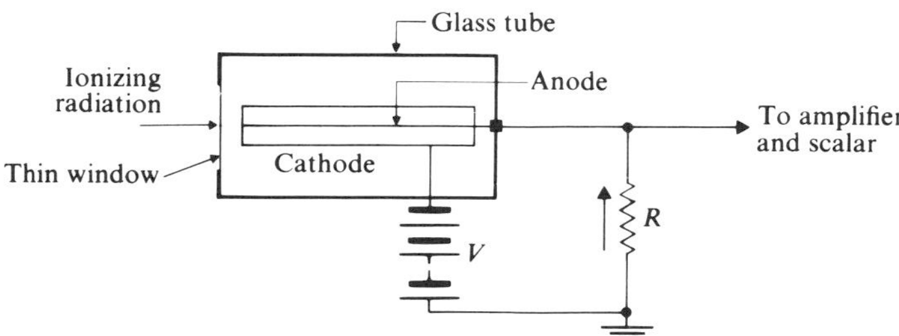

Fig. 13.11 Typical gas-filled detector with a thin mica window so that α or β particles may be counted. Depending on the gas pressure and the applied voltage, the detector may be operated as proportional counter or Geiger counter.

the output pulse signal is independent of the primary ionization, type, and energy of the particles (recall Fig. 13.10). This property makes the Geiger counter well-suited for the detection of individual particles. An individual charged particle passing through the gas ionizes a gas molecule. The central wire (Fig. 13.11) attracts the electron, while the cylindrical electrode attracts the positive ion. Because there is a very high electric field near the central wire, the approaching electrons gain high kinetic energy and cause further ionization of molecules. The multiplication of ions by this process continues until there is a complete discharge along the whole length of the electrode, which gives rise to a sudden pulse of current in the external circuit. This current creates a potential difference across a large resistor R, which in turn reduces the potential of the central wire, thereby extinguishing the discharge and getting the tube ready for the next event.

The flat portion of the Geiger region is called a *plateau*. Operating the counter in this region eliminates the need for a highly regulated (stabilized) power supply, which is the biggest advantage of the Geiger counter.

These detectors are commonly used to detect x-rays, low-energy gamma rays, and beta rays. That is why they are so widely used in prospecting for uranium, as well as in protecting personnel in atomic power plants, and similar applications.

Example 13.3. A gamma ray of energy 2 MeV produces an electron–positron pair, in which electron and positron move in opposite directions with equal speeds. The electron enters a helium-filled detector and loses all its kinetic energy in producing ion pairs. Calculate the number of ion pairs produced, given that the average ionization energy per ion pair in He is 42.6 eV.

According to Eq. (13.16), assuming $K_- = K_+ = K$,

$$h\nu = 2m_0c^2 + K_- + K_+ = 2m_0c^2 + 2K$$

or

$$K = \frac{h\nu - 2m_0c^2}{2} = \frac{2 \text{ MeV} - 2(0.51) \text{ MeV}}{2}$$

$$= 0.49 \text{ MeV} = 0.49 \times 10^6 \text{ eV}.$$

Thus the number of ion pairs produced in helium is

$$n = \frac{0.49 \times 10^6 \text{ eV}}{42.6 \text{ eV/ion pair}} = 13850 \text{ ion pairs}.$$

The total charge of either sign collected by the electrodes is $Q = ne$. Given that the capacitance of the detector is C, the size of the voltage pulse V from the detector will be

$$V = \frac{Q}{C} = \frac{ne}{C}.$$

If $C = 10\ \mu\mu\text{F} = 10 \times 10^{-12}$ F,

$$V = \frac{13850 \times 1.6 \times 10^{-19}\ \text{coulomb}}{10 \times 10^{-12}\ \text{F}}$$

$$= 2.08 \times 10^{-4}\ \text{volt.}$$

II. Cloud chambers

The cloud chamber or *expansion chamber*, invented by C. T. R. Wilson[13] in 1912, makes visible the paths of charged particles passing through it. The cloud chamber works on the principle that a supersaturated vapor condenses preferentially on charged particles. Figure 13.12 shows a cloud chamber, *C*, which is filled with dustfree air and saturated water vapor at room temperature. The piston *P* is allowed to fall freely, resulting in a sudden expansion of the mixture of air and water vapor, which leads to a fall in the temperature. This causes the mixture to become supersaturated. If at this moment a charged particle passes through the chamber, producing ion pairs, the supersaturated vapor condenses on the ions, thus leaving a trail of droplets along the path of the charged particle. One can easily see (and photograph) this path. (See, for example, Fig. 13.6b.)

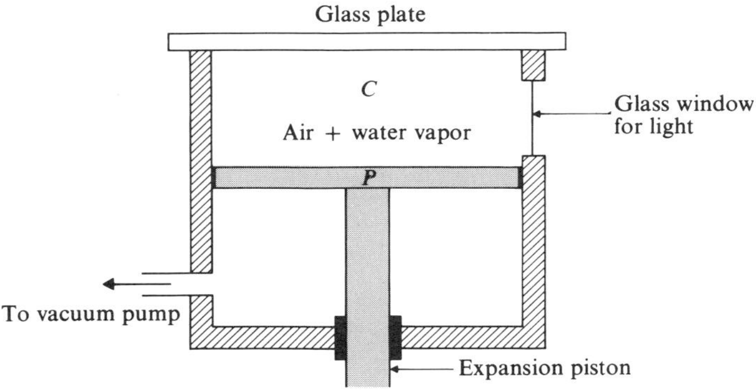

Fig. 13.12 A typical cloud chamber.

The cloud chamber is sensitive only for a very short time (~0.5 sec). There is, however, a diffusion chamber which is continuously sensitive. This device was designed by A. Langsdorf[14] in 1936. The reason it has not been used very often is that it still needs many improvements.

The cloud chamber does a good job of investigating charged particles which have low energy. But for particles with very high energy whose paths in air are very long (because of the low density of air), the cloud chamber is not much use. For particles such as protons, α particles, etc., with energies from 50 MeV to 20 BeV (1 BeV = 1000 MeV), the bubble chamber has been very useful.

III. Bubble chambers

In 1952, Donald A. Glaser,[15] recognizing the instability of superheated liquids, invented the bubble chamber, which takes advantage of the tendency of superheated liquids to form bubbles. Bubbles are formed preferentially along the paths of ion pairs. Similarly, as we have seen, a cloud chamber makes use of the instability of supersaturated vapors (which form droplets).

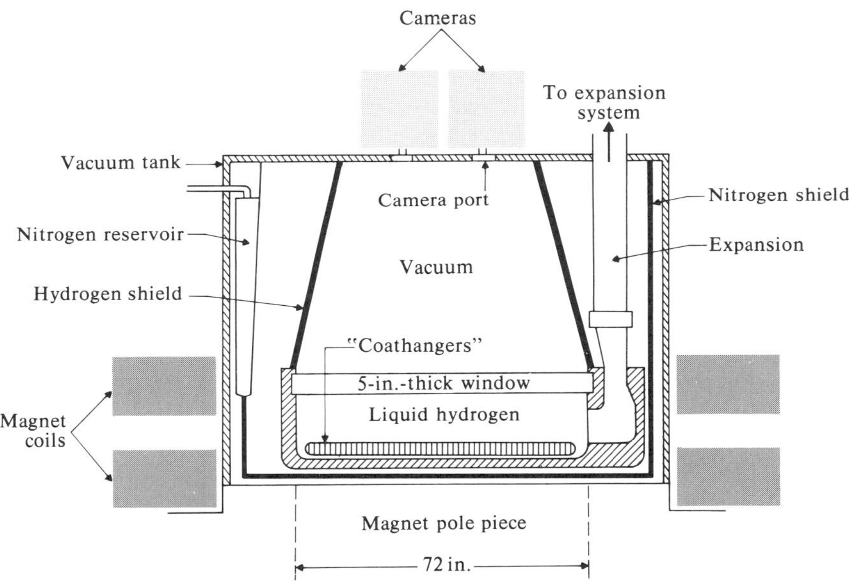

Fig. 13.13 A typical bubble chamber. [Reprinted with permission from D. V. Bugg, *Prog. Nucl. Phys.* **7,** 30 (1959), Pergamon Press]

Figure 13.13 gives an idea of a bubble chamber. The chamber is filled with a liquid such as hydrogen at −246°C, xenon at −20°C, or helium or deuterium. After being heated above its normal boiling point, the liquid is kept in its liquid phase by the application of external pressure. A sudden release of pressure leaves the superheated liquid in an unstable state for a certain length of time. If during this interval a charged particle passes through the chamber and produces ionization, bubbles form along the path of the particle. A photograph of this track may be taken before the boiling starts throughout the liquid. (Figures 15.7 and 15.8 are photograph taken in a bubble chamber of the tracks of charged particles.)

As we said before, because of the great density of the liquid, the bubble chamber is often used in high-energy physics to investigate the interactions of charged particles which have very high energies.

IV. Nuclear emulsions[16,17]

Nuclear emulsions provide permanent visual records of the paths of particles. An ionizing radiation passing through an emulsion affects silver halide grains embedded in a gelatin base. The emulsion is then photographically developed, which changes the affected silver halide grains into black grains of metallic silver. Silver halide grains that are not affected by the ionizing radiation are removed by treating the emulsion chemically in a fixing bath. The black grains of metallic silver therefore provide a permanent image of the paths of the charged particles. These paths can easily be seen with a microscope.

Improved emulsions, from 20 to 600μ thick and coated on glass plates, have been used in high-energy nuclear physics and in the study of cosmic rays, for observing paths of fast-moving particles.

V. Scintillation counters[18]

Such substances as sodium iodide, cesium iodide, anthracene and many others—when struck by a single charged particle, x-ray, or gamma-ray—produce a flash of light. They are called *scintillators.* (Some plastics are also used as scintillators.) When these light flashes are converted into electrical pulses and counted, the arrangement is called a *scintillation counter.*

Figure 13.14 shows how a scintillation counter works. A single crystal of NaI(Tl) (sodium iodide thallium activated) is used as a scintillator. This crystal, which is hygroscopic, is sealed into aluminum foil except for one side, which has optical light-tight contact with a photomultiplier tube. Magnesium oxide coated on the inside of aluminum serves to reflect light.

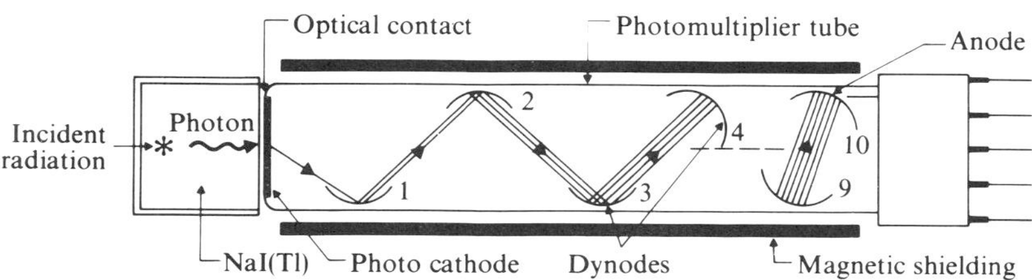

Fig. 13.14 Typical NaI(Tl) scintillation detector. The number of photoelectrons emitted is multiplied by successive dynodes, resulting in an electrical pulse at the anode.

A charged particle or a photon incident on the crystal causes ionization inside the crystal. The electrons produced during the ionization, when combined with the atoms and the molecules of the crystal, emit light in the visible region, with wavelengths from about 3300 Å to about 5000 Å. The NaI crystal is transparent to its own light. Therefore, when this light falls on the photocathode (made of a thin layer of cesium antimony alloy) of the photomultiplier, photoelectrons are emitted. The photomultiplier tube operates at a voltage of anywhere from 700 to 2000 volts, with successive increases in voltage on each dynode. Thus there is a successive multiplication of electrons as they travel from the photocathode to the last dynode (Fig. 13.14). The burst of electrons at the end is an electrical pulse whose size is proportional to the energy of the incident particle or photon. The pulses may be amplified and counted.

NaI(Tl) detectors are very efficient in detecting gamma rays, and are usually cylindrical, 2 × 2 inch or 3 × 3 inch, though very large sizes are also used. A crystal such as anthracene, which is not very sensitive to gamma rays, is used for detection of beta rays.

VI. Solid-state detectors[19,20,21]

In recent years semiconductor materials (discussed in Chapter 10) have been used in several different ways for detecting charged particles and gamma rays. For example, a *p-n* junction diode has been used for detection of alpha particles as explained below.

Let us consider a wafer of *n*-type silicon (made by introducing phosphorus into a silicon crystal as an impurity) and one of *p*-type silicon (made by introducing boron into a silicon crystal as an impurity). These two silicon wafers are fused together to form a *p-n* junction, as explained in Chapter 10. A contact potential is established between the two materials, the *n*-type being at a higher potential than the *p*-type. When the contact is made, the free electrons of the donor go to the acceptor, thus creating a nonconducting region, called the *depletion region* (see Fig. 13.15). One can increase the depletion region still further by applying an external voltage.

Incident radiation entering from the *n*-type side stops in the depletion region, producing free electrons and holes (positive ions). The electrons move toward the *n*-type (positive) layer and the holes toward the *p*-type (negative) layer, resulting in a potential drop across the junction. This potential drop is conveyed to the amplifier. The size of the pulse produced is proportional to the energy of the incident particles.

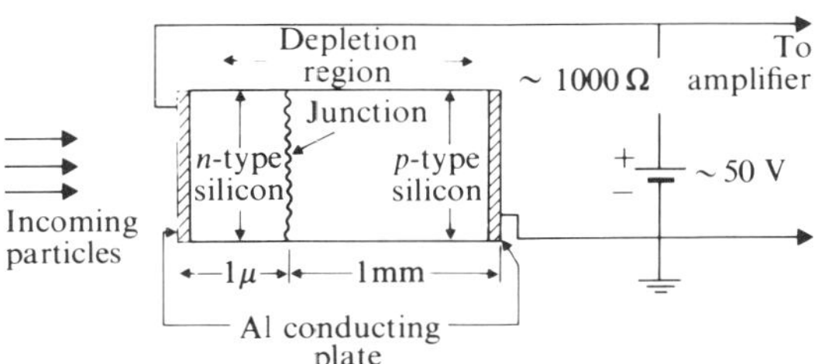

Fig. 13.15 Typical *p-n* junction detector, used to detect charge particles. [From Friedland, Mayer, and Wiggins, *Nucleonics* **18,** 2, 54 (1960)]

Note: The positive ions (the holes) do not move bodily, but contribute to the current by capturing electrons from the neutral atoms which are, say, on their right. A neutral atom which lost an electron to the hole is now a positive ion; i.e., in this process the hole has moved to the right, while the electron has moved to the left. The collection time of the electrons and of the holes is very short, less than 10^{-8} sec.

Such detectors have been used frequently in recent years to detect protons, alpha particles, and other charged particles.[21] If lithium is diffused into a silicon *p-n* junction, it increases the depletion region; such a detector is called a *lithium-drifted silicon*, Si(Li), detector. It is used to detect electrons and beta particles. A *lithium-drifted germanium*, Ge(Li), detector is used to detect gamma rays. The germanium must be kept at the temperature of liquid nitrogen (77°K) at all times, otherwise the lithium will diffuse out of the germanium.

Figure 13.16 shows the gamma spectrum of radioactive cobalt-60 recorded with both a NaI(Tl) crystal and a Ge(Li) detector. We can see that a semiconductor detector affords much better resolution (resolution here means the width of the peak at half-maximum) than a NaI(Tl) detector. However, the semiconductor detector is much *less* efficient in detecting gamma radiation than the NaI(Tl) detector is [its efficiency is about 1% of the efficiency of a NaI(Tl) detector].

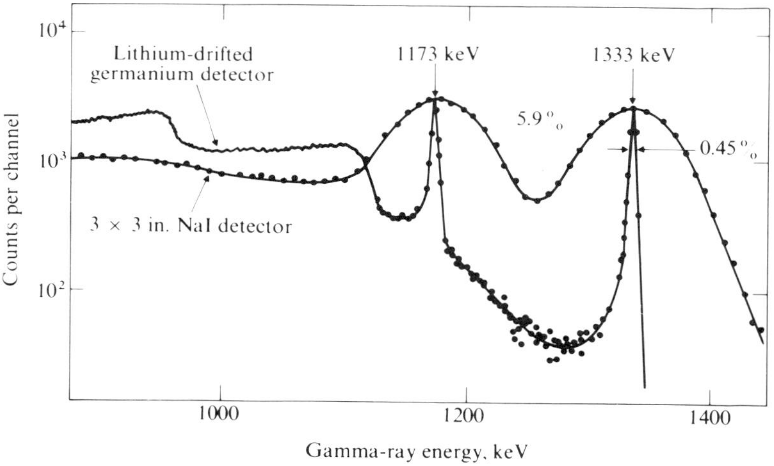

Fig. 13.16 Comparison of gamma ray spectra of ^{60}Co taken with a 3 × 3 in. NaI(Tl) detector and a lithium-drifted germanium detector (at the temperature of liquid nitrogen).

VII. Spark chamber[22,23]

In high-energy investigations, in addition to bubble chambers and nuclear emulsion plates, the spark chamber and Čerenkov detector[24] are also often used.

A typical spark chamber consists of anywhere from 6 to 128 thin metal plates parallel to each other (2 to 20 mm apart) and surrounded by an inert gas such as neon. A very high voltage, 10 to 15 kV is applied to alternate plates. A charged particle passing through the chamber leaves a trail of ionization behind it. The sparks are produced along the path of the ionization, which makes the tracks of the particles visible so that they may be photographed.

The Čerenkov detector makes use of the fact that a particle moving through a dielectric medium with a velocity greater than the phase velocity of light in that medium gives out a very weak visible light. The direction of this light depends on the velocity of the particle through the medium.

13.5 PARTICLE ACCELERATORS

The development of particle accelerators has not only provided nuclear physicists with beams of particles of higher and higher energies so that they can investigate the structure and inner universe of the nucleus, but it has opened a new field: *elementary particle physics* (see Chapter 15). An accelerator is to a nuclear physicist what a microscope is to a biologist and a telescope to an astronomer. Particle accelerators have been developed to such an extent that they may be considered as one branch of a service technology allied with nuclear physics.

Since the first accelerator was developed by John Cockcroft and Ernest Walton,[25] in 1932, many machines based on different principles have been built. Here is a list of them, together with the limits of the energies of the particles they have been able to produce. (One machine not included is the *betatron*,[26] used for obtaining intense beams of electrons.)

a) Direct-voltage accelerators
   1. Cockcroft–Walton accelerator: protons up to 2 MeV
   2. Van de Graaff generator: protons up to 10 MeV
   3. Tandem: protons up to 20 MeV

b) Resonance accelerators
   4. Cyclotron: protons up to 22 MeV

   Linear accelerators
   5. Proton accelerators: protons up to ~68 MeV
   6. Electron accelerators: electrons from 100 MeV to 20 BeV

c) Synchronous accelerators
   7. Synchrocyclotron: protons up to 750 MeV

   Synchrotrons
   8. Proton synchrotrons: protons up to 200 BeV
   9. Electron synchrotrons: electrons from 50 MeV to 1.2 BeV

d) Alternating-gradient accelerators
   Protons up to 30 BeV

e) Colliding-beam accelerators
   Protons with energy equivalent of 1500–100,000 BeV

Now let us deal with some of these particle accelerators one by one.

a) Direct-voltage accelerators

Cockcroft–Walton accelerator. The three main components of this one are (i) ion source, (ii) accelerating tube, and (iii) source of high-voltage supply. The first machine, built by Cockcroft and Walton,[25] used a voltage-multiplier circuit developed by Greinacher[27] as a source of high potential difference. It caused a

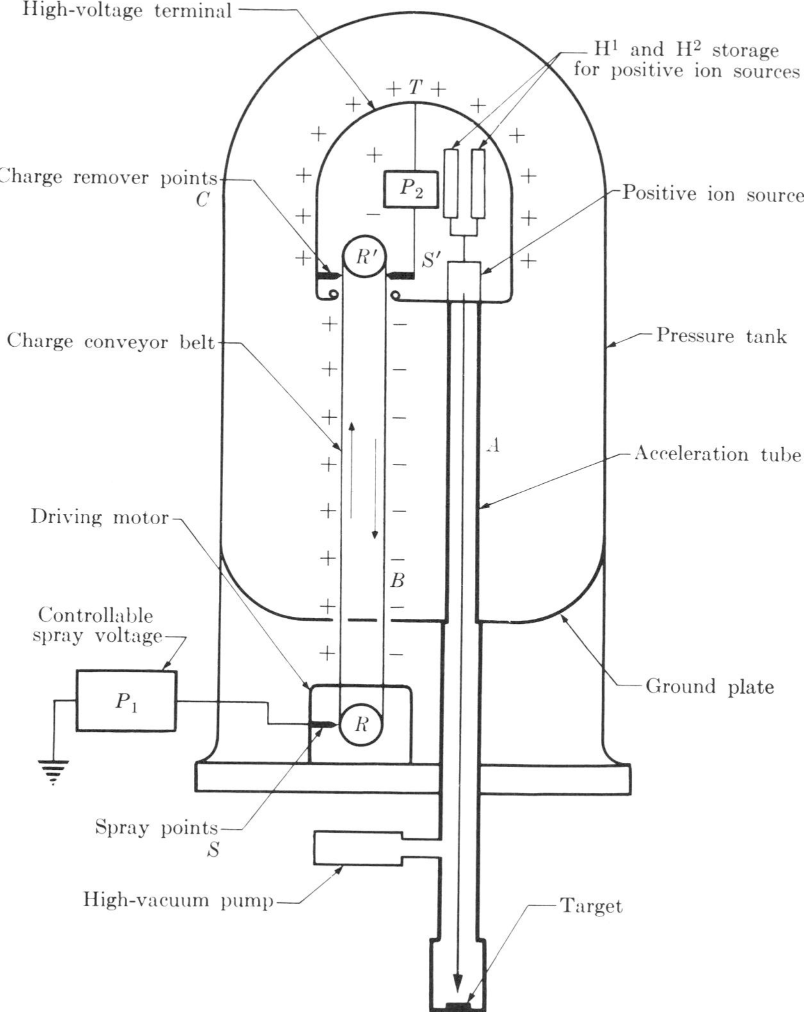

Fig. 13.17 Typical Van de Graaff accelerator. P_1 and P_2 are high-voltage power supplies, A is an accelerating tube, T is a spherical terminal of high voltage, B is the conveyor belt, S, S' are spray points, and C is a collector point.

particle of charge q to fall through a potential difference ΔV. The kinetic energy K of the particle is given by

$$K = q\,\Delta V. \tag{13.22}$$

In the first machine, they were able to accelerate protons up to energies of 800 keV. With further improvements, this limit has been raised to 2 MeV.

Van de Graaff Accelerator. This electrostatic generator was first built by R. J. Van de Graaff[28,29] in 1930. It works on the following principle. When a charged conductor is brought into a hollow conductor and makes an internal contact with it, all the charge on the first conductor is transferred to the second, irrespective of the charge and the potential the second conductor may have.

Figure 13.17 shows a typical Van de Graaff accelerator. Instead of charged bodies being inserted into a hollow conductor one after another, charge is carried in continuously by a belt B of insulated fabric running between two rollers, R and R'. There are two sets of sharp spray points S and S', called *corona points*, and a set of sharp collector points C. The sharp points, S, connected to the controllable spray voltage P_1, cause the ionization of the air and repel positive ions. The positive ions attach themselves to the belt, and when the belt reaches the collector points, C, the positive charges are collected by C. Then C conveys the charges to a sphere, which is the high-voltage terminal. A second set of spray points S' sprays negative charge on the belt, so that the charge-carrying capacity of the belt is doubled. The charge—and hence the potential—on the high terminal keeps on increasing as the belt keeps on transferring the charge. A limit is reached only when the insulation breaks down. The tank is filled with air or nitrogen, containing 3 to 10% freon at a pressure of $\sim$15 atm, so that it is possible to obtain a proton beam of energy 8 MeV and current up to about 100μ amperes.

The Tandem Van de Graaff accelerator[30] is an ingenious modification of the conventional Van de Graaff generator. Two insulating columns are used back to back in one pressure tank, so that the particles are accelerated to energies twice as high as those in conventional machines. The latest design of this machine (by High Voltage Engineering Corporation) is the so-called "Emperor" tandem Van de Graaff (Fig. 13.18), capable of accelerating protons to 20 MeV and attaining electrical currents of $\sim$1.5μ amperes.

b) Resonance accelerators

For years, because of electrical breakdown, electrostatic generators could not be used to accelerate particles to very high energies. But then the discovery of an altogether new principle—resonance—removed this limitation. According to this principle, particles are accelerated in steps by the repeated application of a relatively small voltage. Two types of accelerators now use resonance: *cyclic accelerators* (or *cyclotrons*) and *linear accelerators.*

Fig. 13.18 Tandem Van de Graaff accelerator model MP "Emperor," developed by High Voltage Engineering Corporation (by courtesy of High Voltage Engineering Corporation).

Cyclotron. Ernest Lawrence and M. Stanley Livingston[31] constructed the first cyclotron in 1932. Figure 13.19 is a sketch of a typical one. It consists of two hollow metal chambers D_1 and D_2 (called the *dees* because of their shapes), with open sides which are parallel and slightly displaced from each other. The dees are connected to the terminals of an alternating high-frequency voltage source which produces an alternating electric field in the gap between the dees. That is, when one dee is positive, the other is negative, and vice versa. Because of the electrical shielding effect of the dees, the electric field inside them is zero, and they are in an evacuated chamber from which they are insulated. The whole apparatus is placed between pole faces of a powerful electromagnet, which provides a strong magnetic field in a direction perpendicular to the cross-sectional area of the dees. The source which produces protons or heavy hydrogen (deuteron) ions is placed at S between the dees (see Fig. 13.19).

Suppose an ion of charge $+q$ and mass m is emitted from the source S at the instance when D_1 is positive and D_2 negative. The ion travels toward D_2, accelerated by the electric field, and enters the electric-field-free region in D_2. However,

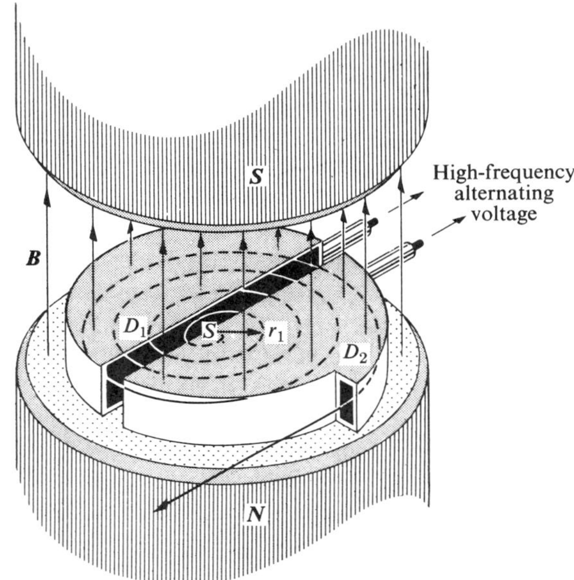

Fig. 13.19 A typical cyclotron.

the magnetic field makes the ion go into a circular path. If the frequency of the alternating voltage is such that when the ion is coming out of D_2 and ready to enter the gap again, D_1 is negative and D_2 positive, the ion is again accelerated while going toward D_1. This process of acceleration in steps continues each time the ion crosses the gap. As the velocity of the ion increases, the magnetic field makes it move into bigger and bigger orbits. At the end, the beam of ions is extracted from the chamber by means of a deflecting magnetic field.

We can calculate the frequency of the oscillating field by equating the magnetic force with the centripetal force. That is,

$$qvB = \frac{mv^2}{r}, \tag{13.23}$$

where B is the flux density of the magnetic field, r is the radius of the circular orbit, and v is the velocity of the ion. Thus the angular frequency ω and the frequency f are given by

$$\omega = \frac{v}{r} = \frac{qB}{m} \tag{13.24}$$

and

$$f = \frac{1}{2\pi}\left(\frac{q}{m}\right)B. \tag{13.25}$$

The ions attain maximum energy when $r = R$. That is,

$$K_{max} = \frac{1}{2} mv_{max}^2 = \frac{1}{2} \frac{(qBR)^2}{m}. \tag{13.26}$$

In a typical cyclotron, $B \sim 2$ webers/m^2, while the alternating voltage applied to the dees is $\sim$200 kV with a frequency of $\sim$10–12 megacycles per second. The maximum proton energy attainable in a cyclotron is $\sim$22 MeV.

As the energies of the ions increase, the relativistic change in mass becomes important and the resonance condition given by Eq. (13.23) does not hold. For this reason, a cyclotron cannot accelerate heavy ions to very high energies. For electrons, the relativistic effect becomes important at much lower energies, and hence a cyclotron cannot be used to accelerate electrons.

Example 13.4. The radius of a cyclotron dee is 0.4 m and the magnetic intensity is 1.5 weber/m^2. What is the maximum energy of a beam of protons?

According to Eq. (13.26), using the non-relativistic expression for K, we obtain

$$K_{max} = \tfrac{1}{2} mv_{max}^2 = \frac{1}{2} \frac{q^2B^2R^2}{m},$$

where m = mass of proton = 1.67×10^{-27} kg, $q = 1.6 \times 10^{-19}$ C, $B = 1.5$ weber/m^2 = 1.5 N/amp-m, and $R = 0.4$ m.

Therefore

$$\begin{aligned} K_{max} &= \frac{1}{2} \frac{(1.6 \times 10^{-19}\ \text{C})^2 (1.5\ \text{weber/m}^2)^2 (0.4\ \text{m})^2}{1.67 \times 10^{-27}\ \text{kg}} \\ &= 2.75 \times 10^{-12}\ \text{joule} \\ &= 2.75 \times 10^{-12} \times (1/1.602 \times 10^{-19}\ \text{joule/eV}) \\ &= 17.3 \times 10^6\ \text{eV} = 17.3\ \text{MeV}. \end{aligned}$$

The frequency of the alternating voltage applied to the dees of such a cyclotron, according to Eq. (13.25), is

$$\begin{aligned} f &= \frac{1}{2\pi} \left(\frac{q}{m}\right) B \\ &= \frac{1}{2\pi} \left(\frac{1.6 \times 10^{-19}\ \text{C}}{1.67 \times 10^{-27}\ \text{kg}}\right) (1.5\ \text{weber/m}^2) \\ &= \frac{1}{2\pi} \frac{1.6}{1.67} \times 10^8 \times 1.5 (\text{C/kg})(\text{N/amp-m}) \\ &= 2.28 \times 10^7\ \text{cycles/sec} = 22.8\ \text{megahertz}. \end{aligned}$$

For a deuteron, the (q/m) is half that for the proton, and hence f will be equal to 11.4 megahertz.

Linear accelerators. As the name indicates, particles are accelerated in straight-line paths, according to the resonance principle first suggested by G. Ising[32] in 1924.

Figure 13.20 diagrams a proton linear accelerator, also called a drift-tube accelerator, or *linac.* A number of cylindrical metal tubes, called *drift tubes*, are placed along the axis of a large vacuum tank. The adjacent tubes are connected to the opposite terminals of a radiofrequency oscillator. It works like this: When a positive ion leaves tube 1, tube 2 is negative, and the ion (proton) is accelerated toward it. Inside tube 2, the ion is electrically shielded, and hence moves with uniform velocity. The length of drift tube 2 is such that the positive ion takes the same time to travel the length of the tube as the oscillator takes to change its polarity, so as the ion leaves tube 2, tube 3 is negative. Thus the ion is accelerated again as it crosses the gap between tubes 2 and 3, and so on. As the ions are accelerated, they move faster and faster through the drift tubes. Because the frequency of the oscillator is fixed, in order to satisfy the resonance condition (i.e., so that the ions will cross each gap at the correct time), the drift tubes must be of increasing lengths. Once the ions have reached velocities near the velocity of light, the lengths of the drift tubes become almost constant.

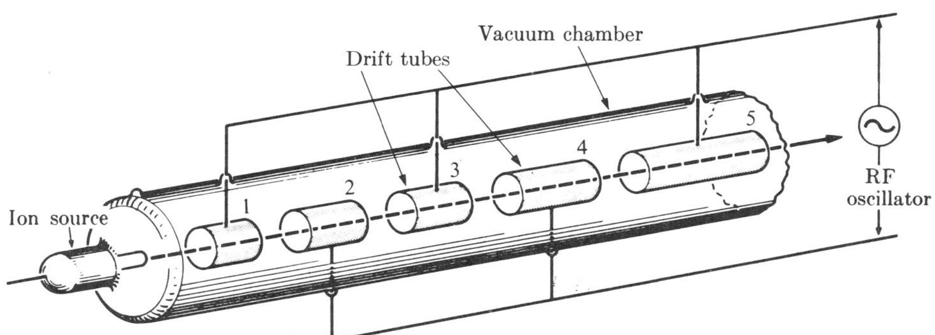

Fig. 13.20 A proton linear accelerator (also called linac or drift tube accelerator).

Figure 13.21 is an overall view of the proton linear accelerator (proton linac) at Brookhaven National Laboratory, on Long Island in New York, which is capable of boosting the energy of protons up to 50 MeV. The largest proton linac, which is at the University of Minnesota, produces 68-MeV protons.

The design of an electron linear accelerator is quite different from that of the proton linear accelerator, since electrons reach the velocity of light at very small energies (for electrons, at 2 MeV, $v = 0.98c$). The design of an electron linear accelerator is based on the traveling wave established inside the resonant cavities, as contrasted to standing waves in the proton linac. Drift tubes are eliminated

Fig. 13.21 The proton linear accelerator (proton linac) at Brookhaven National Laboratory, capable of boosting the energy of protons up to 50 MeV. (By courtesy of Brookhaven National Laboratory)

in the electron accelerator and its overall size is small. The electron accelerators at Stanford University and in the USSR both produce 100-MeV electrons and have lengths of 79 m and 100 m, respectively. The two-mile linear accelerator at Stanford, called SLAC (Stanford Linear Accelerator Center) has an overall length of 3000 m ($\sim$ 2 miles) and produces a 20-BeV beam of electrons.

c) Synchronous accelerators

If we take into account the variation of mass with velocity, the resonance condition for cyclotrons given by Eq. (13.25) takes the form

$$f = \frac{1}{2\pi} \frac{q}{m_0} \sqrt{1 - \frac{v^2}{c^2}}\, B. \tag{13.27}$$

This means that, if particles are to be accelerated to very high energies, in order to maintain the resonance condition one has to either decrease the applied frequency or increase the magnetic field. The application of these methods to accelerate particles to very high energies was made possible by the discovery (by

E. McMillan[33] in the United States and V. Veksler[34] in the USSR) of the principle of *phase stability*. It was pointed out that phase stability may be achieved by slowly decreasing f or increasing B, or both. Machines in which B is kept constant while f is varied are called *synchrocyclotrons*, or *frequency-modulated* (FM) *cyclotrons*. Machines in which B is varied while f may or may not be varied are called *synchrotrons*. In electron synchrotrons, f is kept constant and B varied, while in proton synchrotrons, both f and B are varied.

Of the existing two dozen FM cyclotrons in the world,[35] the machine at Berkeley, California, produces 750-MeV protons, the one at Dubna (USSR) produces 680-MeV protons, and the machine at CERN (the European Organization for Nuclear Research in Geneva, Switzerland) produces 600-MeV protons.

Several electron synchrotrons have been constructed which can produce beams of electrons from 50 MeV to ~1.2 BeV.

A typical proton synchrotron,[36–38] also called a *cosmotron*, located at Brookhaven National Laboratory, is diagrammed in Fig. 13.22. Figure 13.23 is an actual photograph of it. This machine has a magnet which weighs 2000 tons and has a maximum diameter of 75 feet. The magnetic field changes from 300 gauss to

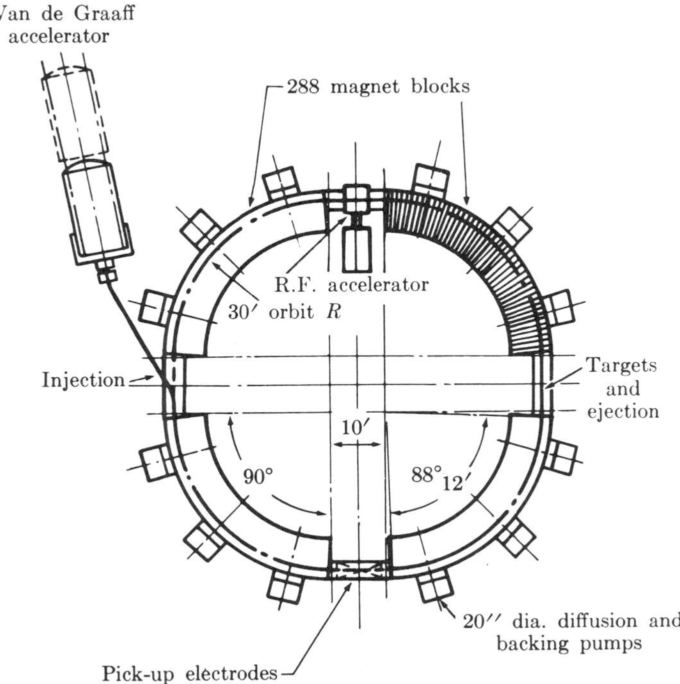

Fig. 13.22 A proton synchrotron, also called a *cosmotron* at Brookhaven. [From M. S. Livingston, J. P. Blewett, G. K. Green, and L. J. Haworth, *Rev. Sci. Int.* **21**, 7 (1950)]

Fig. 13.23 The cosmotron at Brookhaven National Laboratory, which can accelerate protons up to 3 BeV. (By courtesy of Brookhaven National Laboratory)

14,000 gauss, while the radiofrequency source is modulated from 0.37 megacycle to 4.20 megacycles per second. After making 3 million revolutions, the machine obtained protons with energies of 3 BeV. The proton synchrotron at Dubna, USSR produces 10-BeV protons.

d) Alternating-gradient accelerators

Synchrotrons use constant-gradient magnetic fields. The principle of fixed field but alternating gradient was proposed by Ernest Courant, M. Stanley Livingston, and H. Snyder[39] at the Brookhaven National Laboratory in 1952. The AG synchrotron uses a sequence of magnet sectors shaped in such a way that the gradient of the magnetic field is stronger on one side and weaker on the other. The net effect of this alternating gradient is strong focusing, so that the particles oscillate with much smaller amplitude about their equilibrium orbit. This permits one to use smaller magnets and vacuum chamber, thereby making it economically feasible to construct synchrotrons with much larger orbits for obtaining much higher energies.

The first AG proton synchrotron went into operation at CERN,[38] near Geneva, Switzerland, in 1959. Scientists at CERN, starting with 50-MeV pre-accelerated protons, directed them into the AG synchrotron and produced a beam of 28-BeV protons. In 1960, the AG synchrotron at Brookhaven[38] produced a 30-BeV proton beam. Figure 13.24 is a diagram of this synchrotron. The protons were pre-accelerated to 50 MeV by means of a proton linac. The diameter of the orbit is 842.9 ft (circumference $\sim\frac{1}{2}$ mile). Each of the 240 units of the main

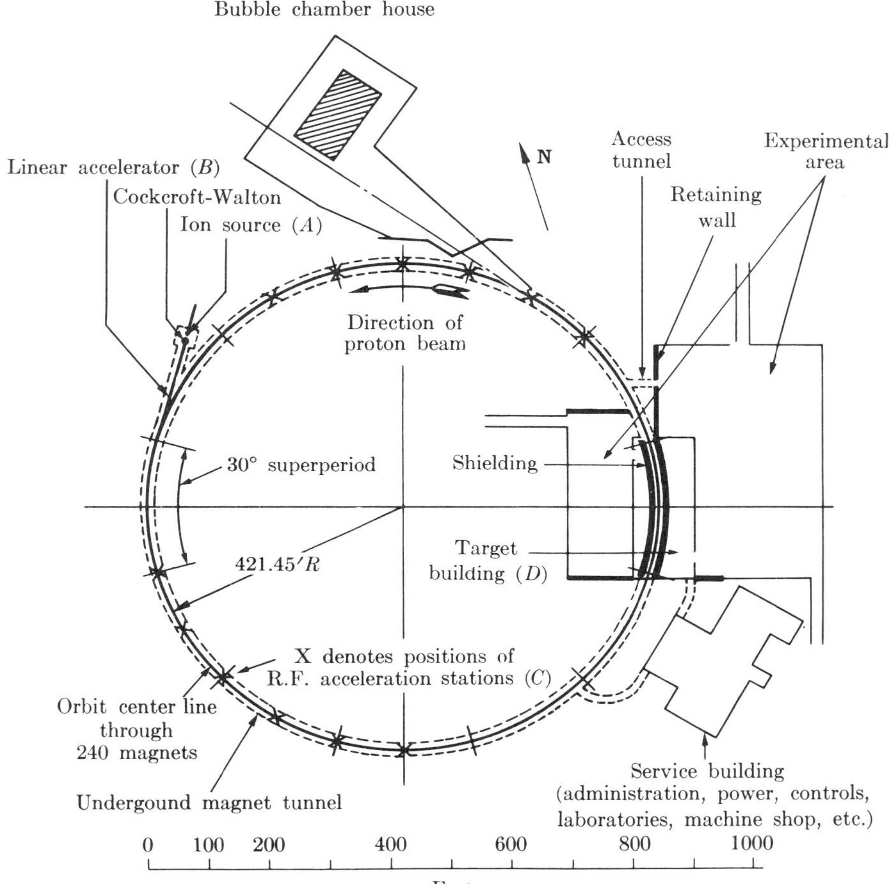

Fig. 13.24 Diagram of the alternating-gradient synchrotron (AGS) at Brookhaven National Laboratory.

magnet weighs 60 tons. The proton synchrotron operating at Serpukhov in the USSR produces 70-BeV protons; while the proton synchrotron currently in operation at the National Accelerator Laboratory in Batavia, Illinois, is rated at 200 BeV.

e) Colliding beam-accelerators

Physicists are always trying to create machines with still higher energy. When the proton synchrotron under construction at Batavia, Illinois, is completed (at a cost of $250 million), it will be able to provide a beam of energy up to 500 BeV.

Scientists at CERN have put one imaginative idea to work: the so-called *colliding beams*. Their machine, based on *intersecting storage rings* (ISR), produces beams with the highest energy achievable so far, and was built at a cost of only \$80 million. Figure 13.25 is a diagram of this accelerator. CERN's synchrotron, after accelerating protons to 28 BeV, deflects them with powerful magnets into two large concentric rings. The protons go alternately in clockwise and counter-clockwise directions in the interlaced vacuum tunnels. When the two opposing beams of protons, each having energy of 28 BeV, collide, their collision produces energy equivalent to that yielded by a conventional accelerator of 1500 BeV. The only problem is that the intensity is extremely low. (Note that in a conventional accelerator producing 100-BeV protons, only about 15 BeV is useful energy. The rest of the energy goes into pushing the target particles aside.) Physicists at Brookhaven, using this colliding-beam technique, hope to achieve energies equivalent to 100,000 BeV!

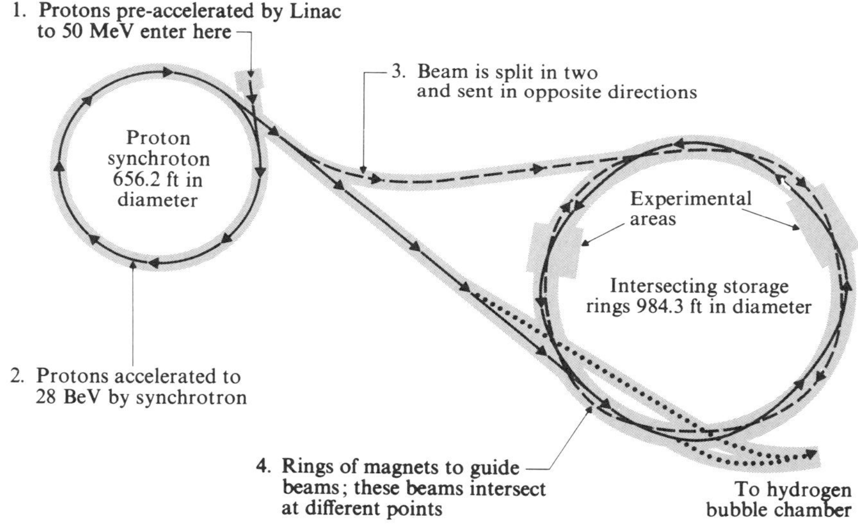

Fig. 13.25 Principle of operation of colliding-beam accelerator.

SUMMARY

Heavy charged particles traveling through a medium are stopped when they lose energy on account of ionization. The *average ionization energy* $\bar{I}$ is the average energy needed to cause excitation or ionization of an atom of the medium. The *mean range* is the distance a particle travels through a medium from its source to the point at which its kinetic energy is almost zero. *Stopping power* is the amount of energy lost per unit length by the incident particle in a given medium; that is, $S(E) = -dE/dx$. *Specific ionization* is the amount of ionization per unit length of the path of the incident particle. *Equivalent thickness* is the product of range and density.

Fast-moving electrons lose energy mainly by two processes: inelastic collisions and radiation (bremsstrahlung). It is found that

$$\frac{(dE/dx)_{\text{rad}}}{(dE/dx)_{\text{coll}}} \approx \frac{EZ}{1600mc^2}.$$

The intensity of a beam of gamma rays after passing through thickness x is $I = I_0 e^{-\mu x}$, where μ is the linear absorption coefficient. *Half-thickness*, $x_{1/2}$, is the thickness of the absorber that will reduce the intensity of the incident beam to one-half.

Three factors contribute significantly to the decrease in intensity of incident photons: the photoelectric effect, the Compton effect, and pair production. Thus the total linear absorption coefficient μ is the sum of μ_τ, μ_σ, and μ_κ.

Neutrons, which cause nuclear reactions, are slowed down by both elastic and inelastic scatterings.

When a photon of energy greater than 1.02 MeV strikes a heavy nucleus, the photon disappears. In its place an electron–positron pair is formed. In this process of formation, called pair production, $h\nu \simeq m_0^- c^2 + m_0^+ c^2 + K_- + K_+$.

The positron, which is the antiparticle of the electron, is taken out of circulation by a process called *pair annihilation*, in which the positron, after being slowed down, captures an electron, thus forming a *positronium* atom. In $\sim 10^{-10}$ sec, the positronium atom decays into two photons, of energy 0.51 MeV each, emitted in opposite directions.

According to the hole theory, an electron cannot have E values between $-m_0c^2$ and $+m_0c^2$. This region is called the *forbidden region*.

The ionization chamber, proportional counter, and Geiger–Müller counter work on the principle of the collection of positive ions and electrons by electrodes. These devices, which differ mainly in the voltages at which they operate, are used to detect the presence of beta rays and low-energy gamma rays.

A cloud chamber operates on the principle that a supersaturated vapor condenses preferentially on charged particles. In a cloud chamber the paths of the particles become visible. Because of the low density of the medium, the cloud chamber is used to detect low-energy particles.

Nuclear emulsions, which are used to detect particles with low as well as high energies, provide permanent records of the paths of these particles.

A bubble chamber makes use of the instability of superheated liquids and their tendency to form bubbles. That is, bubbles are formed preferentially along the paths of ion pairs. Because of high density of the liquids used, the bubble chamber is used for high-energy particles.

In a scintillation counter, commonly used for detection of α, β, and γ rays, a flash of light produced by incident radiation is converted into an electrical pulse.

Solid-state detectors use modified forms of p-n junctions. These detectors have excellent resolution for charged particles as well as gamma rays.

In a spark chamber a high voltage is applied to alternate thin metal plates placed very close and parallel to each other. Sparks are produced along the ionization path of the particle passing through the chamber. The spark chamber and the Čerenkov detector are used only for high-energy particles.

Direct voltage accelerators can produce protons up to 20 MeV.

Resonance accelerators can produce protons up to 68 MeV and electrons from 100 MeV to 20 BeV.

Synchronous accelerators can produce protons up to 200 BeV and electrons from 50 MeV to 1.2 BeV.

Alternating-gradient accelerators produce protons up to 30 BeV.

Colliding-beam accelerators can provide protons with energy equivalents from 1500 BeV to 100,000 BeV.

PROBLEMS

1. Calculate the stopping power of air for a 10-MeV proton from Eq. (13.3) and compare this with the value given in Fig. 13.1. Explain the discrepancy.
2. Compare the stopping power for electrons, protons, and alpha particles, all moving with a speed of $0.50c$ in air at standard temperature and pressure.
3. Calculate (from Fig. 13.2) the thickness of Al foil needed to stop 40-MeV alpha particles, and 10-MeV alpha particles.
4. 20-MeV protons and alpha particles pass through an aluminum foil 0.001 cm thick. What are their energies after they pass through the foil?
5. Calculate (from Fig. 13.2) in terms of mg/cm^2 the thickness of an absorber needed to stop 10-MeV alpha particles, 10-MeV protons, and 10-MeV electrons.
6. For alpha particles of energies between 4 MeV and 7 MeV, an empirical expression for range in air is given by
$$R\ (\text{cm}) = 0.31E^{3/2} \qquad \text{for } E \text{ in MeV.}$$
a) What is the range of a 10-MeV α-particle in air?
b) What is its range in terms of mg/cm^2?
7. The range of a certain group of β^- particles in aluminum is 1755 mg/cm^2. From the empirical relation given below, calculate the energy of the beta group.
$$R\ (\text{mg/cm}^2) = 530E - 106.$$
8. For $Z = 50$, what should be the value of electron energy for which $(dE/dx)_{\text{rad}} = (dE/dx)_{\text{coll}}$? Repeat this for $Z = 20$.

9. What will be the percentage error in the range of a 4-MeV electron if we use the empirical relation $R = 412E^{1.265-0.09 \ln E}$ for $E < 2.5$ MeV instead of $R = 530E - 106$ for $E > 2.5$ MeV?

10. How thick should a sheet of lead be to reduce to one-half the intensity of 1-MeV gamma rays? The absorption coefficient of lead for 1-MeV γ-rays is 0.75 cm^{-1}.

11. How thick should a lead sheet be (in terms of absorption coefficient) to reduce the intensity of a monoenergetic gamma ray of 1 MeV energy to $\frac{1}{100}$ of its initial intensity?

12. What thickness of lead (in centimeters) is needed to reduce the intensity of a monoenergetic γ ray whose energy is 1 MeV to $\frac{1}{1000}$ of its initial value? (The absorption coefficient is 0.75 cm^{-1})

13. Show that when a neutron of energy E_0 collides head-on with a nucleus of mass number A at rest, the energy of the inelastically scattered neutron will be minimum (i.e., the loss is maximum and the neutron has been slowed down), and is given by

$$E_{\min} = E_0 \left(\frac{A - 1}{A + 1}\right)^2.$$

What will the energy of the neutron be after the first collision, the second collision, and the nth collision, given that the target nuclei are hydrogen (in the form of water), carbon, and iron?

14. Using the results of previous problem, calculate the number of collisions needed to reduce the 2-MeV neutron to a 0.1-MeV neutron by elastic collisions between the neutron and carbon atoms.

15. Show that the presence of a heavy nucleus is necessary in order to conserve linear momentum in pair production. [*Hint:* Say $E_{e^-} = E_{e^+} = h\nu/2$ and then show that $p_{e^+} + p_{e^-} \neq h\nu/c$. You may also do this graphically by plotting E versus p.]

16. In pair production, under what conditions will the electron and the positron carry away the maximum amount of momenta?

17. An α particle is stopped in an ionization chamber, in which it produces 150,000 ion pairs. Each time the α particle produces an ion pair, it loses 35 eV of energy. What is the kinetic energy of the α particle? Calculate the amount of charge collected by each plate.

18. Calculate the rate at which charge is collected at the plates of an ionization chamber filled with air when α particles of energy 5 MeV are entering at the rate of 100/sec. Assume that each α particle loses all its energy in the chamber, and that 35 eV of energy are needed to produce an ion pair.

19. In a Geiger counter, the inner wire is of radius a, the outer cylinder is of radius b, and the voltage applied between the electrodes is V. Show that the electric field E at a distance r from the center is

$$E = \frac{V}{r \ln (b/a)}.$$

20. Using the results of Problem 19, calculate E for $a = 0.02$ cm, $b = 5$ cm, $V = 1200$ volts, and for $r = 0.1$ cm, 1 cm, and 3 cm.

21. A radioactive source placed inside a cloud chamber emits 1-MeV electrons. What should the applied magnetic field be so that the radius of curvature of the path of the electrons is 10 cm?

22. A radioactive source placed inside a cloud chamber emits 5-MeV α particles. What should the applied magnetic field be so that the radius of curvature of the path of the alpha particles is 10 cm?

23. Show that the relativistic mass of a particle of rest mass m_0 and charge q, which has been accelerated through a potential difference of V volts, is given by

$$m = m_0(1 + kV),$$

where k is a constant expressed in terms of m_0, q, and c.

24. The sphere of an electrostatic generator has a capacity of 250 $\mu\mu$F and operates at 0.4 MeV. Calculate the amount of charge it carries.

25. Calculate the power needed to drive the belt of a Van de Graaff generator machine used for a 1-MeV deuteron beam of 100μA.

26. Protons are being accelerated in a 40-in. cyclotron which has an oscillator of frequency 8 megacycles. Calculate the magnetic field needed and the maximum energy to which the protons may be accelerated.

REFERENCES

1. M. G. Holloway and M. S. Livingston, *Phys. Rev.* **54,** 18 (1938)
2. Niels Bohr, *Phil. Mag.* **24,** 10 (1913); **30,** 581 (1915)
3. Hans Bethe, *Ann. Physik* **5,** 325 (1930); *Z. Physik* **76,** 293 (1932)
4. F. Bloch, *Ann. Physik* **16,** 285 (1933)
5. Hans Bethe and J. Ashkin, *Experimental Nuclear Physics*, edited by E. Segré, Vol. I, New York: John Wiley, 1952
6. C. M. Davisson and R. D. Evans, *Revs. Mod. Phys.* **24,** 79 (1952)
7. B. T. Feld, *Experimental Nuclear Physics*, Vol. II, New York: John Wiley, 1953
8. P. A. M. Dirac, *Proc. Roy. Soc.* **126,** 360 (1930); **133,** 61 (1931)
9. C. D. Anderson, *Phys. Rev.* **43,** 491 (1933)
10. D. R. Carson and R. R. Wilson, *Rev. Sci. Inst.* **19,** 207 (1948)
11. B. B. Rossi and H. H. Staub, *Ionization Chambers and Counters*, New York: McGraw-Hill, 1949
12. H. Geiger and W. Müller, *Z. Physik* **29,** 839 (1928); **30,** 483 (1929)
13. C. T. R. Wilson, *Proc. Roy. Soc.* **87,** 277 (1912); **104,** 1 (1923)
14. A. Langsdorf, *Phys. Rev.* **49,** 422 (1936)
15. D. A. Glaser, *Phys. Rev.* **91,** 762 (1953)
16. M. M. Shapiro, *Revs. Mod. Phys.* **13,** 58 (1941)

17. J. Rotbalt, *Prog. Nucl. Phys.* **1,** 37 (1950)
18. K. Siegbahn, editor, *Alpha- Beta- and Gamma-Ray Spectroscopy*, Amsterdam: North Holland Publishing Company, 1968
19. P. J. Van Heerden, *The Crystal Counter*, University of Utrecht, Dissertation, 1945
20. D. E. Wooldridge, A. J. Akearn, and J. A. Burton, *Phys. Rev.* **71,** 913 (1947)
21. S. S. Friedland, J. W. Mayer, and J. S. Wiggins, *Nucleonics* **18,** 2, 54 (1960)
22. G. K. O'Neill, *Sci. Am.*, August 1962, page 37
23. E. F. Beall, B. Cork, P. G. Murphy, and W. A. Wenzel, *Nuovo Cimento* **20,** 502 (1961)
24. J. V. Jelley, *Prog. Nucl. Phys.* **3,** 84 (1953)
25. J. D. Cockcroft and E. T. S. Walton, *Proc. Roy. Soc.* **A137,** 229 (1932); **A136,** 619 (1932)
26. D. W. Kerst, G. D. Adams, H. W. Koch, and C. S. Robinson, *Rev. Sci. Inst.* **21,** 462 (1950)
27. H. Greinacher, *Z. Physik* **4,** 195 (1921)
28. R. J. Van de Graaff, *Phys. Rev.* **38,** 1919 (1931)
29. R. J. Van de Graaff, J. J. Trump, and W. W. Buechner, *Rev. Prog. Phys.* **XI,** 1 (1946)
30. R. J. Van de Graaff, *Nucl. Inst. and Meth.* **8,** 195 (1960)
31. E. O. Lawrence and M. S. Livingston, *Phys. Rev.* **40,** 19 (1932)
32. G. Ising, *Arkiv Mat. Astron. Fysik* **18,** No. 30, 1 (1924)
33. E. M. McMillan, *Phys. Rev.* **68,** 143 (1945)
34. V. I. Veksler, *J. Physics*, USSR, **9,** 153 (1945)
35. M. S. Livingston, ICSU, *Review of World Science* **6,** 44 (1964)
36. M. S. Livingston, J. P. Blewett, G. K. Green, and L. J. Haworth, *Rev. Sci. Instr.* **21,** 7 (1950)
37. M. H. Blewett, *Rev. Sci. Instr.* **24,** 725 (1953)
38. Proceedings of International Conference on High-Energy Accelerators and Instrumentation of 1959, Geneva, Switzerland
39. E. D. Courant, M. S. Livingston, and H. S. Snyder, *Phys. Rev.* **88,** 1190 (1950)

Suggested Readings

1. E. Segré, editor, *Experimental Nuclear Physics*, Vol. III, Parts IX, X, and XI, New York: John Wiley, 1959
2. Atam P. Arya, *Fundamentals of Nuclear Physics*, Chapters VII, VIII, IX, III, and XIV, Boston: Allyn and Bacon, 1966
3. H. Staub, *Detection Methods in Experimental Nuclear Physics*, edited by E. Segré, Vol. I, Part I, New York: John Wiley, 1953
4. J. W. Price, *Nuclear Radiation Detection*, New York: McGraw-Hill, 1958

5. M. J. Morant, *Introduction to Semiconductor Devices*, Reading, Mass.: Addison-Wesley, 1964
6. M. S. Livingston and J. P. Blewett, *Particle Accelerators*, New York: McGraw-Hill, 1962
7. M. S. Livingston, editor, *The Development of High-Energy Accelerators*, New York: Dover Publications, 1966

CHAPTER 14
NUCLEAR REACTIONS

Fuel assembly apparatus for commercial nuclear power reactor. This unit contains 204 fuel rods with 41,412 uranium fuel pellets. (Courtesy Westinghouse Electric Company)

14.1 INTRODUCTION

Suppose that you bombard a target of a given material with fast-moving particles such as protons, neutrons, electrons, deuterons, or α particles. Depending on the type of target material and the energy and nature of the incident particles, many different things can happen when the incident particles come close enough to interact with the target nuclei. The incident particle may simply change direction; or it may lose some of its energy; or it may be completely absorbed by the nucleus. Perhaps particles which are altogether different may be knocked out of the nucleus, or the incident particle may be captured and a gamma ray emitted. The target nuclei after the bombardment are usually different from what they were before. They may change mass number or atomic number, or both. Such a change in the target nucleus resulting from its interaction with an incident particle is called a *transmutation*. The process itself is called a *transmutation reaction* or *nuclear reaction*.

Every nuclear reaction has a reaction equation which, in a typical case, may be written as

$$\mathrm{x} + \mathrm{X} \rightarrow \mathrm{Y} + \mathrm{y}, \tag{14.1}$$

which states that an incident particle x interacts with the target nucleus X leading to a nuclear reaction, the outcome of which is a recoil nucleus Y and an outgoing particle y. This nuclear reaction may be written in short as X(x, y)Y.

Ernest Rutherford,[1] in 1919, triggered the first nuclear reaction when he bombarded a nitrogen target with α particles obtained from a natural radioactive source. The reaction produced oxygen as a recoil nucleus and a fast-moving proton as an outgoing particle. That is,

$${}^{4}_{2}\mathrm{He} + {}^{14}_{7}\mathrm{N} \rightarrow {}^{17}_{8}\mathrm{O} + {}^{1}_{1}\mathrm{H}, \tag{14.2}$$

which we may write in short form as ${}^{14}_{7}\mathrm{N}(\alpha, \mathrm{p}){}^{17}_{8}\mathrm{O}$. In 1930 John Cockcroft and Ernest Walton[2] performed the following reaction for the first time, using artificially accelerated protons (p or ${}^{1}_{1}\mathrm{H}$):

$${}^{7}_{3}\mathrm{Li} + \mathrm{p} \rightarrow {}^{4}_{2}\mathrm{He} + \alpha \tag{14.3}$$

or

$${}^{7}_{3}\mathrm{Li}\ (\mathrm{p}, \alpha)\ {}^{4}_{2}\mathrm{He}.$$

Today the list of nuclear reactions has grown into thousands. In most experiments with a given type of incident particle, more than one type of reaction take place. The recoiling nucleus is often unstable (or radioactive), and usually becomes stable by emitting beta and/or gamma rays. The outcome of the nuclear reaction also depends on the energy of the incident particles. For energies $\gtrsim$ 100 MeV, the product of the nuclear reaction may contain more than one outgoing particle. Let us limit our discussion of nuclear reactions to those involving incident particles with energies of less than 100 MeV.

We shall investigate nuclear reactions in detail so that we can obtain information about such properties of nuclei as size, charge distribution, and nuclear forces. In this chapter we shall tackle the subject from three points of view.

a) *Nuclear reaction dynamics.* By applying the laws of conservation of momentum and energy, we shall find the condition necessary for the nuclear reaction to be energetically possible.

b) *Nuclear cross section, σ.* To determine quantitatively the rate of nuclear reactions, we must know the probability that an incident particle will cause a nuclear reaction; i.e., we must know the *cross section* of the nucleus.

c) *Mechanism of nuclear reactions.* We shall investigate theories that enable us to explain nuclear reactions, both qualitatively and quantitatively.

14.2 CONSERVATION LAWS IN NUCLEAR REACTIONS

A bombarding particle of rest mass m_x and kinetic energy K_x is incident on a target containing nuclei of rest mass M_X and kinetic energy K_X. The collision results in the following nuclear reaction:

$$x + X \rightarrow Y + y. \tag{14.1}$$

The recoil nucleus Y of rest mass M_Y is emitted with kinetic energy K_Y. A light particle y of rest mass m_y is also emitted, with kinetic energy K_y, as shown in Fig. 14.1. The law of conservation of energy requires that the total initial energy E_i, which is the sum of the rest-mass energies and kinetic energies, must be equal to the total final energy, E_f. That is,

$$E_i = E_f \tag{14.4}$$

or

$$m_x c^2 + K_x + M_X c^2 + K_X = M_Y c^2 + K_Y + m_y c^2 + K_y. \tag{14.5}$$

This may be written as

$$[(K_Y + K_y) - (K_X + K_x)] = [(M_X + m_x)c^2 - (M_Y + m_y)c^2] \tag{14.6}$$

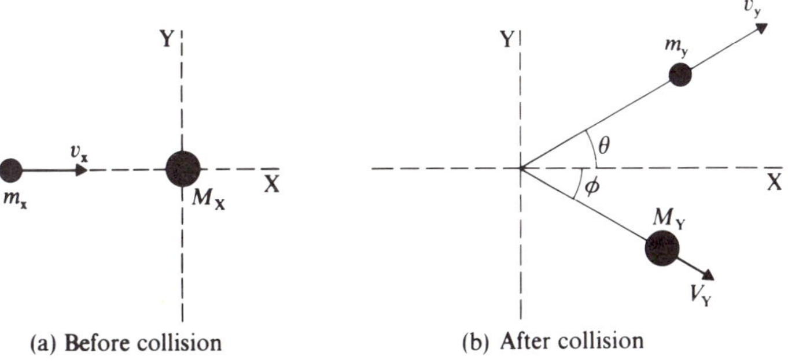

Fig. 14.1 (a) Incident particle of mass m_x and velocity v_x approaching a target nucleus of mass M_X at rest. (b) Outgoing particle of mass m_y makes an angle θ with the direction of the incident particle, and the recoiling nucleus of mass M_Y makes an angle ϕ.

or

$$K_f - K_i = M_i c^2 - M_f c^2, \tag{14.7}$$

where K_i (= $K_X + K_x$) and K_f (= $K_Y + K_y$) are the total initial and final kinetic energies and M_i (= $M_X + m_x$) and M_f (= $M_Y + m_y$) are the total initial and final rest masses, respectively. Equation (14.6) or (14.7) states that the net increase in kinetic energy is equal to the net decrease in rest-mass energy. The *disintegration energy* or *Q-value* of a nuclear reaction is a net change in the kinetic energy which is also equal to the net change in the rest-mass energies. That is,

$$\boxed{Q = K_f - K_i = M_i c^2 - M_f c^2.} \tag{14.8}$$

The nuclear reaction is an *exoergic reaction* if the Q-value is positive, in which case $K_f > K_i$ or $M_i c^2 > M_f c^2$. The nuclear reaction is an *endoergic reaction* if the Q-value is negative, in which case $K_f < K_i$ or $M_i c^2 < M_f c^2$.

If the target nucleus is at rest, which it usually is, its kinetic energy $K_X = 0$. Hence the Q-value of a reaction is given by

$$\begin{aligned} Q &= (K_Y + K_y) - K_x \\ &= [(M_X + m_x) - (M_Y + m_y)]c^2. \end{aligned} \tag{14.9}$$

Thus, knowing K_x, K_y, and K_Y, we can find the Q-value of the nuclear reaction. Since the mass of the recoiling nucleus Y is very large compared to that of the light particle y, the kinetic energy K_Y of Y is extremely small, and hard to measure accurately. By using the law of the conservation of linear momentum, we can calculate the Q-value that is independent of K_Y. Since we are concerned with low energies, we use nonrelativistic expressions.

As Fig. 14.1 shows, the particle y is emitted with velocity v_y at an angle θ, and the recoiling nucleus Y is emitted with a velocity V_Y at an angle ϕ. The conservation of linear momentum in the plane of the paper yields the equations

$$m_x v_x = m_y v_y \cos\theta + M_Y V_Y \cos\phi, \tag{14.10}$$

$$0 = m_y v_y \sin\theta - M_Y V_Y \sin\phi. \tag{14.11}$$

Rearranging gives

$$M_Y V_Y \cos\phi = m_x v_x - m_y v_y \cos\theta, \tag{14.12}$$

$$M_Y V_Y \sin\phi = m_y v_y \sin\theta. \tag{14.13}$$

Squaring and adding these equations, we obtain

$$M_Y^2 V_Y^2 = m_x^2 v_y^2 + m_y^2 v_y^2 - 2m_x m_y v_x v_y \cos\theta. \tag{14.14}$$

Using the kinetic-energy relations,

$$K_x = \tfrac{1}{2} m_x v_x^2, \qquad K_y = \tfrac{1}{2} m_y v_y^2, \qquad K_Y = \tfrac{1}{2} M_Y V_Y^2 \tag{14.15}$$

we get the following expression for K_Y:

$$K_Y = \frac{m_x K_x}{M_Y} + \frac{m_y K_y}{M_Y} - \frac{2}{M_Y}(m_x m_y K_x K_y)^{1/2} \cos\theta. \tag{14.16}$$

Substituting this expression for K_Y into Eq. (14.9) yields

$$Q = K_y\left(1 + \frac{m_y}{M_Y}\right) - K_x\left(1 - \frac{m_x}{M_Y}\right) - \frac{2}{M_Y}(m_x m_y K_x K_y)^{1/2} \cos\theta. \tag{14.17}$$

Note that the above expression for the Q-value is independent of (a) the mass of the target nucleus, and (b) the kinetic energy of the recoiling nucleus.

Example 14.1. The Q-value of a given reaction $^{19}F(n, p)^{19}O$ is -3.9 MeV, and the energy of the incident neutrons is 10 MeV. What is the energy of the emitted protons that are observed at an angle of 90° to the direction of the incident neutrons?

For $\theta = 90°$, $\cos 90° = 0$, Eq. (14.17) reduces to

$$Q = K_y\left(1 + \frac{m_y}{M_Y}\right) - K_x\left(1 - \frac{m_x}{M_Y}\right),$$

where

$Q = -3.9$ MeV

K_x = kinetic energy of incident neutron = 10 MeV

m_x = mass of incident neutron = 1.0087 u

m_y = mass of emitted proton = 1.0078 u

M_Y = mass of ^{19}O nucleus = 18.99 u

Thus, from the above equation, we derive the following:

$$K_y = \frac{Q + K_x(1 - (m_x/M_Y))}{(1 + (m_y/M_Y))} = \frac{-3.9\text{ MeV} + 10\text{ MeV}(1 - (1.0087/18.99))}{(1 + (1.0078/18.99))}$$

$$= \frac{-3.9\text{ MeV} + 9.47\text{ MeV}}{1.053} = \frac{5.57\text{ MeV}}{1.053} = 5.3\text{ MeV}.$$

It is obvious from the definition of the Q-value in Eq. (14.9) that, for an exoergic reaction, Q is positive and the final kinetic energy is the sum of Q and K_x. If $K_x = 0$, $K_Y + K_y = Q > 0$. That is, an exoergic reaction is energetically possible even if the bombarding particle has zero kinetic energy.

Let us consider endoergic reactions: If $K_x = 0$, $K_Y + K_y = Q < 0$. That is, the final kinetic energy is negative, which is physically not possible! Hence K_x cannot be equal to zero. $K_Y + K_y$ is positive only if $K_x \geq |Q|$. If $K_x = |Q|$,

$K_Y + K_y = 0$. That is, the reaction products are produced at rest. But this is not possible, because initially the momentum of the incident particle is not zero, while the final momentum of the system is zero. This would mean that there would be nonconservation of linear momentum.

Thus, for the endoergic reaction to be energetically possible, energy K_x greater than $|Q|$ is needed. The minimum kinetic energy of the incident particle which will initiate an endoergic reaction is called the *threshold energy* for an endoergic reaction. We can calculate the threshold energy by using the center-of-mass coordinate system (CMCS). For any collision or reaction in CMCS, the linear momentum is always zero before and after the reaction. Suppose that K_x' is the kinetic energy of the incident particle in CMCS. The endoergic reaction is energetically possible only if

$$K_x' \geq |Q|. \tag{14.18}$$

But $K_x' = \frac{1}{2}m_{red}v_x^2$, where m_{red} is the reduced mass of the incident particle and the target nucleus, that is, $m_{red} = m_x M_X/(m_x + M_X)$. Thus

$$\frac{1}{2}\frac{m_x M_X}{M_X + m_x}\, v_x^2 \geq |Q| \tag{14.19}$$

or

$$\tfrac{1}{2}m_x v_x^2 \geq \left(\frac{M_X + m_x}{M_X}\right)|Q|. \tag{14.20}$$

$$\boxed{\text{Threshold energy} = (K_x)_{min} = \left(1 + \frac{m_x}{M_X}\right)|Q|.} \tag{14.21}$$

Thus the threshold energy—the minimum energy needed for endoergic reaction to take place—is greater than the Q-value by a factor of $(1 + m_x/M_X)$.

Example 14.2. In Example 14.1, the Q-value of the reaction $^{19}F(n, p)^{19}O$ is -3.9 MeV. What is the threshold energy for this reaction? The Mass of ^{19}F is 18.9984 u.

According to Eq. (14.21),

$$\text{Threshold energy} = (K_x)_{min} = \left(1 + \frac{m_x}{M_X}\right)|Q|$$

$$= \left(1 + \frac{1.0087\ \text{u}}{18.9984\ \text{u}}\right) \times 3.9\ \text{MeV}$$

$$= (1 + 0.053) \times 3.9\ \text{MeV} = 4.1\ \text{MeV}.$$

That is, the reaction will not start until the energy of the incident neutrons is 4.1 MeV.

14.3 CROSS SECTIONS

Even though a nuclear reaction may be possible energetically, every particle incident on a target doesn't necessarily produce a nuclear reaction. As a matter of fact, only a very small fraction of the beam of incident particles produces nuclear reactions. We can't say which incident particle will produce a nuclear reaction and which will not. In the field of microscopic physics, theories do not predict certainties. We speak of probabilities of things happening. In the case of nuclear reactions, we want to know what the probability is that a single incident particle will interact with an individual target nucleus (interact means that particles will be scattered, absorbed, or cause any other type of nuclear reaction). The probability of such a reaction occurring is called the *cross section* or *microscopic cross section*, also called the *nuclear cross section* σ. If we know the cross section, we can calculate the fraction of the incident beam that will be lost while the incident particles are passing through the target.

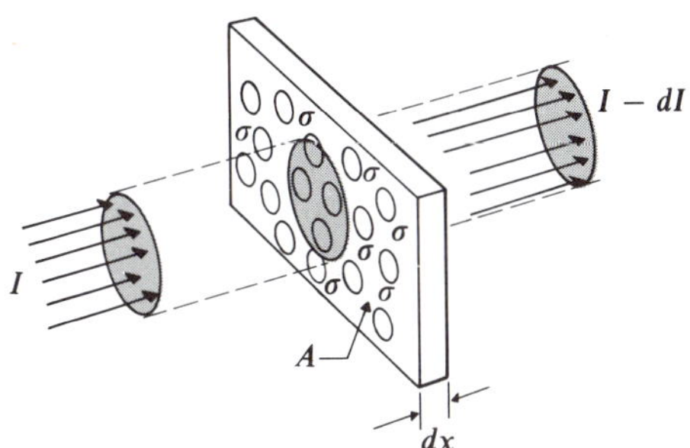

Fig. 14.2 A beam of intensity I is incident on a target of thickness dx. Each small circle indicates a sensitive area surrounding each nucleus. If an incident particle falls within this area, a nuclear reaction will take place.

A beam of monoenergetic particles of intensity I (*intensity* is the number of particles crossing a unit area per second) is incident on a target in the form of a thin foil of thickness dx and face area A (Fig. 14.2). With each target nucleus there is associated a *sensitive area* σ such that if the incident particle hits the nucleus within this area, the particle is absorbed or scattered. That is, there is a nuclear reaction. [*Note:* σ is just a fictitious area introduced to help you grasp this idea. It has nothing to do with the geometrical cross-sectional area, πR^2, of the target nucleus.] Assume that the foil is so thin that the cross-sectional area σ surrounding any nucleus in the target is not shadowed by the nuclear cross section of another target nucleus. If there are n nuclei per unit volume of the target foil, there will be $n\,dx$ nuclei per unit area of the face of the foil, and

$$n\,dx\,A = \text{total number of nuclei in face area } A.$$

Since with each nucleus there is associated a sensitive area σ, therefore

$$n\,dx\,A\sigma = \text{total sensitive area.} \tag{14.22}$$

The fractional sensitive area f is given by

$$f = \frac{\text{total sensitive area}}{\text{total face area}} = \frac{n\,dxA\sigma}{A} = n\sigma\,dx. \tag{14.23}$$

But this fractional sensitive area must be equal to the fractional change in the intensity of the beam as it goes through the foil. That is,

$$f = -\frac{dI}{I} = n\sigma\,dx, \tag{14.24}$$

where dI is the change in intensity. The negative sign means that I decreases as x increases.

Applying the condition that $I = I_0$ at $x = 0$, we integrate Eq. (14.24) to obtain

$$\boxed{I = I_0 e^{-n\sigma x},} \tag{14.25}$$

which expresses the intensity I of the beam after it has passed through a foil of thickness x. Since the number of particles in a beam is proportional to the intensity of the beam, we can write Eq. (14.25) as

$$\boxed{N = N_0 e^{-n\sigma x},} \tag{14.26}$$

where N_0 is the number of particles incident on the foil and N is the number left after the beam has passed through a foil of thickness x.

[We have assumed that the foil is very thin, which is true only so long as either σ is small or x is small, so that $e^{-n\sigma x} \simeq 1 - n\sigma x$, and Eq. (14.25) takes the form $\Delta I = I - I_0 = -I\,n\sigma x$, which is the same as Eq. (14.24).]

The unit of microscopic cross section is a *barn*, denoted by b and defined as

$$1\ \text{b} = 10^{-24}\ \text{cm}^2.$$

Its subunit, the millibarn, 1 mb, is

$$1\ \text{mb} = 10^{-3}\ \text{b}.$$

Another quantity of interest is the *macroscopic cross section*, Σ, defined as the product of n and σ:

$$\boxed{\Sigma = n\sigma.} \tag{14.27}$$

If we are dealing only with absorption, we make use of the term *absorption coefficient* α instead of Σ. That is, in an absorption-type reaction,

$$\boxed{\alpha = n\sigma.} \tag{14.28}$$

A beam of a given type of particle can induce many different types of nuclear reactions in a particular target. Let $\sigma_1, \sigma_2, \sigma_3, \ldots, \sigma_i$ be the microscopic cross sections for the production of different nuclear reactions. The total cross section σ_{tot} is defined as the sum of the partial cross sections $\sigma_1, \sigma_2, \sigma_3, \ldots,$ etc. That is,

$$\sigma_{tot} = \sum_i \sigma_i. \tag{14.29}$$

14.4 REACTION RATE

Another quantity of interest in nuclear reactions is the *reaction rate*, RR, defined as the number of nuclear reactions taking place in a unit time. Let ϕ be the *flux*, or the number of particles crossing a unit area in a unit time, and let A be the area of a foil. The total number of particles incident on this foil per unit time will be ϕA. The probability that each particle will produce a nuclear reaction is equal to the fractional sensitive area $n\sigma x$. For ϕA particles, we may write the RR as

$$\text{Reaction rate} = \text{RR} = (\phi A)(n\sigma x)/\text{sec}. \tag{14.30}$$

Since $xA = V$, the volume of the foil, and $n\sigma = \Sigma$, we may also write Eq. (14.34) as

$$\text{RR} = \phi n\sigma V \tag{14.31}$$

or

$$\text{RR} = \phi\Sigma V \tag{14.32}$$

Example 14.3. 0.1 g of cobalt-59 is exposed to a thermal neutron beam of flux 10^{14} neutrons/cm^2/sec for a period of 1 minute. Calculate the number of radioactive atoms of cobalt-60 produced and the activity of the sample immediately after irradiation, assuming that no nuclei decay during the irradiation process.

^{60}Co is produced by the reaction

$$\text{n} + {}^{59}\text{Co} \rightarrow {}^{60}\text{Co} \xrightarrow[t_{1/2}=5.26y]{\beta^-} {}^{60}\text{Ni} + \beta^- + \bar{\nu}.$$

According to Eq. (14.31),

$$\text{RR} = \phi n\sigma V$$

where

$$\phi = 10^{14} \text{ neutrons/cm}^2\text{/sec}$$

$$\sigma = 19 \text{ barns} = 19 \times 10^{-24} \text{ cm}^2$$

$$n = \text{number of nuclei per unit volume}$$

$$= \frac{\rho N_A}{A} = \frac{(8.71 \text{ g/cm}^3) \times 6.02 \times 10^{23}\text{/mole}}{58.933 \text{ g/mole}}$$

$$= 0.915 \times 10^{23}\text{/cm}^3$$

$$V = M/\rho = 0.1 \text{ g}/(8.71 \text{ g/cm}^3) = 0.0115 \text{ cm}^3$$

Hence

$$\text{RR} = (10^{14}\text{/cm}^2\text{/sec}) \times (0.915 \times 10^{23}\text{/cm}^3) \times (19 \times 10^{-24} \text{ cm}^2) \times (0.0115 \text{ cm}^3)$$

$$= 19.95 \times 10^{11}\text{/sec.}$$

In one minute the total number of radioactive nuclei N produced (assuming no decay) will be

$$N = \text{RR} \times \text{time} = 19.95 \times 10^{11}\text{/sec} \times 60 \text{ sec}$$

$$= 12 \times 10^{13} \text{ radioactive Co-60 nuclei.}$$

The disintegration rate, or activity, is given by

$$\left|\frac{dN}{dt}\right| = \lambda N = \frac{0.693}{t_{1/2}} N$$

$$= \frac{0.693}{5.26 \times 31.4 \times 10^6 \text{ sec}} \times 12 \times 10^{13} = 5 \times 10^5 \text{ disintegrations/sec.}$$

14.5 THEORIES OF NUCLEAR REACTIONS

Theories of nuclear reactions are supposed to explain the qualitative as well as quantitative aspects of these reactions. It would be ideal if we could find a single theory or model of the nucleus which would explain not only the nuclear reactions, but other nuclear properties as well. Not knowing the exact form of nuclear forces has led to the development of many different models to explain nuclear reactions and other nuclear properties: (i) the shell model, (2) the collective model, (3) the liquid-drop model, (4) the compound-nucleus model, (5) the direct-interaction model, (6) the optical model, and (7) the Fermi gas model. We discussed some of these in previous chapters. The model most commonly used in connection with nuclear reactions at low energies is the *compound nucleus model.*

The compound-nucleus model

A plot of the cross section versus the energy of an incident particle shows that variations in cross sections are quite complicated. From Fig. 14.3, we observe that at certain energies the cross section is very high compared to those of its neighboring energies. These sharp peaks are called *resonances* in the cross section. (The position of these resonances reveals the excited levels of the nucleus, and hence the level structure of the nucleus.) At low energies these resonances are very sharp, only a few tenths of an electron volt wide, and only a few electron volts apart from neighboring resonances. Niels Bohr[3] in 1936 explained the existence of these sharp resonances and some other features of nuclear reactions at low energies (from a few eV to less than 50 MeV) by proposing the *compound-nucleus hypothesis.*

According to Bohr, nuclear reactions take place in two stages. In the *first* stage, the incident particle interacts very strongly with the target nucleus. In the process of collisions with the nucleons of the target, the incident particle loses all its energy, thereby losing all its identity and becoming a part of the target. The nucleus thus formed is called a compound nucleus, and is in a highly excited state.

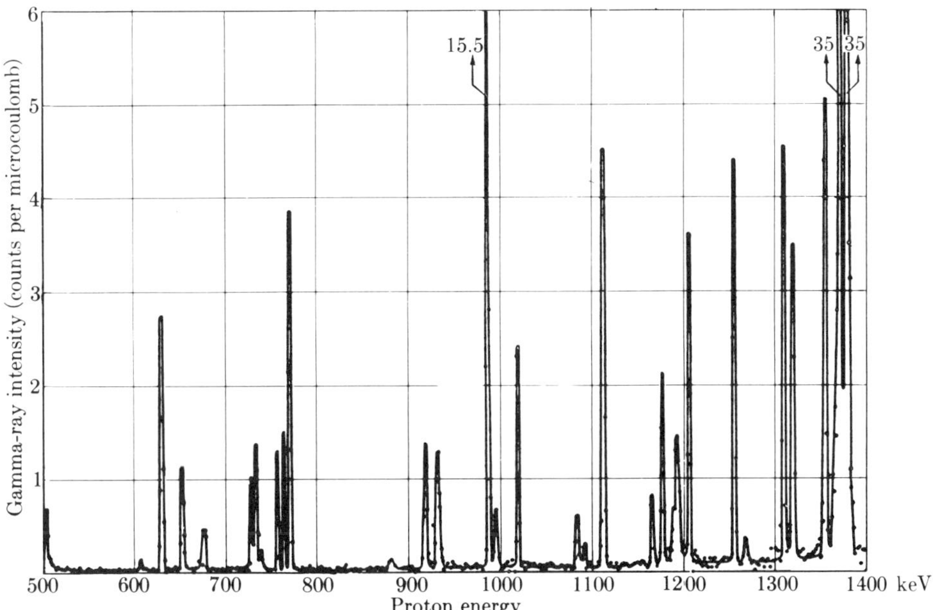

Fig. 14.3 Plot of cross section (proportional to the intensity of the gamma rays) versus energy of incident protons in the radiative capture reaction $^{27}Al(p, \gamma)^{28}Si$. Note sharp increases in intensity at certain energies. These are the so-called resonances, which are sharp, narrow, and well separated in this reaction. [From K. J. Brostrom, T. Huss, and R. Tangen, *Phys. Rev.* **71,** 661 (1947)]

In the *second* stage, if in due course, by statistical fluctuations, one (or more) nucleons happens to be at the surface of the compound nucleus with enough energy to escape, this leads to the decay of the compound nucleus. We can represent these two stages as follows:

$$\text{Incident particle} + \text{target nucleus} \to \text{compound nucleus}$$

$$\text{Compound nucleus} \to \text{recoil nucleus} + \text{outgoing particle}$$

or

$$\mathrm{x} + \mathrm{X} \to \mathrm{CN}, \qquad \mathrm{CN} \to \mathrm{Y} + \mathrm{y} \tag{14.33}$$

The formation of the compound nucleus as an intermediate state implies that its decay mode is independent of the mode of its formation. That is, the compound nucleus does not "remember" the way it was formed. This is possible only if the decay time of the compound nucleus is much longer than the *natural nuclear time*. (The nuclear time is the time it takes the incident particle to travel across the diameter of the nucleus.) For example, a neutron moving with velocity $\sim 10^9$ cm/sec (equivalent to a neutron of kinetic energy ~ 1 MeV), when crossing a nucleus of $\sim 10^{-12}$ cm diameter, will take $\sim 10^{-12}$ cm/(10^{+9} cm/sec) $\sim 10^{-21}$ sec, which is much shorter than the decay time of the compound nucleus ($\sim 10^{-14}$ sec). Once the compound nucleus is formed, there are many modes by which it may decay [see Eq. (14.34) below]. Figure 14.4 is an energy-level diagram of the formation and decay of a compound nucleus.

The experiments of S. Ghoshal[4] give the best test of the hypothesis that the decay of the compound nucleus is independent of the mode of its formation. The compound nucleus $({}^{64}_{30}\mathrm{Zn})^*$ (* indicates that the nucleus is in the excited state) may be produced by two different methods, and then decays as shown below.

$$\begin{aligned} {}^{1}_{1}\mathrm{H} + {}^{63}_{29}\mathrm{Cu} \to ({}^{64}_{30}\mathrm{Zn})^* &\to {}^{63}_{30}\mathrm{Zn} + {}^{1}_{0}\mathrm{n} \\ &\to {}^{62}_{30}\mathrm{Zn} + {}^{1}_{0}\mathrm{n} + {}^{1}_{0}\mathrm{n} \\ &\to {}^{62}_{29}\mathrm{Cu} + {}^{1}_{0}\mathrm{n} + {}^{1}_{1}\mathrm{H} \end{aligned} \tag{14.34}$$

or

$$\begin{aligned} {}^{4}_{2}\mathrm{He} + {}^{60}_{20}\mathrm{Ni} \to ({}^{64}_{30}\mathrm{Zn})^* &\to {}^{63}_{30}\mathrm{Zn} + {}^{1}_{0}\mathrm{n} \\ &\to {}^{62}_{30}\mathrm{Zn} + {}^{1}_{0}\mathrm{n} + {}^{1}_{0}\mathrm{n} \\ &\to {}^{62}_{29}\mathrm{Cu} + {}^{1}_{0}\mathrm{n} + {}^{1}_{1}\mathrm{H} \end{aligned} \tag{14.35}$$

If the decay of the compound nucleus really is independent of the mode of its formation, the ratio of the cross sections obtained for different modes of decay from Eq. (14.34) should be the same as that from Eq. (14.35). That is,

$$\sigma(\mathrm{p, n}):\sigma(\mathrm{p, nn}):\sigma(\mathrm{p, pn}) = \sigma(\alpha, \mathrm{n}):\sigma(\alpha, \mathrm{nn}):\sigma(\alpha, \mathrm{np}). \tag{14.36}$$

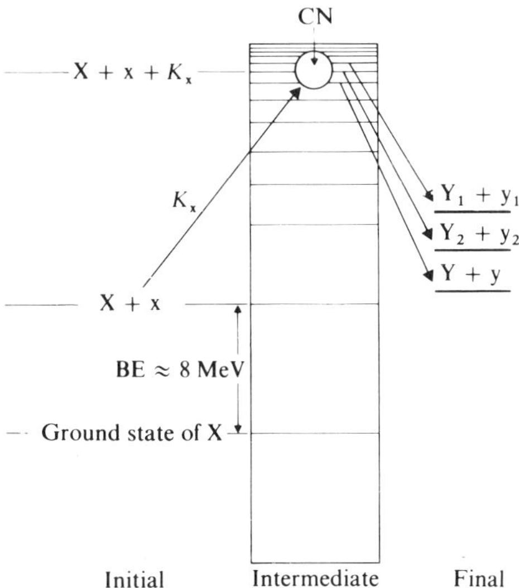

Fig. 14.4 A nuclear reaction according to the compound-nucleus model: energy-level diagram for the formation and decay of a compound nucleus. Such a nucleus may decay by emission of gamma rays or particles.

Scientists have experimentally tested these results by measuring the cross sections, thus verifying the hypothesis of the compound nucleus.

This compound-nucleus model has been very successful in explaining nuclear reactions for incident particles of energies up to ~15 MeV and for targets of mass numbers $A > 10$. For energies greater than 15 MeV but less ~50 MeV, it has some success. But above 50 MeV, we must look to other models to explain nuclear reactions. For incident particles of energies between 15 MeV and 50 MeV, the *direct interaction model*[5] and the *optical model*[6,7] are quite successful.

The picture of nuclear reactions for incident energies greater than 50 MeV becomes very complicated, so naturally the theoretical explanation becomes complicated too. If the energies of the incident particles are greater than 300 MeV and the target is of high mass number, π-mesons are produced. The production of pions and other elementary particles takes place with increasing energy, as we shall see later.

14.6 FISSION AND NUCLEAR REACTORS

Fission is a special type of nuclear reaction in which a nucleus breaks into two or more fragments of comparable size. Each time there is a fission, a large amount of energy is emitted. In this section we shall talk about a special type of fission in which a heavy nucleus absorbs a slow-moving neutron and breaks into two almost-equal-size fragments. In *each* such fission, ~200 MeV of energy is released!

Otto Hahn and Fritz Strassmann[8] discovered fission in 1939 when they bombarded uranium with slow neutrons and found that the product nuclei were isotopes of barium (^{139}Ba) and lanthanum (^{140}La). They found further that the isotope uranium-235 was the one that absorbed the slow neutron and then underwent fission. Today we know that there are many other heavy elements that undergo fission, and that fission can be caused not only by slow neutrons, but also by fast neutrons, charged particles, and gamma rays. In addition to Ba and La, many other isotopes of light elements in the range of $Z = 30$ (zinc) to $Z = 65$ (terbium) are emitted in the fission process.

a) *Fission by thermal neutrons.* There are three nuclei—^{233}U, ^{235}U, and ^{239}Pu—in which fission may be produced by absorption of thermal neutrons (neutrons with energies $\sim$0.025 eV). When these nuclei undergo fission, not only do they split into two large fragments, but they also emit neutrons and gamma rays; usually on the average of 2.51, 2.44, and 2.89 neutrons per fission of ^{233}U, ^{235}U, and ^{239}Pu, respectively. ^{235}U is a naturally occurring isotope with a long half-life, which is why it is used for fission by thermal neutrons, and because the fission behavior of ^{233}U and ^{239}Pu is very similar to that of ^{235}U, we shall limit our discussion to the fission of ^{235}U. Natural uranium contains only 0.72% of the ^{235}U isotope. But to separate it completely from other uranium isotopes would be too difficult and costly, so natural uranium enriched in the isotope ^{235}U is used for nuclear reactors and other purposes.

The fission of ^{235}U by slow or thermal neutrons results in the emission of many different products between $A = 70$, $Z = 30$, and $A = 160$, and $Z = 65$, as shown in Fig. 14.5, in which the fission yield, $Y(A)$, resulting from the reaction

$$^{235}\text{U} + {}^{1}_{0}\text{n} \rightarrow {}^{236}\text{U} \rightarrow \text{fission fragments} + \text{radiation} + \text{neutrons}$$

is plotted against A. The fission yield $Y(A)$ is defined as

$$Y(A) = \frac{\text{number of nuclei of mass number } A \text{ formed in fission}}{\text{total number of fissions}} \times 100\%$$

or

$$\boxed{Y(A) = \frac{N_A}{N_0} \times 100\%.} \tag{14.37}$$

There are more than 30 different modes of fission, in each of which a different pair of nuclei is formed. Some of the common nuclei are barium, lanthanum, bromine, molybdenum, rubidium, antimony, tellurium, krypton, iodine, xenon, and cesium. The fission fragments have too many neutrons to maintain stability, and hence are unstable. To get rid of their excess neutrons, most of the fission fragments

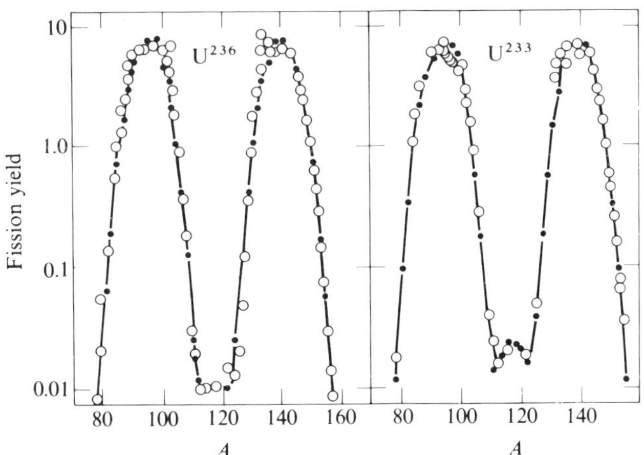

Fig. 14.5 Mass yield curves for fission of ^{236}U and ^{233}U nuclei resulting from the absorption of neutrons by ^{235}U and ^{232}U nuclei. [From W. H. Newson, *Phys. Rev.* **122,** 1224 (1961)]

decay by β^- emission. Fragments often go through many β^- decays before they become stable isotopes. A series of products with the same mass number A is a *fission chain.* There are 60 such chains. One of the longest and one of the shortest are

$$^{143}_{54}\text{Xe} \xrightarrow[1\text{ sec}]{\beta^-} {}^{143}_{55}\text{Cs} \xrightarrow[<1\text{ sec}]{\beta^-} {}^{143}_{56}\text{Ba} \xrightarrow[<0.5\text{min}]{\beta^-} {}^{143}_{57}\text{La} \xrightarrow[\sim 19\text{ min}]{\beta^-}$$

$$^{143}_{58}\text{Ce} \xrightarrow[33\text{ hr}]{\beta^-} {}^{143}_{59}\text{Pr} \xrightarrow[13.8\text{ da}]{\beta^-} {}^{143}_{60}\text{Nd (stable)}$$

and

$$^{147}_{60}\text{Nd} \xrightarrow[11\text{ da}]{\beta^-} {}^{147}_{61}\text{Pm} \xrightarrow[4\text{ yr}]{\beta^-} {}^{147}_{62}\text{Sm } (\sim 10^{11}\text{ yr})$$

Some idea of the amount of energy released in fission may be obtained from the following: The binding energy of each nucleon in uranium is ~7.6 MeV (see the BE/A versus A curve in Fig. 11.3). The binding energy of the fission fragments (A ~118) is ~8.5 MeV/nucleon. Thus the energy released in a single fission is

$$(2 \times 118 \times 8.5 - 236 \times 7.6) \simeq 212 \text{ MeV}.$$

Accurate calculations yield ~200 MeV/fission, which agrees well with the measured values. This energy appears in the form of kinetic energy of fission fragments, neutrons, β rays, γ rays, and neutrinos, as shown in Table 14.1.

Table 14.1

Energy Distribution after Fission of ^{235}U by Thermal Neutrons

Kinetic energy of:	MeV
Fission fragments	162
Fission neutrons	
(2.5 neutrons per fission carry about 2 MeV each)	5
Gamma rays emitted in fission	7
Gamma rays, beta rays, and neutrinos emitted	
in the decay of fission fragments	21
Total energy per fission	195

b) *Theory of fission.* No theory exists at present that can explain *all* the different aspects of fission. However, Niels Bohr and John Wheeler[9] developed a theory of fission based on the liquid-drop model which accounts for several different aspects of it. According to this model, a fissionable nucleus such as ^{235}U in the ground state may be treated as a perfectly spherical liquid drop. The two terms that are responsible for the shape of the nucleus are the *surface tension* and the *coulomb repulsion.* For a nucleus of radius R, the total energy E due to these two effects is

$$E = 4\pi R^2 O + \frac{3Z^2e^2}{5R}, \tag{14.38}$$

where O is the coefficient of surface tension. Coulomb repulsion leads to deformation of the shape of the nucleus. Surface tension overcomes coulomb repulsion and keeps the nucleus in a spherical shape. When a spherical nucleus absorbs a particle or a gamma ray, this increases the energy of the nucleus, causing it to deform, and we say that the nucleus is in the *excited state.* As the amount of excitation energy increases, the deformation of the nucleus increases progressively, as shown in Fig. 14.6.

Consider the state of the nucleus in Fig. 14.6(c). Two things can happen. If the effect of the surface tension is greater than that of the coulomb repulsion, the nucleus gets rid of the extra energy by emitting a gamma ray, and then adopts a spherical shape again and becomes stable. If, on the other hand, the effect of the coulomb repulsion is greater than that of the surface tension, the excited nucleus breaks into two; i.e., fission occurs. Of course, for the degree of deformation shown in Fig. 14.6(b) and (c), the nucleus always gets rid of its extra energy and no fission takes place.

Some heavy nuclei undergo fission as a result of bombardment by thermal neutrons, others after bombardment by fast or high-energy neutrons or charged particles, and others undergo fission only if the energies of incident particles are as high as 20 or 30 MeV. What one has to do is to supply enough energy so that

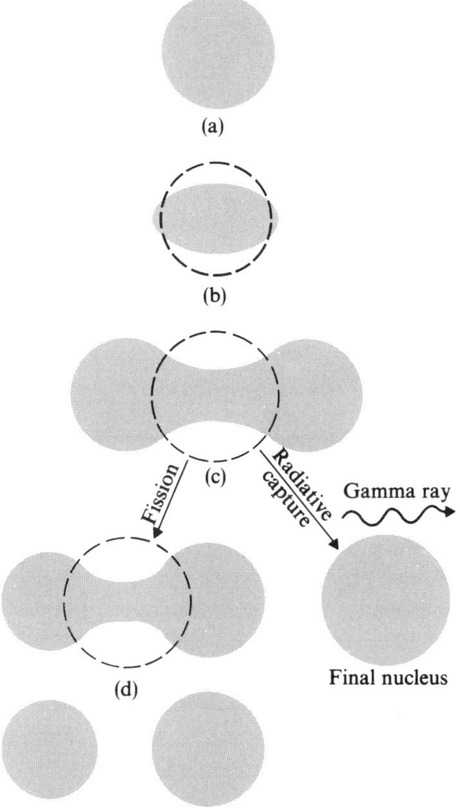

Fig. 14.6 A fissionable nucleus takes different shapes depending on the degree of excitation resulting from the absorption of an incident particle. (a) Spherical nucleus before collision with incident particle. (b) After collision, nucleus undergoes deformation to an ellipsoid. (c) Nucleus still further deformed. (d) If the excitation is high, the nucleus will undergo fission. Otherwise the particle is captured and a gamma ray is emitted.

the absorbing nucleus takes the shape shown in Fig. 14.6(d). For nuclei with mass number $A \sim 200$, very high excitation energies are needed to cause fission. For those with mass numbers between $A = 230$ and $A = 240$, energies of only a few MeV are needed to produce fission, while three nuclei—^{233}U, ^{235}U, and ^{239}Pu—are a special case, in which the shape of Fig. 14.6(d) is reached by these nuclei when they absorb neutrons with almost zero kinetic energy. That's why they are so useful as target materials for fission.

c) *Chain reactions and nuclear reactors.* Nuclear fission differs from other nuclear reactions in two ways: (1) For each neutron absorbed, *more* than one neutron is emitted (on the average 2.5 neutrons in the fission of ^{235}U). (2) A large amount of energy is released in each fission, and is available in the form of heat for power production. Nuclear-fission reactors utilize these two characteristics for the production of power.

Suppose that we bombard a small amount of uranium by 100 neutrons. Say that 40 of these neutrons produce fission, and the other 60 are either absorbed without producing fission or escape from the uranium target. After the fission, we still have $40 \times 2.5 = 100$, that is, 100 neutrons. Of these 100, again 40 will produce fission, resulting in 100 new neutrons. Thus no neutrons are lost, and the process of fissioning continues without our supplying additional neutrons from outside. This *self-sustained chain reaction* is possible only if the number of neutrons produced in a given generation is equal to, or more than, the previous generation. The *reproduction factor* or *multiplication factor* k is defined as

$$k = \frac{\text{number of neutrons in the } (n+1)\text{th generation}}{\text{number of neutrons in the } n\text{th generation}}. \tag{14.39}$$

If $k = 1$, the number of neutrons in a given generation is equal to the number in the previous generation, and the system is said to be *critical.* If $k < 1$, the number of neutrons in a given generation is less than that in the previous generation, and the system is said to be *subcritical* or *convergent.* If $k > 1$, the number of neutrons in a given generation is more than that in the previous generation, the system is said to be *supercritical* or *divergent.*

Each fission produces energy. If we are to obtain a continuous, uniform supply of power from a nuclear fission reactor, the system must be critical, that is, $k = 1$. This is the prerequisite for running a nuclear power plant. In a supercritical system, the power continuously increases, finally becoming uncontrollable, i.e., resulting in an explosion, as in the case of an atomic bomb. If the system is subcritical, the number of neutrons decreases continuously, and hence the power decreases continuously, finally reaching a power level of zero. So the system will shut itself down.

To start up a nuclear reactor, one makes k slightly greater than unity, and after the desired power level is reached, k is maintained at unity. If natural uranium is used as a fuel, one cannot attain a critical system, because natural uranium consists of ^{235}U and ^{238}U in the ratio of 1:138. That is, the abundance of ^{235}U is only 0.72%. When thermal neutrons are incident on a system containing natural uranium, most of the neutrons are lost by being absorbed in the ^{238}U in nonfission processes, and by escaping from the system. To increase the probability of fission, one has to use uranium enriched in the ^{235}U isotope as fuel in a nuclear reactor.

Figure 14.7 shows a typical arrangement of a nuclear power plant. The core consists of fuel elements made of uranium that has been enriched by ^{235}U. These elements are surrounded by moderator and shielding, as shown in part (b). The moderator (which may be water or heavy water) slows down the fast neutrons

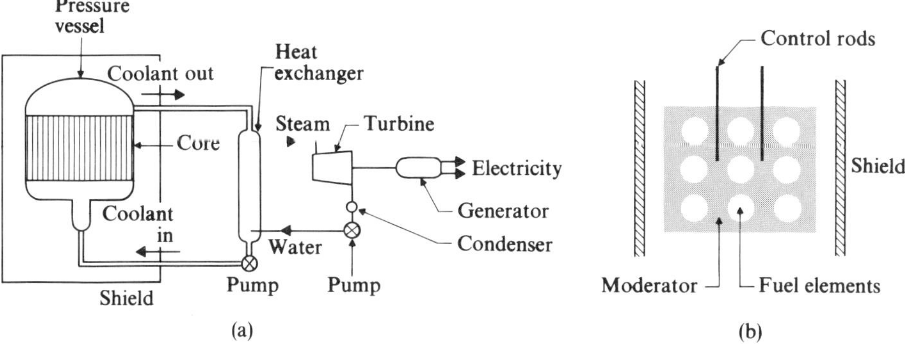

Fig. 14.7 (a) A typical nuclear power plant. (b) Arrangement of fuel elements in the core of a nuclear reactor. [From A. P. Arya, *Fundamentals of Nuclear Physics*, Boston: Allyn and Bacon, 1966]

produced in the fission, so as to increase the probability that they will be captured and produce fission. Control rods made of cadmium, which has a high absorption cross section for thermal neutrons, are used to increase the power level (this is achieved by pulling them out) or decreasing it (achieved by pushing them in); i.e., they are used to control the power level. The heat produced by fission in the reactor core is taken away by the coolant, which in turn is used to produce steam. The steam runs a turbine, and electricity is made available through the generator.

The staggering amount of power we can get from a nuclear reactor is apparent from the following example.

Example 14.4. Calculate the amount of energy available if one gram of ^{235}U is completely fissioned.

$$\text{Energy released from one fission of } ^{235}\text{U atom} = 200 \text{ MeV}$$
$$= 3.2 \times 10^{-11} \text{ watt-sec.}$$

$$\text{One gram } ^{235}\text{U contains } \frac{6.02 \times 10^{23}}{235} = 2.16 \times 10^{21} \text{ atoms.}$$

Therefore one gram of ^{235}U will release energy in fission

$$= 3.2 \times 10^{-11} \text{ watt-sec} \times 2.16 \times 10^{21}$$
$$= 8.2 \times 10^{10} \text{ watt-sec}$$
$$\approx 1 \text{ MWD (megawatt-day).}$$

By comparison, one ton of coal yields $\simeq$ 0.36 MWD of heat. In other words, as far as production of power is concerned,

$$1 \text{ ton of uranium} = 2.7 \times 10^6 \text{ tons of coal.}$$

14.7 FUSION AND THERMONUCLEAR ENERGY

At present oil and coal—they are called *fossil fuels*, for obvious reasons—are the main types of fuels used for the production of power. However, the rate of consumption of power is increasing so fast that it is possible that the world's reserves of fossil fuels will be depleted entirely in less than a century. So, in the last decade, many power plants have been built which use fissionable fuels to produce power. The production of power by nuclear *fusion*, or *thermonuclear reaction*, if mankind could successfully achieve it, would satisfy the power needs of the world's increasing population for billions of years, even if the consumption rate went as high as 1000 times the present level.

Nuclear fusion is a process in which two very light nuclei ($A \leq 8$) combine to form a heavy nucleus, which releases an enormous amount of energy. As is evident from the plot of the BE/A versus A curve (Fig. 11.3), the binding energy of this heavy nucleus is much more than the combined binding energies of the two light nuclei. That is, the rest mass of the heavy nucleus is less than the sum of the individual masses of the light nuclei. This decrease in the rest mass appears as energy, which one can use to produce electric power. The process of fusion is opposite to that of fission; the only thing the two have in common is that both produce energy.

The energy released by the sun and other stars is due to the fusion reactions taking place inside them. The temperature inside the sun and the stars is so high ($\sim 10^8$ or 1 million °K) that the atoms of different elements are completely ionized. A collection of the bare nuclei and electrons is called a *plasma*. The nuclei in these plasmas move with very high velocities and have a very high likelihood of combining with each other when they collide. And when they do, since the reactions are exoergic, large amounts of energy are released. Such reactions, which are made possible by virtue of high temperatures, are called *thermonuclear reactions*. Many thermonuclear reactions take place in the stars, but these two are the most common:

Proton–proton cycle

$$\begin{aligned}
{}^{1}\mathrm{H} + {}^{1}\mathrm{H} &\rightarrow {}^{2}\mathrm{H} + \beta^{+} + \nu + 0.42\ \mathrm{MeV} \\
{}^{2}\mathrm{H} + {}^{1}\mathrm{H} &\rightarrow {}^{3}\mathrm{He} + \gamma + 5.49\ \mathrm{MeV} \\
{}^{3}\mathrm{He} + {}^{3}\mathrm{He} &\rightarrow {}^{4}\mathrm{He} + 2\,{}^{1}\mathrm{H} + 12.86\ \mathrm{MeV} \\
\text{Total energy released} &= 18.77\ \mathrm{MeV}
\end{aligned} \tag{14.46}$$

Carbon cycle

$$\begin{aligned}
{}^{12}\mathrm{C} + {}^{1}\mathrm{H} &\rightarrow {}^{13}\mathrm{N} + \gamma + 1.95\ \mathrm{MeV} \\
{}^{13}\mathrm{N} &\rightarrow {}^{13}\mathrm{C} + \beta^{+} + \nu + 1.20\ \mathrm{MeV} \\
{}^{13}\mathrm{C} + {}^{1}\mathrm{H} &\rightarrow {}^{14}\mathrm{N} + \gamma + 7.55\ \mathrm{MeV} \\
{}^{14}\mathrm{N} + {}^{1}\mathrm{H} &\rightarrow {}^{15}\mathrm{O} + \gamma + 7.34\ \mathrm{MeV} \\
{}^{15}\mathrm{O} &\rightarrow {}^{15}\mathrm{N} + \beta^{+} + \nu + 1.68\ \mathrm{MeV} \\
{}^{15}\mathrm{N} + {}^{1}\mathrm{H} &\rightarrow {}^{12}\mathrm{C} + {}^{4}\mathrm{He} + 4.96\ \mathrm{MeV} \\
\text{Total energy released} &= 24.68\ \mathrm{MeV}
\end{aligned} \tag{14.47}$$

The following thermonuclear fusion reactions, which involve the isotopes of hydrogen (deuterium, ^{2}H, and tritium, ^{3}H), have been studied in laboratories:

$$\begin{aligned} {}^2_1\text{H} + {}^3_1\text{H} &\to {}^4\text{He} + \text{n} + 17.6 \text{ MeV} \\ {}^2_1\text{H} + {}^2_1\text{H} &\to {}^3\text{He} + \text{n} + 3.2 \text{ MeV} \\ {}^2_1\text{H} + {}^2_1\text{H} &\to {}^3\text{H} + {}^1_1\text{H} + 4.0 \text{ MeV} \end{aligned} \tag{14.48}$$

Large quantities of deuterium are available in the world's oceans, so it would seem that obtaining fuel for a nuclear-fusion power plant would be no great problem. However, in order to harness nuclear-fusion energy for useful purposes, we have yet to overcome many problems. The main difficulty is the containment of the plasma at temperatures of $\sim 10^8$°K. At present there isn't any material which will stand up to 1,000,000°K. At such high temperatures, any material used for a container vaporizes, and the plasma disperses!

An alternative that is being tried instead of putting the plasma in a container, is confining it, by means of external magnetic fields, at the center of a tube of very large diameter. One such technique utilizes the so-called *pinch effect*. Other magnetic systems that are being used are the *stellarator*, the *magnetic mirror*, and the *astron*.[10]

SUMMARY

A typical nuclear reaction is represented as x + X → Y + y or X(x, y)Y. The disintegration energy or *Q-value* of a nuclear reaction is defined as a net change in the kinetic energy ($K_f - K_i$), which is also equal to the net change in rest-mass energy ($M_i c^2 - M_f c^2$). If the Q-value is positive, a reaction is said to be *exoergic*; if the Q-value is negative, the reaction is said to be *endoergic*.

The minimum kinetic energy of an incident particle which will initiate an endoergic reaction is called the *threshold energy* for an endoergic reaction. Threshold energy $= (K_x)_{\min} = (1 + m_x/M_X)|Q|$.

The probability that a single incident particle will interact with an individual target nucleus is called the *cross section* or *microscopic cross section*, σ. The intensity of a beam after it passes through a material of thickness x is $I = I_0 e^{-n\sigma x}$. The unit of microscopic cross section is the *barn*, where 1 b $= 10^{-24}$ cm^2.

The macroscopic cross section is defined as $\Sigma = n\sigma$ and the absorption coefficient $\alpha = n\sigma$.

The reaction rate, RR, is the number of nuclear reactions taking place in a unit time; that is, RR $= \phi n \sigma V = \phi \Sigma V$.

The model of the nucleus most commonly used in connection with nuclear reactions at low energies, the compound-nucleus model, was suggested by Niels Bohr in 1936. According to Bohr's hypothesis, nuclear reaction takes place in two stages:

$$\text{x} + \text{X} \to \text{CN} \to \text{Y} + \text{y}.$$

That is, the decay of a compound nucleus is independent of the mode of its formation.

In nuclear fission, a nucleus breaks into two or more fragments of about equal size, and a large amount of energy is emitted. The isotope U-235 is particularly favorable to the absorption of thermal neutrons. When it does so, it breaks into two fragments of approximately equal size and releases $\sim$200 MeV of energy in the form of the kinetic energy of fission fragments, neutrons, gamma rays, beta rays, and neutrinos.

According to the theory of fission based on Bohr's and Wheeler's liquid-drop model, the nucleus that absorbs a neutron changes in shape from a sphere to an ellipsoid. If the coulomb repulsion force is greater than the surface-tension energy, the excited nucleus breaks into two fragments.

A self-sustained nuclear-fission reaction is possible only if the number of neutrons produced in a given generation is equal to, or more than, that produced in the previous generation:

$$\text{Reproduction factor } k = \frac{\text{number of neutrons in the } (n+1)\text{th generation}}{\text{number of neutrons in the } n\text{th generation}}.$$

If $k = 1$, the system is critical; if $k < 1$, it is subcritical; if $k > 1$, it is supercritical. In a nuclear reactor, to begin with, k is made slightly larger than 1. But when the desired power level is reached, the system is adjusted so that $k = 1$, and is maintained at that level.

Nuclear fusion is a process in which two light nuclei ($A \leq 8$) combine to form a heavy nucleus, with the release of energy. Light nuclei have to be at very high temperatures ($\sim 10^8$°K) to produce fusion. Such reactions are therefore called *thermonuclear reactions.*

PROBLEMS

1. A beam of protons is incident on a $^{31}_{15}P$ target. Write the complete equation for the nuclear reaction, given that the emitted particle is (a) a neutron, (b) a proton, (c) a deuteron, and (d) an alpha particle.
2. A beam of deuterons is incident on a $^{27}_{13}Al$ target. Write the complete equation for the nuclear reaction, given that the emitted particle is (a) a neutron, (b) a proton, and (c) an alpha particle.
3. Calculate the Q-values for the following reactions. Which are exoergic and which endoergic?

$$^{10}B(\alpha, p)^{13}C \qquad ^{27}Al(\alpha, d)^{29}Si$$
$$^{9}Be(d, n)^{10}B \qquad ^{15}N(d, n)^{16}O$$

4. A proton of 20-MeV kinetic energy strikes a ^{11}B nucleus at rest. An α particle is given out in the same direction as the original direction of the proton. Calculate the Q-value of the nuclear reaction and the kinetic energy of the emitted proton.

5. A beam of fast neutrons can be produced in the laboratory by using a 400-keV deuteron beam in the following reaction:

$$ {}^2_1\text{H} + {}^3_1\text{H (tritium)} \rightarrow {}^4_2\text{He} + {}^1_0\text{n} $$

a) Calculate the Q-value of the reaction.
b) What are the kinetic energies of the alpha particles and the neutrons, assuming that the kinetic energies of ${}^2_1\text{H}$ and ${}^3_1\text{H}$ are negligible?
c) Make a plot of the kinetic energy of the neutron versus θ in steps of 30° from $\theta = 0°$ to $\theta = 180°$.

6. The Q-value of the reaction ${}^{14}\text{N}(\alpha, \text{p}){}^{17}\text{O}$ is -1.16 MeV.
a) Calculate the threshold energy of the reaction.
b) What is the energy of the proton at the threshold?

7. The Q-value of the reaction ${}^{27}\text{Al}(\alpha, \text{n}){}^{30}\text{P}$ is -2.9 MeV. Calculate the threshold energy of the nuclear reaction.

8. Suppose that, in a given nuclear reaction, the kinetic energy of the particle emitted at an angle θ_1 is K_{y1} and emitted at an angle θ_2 is K_{y2}. Calculate the energy of the incident particle in terms of K_{y1}, K_{y2}, and the rest masses.

9. The Q-value of the nuclear reaction ${}^{19}\text{F}(\text{n}, \text{p}){}^{19}\text{O}$ is -3.9 MeV. (a) Calculate the threshold energy of the reaction. (b) Calculate the difference in mass between ${}^{19}\text{F}$ and ${}^{19}\text{O}$.

10. The intensity of a beam of thermal neutrons (i.e., a beam of energy ~0.025 eV) is reduced to one-half when it passes a layer of water 5 cm thick. Assume that the beam is absorbed by the hydrogen atoms only, because the absorption by the oyxgen atoms is negligible.
a) Calculate the absorption cross section.
b) Calculate the absorption coefficient.
c) What thickness of water is needed to reduce the intensity of the incident beam to one-tenth of its initial intensity?

11. What thickness of aluminum is needed to reduce the intensity of an incident beam of 0.025-eV neutrons to (a) one-tenth, (b) one-hundredth, (c) one-thousandth of its initial intensity? The density of aluminum is 2.7 g/cm³, and its cross section for the 0.025-eV neutrons is 0.23 b.

12. By definition, the mean free path $\bar{x}$ of a particle is given by

$$ \bar{x} = \frac{\int_0^{N_0} x \, dN}{\int_0^{N_0} dN}, $$

where dN is the number of particles that travel a distance x before being absorbed. Show that

$$ \bar{x} = \frac{1}{n\sigma} = \frac{1}{\Sigma}. $$

13. Using the expression for the mean free path $\bar{x} = 1/n\sigma = 1/\Sigma$, calculate the mean free path of thermal neutrons in (a) water, for which $\sigma = 0.33$ b and density $=$ 1 g/cm³, and (b) graphite, for which $\sigma = 2.6$ b and density $= 2250$ kg/m³.

14. Suppose that 0.1 g of ^{59}Co is bombarded with a flux of 2×10^{13} thermal neutrons/cm^2/sec for 1 hour. The capture cross section is 19 b. Calculate, in Millicuries, the amount of radioactive ^{60}Co ($t_{1/2} = 5.27$ years) produced. Neglect any decay during production.

15. One gram of natural neodymium is exposed for four days in a nuclear reactor in which the thermal neutron flux is 10^{13} neutrons/cm^2/sec. The natural abundance of Nd-146 is 17.22% and its cross section for thermal neutrons is 2 b. What is the amount of Nd-147 in the following reaction? (Density of Nd is 6.9 g/cm^3)

$$^{146}\text{Nd} + {}^1_0\text{n} \rightarrow {}^{147}\text{Nd} \xrightarrow[11.1\,\text{d}]{\beta^-} {}^{147}\text{Pm}.$$

What is the activity soon after the irradiation? Neglect any decay during production.

16. A thin gold foil weighing 0.2 g is irradiated for 2 min in the thermal neutron flux of a nuclear reactor. ^{198}Au is produced by the reaction

$$^{197}\text{Au} + {}^1_0\text{n} \rightarrow {}^{198}\text{Au} \xrightarrow[64.8\,\text{hr}]{\beta^-} {}^{198}\text{Hg} + \gamma.$$

The neutron capture cross section is 98 b. The activity of the foil right after it is taken out of the reactor is 2 Millicuries. Calculate the thermal neutron flux.

17. In the formation of a compound nucleus, the widths of the resonance levels are 1 eV, 1 keV, and 1 MeV. What are the corresponding lifetimes of these states? What do you conclude from these data?

18. Consider the reaction ^{63}Cu(p, n)^{63}Zn. What are the compound and the residual nuclei?

19. An alpha particle is absorbed by $^{27}_{13}$Al. What compound nucleus is formed? What are the different modes by which this compound nucleus may decay? What will the corresponding residual nuclei be?

20. The excitation energy, E_c, of a compound nucleus is given by

$$E_c = E'_x + E_B,$$

where E_B is the binding energy of the particle x when the compound nucleus is in the ground state ($\approx$ 8 MeV/nucleon), and E'_x is the fraction of the energy of the incident particle that is used in exciting the nucleus. Use the law of the conservation of linear momentum to show that

$$E'_x = E_x\left(\frac{M_X}{M_X + m_x}\right)$$

where E_x is the energy of the incident particle. (You may use nonrelativistic approximation.)

21. Using the expression given in Problem 20, calculate the excitation energy for the formation of the compound nucleus formed by the capture of 8-MeV alpha particles by ^{27}Al.

22. Calculate the binding energy of thermal neutrons added to the following nuclei:

$$^{227}\text{Th}, \quad ^{233}\text{U}, \quad ^{235}\text{U}, \quad ^{239}\text{Pu}, \quad ^{242}\text{Pu}$$

Which of these will be classified as fissionable by thermal neutrons?

23. In the fission of ^{235}U, the mass ratio of the two fission fragments produced is 1.5. What is the ratio of the velocities of these two fragments?
24. Calculate the surface energies and coulomb energies of the following nuclei:

$$^{228}\text{Th}, \quad ^{234}\text{U}, \quad ^{236}\text{U}, \quad ^{240}\text{Pu}, \quad ^{243}\text{Pu}$$

From your calculations, do you get any hint as to which of these nuclei will undergo fission more easily than the others? (The coefficient O $\simeq 10^{10}$ tons/mm.)

25. Calculate the rate of consumption of ^{235}U as fission fuel in a nuclear reactor operating at a power level of 500 megawatts of electricity. Note that the efficiency for conversion from heat to electricity is only 5%.
26. Calculate the total energy liberated in (a) proton-proton cycle, and (b) carbon–nitrogen cycle, when 1 g of material undergoes complete fusion.
27. How much energy is liberated when 1 g of hydrogen atoms is converted into helium atoms by *fusion*? Compare this with the energy liberated in the *fission* of 1 g of ^{235}U.

REFERENCES

1. Ernest Rutherford, *Phil. Mag.* **37,** 537 (1919)
2. J. D. Cockcroft and E. T. S Walton, *Proc. Roy. Soc.* **137,** 229 (1932)
3. Niels Bohr, *Nature* **137,** 344 (1936)
4. S. N. Ghoshal, *Phys. Rev.* **80,** 939 (1950)
5. S. T. Butler, *Proc. Roy. Soc.* **A208,** 559 (1951); *Phys. Rev.* **106,** 272 (1957)
6. R. Serber, *Phys. Rev.* **72,** 1114 (1947)
7. H. Feshbach, C. E. Porter, and V. F. Weisskopf, *Phys. Rev.* **77,** 606 (1950)
8. O. Hahn and F. Strassmann, *Naturwissenschaften* **27,** 11, 89 (1939)
9. Niels Bohr and J. A. Wheeler, *Phys. Rev.* **56,** 426 (1939)
10. A. S. Bishop, *Project Sherwood—The U. S. Program in Controlled Fusion*, Reading, Mass.: Addison-Wesley (1958)

Suggested Readings

1. Atam P. Arya, *Fundamentals of Nuclear Physics*, Chapters IV, XI, and XIII, Boston: Allyn and Bacon, 1966
2. M. S. Livingston and H. A. Bethe, *Revs. Mod. Phys.* **9,** 245–390 (1937)
3. P. Morrison, *Experimental Nuclear Physics*, edited by E. Segré, Vol. II. New York: John Wiley, 1953

4. J. M. Blatt and V. F. Weisskopf, *Theoretical Nuclear Physics*, New York: John Wiley, 1952
5. S. Glasstone and M. C. Edlund, *The Elements of Nuclear Reactor Theory*, Princeton, N.J.: Van Nostrand, 1952
6. W. P. Allis, editor, *Nuclear Fusion*, Princeton, N.J.: Van Nostrand, 1960
7. D. J. Rose and M. Clark, Jr., *Plasmas and Controlled Fusion*, Cambridge, Mass.: M.I.T. Press, 1961
8. B. L. Cohen, *Concepts of Nuclear Physics*, New York: McGraw-Hill, 1971
9. W. E. Meyerhof, *Elements of Nuclear Physics*, Chapter 4, New York: McGraw-Hill, 1967
10. I. Kaplan, *Nuclear Physics*, second edition, Chapters 16, 19, and 20, Reading: Addison-Wesley, 1963

CHAPTER 15
ELEMENTARY PARTICLES

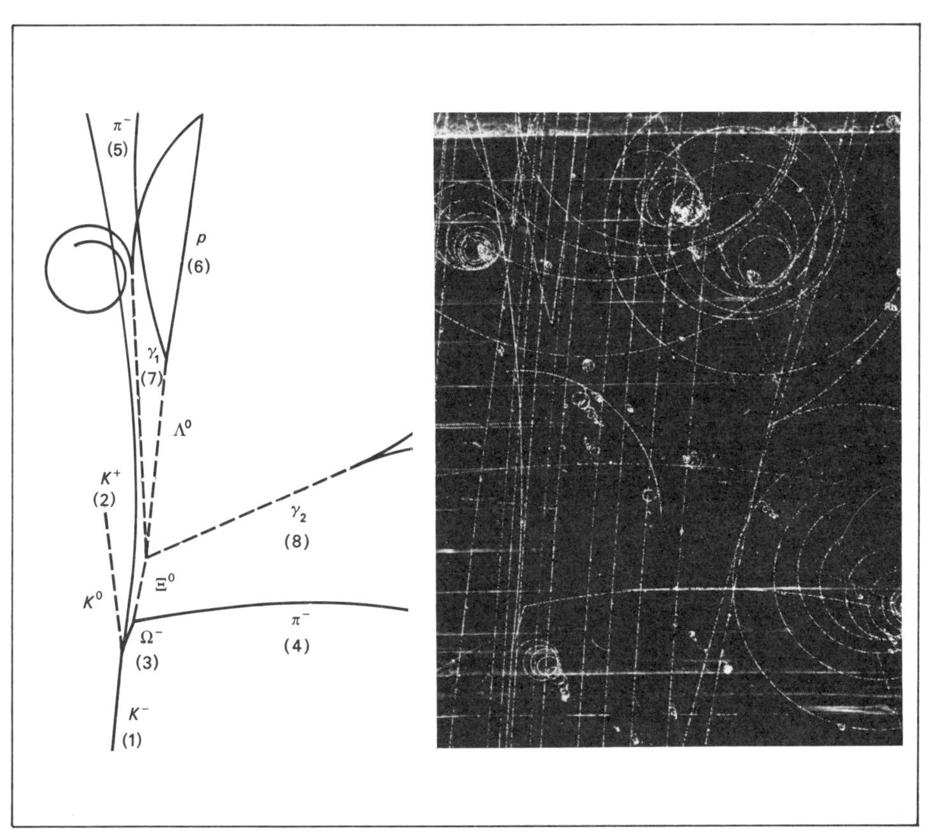

Sketch (left) and bubble-chamber photograph (right) of events showing decay of omega minus, Ω^-. *[From V. F. Barnes* et al., Phys. Rev. Lett. ***12****, 204 (1964)]*

15.1 INTRODUCTION

It has always been one of the main aims of physics to find the basic building blo of which all matter is made, and the forces which hold different parts of mat together. In 1808 Dalton put forth the atomic theory of matter, which held tł the atom was the smallest indivisible constituent of matter. But by 1900, althou people still didn't know much about the atom, they began to be aware that it ha a substructure, so the atom wasn't the smallest constituent of matter after all. The discovery of the electron by J. J. Thomson[1] in 1897, the establishment of the nuclear model of the atom by Rutherford[2] in 1911, and the discovery of the neutron by James Chadwick[3] in 1932 led physicists to believe that all matter was made of electrons, protons, and neutrons. In the early 30's, people knew about only four basic particles: the electron, the proton, the neutron, and the photon (a quantum of electromagnetic radiation). These particles were called the fundamental or elementary particles. By 1947, with further investigations and discoveries, physicists found that the list of elementary particles in our universe had grown to 14; by 1957, it had reached 32 (see Table 15.1). By 1965, this list had been more than doubled, and at present there doesn't seem to be any end to it!

What is a fundamental or elementary particle? You might say that an elementary particle is one which is not a composite of others. But this definition does not hold good because there are many particles which have been called elementary but show composite structure. Even particles like protons and neutrons have been found to have definite structures. At present it seems hard to make rules as to what an elementary particle is. It is hoped that future discoveries in the field of high-energy nuclear physics will be able to answer these questions satisfactorily.

The rapid growth of the list of elementary particles over a period of only 40 years may be atributed to three technological developments: (a) high-energy accelerators, (b) new types of detectors, and (c) data analysis techniques, such as the use of high-speed computers.

The purpose of this chapter is manyfold. First we shall briefly describe the discoveries of the different elementary particles, then classify them, in order to create some order out of chaos. We shall also investigate what types of forces bind these particles when they are brought together. Finally, we shall discuss the predictions for the future in the field of elementary particles.

15.2 THE OLD THIRTY-TWO ELEMENTARY PARTICLES

Table 15.1 is a list of the 32 elementary particles that were known in 1957; these are also called the *old* fundamental particles. All of them had been theoretically predicted and experimentally confirmed before 1957 except for ν_μ and $\bar{\nu}_\mu$. Most of the particles in Table 15.1 were originally noticed in cosmic rays and later on produced in the laboratories. (Cosmic rays are highly penetrating radiation which bombards our earth's atmosphere. It comes from outer space, and is of unknown origin.) Note that almost every particle has an antiparticle, except for a few which are *their own* antiparticles. Furthermore every particle annihilates when it comes in contact with its antiparticle.

Table 15.1

The 32 Elementary Particles of 1957

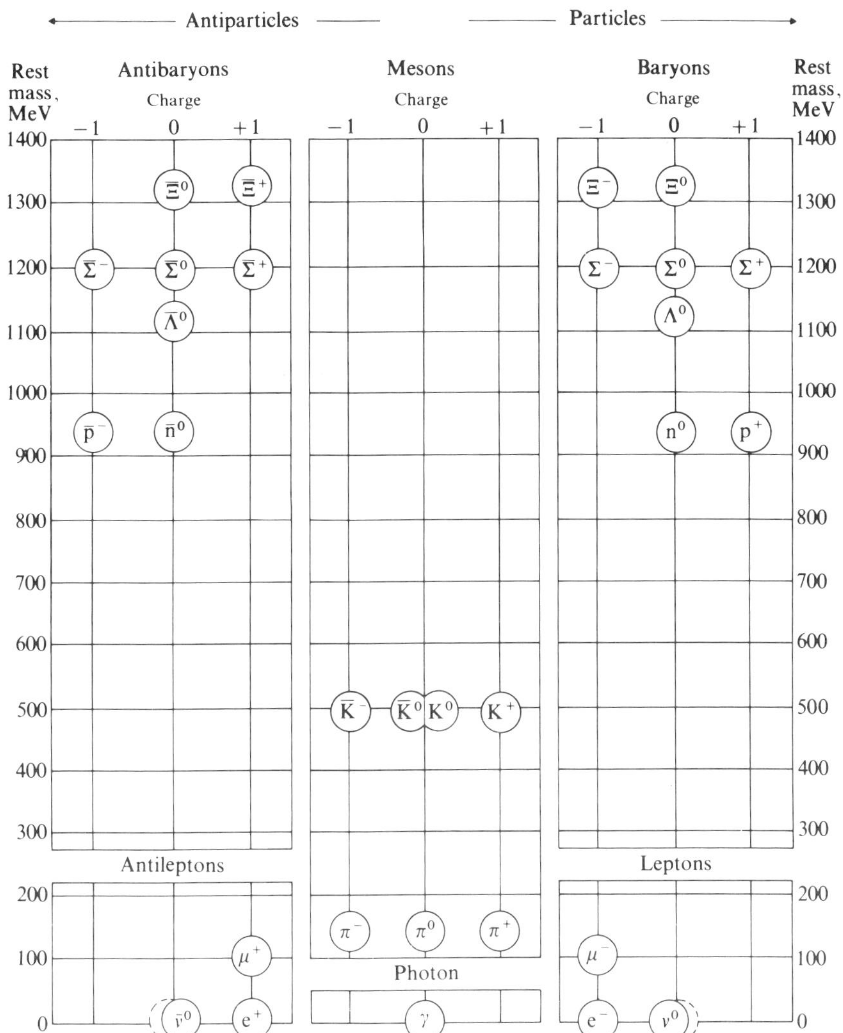

[From "Strongly Interacting Particles" by Geoffrey F. Chew, Murray Gell-Mann and Arthur H. Rosenfeld.]

The 32 particles known in 1957 are conveniently divided into four groups.

a) *Baryons and antibaryons.* There are 16 of these. They are particles with rest masses equal to or greater than the rest mass of a proton, i.e., greater than 938 MeV. There are 8 *baryons*: proton (p^+), neutron (n^0), lambda (Λ^0), sigma plus (Σ^+), sigma zero (Σ^0), sigma minus (Σ^-), xi zero (Ξ^0), and xi minus (Ξ^-). Each baryon has its antiparticle, called an antibaryon. In other words, there are 8 antibaryons, as shown in Table 15.1 and Table 15.2. (Antiparticles are denoted by a horizontal bar over their corresponding particle symbol.) Table 15.2 also gives some of the characteristics of these particles, and indicates their mean lives and decay products. Note that most of them are not stable.

b) *Mesons.* There are 7 mesons. They are lighter than the lightest of the baryons, but heavier than the heaviest of the leptons, i.e., their rest masses are between ~130 and ~500 MeV. Three of them are called *pi mesons* (π^+, π^-, π^0); the other four are called *K-mesons* (K^+, K^-, K^0, $\overline{K}^0$). Table 15.2 lists their mean lives and some other properties (note that they also are unstable).

c) *Leptons.* The 8 leptons have rest masses which are less than those of the mesons, i.e., between 130 MeV and 0 MeV. There are 4 leptons: electron (e^-), mu-minus meson (μ^-), neutrino associated with electron (ν_e^0), and neutrino associated with mu-meson (ν_μ^0). Corresponding to these, there are 4 antileptons (or antiparticles): e^+, μ^+, $\bar{\nu}_e^0$, and $\bar{\nu}_\mu^0$, respectively. Table 15.2 lists these particles plus some of their characteristics.

d) *Photon.* The sole photon is a quantum of electromagnetic radiation with zero rest mass. It is its own antiparticle, and is in a class by itself.

This division into four groups may look very superficial, but it isn't. The division is based on the *types* of forces by which these particles interact with each other (or interact with other particles). Let us now discuss the circumstances and the experimental conditions which led to the discoveries of some of these particles. We can't follow a strict chronological order because of the time lag between theoretical predictions and experimental discoveries.

i) *Electron, proton, neutron, and photon* (e^-, p^+, n, γ). These were the only particles known in the early thirties. The electron (discovered by J. J. Thomson[1] in 1897) has a rest mass of 0.51 MeV, a unit negative charge, and a spin of $\frac{1}{2}$ unit. The proton (discovered by Ernest Rutherford[2] in 1919) has a rest mass of 1836 times that of the electron, a unit positive charge, and a spin of $\frac{1}{2}$ unit. The discovery of the neutron by Chadwick[3] in 1932 led to the establishment of the *neutron–proton model* of the atom (see Chapter 4). The neutron's rest mass is slightly larger (1.5 MeV) than the mass of the proton, but, it has no charge. It has a spin of $\frac{1}{2}$ unit. However, even though the neutron's net charge is zero, it has a negative magnetic moment. This can be true only if the neutron has a charge-bearing substructure. In fact, it has been experimentally verified that both neutron *and*

Table 15.2

Properties of the 32 Elementary Particles

		Particle state	Antiparticle state	Rest mass of particles MeV	Spin	Mean life, sec.	Decay products
Baryons	Xi	Ξ^-	$\overline{\Xi}^+$	1318.4	$\frac{1}{2}$	1.3×10^{-10}	$\Lambda^0 + \pi^-$
		Ξ^0	$\overline{\Xi}^0$	1311	$\frac{1}{2}$	1.5×10^{-10}	
	Sigma	Σ^-	$\overline{\Sigma}^+$	1196	$\frac{1}{2}$	1.6×10^{-10}	$p + \pi^0, n + \pi^+$
		Σ^0	$\overline{\Sigma}^0$	1191.5	$\frac{1}{2}$	$<0.1 \times 10^{-10}$	$n + \pi^-$
		Σ^+	$\overline{\Sigma}^-$	1189.4	$\frac{1}{2}$	0.8×10^{-10}	$\Lambda^0 + \gamma$
	Lambda	Λ^0	$\overline{\Lambda}^0$	1115.4	$\frac{1}{2}$	2.5×10^{-10}	$p + \pi^-, n + \pi^0$
	Neutron	n^0	$\overline{n}^0$	939.5	$\frac{1}{2}$	932 ± 14	$p + e^- + \overline{\nu}_e$
	Proton	p	$\overline{p}^-$	938.2	$\frac{1}{2}$	Stable	
Mesons	K meson	K^0	$\overline{K}^0$	497.8	0	6×10^{-8}, 1×10^{-8}	
		K^+	$\overline{K}^-$	494	0	1.2×10^{-8}	$\mu^+ + \nu, \pi^+ + \pi^0, \pi^+ + \pi^+ + \pi^-,$
							$\pi^+ + \pi^0 + \pi^0, \mu^+ + \nu + \pi^0, e^+ + \nu + \pi^0$
		K_1^0			0		$\pi^+ + \pi^-, \pi^0 + \pi^0, \pi^+ + e^- + \overline{\nu},$
							$\pi^- + e^+ + \nu, \pi^+ + \mu^- + \overline{\nu}$
		K_2^0			0		$\pi^- + \mu^+ + \nu, \pi^+ + \pi^- + \pi^0,$
							$\pi^0 + \pi^0 + \pi^0$
	Pion	π^+	π^-	139.6	0	2.6×10^{-8}	$\mu^+ + \nu_\mu$
		π^0	π^0	135	0	2.3×10^{-16}	$\gamma + \gamma$
Leptons	Muon	μ^-	μ^+	105.66	$\frac{1}{2}$	2.2×10^{-6}	$e^- + \nu_\mu + \overline{\nu}_e$
	Electron	e^-	e^+	0.51	$\frac{1}{2}$	Stable	
	Neutrino	ν_e	$\overline{\nu}_e$	0	$\frac{1}{2}$	Stable	
		ν_μ	$\overline{\nu}_\mu$	0	$\frac{1}{2}$	Stable	
Photon	Photon	γ	γ	0	1	Stable	

proton have charge-bearing substructures. It was the theoretical considerations of the substructure of nucleons that led to the prediction and eventual discovery of new particles.

The electron and the proton are both stable particles. The *bound* neutron in a nucleus is also stable, but a *free* neutron is unstable. It decays (with a half-life of about 15.5 minutes) to a proton, simultaneously emitting an electron and an antineutrino:

$$\mathrm{n} \rightarrow \mathrm{p} + \mathrm{e}^- + \bar{\nu}_{\mathrm{e}}. \tag{15.1}$$

In 1904, Einstein[4] established the existence of the photon, a quantum of electromagnetic radiation which is a stable particle of rest mass zero and spin of 1 unit. The photon is its own antiparticle.

ii) *Positron* (e^+). As explained in Chapter 13, the positron was theoretically predicted by Dirac[5] and discovered experimentally by Anderson[6] in 1932 (Anderson observed electron–positron pairs in cloud-chamber photographs). The positron is identical to the electron in mass and spin, except that it has a unit positive charge. In the absence of an electron, the positron is stable. Otherwise it combines with an electron and annihilates, thereby producing 0.51-MeV photons. Thus the positron is an antiparticle of an electron.

iii) *Neutrino and antineutrino associated with the electron* (ν_{e}, $\bar{\nu}_{\mathrm{e}}$). In 1930 Wolfgang Pauli, in order to explain the conservation of linear momentum, angular momentum, and energy in beta decay, introduced these particles. In 1934 Enrico Fermi[7] used them to develop a satisfactory theory of beta decay. These particles participate in beta decay according to the following equations:

$$\begin{aligned} \mathrm{n} &\xrightarrow{\beta^-\text{-decay}} \mathrm{p} + \mathrm{e}^- + \bar{\nu}_{\mathrm{e}} \\ \mathrm{p} &\xrightarrow{\beta^+\text{-decay}} \mathrm{n} + \mathrm{e}^+ + \nu_{\mathrm{e}} \\ \mathrm{p} + \mathrm{e}^- &\xrightarrow{\text{EC decay}} \mathrm{n} + \nu_{\mathrm{e}} \end{aligned} \tag{15.2}$$

Both the neutrino and the antineutrino have zero rest mass, no charge, and a spin of $\frac{1}{2}$ unit. The neutrino is a particle; the antineutrino is its antiparticle. Though postulated in 1930, the existence of these two particles was firmly established only in 1957 by the experiments of Frederick Reines and Clyde Cowan[8]. The reaction they investigated is the reverse of neutron decay,

$$\mathrm{p}^+ + \bar{\nu}_e \rightarrow {}^1_0\mathrm{n} + \beta^+. \tag{15.3}$$

They obtained a beam of $\bar{\nu}_{\mathrm{e}}$ from a powerful nuclear reactor, and used it to bombard five large tanks filled with water which had $CdCl_2$ dissolved in it. Cd absorbs a neutron and emits a gamma ray. When they found that neutrons and positrons were being emitted simultaneously at the rate of (2.88 ± 0.22) particles/hour, they realized that this confirmed the existence of the antineutrino.

iv) *Mu mesons* (μ^+, μ^-); *neutrinos* (ν_μ) *and antineutrinos* ($\bar{\nu}_\mu$) *associated with mesons.* Hideki Yukawa,[9] in 1935, using Heisenberg's suggestions, postulated that the nuclear force between two nucleons (two neutrons, two protons, or a neutron and a proton) was due to the exchange of a new type of particle, called a *meson.* [This is in analogy with the electromagnetic force, in which charged particles exchange a photon, with the result that there is a force between the two charged particles.] As we said in Chapter 11, the mass of this particle was calculated to be about 270 m_e. Note that Yukawa needed to pinpoint the finite mass of the exchanged particle in order to explain the finite range of the nuclear force. With these ideas in the open, experimental physicists began a vigorous search for this elusive particle, the meson.

Until the late 1940's the only source of radiation of high-energy particles that could be used for investigation was cosmic rays. Cosmic radiation has two components:[10] the so-called *soft components*, consisting of electron–positron showers, and *hard components*, consisting of mesons and other particles. Physicists began to study cosmic radiation by various means. In 1937 S. Neddermeyer and C. Anderson,[11] of the California Institute of Technology, used nuclear emulsions. J. Street and E. Stevenson,[12] at Harvard, used a cloud chamber. These scientists observed particles that ionized less readily than protons and lost less energy than high-energy electrons. On further investigation, they concluded that these particles had either unit positive charge or unit negative charge, and that their masses were about 207 m_e, that is, masses which were between the mass of the proton and that of the electron. They originally called these particles mesons, but when they discovered other mesons (pi mesons), they called these particles *mu mesons* or *muons.* [Until the discovery of the pi meson in 1947, the mu meson (with its mass of 207 m_e as compared to that of the pi meson, 270 m_e) was mistakenly thought to be the particle that was exchanged between two nucleons which resulted in the nuclear force predicted by the Yukawa theory. But μ mesons do not interact strongly with nucleons, while π mesons do.] Also there are two types of mu mesons, μ^+ and μ^-.

Mu mesons are unstable; they decay with a mean life of 2.2×10^{-6} sec. One of the decay products of the mu meson is an electron. The energies of the electrons so produced show a range of energy of 9 to 55 MeV. This means that—as in beta decay—in mu meson decay the situation is a three-body one. The decay of mu mesons takes place according to the reaction

$$\mu^{\pm} \rightarrow e^{\pm} + \nu_\mu + \bar{\nu}_e. \tag{15.4}$$

Note that the neutrino ν_μ and the antineutrino $\bar{\nu}_e$ belong to two different classes: one belongs to the meson type and the other to the electron type. This distinction between the two types of neutrinos and antineutrinos, that is, (ν_e, $\bar{\nu}_e$) and (ν_μ, $\bar{\nu}_\mu$), is real. If it weren't, the neutrino and antineutrino in Eq. (15.4) would annihilate each other, thereby producing a gamma ray:

$$\mu^- \rightarrow e^- + (\bar{\nu} + \nu) \overset{?}{\rightarrow} e^- + \gamma. \tag{15.5}$$

The fact that this annihilation reaction does not take place has been confirmed by further sophisticated experiments.[13] The masses of ν_μ and $\bar{\nu}_\mu$ are almost zero.

The mu mesons, μ^+ and μ^-; the electron and positron, e^- and e^+; the neutrino and antineutrino associated with the electrons and positron, ν_e, $\bar{\nu}_e$, ν_μ, and $\bar{\nu}_\mu$, each have a spin of $\frac{1}{2}$ unit. All these 8 particles are called *leptons*, and a system of any of them is governed by Fermi–Dirac statistics.

v) *Pi mesons* (π^+, π^-, π^0). In 1947, C. Lattes, G. Occhialini, and C. Powell[14] exposed nuclear emulsion plates to cosmic rays at high altitude and observed tracks of the type shown in Fig. 15.1. Part (a) is a schematic line drawing of the decays, and part (b) is an actual photograph of the tracks. From the rate of increase of grain density, it was obvious that the first part of the track was due to a charged

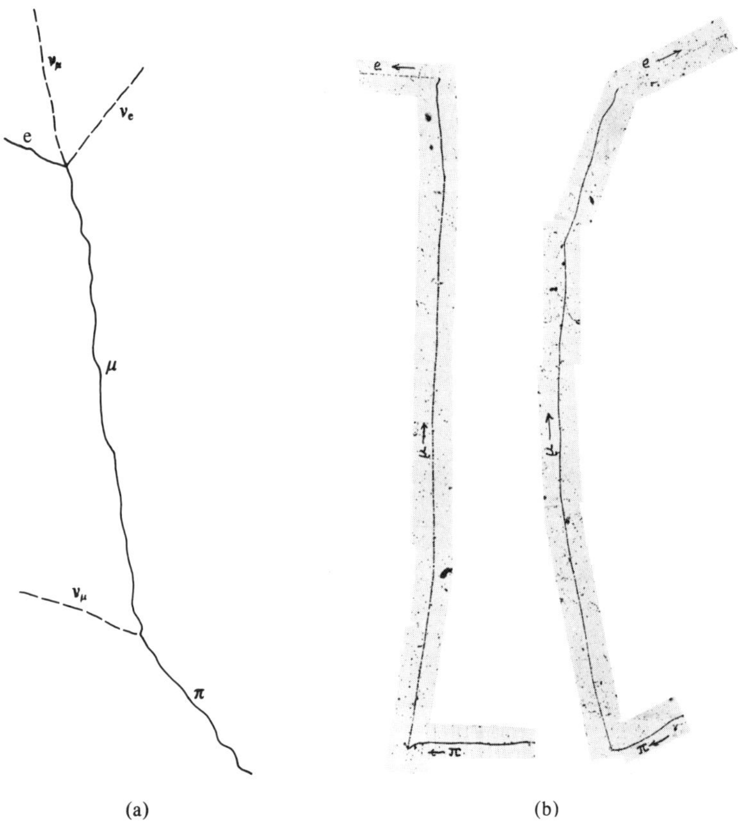

Fig. 15.1 Sequence of pion–muon–electron (π–μ–e) decay. The positive pion, after coming to rest, decays into a positive muon and a neutrino. After the muon comes to rest, it decays into an electron and two neutrinos. (a) Diagram of the decays. (b) Photograph of tracks of particles (courtesy University of Bristol).

particle which had a mass several hundred times as great as that of an electron. This particle decays into another charged particle, which also has a mass several hundred times the mass of the electron, and leaves a second track before decaying into an electron (see Fig. 15.1). They concluded that the first particle was a "heavy" meson and the second a "light" meson resulting from the decay of the first. They further confirmed that the light meson was the mu meson discussed above. They called the heavier meson a *pi meson* or *pion*, and established its mass as $\sim 273\ m_e$. Pi mesons are the particles postulated by Yukawa[9] in 1935 to explain nuclear force.

Pi mesons come positively charged, π^+; negatively charged, π^-; and neutral π^0. The masses of π^+ and π^- are 273.7 m_e. The mass of π^0 is 265 m_e. All pi mesons have zero spins. The mean life of π^+ and π^- is 2.6×10^{-8} sec, while that of π^0 is 2.3×10^{-16} sec. The π^+ and π^- mesons decay into μ^+ and μ^- mesons, respectively, but they themselves never decay directly into electrons. Since no photons are emitted in the decay of π^+ or π^- mesons, conservation of momentum and energy requires that another particle be emitted at the same time as the decay of each charged pion. These particles are ν_μ and $\bar{\nu}_\mu$. Hence the decay processes are

$$\pi^+ \to \mu^+ + \nu_\mu, \qquad \pi^- \to \mu^- + \bar{\nu}_\mu, \tag{15.6}$$

while μ^+ and μ^- decay as

$$\mu^\pm \to e^\pm + \nu_\mu + \bar{\nu}_e. \tag{15.7}$$

Negative pions and muons are easily captured by matter; the result is the formation of mesonic atoms,[15] which disintegrate in short times. The energy released in such captures produces "star"-type patterns when stopped in nuclear emulsions. The neutral pion, π^0, decays into two gamma rays (photons):

$$\pi^0 \to \gamma + \gamma. \tag{15.8}$$

In 1950, Stellar, Steinberger, and Panofsky at the University of California in Berkeley produced pions by bombarding hydrogen or beryllium targets with a 330-MeV x-ray beam which they obtained from a synchrotron. They found that some of the reactions by which charged pions are produced are

$$\begin{aligned} h\nu + p &\to \pi^0 + p \\ h\nu + p &\to \pi^+ + n \\ h\nu + n &\to \pi^- + p \end{aligned} \tag{15.9}$$

Strange particles

In 1947, G. Rochester and C. Butler,[16] who were members of a group studying cosmic rays at the University of Manchester in England, detected some new particles with confusing properties. They singled out these particles in the showers of particles produced when cosmic rays passed through lead plates placed in cloud

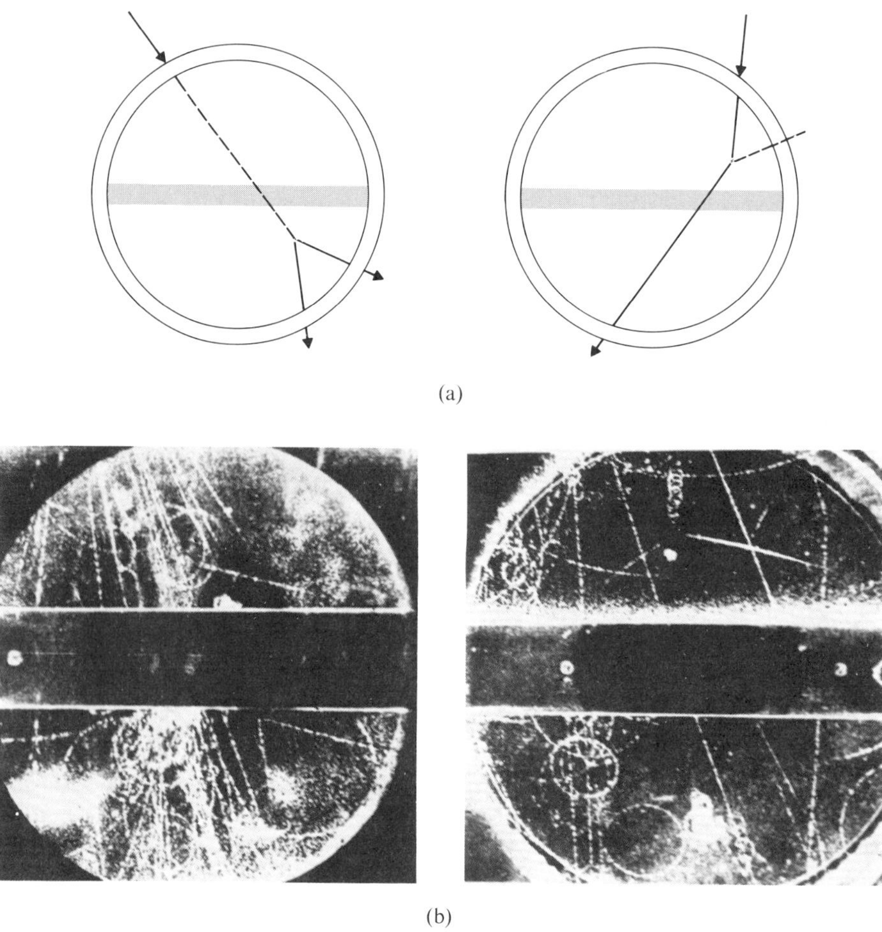

(a)

(b)

Fig. 15.2 (a) Diagram of decay of lambda particle $\Lambda^0 \rightarrow p^+ + \pi^-$ and of sigma particle $\Sigma^+ \rightarrow n + \pi^+$. (b) Cloud-chamber photographs. [From G. D. Rochester and C. C. Butler, *Nature*, **160,** 855 (1947).]

chambers. The particles formed curious two-prong or V-shaped patterns, so they called them *V-particles.*

Figure 15.2 presents photographs of the tracks of two such particles. The first photograph they attributed to a neutral particle which did not leave any track in the cloud chamber, but which decayed into two charged particles. They called this neutral particle a *lambda particle*, and found that it decayed into a proton and a pi meson:

$$\Lambda^0 \rightarrow p^+ + \pi^-. \tag{15.10}$$

The tracks in the second photograph of Fig. 15.2 were attributed to an unknown charged particle which decayed into another charged particle plus a neutral particle. They called this particle a *positive sigma particle*, and found that it decayed into a positive pi meson and a neutron:

$$\Sigma^+ \to \pi^+ + \text{n}. \tag{15.11}$$

Further investigations by scientists all over the world led to the discovery of many new particles. These were classified into two groups on the basis of their masses. (1) *Hyperons*, particles with masses greater than the mass of the proton. They are the lambda (Λ^0), the sigma (Σ^+, Σ^-, Σ^0), the negative and neutral xi (Ξ^-, Ξ^0), and the antiparticles of all these, i.e., the *antihyperons.* (2) *Heavy mesons*, particles heavier than pi mesons and light than protons. These are called *K-mesons*, and are classed as K^+, K^-, K^0, and $\overline{K}^0$. Before we talk about the discoveries and properties of these particles, let us explain why they are called *strange particles.*

The decay times of these newly discovered particles have a peculiar property. On the basis of theory, we would expect these particles, which are manufactured by nuclear forces (called *strong interactions*), to have a time scale of $\sim 10^{-23}$ sec. But the lifetimes of these particles range from $\sim 10^{-8}$ to $\sim 10^{-10}$ sec, which is the time scale of particles manufactured by *weak interactions* (forces which control the production and decay of leptons). So here we have a contradictory situation: These particles are produced by *strong* interactions, but they decay as if they had been produced by *weak* interactions; i.e., they live 100,000 billion times longer than they should! So they have been called *strange* or *queer* particles, because, according to the principle of reversibility, a particle produced by strong interaction should decay by strong interaction.

A. Pais[17] and others explained this paradox by introducing a concept known as *associated production.* They postulated that more than one strange particle at a time is manufactured by strong interaction. Thus (K-particle, anti-K-particle) or (K-particle, hyperon) or (anti-K-particle, antihyperon) pairs should be produced together. After they have been produced, the probability of one strange particle locating its mate before it decays is extremely small; and an individual strange particle produced by strong interaction does not have enough energy to decay by strong interaction. Hence the individual particles move away from each other and eventually decay by weak interactions, which explains why they have longer lifetimes than by all rights they should have.

vi) *K mesons* (K^+, K^-, K^0, $\overline{K}^0$). Charged K mesons were first observed in 1949 by Cecil Powell[14] and a group of his colleagues at the University of Bristol in England. Figure 15.3 shows a typical photograph (taken in an emulsion) of the tracks. Charged K mesons decay into three pi mesons according to the following reaction:

$$K^\pm \to \pi^\pm + \pi^\pm + \pi^\mp \tag{15.12}$$

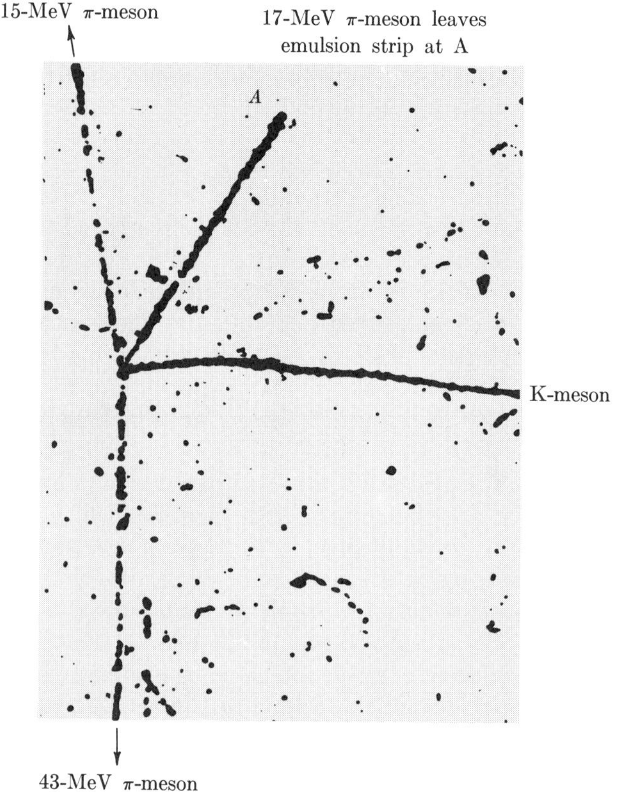

Fig. 15.3 The decay of the K^- meson by the τ mode into three mesons. [Courtesy Brookhaven National Laboratory, Upton, New York]

The K^+ is called the *positive* K *meson*, and the K^- is called the *negative* K *meson.* Their masses are $\sim 964\ m_e$. The K mesons decay by many different modes, some of which are given in Table 15.2. Originally the K^+ meson was called a *tau* (τ) *meson* and K^- was called a *theta* (θ) *meson.* Actually, these are two different modes of decay of K-mesons:

$$K^+ \to \tau \to \pi^+ + \pi^- + \pi^+, \tag{15.13a}$$

$$K^- \to \theta \to \pi^- + \pi^0. \tag{15.13b}$$

The spins of the charged as well as neutral K mesons are zero; their mean lifetimes are of the order of 10^{-8} sec.

The neutral K meson, K^0, is very strange and interesting. Its counterpart, the anti-K-zero meson, $\overline{K}^0$, is not identical to it! The reason for this discrepancy was explained in 1947 by G. Rochester and C. Butler.[16] Both K^0 and $\overline{K}^0$ are different mixtures of two other particles, K_1^0 and K_2^0. K_1^0 decays with a lifetime of 1×10^{-10} sec and K_2^0 with a lifetime of 6.1×10^{-8} sec. The amounts of K_1^0 and K_2^0 mixed in K^0 and $\overline{K}^0$ were predicted by Gell-Mann.[18]

vii) *Hyperons: lambda* (Λ^0), *sigma* (Σ^+, Σ^-) *and xi* (Ξ^0, Ξ^-). These particles were originally observed in cosmic radiation, and were identified in 1956 only after Murray Gell-Mann[18] and Kazuhiko Nishijima[19] independently postulated them theoretically. Table 15.2 lists the properties of these particles; note that all hyperons are heavier than nucleons, and xi minus is the heaviest of them all, with a mass $\sim 2585\ m_e$. The xi particles are also called *cascade particles* because they go through several decays before becoming stable. The spins of all the hyperons are $\frac{1}{2}$, and they are all strange particles. That is, they decay by weak interactions, except Σ^0, which also decays by electromagnetic interaction, by the reaction

$$\Sigma^0 \to \Lambda^0 + \gamma. \tag{15.14}$$

Physicists finally succeeded in producing these particles in laboratories by using high-energy accelerators. In the beginning, they produced hyperons in conjunction with K mesons, but when accelerators capable of producing particles with still higher energy became available, they found that they could produce hyperon pairs instead of hyperon–K-meson pairs.

The lambda particle, first produced in the laboratory in 1953 by William Fowler, Ralph Shutt, and Alan Thorndike,[20] follows the reaction path

$$\pi^- + p \to \Lambda^0 + K^0. \tag{15.15}$$

Fowler *et al.* allowed a beam of 1.5-BeV pi mesons to enter a cloud chamber containing hydrogen at 18 atm. The Λ^0 produced traveled 0.65 cm before it decayed into a π^- meson and a proton:

$$\Lambda^0 \to p + \pi^-.$$

Similarly, by using a beam of pi mesons, they produced Σ by the reaction path

$$\pi^+ + p \to \Sigma^+ + K^+. \tag{15.16}$$

Note that lambda particles exhibit only neutral charge, while xi hyperons exist as two types: negative and neutral.

viii) *Antinucleons: antiproton* ($\bar{p}$) *and antineutron* ($\bar{n}$). P. A. M. Dirac predicted theoretically that an electron has an antiparticle—the positron. This was experimentally observed by Carl D. Anderson in 1932. Dirac furthermore postulated that *every* particle has an antiparticle. According to theory, the antiparticle of the proton—the antiproton—should have the same mass as a proton, but an opposite charge, and the same spin as a proton, but an opposite magnetic moment. The antiproton should be stable, and it and a proton should mutually annihilate each other. The antiproton should be generated as one member of a proton–antiproton pair.

The 20-year search for the antiproton ended in 1955 when O. Chamberlain, E. Segré, C. Wiegand, and T. Ypsilantis[21] using a beam of 6-BeV protons obtained from the bevatron at the University of California at Berkleley, produced antiprotons by proton–proton collisions:

$$\mathrm{p} + \mathrm{p} \rightarrow \mathrm{p}^+ + \mathrm{p}^+ + \mathrm{p}^+ + \bar{\mathrm{p}}^-. \tag{15.17}$$

There are two important aspects of antiproton production: (1) The antiproton in the $\mathrm{p}^+ - \bar{\mathrm{p}}^-$ pair must be formed with a new proton. (2) It may look as though the energy needed to produce an antiproton should be twice the rest-mass energy of the proton, that is, $2 \times 938 = 1876$ MeV. But this is not true. In order to conserve momentum, the minimum initial kinetic energy needed is ~6 BeV, as we see in Example 15.1.

Example 15.1. Show that the amount of energy needed to produce an antiproton by the reaction $\mathrm{p} + \mathrm{p} \rightarrow \mathrm{p}^+ + \mathrm{p}^+ + \mathrm{p}^+ + \bar{\mathrm{p}}^-$ is about 6 BeV.

Suppose that the initial momentum of the system is p_i. After the reaction it is equally distributed between the four particles:

$$p_{1f} = p_i/4 \tag{a}$$

and

$$K_i = 4K_{1f}, \tag{b}$$

which is related to the minimum initial bombarding energy K_i of the incident proton by the relation

$$K_i = 4K_{1f} + 2M_0c^2, \tag{c}$$

where M_0 is the rest mass of the proton.

But the final momentum p_{1f}, kinetic energy K_f, and total energy E_T of the product particle is given by the relativistic relation

$$E_T^2 = (K_{1f} + M_0c^2)^2 = p_{1f}^2c^2 + M_0^2c^4 \tag{d}$$

or

$$p_{1f}^2c^2 = (K_{1f} + M_0c^2)^2 - M_0^2c^4. \tag{e}$$

Similarly,

$$p_{1i}^2c^2 = (K_i + M_0c^2)^2 - M_0^2c^4. \tag{f}$$

Substituting Eqs. (e) and (f) in Eq. (a), we get

$$(K_i + M_0c^2)^2 - M_0^2c^4 = 16[(K_{1f} + M_0c^2)^2 - M_0^2c^4] \tag{g}$$

Substituting for K_i from Eq. (c) into Eq. (g) and solving for K_{1f}, we get

$$K_{1f} = M_0c^2. \tag{h}$$

Thus, from Eq. (c), using Eq. (h), we obtain

$$K_i = 4M_0c^2 + 2M_0c^2 = 6M_0c^2 = 6(938 \text{ MeV}) = 5628 \text{ MeV}.$$

That is, the incident proton needs about 6 BeV of energy to produce an antiproton.

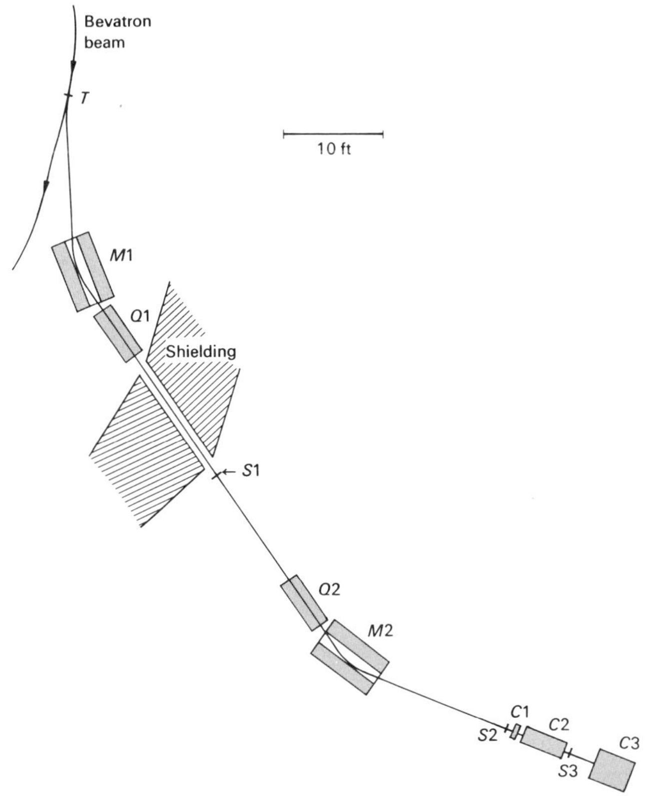

Fig. 15.4 The mass spectrograph uses a system of magnets, scintillation detectors, and Čerenkov detectors to detect the presence of antiprotons. [After Chamberlain *et al.*, 1955]

Figure 15.4 shows an experimental arrangement used by Chamberlain *et al.*[21] to produce an antiproton. When the protons collided with other protons, for each proton–antiproton pair produced, there were also ~100,000 mesons produced at the same time! Using a system of magnets (M), scintillation detectors (S), and Čerenkov detectors (C), they were able to count particles with different rest masses, charges, and kinetic energies. Figure 15.5 shows the results of their experiments. The antiproton appears at the correct position on the mass scale.

The mutual annihilation of an antiproton and a proton was investigated at Berkeley. Observers were able to distinguish 20 tracks showing annihilations.

It is much harder to detect an antineutron ($\overline{\text{n}}$) because it has no charge. But when it collides with a neutron and annihilates, it produces tracks, and such tracks were observed in propane bubble-chamber photographs.[22] Like the

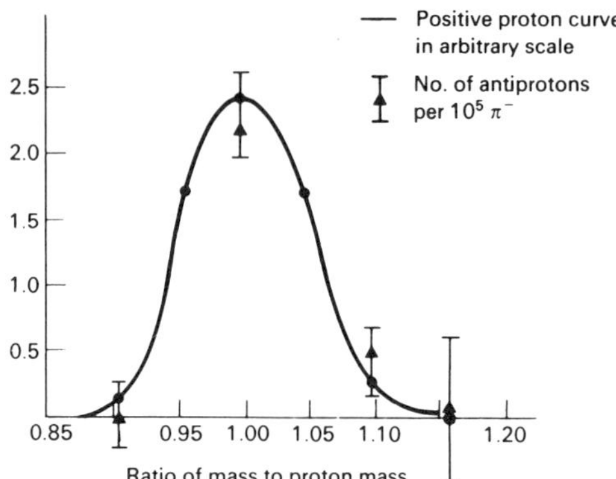

Fig. 15.5 Appearance of a peak at the right place in the plot of the experimental counting rate of antiprotons versus the ratio of mass to mass of protons confirms the existence of the antiproton. [From O. Chamberlain, E. Segré, C. Wiegand, and T. Ypsilantis, *Phys. Rev.* **100,** 947 (1955)]

neutron, the antineutron has no charge. It decays by emitting a positron with the same lifetime as the neutron.

ix) *Antihyperons.* Of the many antihyperons, we mention only two: *anti-xi-plus* ($\bar{\Xi}^+$) and *anti-xi-zero* ($\bar{\Xi}^0$). Others are listed in Tables 15.1 and 15.2.

Anti-xi-plus is also written as anti-xi-minus because $\bar{\Xi}^+$ is the antiparticle of Ξ^-. It is the next-to-the-last strange particle predicted by Gell-Mann and Nishijima. It was discovered[23] in 1962 in experiments using the 30-BeV accelerator at Brookhaven in New York and the accelerator at CERN in Geneva. When a high-speed antiproton collides with a proton, it produces an anti-xi-plus together with its antiparticle, the xi-minus.

$$\bar{p} + p \rightarrow \Xi^- + \bar{\Xi}^+. \tag{15.18}$$

In about one-billionth of a second, the anti-xi-plus decays into an antilambda and a pi meson. These particles then decay further. The complex decay that follows is shown in Fig. 15.6. The lengthy decays of xi particles have earned them the name *cascade particles.*

A group of physicists from Yale investigated 34,000 photographs of bubble-chamber tracks. Of these, 14 showed tracks of antiprotons, and only one track was a picture of an anti-xi-minus ($\bar{\Xi}^+$) event.

Anti-xi-zero ($\bar{\Xi}^0$) particles, resulting from the interaction of antiprotons with protons, were produced by the 30-BeV synchrotron at Brookhaven.[24] Since the anti-xi-zero is a neutral particle, it is detected by its decay products. Out of 300,000 photographs observed by the Yale group, only three showed the rare anti-xi-zero particle.

Tables 15.1 and 15.2 show all the particles discovered up until the late 1950's.

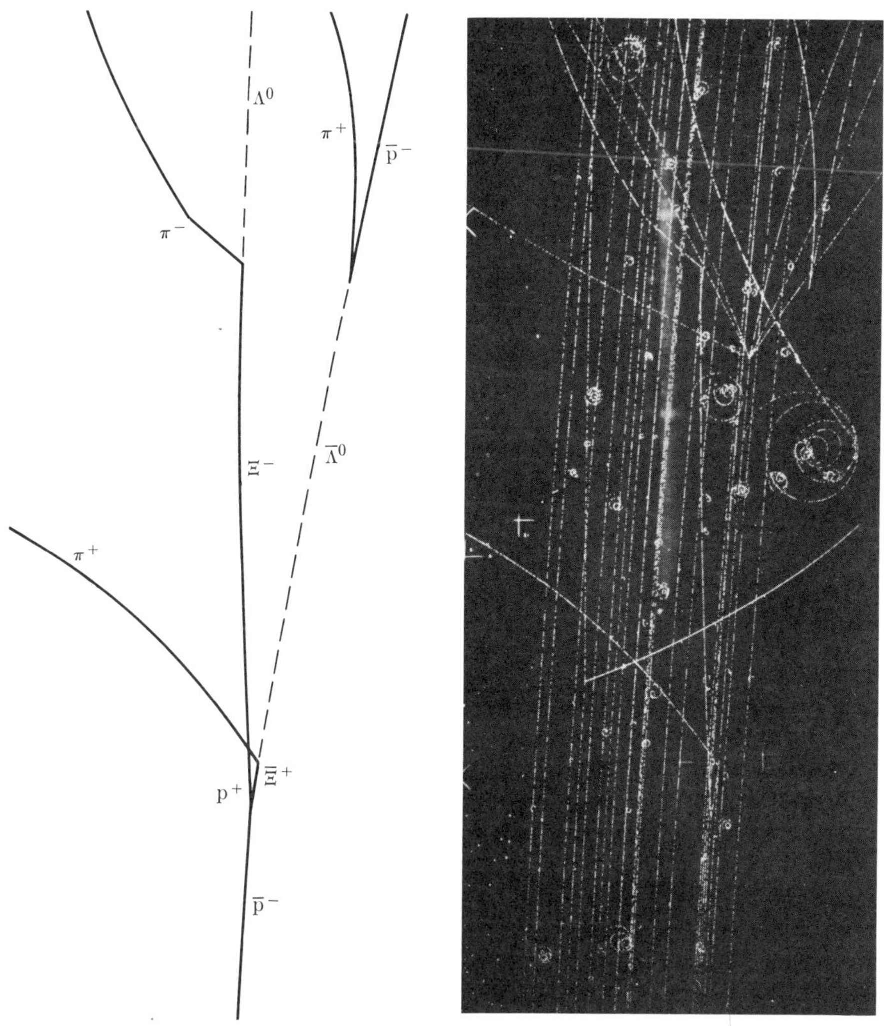

Fig. 15.6 The decay products of an anti–xi–minus particle and its antiparticle produced by the reaction $\overline{p} + p \rightarrow \Xi^- + \overline{\Xi}^+$. (a) Schematic of the decays. (b) Bubble-chamber photograph. [From H. N. Brown, *et al., Phys. Rev. Letters* **8**, 255 (1962)]

As we said before, these particles have been classified into four groups. For a while, physicists thought that all the elementary particles had been discovered. But very shortly another new class of particles started emerging, which we shall discuss in Section 15.4. For the time being, we shall see how to classify these 32 elementary particles.

15.3 BASIC INTERACTIONS AND CONSERVATION LAWS

The 32 "old" fundamental particles which we discussed in Section 15.2 were classified into four groups—baryons, leptons, mesons, and photon. As we shall see, this classification is based on the laws of conservation. The production and decay of any particle are caused by a *force*. At present there are only four known basic forces or interactions and one (or more) of these four interactions governs the production or decay of each of these 32 particles. Actually we may say that not only microscopic systems, but even macroscopic systems, are governed by one (or more) of these four forces. So let's review these four forces briefly. Table 15.3 lists them, together with their relative strengths and reaction time scales.

Table 15.3

Four Basic Interactions

Types of interactions	Relative strength	Range	Reaction time scale
Strong	1	$\sim 10^{-13}$ cm	10^{-23} sec
Electromagnetic	$(\frac{1}{137}) \sim 10^{-2}$	∞	10^{-21} sec
Weak	$\sim 10^{-13}$	Almost zero	10^{-8} sec
Gravtitational	$\sim 10^{-39}$	∞	

a) *Strong interactions.* The strong interactions, also called *nuclear interactions* or *nuclear forces*, are strongest of all, and are responsible for holding nuclear matter (nucleons) together. They also are responsible for the production of baryons and K-mesons. As explained earlier, nuclear interactions are the result of virtual emission and absorption of pions. The range of strong interactions is $\sim 10^{-13}$ cm. They are called strong because their reaction time (the time it takes two particles to interact or feel interaction) is very small: $\sim 10^{-23}$ sec.

b) *Electromagnetic interactions.* Electromagnetic forces, which are only 1/137th as strong as nuclear forces, act between charged particles and particles with electric and magnetic moments. The electromagnetic force is the result of virtual emission and absorption of the photons by charged particles. Because photons have zero rest masses, the range of the electromagnetic interaction is infinite. Their time scale for interaction is $\sim 10^{-21}$ sec. Electromagnetic forces are what cause the binding of electrons in atoms, and the formation of molecules.

c) *Weak interactions.* The strength of weak interactions is only 10^{-13} the strength of nuclear interactions. The range of weak interactions is almost zero. These weak forces govern the properties of leptons, and are responsible for beta decay, as well as the decay of some heavier particles, such as strange particles.

d) *Gravitational interactions.* These are the weakest of all the basic forces. On a microscopic scale, their effect is negligible. Gravitational forces are always attractive, and their effect is felt quite clearly between large masses, such as planets.

The classification of the 32 elementary particles into four groups—baryons, mesons, leptons, and photons—does not mean that each group acts with one particular type of interaction. As a matter of fact, an elementary particle may exhibit some or all four types of interaction. For example, a proton, when inside the nucleus, is bound by the influence of strong interactions. But because of its charge it acts by electromagnetic interaction. The production of a proton by the decay of a neutron is governed by weak interaction. And, of course, since the proton has a mass, it exhibits gravitational interaction.

Investigation of the properties of the elementary particles plus application of the laws of conservation has been very helpful in grouping and naming new particles. All four interactions obey the well-known conservation laws—conservation of linear momentum, angular momentum, and energy. Then there is a group of quantities with which we are not so familiar, and which are not exhibited by all the interactions: parity, baryon number, lepton number, isospin and its Z-component, and strangeness (related to hypercharge and average charge). Table 15.4 lists these quantities. In the language of quantum mechanics, there is a quantum number corresponding to each conserved quantity. Let us now take a look at some of these conserved quantities.

Using the Mössbauer effect, scientists have verified the conservation of energy and momenta to an accuracy of one part in 1000. They have verified the conservation of charge to one part in 10^{17}. Different reactions may be allowed by some conservation laws and forbidden by others. For example, the laws of conservation

Table 15.4

Conservation Laws in Different Interactions

Conserved quantity	Strong	Electromagnetic	Weak	Gravitational
Linear momentum	Yes	Yes	Yes	Yes
Angular momentum	Yes	Yes	Yes	Yes
Energy	Yes	Yes	Yes	Yes
Charge	Yes	Yes	Yes	—
Baryon number	Yes	Yes	Yes	—
Lepton number	Yes	Yes	Yes	—
Muon number	Yes	Yes	Yes	—
Parity	Yes	Yes	No	—
Total isotopic spin	Yes	No	No	—
Z-component of isotopic spin	Yes	Yes	No	—
Strangeness	Yes	Yes	No	—

of energy and momentum allow the decay of an electron into a gamma ray and a neutrino, but the law of conservation of charge forbids such a reaction.

The conservation of parity (right-left symmetry) holds only approximately. It is conserved in strong and electromagnetic interactions, but does not hold in weak interactions.

i) *The conservation of baryon number, lepton number, and muon number*. The baryon number, or the atomic mass number A, is defined as the number of baryons minus the number of antibaryons. For baryons $A = +1$, for antibaryons $A = -1$, while for mesons $A = 0$. The conservation of baryon number means that in a given system the value of A remains constant. That is, no matter what interaction governs a given reaction, the value of A does not change. This implies that baryons cannot be created or destroyed singly, but only in baryon–antibaryon pairs. For example, a proton does not decay into a positron and a gamma ray because the conservation of baryon number forbids it, even though charge, energy, and momentum may be conserved.

For leptons, the law of conservation of lepton number, and for muons, the law of conservation of muon number, have been found to hold good. There is not much physical justification for these laws except that they do forbid certain reactions.

ii) *The conservation of isotopic spin*. This is one of the most important quantities in classifications of new elementary particles. It also explains some of the mysteries associated with the properties of elementary particles. The name of this quantity is misleading, because isotopic spin (or isospin) I has nothing to do with spin *or* angular momentum; it is purely a mathematical quantity. But using the concept of the symmetry of isotopic spin in analogy with the real spin or angular momentum helps us classify baryons and mesons, i.e., for strongly interacting particles.

According to the concept of isotopic spin, the proton and neutron are regarded as two different states of the nucleon. The nucleon is assigned an isotopic spin of $I = \frac{1}{2}$, and its projection on the Z axis results in two components, $I_Z = +\frac{1}{2}$, which corresponds to the positively charged state, the proton; and $I_Z = -\frac{1}{2}$, which corresponds to the neutral charged state, the neutron. Thus a nucleon with $I = \frac{1}{2}$ is a charged doublet. Similarly, pions with $I = 1$ constitute a charged triplet, corresponding to $I_Z = 0, \pm 1$, where $I_Z = 0$ is the neutral π^0 meson, $I_Z = +1$ is the π^+ meson, and $I_Z = -1$ is the π^- meson. Thus we can deduce that the multiplicity M of isotopic states is given by $M = 2I + 1$.

If strong interactions were the only interactions present, the two nucleons would be identical. We state this fact by saying that "the two states of the nucleon are alike, and are related to each other by the symmetry of the isotopic spin." But the presence of the electromagnetic interaction, which violates the symmetry of the isotopic spin (i.e., electromagnetic interactions do not conserve isotopic spin), leads to the fact that the rest mass of the neutron is 1.3 MeV more than that of the proton. If the electromagnetic interaction could be turned off in the laboratory, the masses of the proton and neutron would be the same. Similar remarks apply to pions. Table 15.5 lists the isotopic spins assigned to different baryons and mesons.

Table 15.5

Quantum Numbers Associated with Baryons and Mesons

Particle	Symbol	Baryon number A	Charge Q	Isotopic spin I	Z-component of I I_Z	Average charge $\bar{Q}$	Strangeness S
Nucleon	p	+1	+1	$\frac{1}{2}$	$+\frac{1}{2}$	$+\frac{1}{2}$	0
	n	+1	0		$-\frac{1}{2}$		0
Antinucleon	$\bar{\text{p}}$	−1	−1	$\frac{1}{2}$	$-\frac{1}{2}$	$-\frac{1}{2}$	0
	$\bar{\text{n}}$	−1	0		$+\frac{1}{2}$		0
Lambda	Λ^0	+1	0	0	0	0	−1
Antilambda	$\bar{\Lambda}^0$	−1	0	0	0		+1
Sigma	Σ^+	+1	+1	1	+1	0	−1
	Σ^0	+1	0		0		−1
	Σ^-	+1	−1		−1		−1
Antisigma	$\bar{\Sigma}^+$	−1	+1	1	+1	0	+1
	$\bar{\Sigma}^0$	−1	0		0		+1
	$\bar{\Sigma}^-$	−1	−1		−1		+1
Xi	Ξ^-	+1	−1	$\frac{1}{2}$	$-\frac{1}{2}$	$-\frac{1}{2}$	−2
	Ξ^0	+1	0		$+\frac{1}{2}$		−2
Anti-xi	$\bar{\Xi}^+$	−1	+1	$\frac{1}{2}$	$+\frac{1}{2}$	$+\frac{1}{2}$	+2
	$\bar{\Xi}^0$	−1	0		$-\frac{1}{2}$		+2
Pion	π^+	0	+1	1	+1	0	0
	π^0	0	0		0		0
	π^-	0	−1		−1		0
K-meson	K^+	0	+1	$\frac{1}{2}$	$+\frac{1}{2}$	$+\frac{1}{2}$	+1
	K^0	0	0		$-\frac{1}{2}$		+1
Anti-K	$\bar{\text{K}}^-$	0	−1	$\frac{1}{2}$	$-\frac{1}{2}$	$-\frac{1}{2}$	−1
	$\bar{\text{K}}^0$	0	0		$+\frac{1}{2}$		−1

iii) *The conservation of strangeness.* The strangeness number S is assigned to all the strongly interacting particles. It is defined as

$$S = Y - A = 2\bar{Q} - A, \tag{15.19}$$

where A is the baryon number, $\bar{Q}$ is the average charge of the isotopic-spin multiplet, and Y $(= 2\bar{Q})$ is the hypercharge.

For example, for the nucleon multiplet, the average charge (remembering that the charge of the proton is $+1$ and that of the neutron is 0) is

$$\bar{Q} = \frac{1 + (0)}{2} = \frac{1}{2},$$

and $A = 1$. Therefore $S = 2(\frac{1}{2}) - 1 = 0$. That is, the nucleons are assigned a strangeness of zero. For pions, the average charge

$$\bar{Q} = \frac{1 + 0 + (-1)}{3} = 0$$

and $A = 0$. Hence $S = 2(0) - 0 = 0$. That is, the strangeness of the pions is zero. Similarly, for lambda zero, Λ^0, the average charge of the multiplet is zero, and $A = 1$. Hence $S = 2(0) - 1 = -1$. That is, the strangeness of the Λ^0 is -1. Table 15.5 lists the strangeness number assigned to other strongly interacting particles.

The conservation of strangeness number is an experimental law. Its conservation explains "associated production," i.e., the strange particles are produced by ordinary particles which have zero strangeness. The product particles must contain at least two strange particles with equal and opposite strangeness numbers, so that the total strangeness of the final products is also zero.

The quantum numbers associated with strongly interacting particles are listed in Table 15.5.

15.4 RESONANCE PARTICLES

Most of the elementary particles discussed in the previous sections have a short mean lifetime of $\sim 10^{-10}$ sec. But this time is sufficient to enable the particles to travel long enough distances in the detectors so that their tracks are visible. In the late 1950s and 1960s, a new class of particles began to be observed. Their common characteristic was that they had mean lifetimes of $\sim 10^{-23}$ sec. Since they did not leave observable tracks, the only way to identify them was through their decay products!

Did these new particles really exist autonomously before decaying, or were they just a group of particles which traveled together for about 10^{-23} sec and then decayed (or separated)? Physicists have avoided answering this question by calling this new type of particles *resonance particles* or *resonance states.*

The first resonance particle, N*, was discovered by Enrico Fermi in 1952, but was not so named. In 1960, at the Lawrence Radiation Laboratory, Luis W. Alvarez *et al.*,[25] using high-energy K-minus mesons aimed at liquid hydrogen (protons) in a bubble chamber, observed the following reaction:

$$\overline{K}^- + p \rightarrow \Lambda^0 + \pi^+ + \pi^-. \tag{15.20}$$

They analyzed the visible tracks of π^+ and π^- for their momentum and energy by means of a computer. Every once in a while it looked as if one pion was recoiling from one other particle and not two. The implication was that the other pion and the lambda particle did not break apart immediately, but remained together long enough for the pion to recoil from the lambda particle. That is, $\pi^\pm$ was recoiling from $(\Lambda^0 + \pi^\mp)$. They called this combination $(\Lambda^0 + \pi^\mp)$ by the name $Y^{*\pm}$ resonance. They assumed that the reaction was taking place in two stages:

$$\overline{K}^- + p^+ \rightarrow Y^{*\pm} + \pi^\mp, \tag{15.21a}$$

$$Y^{*\pm} \rightarrow \Lambda^0 + \pi^\pm. \tag{15.21b}$$

Figure 15.7 shows the tracks obtained from these reactions. The mass of the Y* resonance was found to be 1384 MeV and the lifetime 10^{-23} sec.

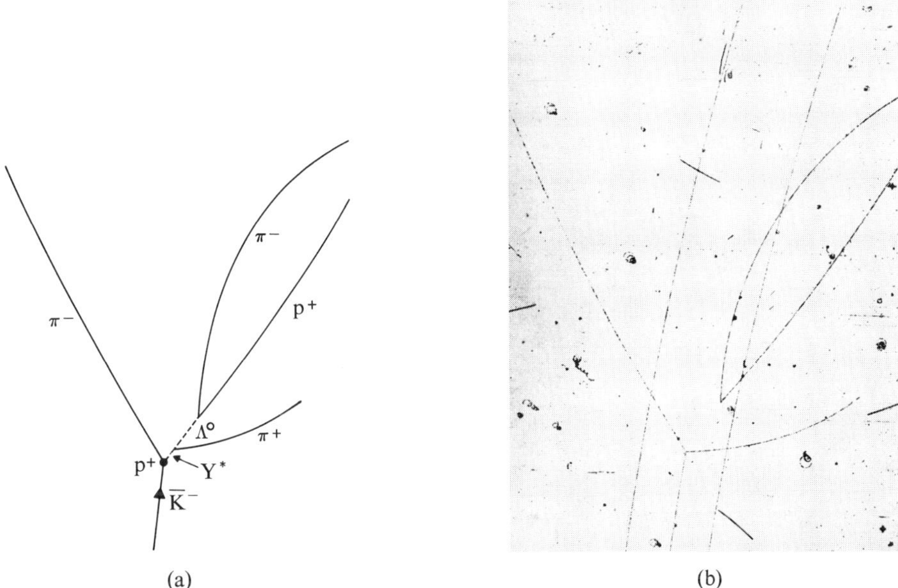

Fig. 15.7 (a) Schematic, and (b) bubble-chamber photograph of a Y^* resonance produced by the reaction $\overline{K}^- + p^+ \rightarrow Y^* + \pi^-$. Broken line in (a) representing no track of Y^* before it disintegrates into Λ^0 and π^+. Λ^0 further decays into p^+ and π^-. [Courtesy L. W. Alvarez, Lawrence Radiation Laboratory.]

Table 15.6

Strongly Interacting Particles as of 1964

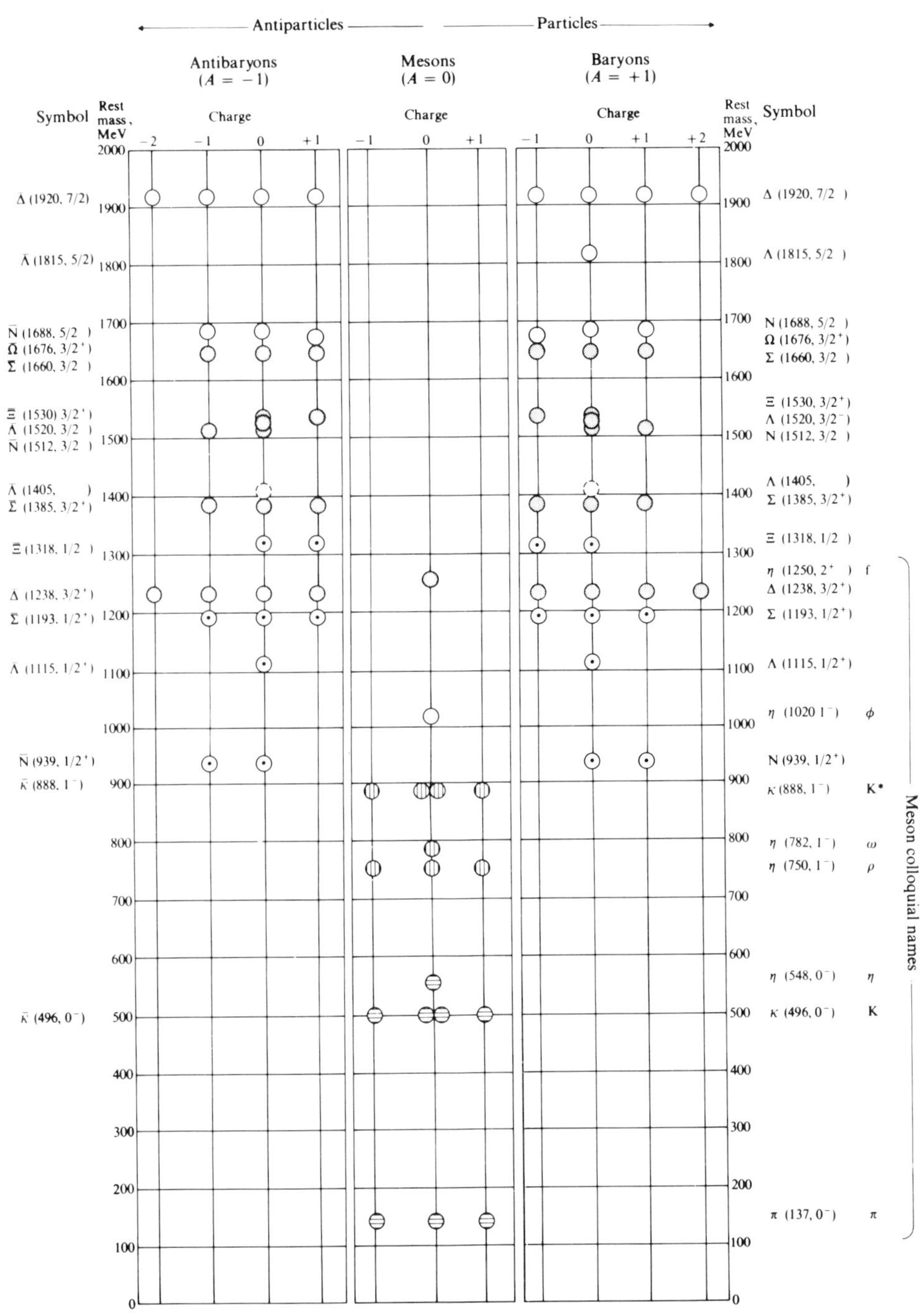

The establishment of Y* resonance led to the discovery of many more resonances. For example, N* resonance was found to be a combination of a pion and a nucleon, with a mass of 1237 MeV and a lifetime slightly shorter than that of Y*. Incidentally, the discovery of *rho* (ρ) and *omega* (ω) *resonances* helped explain the interior structure of nucleons,[26] an important step forward.

The list of resonance particles grew very fast; Table 15.6 shows a list of them as of 1964.

We shall conclude our discussion with one of the most important resonance particles: *omega minus* (Ω^-).

One of the best ways to classify particles is according to the so-called *eightfold way*. This predicts that in order for a supermultiplet of 10 particles sharing the same basic characteristic to be completed (see Section 15.5), there should exist a particle with a mass between 1676 and 1680 MeV. Gell-Mann and Susumu Okubo named this particle omega minus (Ω^-). They expected that it would decay by one of these reactions:

$$\Omega^- \to \Xi^0 + \pi^-, \tag{15.22a}$$

$$\Omega^- \to K^- + \Lambda^0, \tag{15.22b}$$

$$\Omega^- \to \Xi^- + \pi^0. \tag{15.22c}$$

The Brookhaven team, using the AGS (alternating gradient synchrotron) machine, produced 33-BeV protons which, when they collided with a tungsten target, produced 5-BeV K^- mesons. They separated these K^- mesons from the main beam by means of 450 feet of bending and focusing magnets, collimators, and separators. When these 5-BeV K^- mesons entered the 80-inch hydrogen bubble chamber, they reacted with protons. The Brookhaven group took 100,000 photographs, and finally were able to observe all three reactions in Eq. (15.22): in January 1964, February 1964, and November 1964, respectively.[27] Figure 15.8 shows a photograph and a line drawing of reaction (15.22a). The mass of the omega-minus was found to be 1686 $\pm$ 12 MeV, and its lifetime 0.7×10^{-10} sec. (Even though Ω^- is classified as a resonance state, it is a baryon.)

◀ Strongly interacting particles are presented with a new naming system. The 82 particles and antiparticles in this chart include only those with a rest mass below 2000 million electron volts (MeV) and with atomic mass number (A) of -1,0 or +1. Their existance has been predicted by the "eightfold way." The mass assignments are averages for the members of a charge multiplet, or family of states, differing only in their electric charge. Multiplets that have the same spin angular momentum (J) and parity (P) are assigned a common mark. The same mark is also assigned to "recurrences" of these particles. Recurrences are particles with 2, 4, 6 (and so on) more units of spin than their "ground" states of lowest mass. Baryons: Ξ xi, Σ sigma, Λ lambda, N nucleon, Δ delta, Ω omega; Mesons: η eta, π pi, $\bar{\kappa}$ antikappa, κ kappa; Colloquial names: ϕ phi, ω omega, ρ rho. Letter A = atomic mass, or baryon, number. Parentheses contain: (rest mass in million electron volts, spin angular momentum, parity). [From "Strongly Interacting Particles" by Geoffrey F. Chew, Murray Gell-Mann and Arthur H. Rosenfeld.]

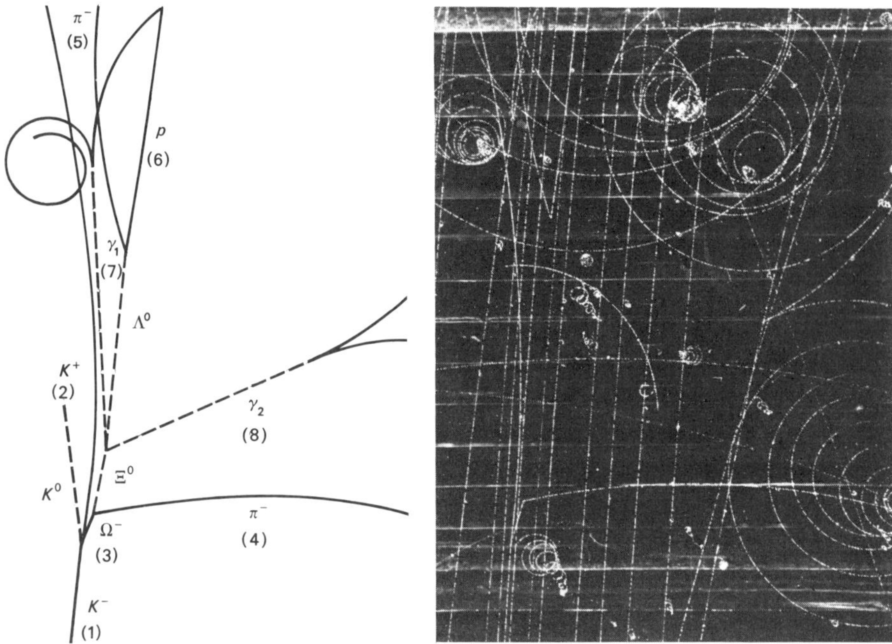

Fig. 15.8 Sketch (left) and bubble-chamber photograph (right) of events showing decay of omega minus, Ω^-. [From V. F. Barnes *et al.*, *Phys. Rev. Lett.* **12**, 204 (1964)]

15.5 THE EIGHTFOLD WAY AND CLASSIFICATION OF STRONGLY INTERACTING PARTICLES

Classifying the 32 elementary particles discovered up to 1957 into four groups—baryons, mesons, leptons, and photon—was a logical and reasonably simple matter. But the discovery of some 100 resonance states (or particles) by the middle 1960s forced physicists to come up with new classifications for strongly interacting particles, which would include not only the newly discovered resonance states but be able to predict resonance states which may be produced in the future with the beams of higher-energy particles which will become available when bigger accelerators are built.

The new classifications are based on five quantities (or quantum numbers) which are conserved in strong interactions.

A, the atomic mass number or baryon number
J, the spin angular momentum of the particle
P, the parity of the state
I, the isotopic spin
Y, the hypercharge [or one may use S or $\bar{Q}$, since these three are related by Eq. (15.19)]

The baryon number A may take the values 0 or ± 1, where $A = 0$ stands for mesons, $A = +1$ for baryons with mass number 1, and $A = -1$ for antibaryons with mass number 1. The state with spin angular momentum J and parity P is denoted by J^P. A state with isotopic spin I has a multiplicity of $M = 2I + 1$.

The eightfold way[28]

Of all the new schemes, the one which has been most useful not only in classifying the existing particles but also in predicting still-undiscovered ones is the so-called *eightfold way*. This new idea of classification, introduced in 1961 independently by M. Gell-Mann and Y. Neeman, makes use of eight operators corresponding to eight quantum numbers. Hence the name eightfold way. Or perhaps the name is derived from Buddha's "eightfold way of life": (1) right views, (2) right intentions, (3) right speech, (4) right actions, (5) right living, (6) right effort, (7) right mindfulness, and (8) right concentration.

The theory is based on the idea that a group of particles which have the same values of J and P but different rest masses and different Y and I must be interrelated. For example, four members—N, Λ, Σ, and Ξ—of the baryon family all have the same values of $J^P = \frac{1}{2}^+$ but different values of rest masses and of Y and I. We say that there exists a *supermultiplet* which is split into four multiplets N, Λ, Σ, and Ξ. Each multiplet may further be split into different states. For example (with masses given in parentheses), N splits into a doublet, n (939.5 MeV) and p (938.2 MeV); Σ into a triplet, Σ^+ (1189.4 MeV), Σ^0 (1191.5 MeV), and Σ^- (1196 MeV); Ξ splits into a doublet, Ξ^0 (1311 MeV) and Ξ^- (1318.4 MeV); and Λ is a singlet Λ^0 (1115.4 MeV). An important characteristic is that the mass differences *between the multiplets* are much larger (by a factor of ten or more) than the mass differences *within the multiplets*, as illustrated in Fig. 15.9.

We have already shown how splitting within a multiplet can be explained. For example, the difference in mass between the proton and neutron was regarded as a splitting caused by the nonconservation of total isotopic spin in the electromagnetic interaction. That is, the violation of isotopic spin by the electromagnetic interaction results in differences in mass within a multiplet. One can extend these ideas further and say that the violation of some other conservation laws of symmetry (i.e., the violation of some other quantum numbers) may be the cause of a difference in mass between the multiplets themselves. The idea is to divide strongly interacting particles into supermultiplets with the same values of J^P and different values of Y and I. The violation of the law of symmetry (corresponding to the violation of some new quantum number) results in the difference in mass between multiplets of a supermultiplet.

The mathematical basis of the theory of the eightfold way (involving eight operators corresponding to eight quantum numbers) is the algebra of *Lie groups*. The Lie groups involved here are a *special unitary* group for 3×3 arrays [hence the name SU_3 or SU(3)], with a condition that reduces the nine components to eight. Each component stands for some conserved quantity. The eight components

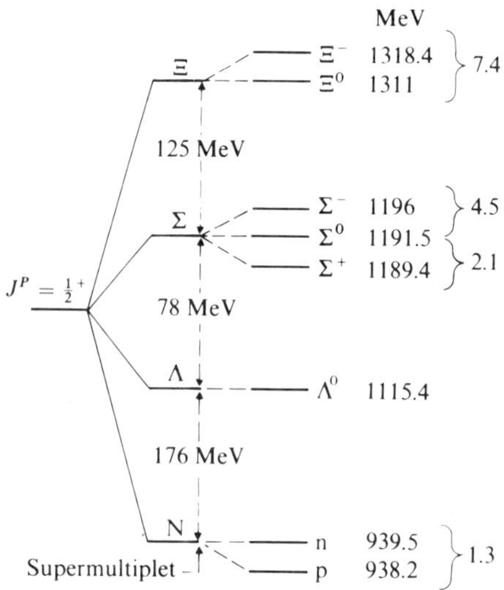

Fig. 15.9 Four member states Ξ, Σ, Λ, and N of a supermultiplet splitting into multiplet states of 2, 3, 1, and 2, respectively.

of SU(3) are: The three components of the isotopic spin; one component corresponding to hypercharge Y; two operators that change Y from up or down by one unit without changing the electric charge; and two more operators that change both Y and the charge by one unit. Violation of the last four operators (or components) by part of the strong interaction is what changes the masses of the multiplets and causes them to form a supermultiplet. Figure 15.10 shows the formation of five supermultiplets on the basis of SU(3). As usual, many particles were predicted before they were discovered experimentally. One outstanding example is omega minus, Ω^-, the resonance state shown in Fig. 15.10(c).

The theory of the eightfold way has been very successful in predicting new particles and in classifying particles, but it is not without its shortcomings.

As physicists probed more deeply into the algebra of the eightfold-way theory, they realized that there are three mathematical entities which constitute the first fundamental triplet of SU(3). This led them to some interesting discoveries, as we shall see.

15.6 THE EIGHTFOLD WAY LEADING TO QUARKS[30]

In 1964 Gell-Mann and Zweig independently proposed that there exist three mathematical entities that form the first fundamental triplet of SU(3). The success of the eightfold way in classifying strongly interacting particles lies in the fact that all strongly interacting particles can be constructed by different possible combinations of these three mathematical entities. Gell-Mann said that these three entities were particles, and named them *quarks*, while Zweig called them *aces*. In order

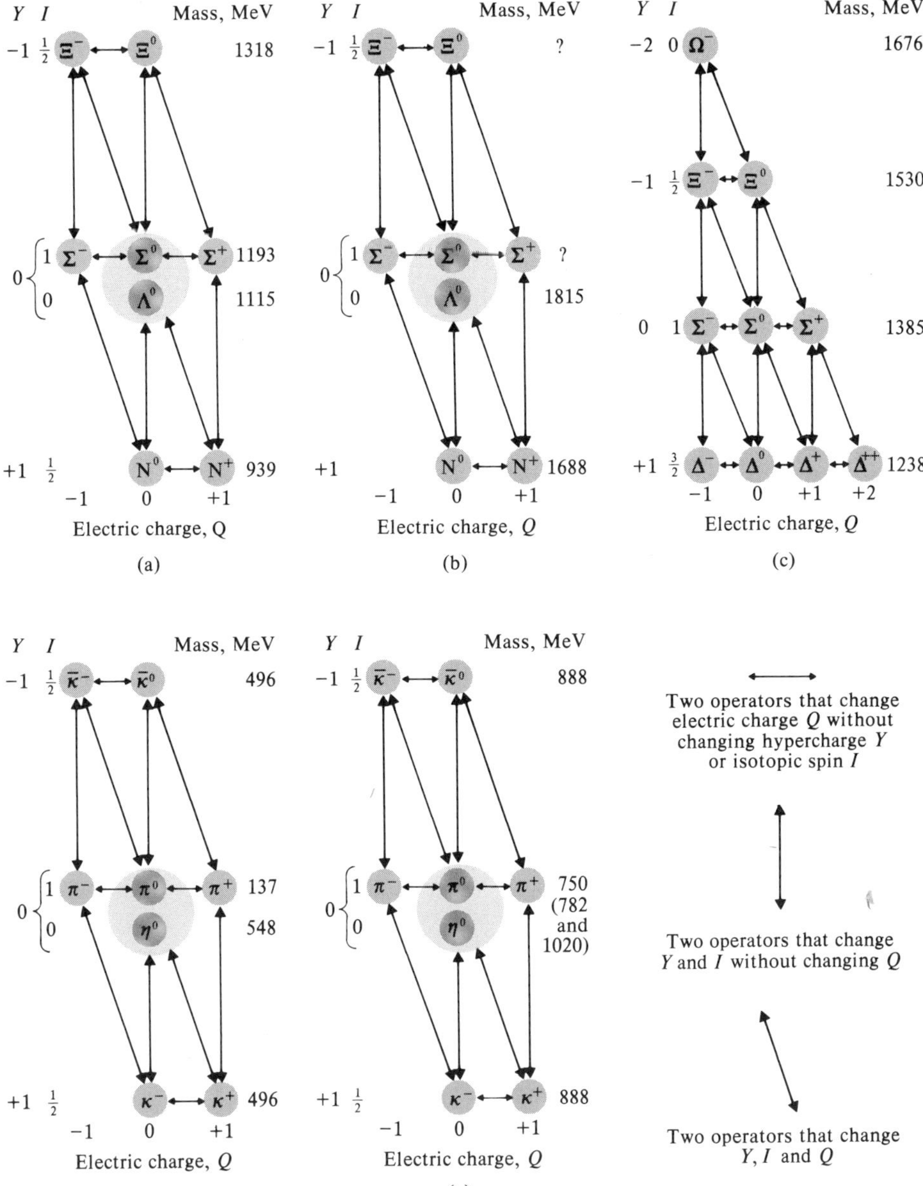

Fig. 15.10 The eightfold way invokes a new system of symmetries to group multiplets of particles into "supermultiplets." The term eightfold refers to a special algebra showing realtions among eight things, in this case eight conserved quantities. The new system of symmetries (*slanting arrows*) connects different values of hypercharge (Y) and isotopic spin (I) in the same way that isotopic-spin symmetry (*horizontal arrows*) connects different values of electric charge. Four of the diagrams (a, b, d, e) show supermultiplets with eight members; another group (c) contains 10 members. Several new particles are predicted by the eightfold way, notably Ω (1,676, $3/2^+$), which appears in (c). Note that the η meson in (e) is given two mass values. [From "Strongly Interacting Particles" by Geoffrey F. Chew, Murray Gell–Mann and Arthur H. Rosenfield. Copyright © 1964 by *Scientific American, Inc.* All rights reserved.]

for strongly interacting particles to be constructed from these three quarks, the quarks must be assigned the charges and quantum numbers shown in Table 15.7.

Table 15.7

Quarks, their charges and Quantum numbers

Symbols for quarks	e_q/e	A	I	I_Z	Y	S	J^P
p	$\frac{2}{3}$	$\frac{1}{3}$	$\frac{1}{2}$	$\frac{1}{2}$	$\frac{1}{3}$	0	$\frac{1}{2}^+$
n	$-\frac{1}{3}$	$\frac{1}{3}$	$\frac{1}{2}$	$-\frac{1}{2}$	$\frac{1}{3}$	0	$\frac{1}{2}^+$
λ	$-\frac{1}{3}$	$\frac{1}{3}$	0	0	$-\frac{2}{3}$	-1	$\frac{1}{2}^+$

Note that the ratio of the charge of the quark to that of the electron, e_q/e, is a fraction. That is, we are assuming that e is no longer the smallest possible fundamental charge. The baryon numbers or mass numbers A of the quarks are also fractional numbers.

Furthermore, it is assumed that there exist three antiparticles of quarks, called *antiquarks.* The quantum numbers A, I_Z, Y, and S, and the charge ratio e_q/e for the antiquarks are opposite to those of the quarks given in Table 15.7. Theoretical calculations have shown that three quarks are necessary to construct a baryon, and a quark–antiquark pair is necessary to construct a meson. These basic triplets of quarks and antiquarks of the SU(3) theory may be shown graphically by plotting Y as a function of I_Z, as shown in Fig. 15.11.

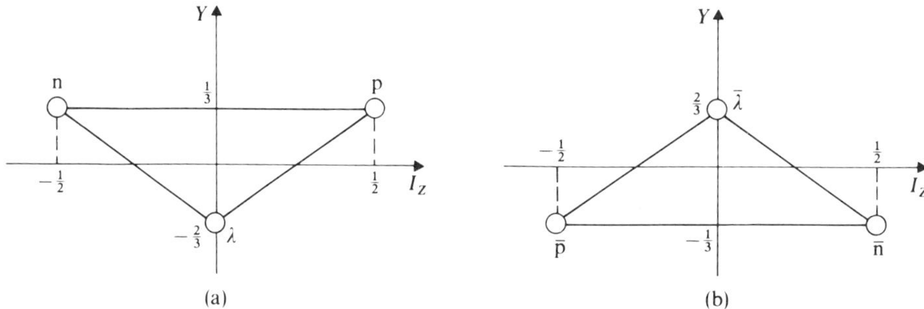

Fig. 15.11 The two basic triplets of (a) quarks and (b) antiquarks, obtained by plotting Y versus I_Z.

The spins of the quarks must be half-integral because there is no way to construct particles with half-integral spins out of particles that have integral spins. Since SU(3) does not involve parity, all quarks have the same parity, assumed to be positive. Theoretically, quarks are supposed to be very heavy, much heavier than nucleons, but according to known properties of the strongly interacting particles, the mass of the quark should be about one-third of the mass of the proton. Two main explanations for this discrepancy are : (1) Quarks are heavy, but, when bound, act like quasi-particles with small mass. (2) Quarks are light, but have some special property that prevents them from being knocked out of baryons and mesons.

Are these three quarks real particles that exist in nature, or are they purely mathematical entities? Since 1964, a number of experiments have been performed to detect the existence of quarks. These experiments may be divided into three categories.

a) Experiments using high-energy accelerators
b) Experiments with cosmic rays
c) Experiments designed to detect quarks as new chemical elements

Attempts to find quarks in nucleon–nucleon collisions, using 30-BeV beams from high-energy accelerators, have had negative results. Either the energies of the nucleons are not high enough, or the production cross section for quarks is extremely low.

Physicists have also tried to study quarks by examining the Lyman series, at wavelengths 2733, 2306, and 2186 Å. These wavelengths arise from quarkium atoms formed in nature. The quarkium atom is formed with a $+\frac{2}{3}e$ quark as a nucleus, plus an electron. Still another place to look for quarks is in the organisms from the sea, sea water, and lake water. Experiments like the Millikan oil-drop experiment may also be helpful.

All these and many other tactics have been tried, but, after some six years of searching for quarks, physicists still have no definite evidence of their existence as real physical particles.

SUMMARY

The 32 elementary particles known by 1957 may be classified in four groups: (a) baryons and antibaryons (16); (b) mesons (7); (c) leptons (8); and photon (1).

Particles that are produced by strong interactions but decay by weak interactions (i.e., they are observed to live 100,000 billion times longer than they should) are called *strange particles.* This behavior of the particles was explained by Pais, who introduced the concept of *associated production.*

Particles with lengthy decays, such as xi particles, are called *cascade particles.*

There are four basic interactions—strong, electromagnetic, weak, and gravitational—with relative strengths of 1, 10^{-2}, 10^{-13}, and 10^{-39}, respectively. Different interactions conserve different quantities. In the language of quantum mechanics, there is a quantum number corresponding to each conserved quantity.

Violation of isotopic-spin symmetry by electromagnetic interactions leads to splitting between states of multiplets.

A strangeness number S is assigned to all strongly interacting particles. It is defined as $S = Y - A = 2\bar{Q} - A$. Strong interactions conserve strangeness.

The "old" 32 particles known in 1957 have lifetimes long enough to leave observable tracks. But scientists have observed a new class of particles with very short lifetimes, $\sim 10^{-23}$ sec, which could be identified only through their decay products. These particles (or states) are called *resonance particles* or *resonance states*. The list of these particles has grown very fast, until today there are hundreds of resonance states.

In order to classify not only the 32 "old" elementary particles, but also the newly discovered resonance states, physicists have introduced many new theories, which are based on five quantities, A, J, P, I, and Y, all of which are conserved by strong interactions. The most important of these is the classification according to the *eightfold way*, suggested independently by M. Gell-Mann and Y. Neeman. This theory makes use of eight operators corresponding to eight quantum numbers. The theory is based on the idea that a group of particles which have the same values of J and P, but different values of rest masses and of Y and I must be interrelated.

The theory of the eightfold way suggests that there exist three mathematical entities, and that all strongly interacting particles can be constructed by different combinations of these entities. These three entities have been called *quarks*. Three *antiquarks* have also been postulated. However, the existence of quarks and antiquarks is still in doubt.

PROBLEMS

1. A mu meson decays by the following reaction:

$$\mu^- \to e^- + \nu_\mu + \bar{\nu}_\mu.$$

The rest mass of the mu meson is 207 m_e and the rest masses of the neutrino and antineutrino are assumed to be zero. What are the maximum kinetic energy and maximum momentum of the electron?

2. A pi meson decays by the following reaction:

$$\pi^- \to \mu^- + \nu.$$

How much momentum and energy are carried away by the mu meson and the neutrino?

3. Suppose that a pi-minus meson at rest interacts with a proton and produces a neutron and a gamma ray:

$$\pi^- + p \to n + \gamma.$$

Calculate the energy of the gamma ray emitted.

4. A pi-zero meson decays by electromagnetic interaction into two gamma rays,

$$\pi^0 \to \gamma + \gamma.$$

What are the momenta and energies of these gamma rays?

5. What is the velocity and momentum of 10-BeV protons?

6. Calculate the magnetic field needed to bend 30-BeV protons through 10°.
7. Calculate the threshold energy for the production of pions by the following reaction:

$$p + p \rightarrow p + n + \pi.$$

8. Calculate the velocity and momentum of a 15-MeV pion. What is the radius of curvature of its path when it is in a magnetic field of 1.5 webers/m^2? Calculate these quantities for a 15-MeV muon.
9. One way of producing positive pions is to bombard a proton target with high-energy photons. Suppose that the pions produced have a kinetic energy of 20 MeV when emitted at an angle of 90° with the direction of the incident photons. What is the energy of the incident photons?
10. When a muon decays at rest into an electron, what are the minimum and maximum energies of the electrons? How much energy is carried away by a neutrino–antineutrino pair in each case?
11. A μ^- meson when at rest replaces an electron in a hydrogen atom, forming a so-called mesonic atom. Assuming a point nucleus, calculate the radius of the first Bohr-type orbit.
12. In Problem 11, suppose that it is a π^- meson that replaces the electron in the hydrogen atom. What is the radius of the first Bohr-type orbit?
13. Calculate the Q values for the decay of a K^- meson by the following reactions:

$$K^- \rightarrow \pi^- + \pi^0 + \pi^0$$
$$K^- \rightarrow \mu^- + \bar{\nu} + \pi^0$$
$$K^- \rightarrow e^- + \bar{\nu} + \pi^0$$

14. Using the reactions given in Eq. (15.22), calculate the values of A, Q, I, I_Z, and S for the omega-minus, Ω^-, particle.
15. Using the reactions given in Eq. (15.22) and the results of the previous problem, calculate the values of A, Q, I, I_Z, and S for $\overline{\Omega}^-$.
16. Explain why the following reactions are forbidden:

$$\pi^0 \rightarrow \mu^+ + e^- + \bar{\nu}$$
$$\Sigma^+ \rightarrow \pi^+ + \pi^0$$
$$\Lambda^0 \rightarrow p + e^-$$

17. What combination of the three quarks p, n, λ leads to the formation of (a) a proton, and (b) a neutron?
18. How would you construct π^+ and π^- mesons from quarks and antiquarks?
19. How do you construct Λ^0 and $\overline{\Lambda}^0$ from quarks and antiquarks?
20. What type of particle results from the following combination of quarks?

(a) $\bar{\lambda}$p (b) $\bar{\lambda}$n (c) $\bar{n}$p (d) $-\bar{p}$n

21. Identify the particle that results from the combination of three quarks $\lambda\lambda\lambda$.

22. Show that Ξ^0 and Ξ^- are formed from the following combination of quarks: $p\lambda\lambda$ and $n\lambda\lambda$.

23. A quarkium atom is formed with $+\frac{2}{3}e$ quark as the nucleus, plus an electron. Calculate the radii of the Bohr-type orbits and the wavelengths of the Lyman series. (Assume the mass of the quark to be $\frac{1}{3}$ of a nucleon.)

REFERENCES

1. J. J. Thomsom, *Phil. Mag.* **44,** 5 (1897)
2. Ernest Rutherford, *Phil. Mag.* **21,** 669 (1911)
3. J. Chadwick, *Proc. Roy. Soc.* **A136,** 692 (1932)
4. Albert Einstein, *Ann. Physik* **17,** 132 (1905)
5. P. A. M. Dirac, *Proc. Roy. Soc.* **126,** 360 (1930); **133,** 61 (1931)
6. C. D. Anderson, *Phys. Rev.* **43,** 491 (1933)
7. Enrico Fermi, *Z. Physik* **81,** 161 (1934)
8. F. Reines and C. L. Cowan, Jr., *Phys. Rev.* **90,** 492 (1953); **113,** 273 (1959)
9. Hideki Yukawa, *Proc. Phys., Math. Soc.* (Japan) **17,** 48 (1935)
10. B. Rossi, *Cosmic Rays*, New York: McGraw-Hill, 1964
11. S. H. Neddermeyer and C. D. Anderson, *Phys. Rev.* **51,** 884 (1937)
12. J. C. Street and E. C. Stevenson, *Phys. Rev.* **51,** 1005 (1937)
13. R. D. Hill, *Sci. Am.*, January 1963
14. C. Lattes, G. Occhialini, and C. Powell, *Nature* **159,** 694 (1947)
15. V. L. Fitch and J. Rainwater, *Phys. Rev.* **92,** 789 (1953)
16. G. D. Rochester and C. D. Butler, *Nature* **160,** 855 (1947)
17. A. Pais, *Phys. Rev.* **86,** 513 (1952)
18. M. Gell-Mann, *Phys. Rev.* **92,** 833 (1953)
19. K. Nishijima, *Prog. Theor. Phys.* (Japan) **12,** 107 (1954); **13,** 285 (1954)
20. W. B. Fowler, R. P. Shutt, and A. M. Thorndike, *Phys. Rev.* **91,** 1287 (1953)
21. O. Chamberlain, E. Segré, C. Wiegand, and T. Ypsilantis, *Phys. Rev.* **100,** 947 (1955)
22. B. Cork, *et al.*, *Phys. Rev.* **104,** 1193 (1956); E. Segré, *Am. J. Phys.* **25,** 363 (1957)
23. H. N. Brown *et al.*, *Phys. Rev. Lett.* **8,** 255 (1962)
24. C. Baltay, *et al.*, *Phys. Rev. Lett.* **11,** 165 (1963)
25. R. D. Hill, *Sci. Am.*, January 1963
26. R. Brown, *et al.*, *Nature* **163,** 82 (1949)
27. V. F. Barnes, *et al.*, *Phys. Rev. Lett.* **12,** 204 (1964)
28. G. F. Chew, M. Gell-Mann, and A. H. Rosenfeld, *Sci. Am.* February 1964
29. R. M. Littauer, H. F. Schopper, and R. R. Wilson, *Phys. Rev. Lett.* **7,** 141 (1961)
30. L. M. Brown, *Physics Today*, p. 44, February 1966.

Suggested Readings

1. K. W. Ford, *Elementary Particles*, New York: Blaisdell, 1963
2. D.H. Frisch and A.M. Thorndike, *Elementary Particles*, Princeton: D.Van Nostrand, 1965
3. D. H. Perkins, *Introduction to High-Energy Physics*, Reading, Mass.: Addison-Wesley, 1972
4. C. E. Swartz, *The Fundamental Particles*, Reading, Mass.: Addison-Wesley, 1965
5. C. N. Yang, *Elementary Particles*, Princeton, N.J., Princeton University Press, 1962
6. R. D. Hill, *Tracking Down Particles*, New York: W. A. Benjamin, 1963
7. The following articles from *Scientific American*: (a) R. E. Marshak, "Pions," January 1957; (b) E. Gell-Mann and E. P. Rosenbaum, "Elementary Particles," July 1957; (c) B. Rossi, "High-Energy Cosmic Rays," November 1959; (d) S. Penman, "The Muon," July 1961; (e) L. Lederman, "Two Neutrino Experiments," March 1963; (f) G. F. Chew, M. Gell-Mann, and A. H. Rosenfeld, "Strongly Interacting Particles," February 1964; (g) R. D. Hill, "Resonance Particles," January 1963; (h) E. Segré and C. E. Wiegand, "The Antiproton," June 1956; (i) W. B. Fowler and N. P. Samios, "The Omega-Minus Experiment," October 1964
8. E. Segré, *Nuclei and Particles*, Chapters XIII, XIV, and XV, New York: W. A. Benjamin (1964)

ANSWERS TO ODD-NUMBERED PROBLEMS

CHAPTER 1

1. $x' = a + (b - v)t + ct^2$; $u'_x = (b - v) + 2ct$; $a'_x = 2c$
3. $t_1 = 2L/\sqrt{c^2 - v^2}$, $t_2 = 2Lc/(c^2 - v^2)$. Calculate v from $t_1 - t_2$.
5. (a) 0.42×10^8 m/sec (b) 2.61×10^8 m/sec
7. 7.90 ft; 63.5°
11. 50.25 sec; 19.9 m
13. 4.59 sec
15. 1.005 sec
19. (a) $1.8c$ (b) $0.995c$
21. $0.75c$; 55.3°
23. (a) 3.45×10^{-27} kg; 1696 MeV/c (b) 1.78×10^{-27} kg; 1000.5 MeV/c
27. (a) 1065 MeV; 0.58 MeV (b) 4502 MeV; 2.45 MeV
29. (a) $0.96c$ (b) $3.57\ m_0$
31. 0; ∞
33. 120 MeV; 0.16 cm
35. 28.8 MeV

CHAPTER 2

1. 2.26 eV
7. 0.54 eV; 0.54 V
9. 2000 Å; 5.31×10^{14} cycles/sec
11. 124,000 V
13. 20,000 Å, 4000 Å; 0.63 eV; 3.1 eV
17. (a) 143 keV (b) 57 keV (c) 51.5°
19. $h\nu[2\alpha/(\alpha^2 + 4\alpha + 2)]$
21. $\cos\theta_0 = [(1 + \alpha)/(2 + \alpha)]$; 17.3°
23. 32 keV
25. 0.74×10^{-14} m
27. 15,000 V
29. 3.5 keV
31. 6.05×10^{26} atoms/kmole
33. 13.1°; 27.7°
35. 8.7°
37. 1.37 eV

CHAPTER 3

1. 12.4×10^8 m/sec
3. $1.005c$
7. 1.05×10^{-28} kg m/sec; 1.16×10^{-4}
11. 9.7 MeV
13. 42 photons/cm^2-sec
15. 0.155 eV
17. 1000, 100, 10, 1 radians
19. $a/2$, $a^3/3(1 - 6/\pi^2)$

21. 37, 37 × 10^2, 37 × 10^4, 37.7 × 10^4 eV
23. $E_n = (n_x^2 + n_y^2 + n_z^2)(\pi^2\hbar^2/2ma^2)$
25. $\hbar^2/2ma^2$ as compared to $\pi^2\hbar^2/2ma^2$
29. 6.22 × 10^{-15} eV

CHAPTER 4

1. 2.46 × 10^{-15} A
3. 0.078 $\cos^2 \theta$
7. (0.55, 0.815, 1.5) × 10^{-19} kg m/sec
9. 1.03 × 10^{-3} cm
11. 9° 40′; 8.3 × 10^{-15} m
13. ~5.5/sec
17. (b) $y_2/y_1 = (M_2/M_1)^{1/2}(x_1/x_2)^2$ (c) $r_1/r_2 = \sqrt{M_1/M_2}$ (d) $(f_1/M_1) = (f_2/M_2)$
19. 65.37
21. Same
23. 0.644 cm

CHAPTER 5

1. $F_e/F_g \simeq 10^{39}$
3. 12.9 Wb/m^2
5. 44, 440, 4400; none
7. $h\nu - [(h\nu)^2/2mc^2]$
9. 13.62 eV; 13.63 eV
13. 6561, 4860.017, 4339.294, 4100.652, 1.79, 1.313, 1.176, 1.088
15. 1215.7, 1215.33, 1215.3 Å
17. 1215.243 Å, 1215.68 Å, 1215.35 Å
19. Wavelengths corresponding to Lyman series; dark lines at the position of the Lyman-series wavelengths.
21. 1.6, 3.2, 4.3, 4.8, 5.9, 6.4, . . . volts
23. 3.7 × 10^{4}°K

CHAPTER 6

1. $2/a^{3/2}$
9. 45°, 90°, 135°; 35° 54′, 66°, 90°, 114°′, 144° 6′; 29° 32′, 54° 32′, 73° 8′, 90°, 106° 52′, 125° 28′, 150° 28′
11. $\frac{3}{2}, \frac{5}{2}$ (a) $\sqrt{15}/2\,\hbar$, $\sqrt{35}/2\,\hbar$ (b) $2\hbar$
13. 60°, 131° 50′
15. 18° 11′, 15° 12′, 12° 50′, 10° 53′
17. −13.5862, −13.6112, −13.6112 eV; last two are degenerate
21. $[-5, +4; -11, +10; -16, +15]a\hbar^2/2$

CHAPTER 7

1. $1s^22s^22p^63s^23p^2$; $1s^22s^22p^63s^23p^5$; $1s^22s^22p^63s^23p^64s^23d^{10}4p^5$; $1s^22s^22p^63s^23p^64s^2$
3. 1; $\frac{1}{2}$; $\frac{1}{2}$ or $\frac{3}{2}$; $2\hbar$; $\sqrt{3}/2\ \hbar$, $\sqrt{15}/2\ \hbar$
5. 128
7. -3.02103, -3.02317, 0.00214 eV
9. $1s^22s^1$; -3.395 eV; 1.74, 1.26
11. Same order of levels and transitions, but different energies, due to different amount of shielding by electrons.
13. $i_n = 1.05 \times 10^{-3}/n^3$ amp, $\mu_n = n\mu_\beta$
15. 8.8×10^{10} rev/sec, 1.6×10^{-23} J
17. 2.8×10^{10} cycles/sec
19. 0.87×10^{-4} eV
25. 8/7
27. 0.7×10^{-22} N; 0.005 cm
29. ~ 10 Å, ~ 3 Å, ~ 1 Å
31. 5.6 keV, 30 keV
33. $\sim 10^7$
35. 0.5×10^8, 0.35

CHAPTER 8

1. -82.4 kcal/mole
3. (a) -5.4 eV (b) -5.13 eV
7. (a) 1.03×10^{-47} kg-m^2 (b) 0.8 Å
9. $I \propto 1/E_V^2$
11. 189 N/m
13. (a) 560 N/m (b) 0.27 eV (c) 0.81, 1.35, 1.89 eV
17. 1000:281:79
19. 2130°K

CHAPTER 9

5. -6.55 eV/ion pair
7. -3.07 eV/ion or -6.14/ion-pair
9. ~ 2 eV as compared to 3 eV
11. 12.1 eV; 7.25 eV
13. $p = 2E/3V$
15. 0.0037, 0.017
17. 1.22×10^6 m/sec
19. 5×10^{-3}
21. ~ 16.3 eV

CHAPTER 10

7. P, As, or Sb impurities in Si or Ge form *n*-type; B, Al, Ga, or In impurities in Si or Ge form *p*-type.
9. In diamond the gap is much wider than the energy of the visible light. Hence visible energy is transmitted without absorption, making the diamond transparent.
11. 6.36 Å

CHAPTER 11

3. 760 *A*, 0.86 μ_N
5. 8.05, 6.75, 7.37 MeV
7. 20.6 12.1, 8.8 MeV
9. 2.014107 u
11. $p_{3/2}$, $d_{5/2}$, $f_{7/2}$
13. (a) $5/2^+$, $1/2^+$, $7/2^+$ (b) $1/2^+$, $3/2^+$, $5/2^+$
15. Yes
17. Because there is a tendency to form proton pairs.

CHAPTER 12

1. (a) 13 hr, 1.48×10^{-5} sec^{-1}, 18.8 hr (b) 25282 dis/sec (c) 1.72×10^9
3. (a) 4.17×10^{-9} sec^{-1}, 7.6 yr (b) 4.2×10^{13} dis/sec; 1130 C; 4.2×10^7*rd* (c) 18.8 J/sec; 4.5 calorie/sec
5. 0.0088 g
9. 10 days; 14.5 days; 8×10^{-7} sec^{-1}
11. 4.79 MeV; 4.60 MeV; 0.19 MeV
13. 0.072 MeV; 2.44×10^5 m/sec
15. 31.4 MeV
19. 789.575 keV; 0.425 keV
21. 6 cal/sec
23. 0.862945 MeV
25. 1.176 MeV; 0.5144 MeV; 661.6 keV; 624.16 keV, 655.61 keV

CHAPTER 13

1. ~0.031 MeV/cm; ~0.045 MeV/cm
3. 0.062 cm, 0.006 cm
5. (16.5, 160, ~6000) mg/cm^2
7. 3.52 MeV
9. ~18%

11. 6.13 cm, or 4.61 μ
13. (0; 0.7, 0.51, 0.36; 0.94, 0.88, 0.83) $\times E_0$
17. 5.25 MeV; 2.4×10^{-14} C
21. 0.0034 Wb/m^2
25. 100 watts

CHAPTER 14

1. ${}^{31}_{16}S$, ${}^{31}_{15}P$, ${}^{30}_{15}P$, ${}^{28}_{14}Si$
3. 4.08, 4.35, −6, 9.85 MeV
5. (a) 18 MeV (b) 3.52 MeV, 14.08 MeV
7. 3.33 MeV
9. ${}^{19}O$ is heavier by 4.88 MeV (or 0.00524 u)
11. (a) 170 cm (b) 340 cm (c) 510 cm
13. (a) 91 cm (b) 3.4 cm
15. 5.6×10^{19} radioactive atoms; 4×10^{13} dis/sec
17. 6.6 (10^{-14}, 10^{-11}, 10^{-8}) sec
19. $({}^{31}_{15}P)^*$; (α, p), (α, d), (α, α'); ${}^{30}_{14}Si$, ${}^{29}_{14}Si$, ${}^{27}_{13}Al$
21. 38.97 MeV
23. 0.667
25. 7.32 g/min
27. 14.1×10^{23} MeV or 22.6×10^{10} joules

CHAPTER 15

1. 105 MeV, 105.4 MeV/c
3. 131 MeV
5. 0.996c
7. 282 MeV
9. 164 MeV
11. 0.005 Å
13. 43 MeV, 78.5 MeV, 182 MeV
15. $A = -1, I = 0, I_Z = 0, Q = +1, S = +3$
17. ppn; nnp
19. $\lambda(np - pn)/\sqrt{2}$
21. Omega minus
23. 0.795 n^2 Å, where $n = 1, 2, 3, \ldots$
    2733, 2306, 2186 Å

APPENDIX

Table A.1

Table of Nuclear Masses

Subscripts preceding symbols denote atomic numbers; asterisks denote radioactive isotopes. (Data from D. T. Goldman and J. R. Roesser, *The Chart of the Nuclides,* ninth edition, distributed by Educational Relations, General Electric Company, Schenectady, N.Y.)

Symbol	Mass number	Mass
${}_{0}$n	1*	1.008665
${}_{1}$H	1	1.007825
${}_{1}$D	2	2.014102
${}_{1}$T	3*	3.016050
${}_{2}$He	3	3.016030
	4	4.002603
${}_{3}$Li	6	6.015125
	7	7.016004
${}_{4}$Be	9	9.012186
	10*	10.013534
${}_{5}$B	10	10.012939
	11	11.009305
${}_{6}$C	12	12.000000
	13	13.003354
	14*	14.003242
${}_{7}$N	14	14.003074
	15	15.000108
${}_{8}$O	16	15.994915
	17	16.999133
	18	17.999160
${}_{9}$F	19	18.998405
${}_{10}$Ne	20	19.992440
	21	20.993849
	22	21.991385
${}_{11}$Na	22*	21.994437
	23	22.989771
${}_{12}$Mg	24	23.985042
	25	24.986809
	26	25.982593
${}_{13}$Al	26*	25.986892
	27	26.981539
${}_{14}$Si	28	27.976929
	29	28.976496
	30	29.973763
	32*	31.974020
${}_{15}$P	31	30.973765
${}_{16}$S	32	31.972074
	33	32.971462
	34	33.967865
	36	35.967089
${}_{17}$Cl	35	34.968851
	36*	35.968309
	37	36.965898
${}_{18}$A	36	35.967544
	38	37.962728
	39*	38.964317
	40	39.962384
	42*	41.963048
${}_{19}$K	39	38.963710
	40*	39.964000
	41	40.961832
${}_{20}$Ca	40	39.962589
	41*	40.962275
	42	41.958625
	43	42.958780
	44	43.955492
	46	45.953689
	48	47.952531
${}_{21}$Sc	45	44.955920
${}_{22}$Ti	44*	43.959572
	46	45.952632
	47	46.951768
	48	47.947950
	49	48.947870
	50	49.944786
${}_{23}$V	50*	49.947164
	51	50.943961
${}_{24}$Cr	50	49.946054
	52	51.940513
	53	52.940653
	54	53.938882
${}_{25}$Mn	55	54.938050

Symbol	Mass number	Mass
$_{26}$Fe	54	53.939616
	55*	54.938299
	56	55.939395
	57	56.935398
	58	57.933282
	60*	59.933964
$_{27}$Co	59	58.933189
	60*	59.933813
$_{28}$Ni	58	57.935342
	59*	58.934342
	60	59.930787
	61	60.931056
	62	61.928342
	63*	62.929664
	64	61.927958
$_{29}$Cu	63	62.929592
	65	64.927786
$_{30}$Zn	64	63.929145
	66	65.926052
	67	66.927145
	68	67.924857
	70	69.925334
$_{31}$Ga	69	68.925574
	71	70.924706
$_{32}$Ge	70	69.924252
	72	71.922082
	73	72.923462
	74	73.921181
	76	75.921405
$_{33}$As	75	74.921596
$_{34}$Se	74	73.922476
	76	75.919207
	77	76.919911
	78	77.917314
	79*	78.918494
	80	79.916527
	82	81.916707
$_{35}$Br	79	78.918329
	81	80.916292
$_{36}$Kr	78	77.920403
	80	79.916380
	81*	80.916610
	82	81.913482
	83	82.914131
	84	83.911503
	85*	84.912523
	86	85.910616
$_{37}$Rb	85	84.911800
	87*	86.909186
$_{38}$Sr	84	83.913430
	86	85.909285
	87	86.908892
	88	87.905641
	90*	89.907747
$_{39}$Y	89	88.905872
$_{40}$Zr	90	89.904700
	91	90.905642
	92	91.905031
	93*	92.906450
	94	93.906313
	96	95.908286
$_{41}$Nb	91*	90.906860
	92*	91.907211
	93	92.906382
	94*	93.907303
$_{42}$Mo	92	91.906810
	93*	92.906830
	94	93.905090
	95	94.905839
	96	95.904674
	97	96.906021
	98	97.905409
	100	99.907475
$_{43}$Tc	97*	96.906340
	98*	97.907110
	99*	98.906249
$_{44}$Ru	96	95.907598
	98	97.905289
	99	98.905936
	100	99.904218
	101	100.905577
	102	101.904348
	104	103.905430
$_{45}$Rh	103	102.905511
$_{46}$Pd	102	101.905609
	104	103.904011

Symbol	Mass number	Mass
	105	104.905064
	106	105.903479
	107*	106.905132
	108	107.903891
	110	109.905164
$_{47}$Ag	107	106.905094
	109	108.904756
$_{48}$Cd	106	105.906463
	108	107.904187
	109*	108.904928
	110	109.903012
	111	110.904188
	112	111.902762
	113	112.904408
	114	113.903360
	116	115.904762
$_{49}$In	113	112.904089
	115*	114.903871
$_{50}$Sn	112	111.904835
	114	113.902773
	115	114.903346
	116	115.901745
	117	116.902958
	118	117.901606
	119	118.903313
	120	119.902198
	121*	120.904227
	122	121.903441
	124	123.905272
$_{51}$Sb	121	120.903816
	123	122.904213
	125	124.905232
$_{52}$Te	120	119.904023
	122	121.903064
	123*	122.904277
	124	123.902842
	125	124.904418
	126	125.903322
	128	127.904476
	130	129.906238
$_{53}$I	127	126.904070
	129*	128.904987
$_{54}$Xe	124	123.906120
	126	125.904288
	128	127.903540
	129	128.904784
	130	129.903509
	131	130.905085
	132	131.904161
	134	133.905815
	136	135.907221
$_{55}$Cs	133	132.905355
	134*	133.906823
	135*	134.905770
	137*	136.906770
$_{56}$Ba	130	129.906245
	132	131.905120
	133*	132.905879
	134	133.904612
	135	134.905550
	136	135.904300
	137	136.905500
	138	137.905000
$_{57}$La	137*	136.906040
	138*	137.906910
	139	138.906140
$_{58}$Ce	136	135.907100
	138	137.905830
	140	139.905392
	142*	141.909140
$_{59}$Pr	141	140.907596
$_{60}$Nd	142	141.907663
	143	142.909779
	144*	143.910039
	145	144.912538
	146	145.913086
	148	147.916869
	150	149.920960
$_{61}$Pm	145*	144.912691
	146*	145.914632
	147*	146.915108
$_{62}$Sm	144	143.911989
	146*	145.912992
	147*	146.914867
	148*	147.914791
	149*	148.917180

Symbol	Mass number	Mass	Symbol	Mass number	Mass
	150	149.917276	$_{71}$Lu	173*	172.938800
	151*	150.919919		175	174.940640
	152	151.919756		176*	175.942660
	154	153.922282	$_{72}$Hf	174*	173.940360
$_{63}$Eu	151	150.919838		176	175.941570
	152*	151.921749		177	176.943400
	153	152.921242		178	177.943880
	154*	153.923053		179	178.946030
	155*	154.922930		180	179.946820
$_{64}$Gd	148*	147.918101	$_{73}$Ta	180	179.947544
	150*	149.918605		181	180.948007
	152*	151.919794	$_{74}$W	180	179.947000
	154	153.920929		182	181.948301
	155	154.922664		183	182.950324
	156	155.922175		184	183.951025
	157	156.924025		186	185.954440
	158	157.924178	$_{75}$Re	185	184.953059
	160	159.927115		187*	186.955833
$_{65}$Tb	159	158.925351	$_{76}$Os	184	183.952750
$_{66}$Dy	156*	155.923930		186	185.953870
	158	157.924449		187	186.955832
	160	159.925202		188	187.956081
	161	160.926945		189	188.958300
	162	161.926803		190	189.958630
	163	162.928755		192	191.961450
	164	163.929200		194*	193.965229
$_{67}$Ho	165	164.930421	$_{77}$Ir	191	190.960640
	166*	165.932289		193	192.963012
$_{68}$Er	162	161.928740	$_{78}$Pt	190*	189.959950
	164	163.929287		192	191.961150
	166	165.930307		194	193.962725
	167	166.932060		195	194.964813
	168	167.932383		196	195.964967
	170	169.935560		198	197.967895
$_{69}$Tm	169	168.934245	$_{79}$Au	197	196.966541
	171*	170.936530	$_{80}$Hg	196	195.965820
$_{70}$Yb	168	167.934160		198	197.966756
	170	169.935020		199	198.968279
	171	170.936430		200	199.968327
	172	171.936360		201	200.970308
	173	172.938060		202	201.970642
	174	173.938740		204	203.973495
	176	175.942680	$_{81}$Tl	203	202.972353

Symbol	Mass number	Mass
	204*	203.973865
	205	204.974442
Ra E″	206*	205.976104
Ac C″	207*	206.977450
Th C″	208*	207.982013
Ra C″	210*	209.990054
$_{82}$Pb	202*	201.927997
	204*	203.973044
	205*	204.974480
	206	205.974468
	207	206.975903
	208	207.976650
Ra D	210*	209.984187
Ac B	211*	210.988742
Th B	212*	211.991905
Ra B	214*	213.999764
$_{83}$Bi	207*	206.978438
	208*	207.979731
	209	208.980394
Ra E	210*	209.984121
Th C	211*	210.987300
	212*	211.991876
Ra C	214*	213.998686
	215*	215.001830
$_{84}$Po	209*	208.982426
Ra F	210*	209.982876
Ac C′	211*	210.986657
Th C′	212*	211.989629
Ra C′	214*	213.995201
Ac A	215*	214.999423
Th A	216*	216.001790
Ra A	218*	218.008930
$_{85}$At	215*	214.998663
	218*	218.008607
	219*	219.011290
$_{86}$Rn		
An	219*	219.009481
Tn	220*	220.011401
Rn	222*	222.017531
$_{87}$Fr		
Ac K	223*	223.019736
$_{88}$Ra		
Ac X	223*	223.018501
Th X	224*	224.020218
Ra	226*	226.025360
Ms Th$_1$	228*	228.031139
$_{89}$Ac	227*	227.027753
Ms Th$_2$	228*	228.031080
$_{90}$Th		
Rd Ac	227*	227.027706
Rd Th	228*	228.028750
	229*	229.031652
Io	230*	230.033087
UY	231*	231.036291
Th	232*	232.038124
UX$_1$	234*	234.043583
$_{91}$Pa	231*	231.035877
UZ	234*	234.043298
$_{92}$U	232*	232.037168
	233*	233.039522
	234*	234.040904
Ac U	235*	235.043915
	236*	236.045637
UI	238*	238.048608
$_{93}$Np	235*	235.044049
	236*	236.046624
	237*	237.048056
$_{94}$Pu	236*	236.046071
	238*	238.049511
	239*	239.052146
	240*	240.053882
	241*	241.056737
	242*	242.058725
	244*	244.064100

Table A.2

Adjusted Values of Physical Constants (MKSA)

Constant	Symbol	Value
Speed of light in vacuum	c	2.997925×10^{8} m/sec
Gravitational constant	G	6.670×10^{-11} N-m²/kg²
Acceleration of free fall	g	9.80665 m/sec²
Avogadro constant	N_A	6.02252×10^{26}/kmole
Faraday constant	F	9.64870×10^{4} coul/mole
Coulomb force constant	$k = \dfrac{1}{4\pi\epsilon_0}$	8.98755×10^{9} N-m²/coul²
Standard volume of ideal gas	V_0	22413.6 cm³/mole
Thermochemical calorie	cal_{th}	4.184 joules = 4.184×10^{7} erg
Gas constant	R_0	8.31434 joules/mole-°K
Boltzmann constant	k	1.38054×10^{-23} joule/°K
Stefan-Boltzmann constant	σ	5.6697×10^{-8} watt/m²-deg⁴
Elementary charge	e	1.60210×10^{-19} coul
Planck constant	h	6.62559×10^{-34} joule-sec
	$\hbar (= h/2\pi)$	1.054494×10^{-34} joule-sec
Mass unit	u	1.66043×10^{-27} kg = 931.478 MeV/c^2
Electron rest mass	m_e	9.10908×10^{-31} kg = 0.000548597 u = 0.511006 MeV/c^2
Proton rest mass	m_p	1.67252×10^{-27} kg = 1.0072766 u = 938.256 MeV/c^2 = 1836.10 m_e
Neutron rest mass	m_n	1.67482×10^{-27} kg = 1.0086654 u = 939.550 MeV/c^2 = 1838.63 m_e
Hydrogen atom rest mass	M_H	1.67343×10^{-27} kg = 1.007825 u
Charge-to-mass ratio for electron	e/m_e	1.758796×10^{11} coul/kg
Electron volt	1 eV	1.60210×10^{-19} joule
Rydberg constant	R_∞	1.0973731×10^{7}/m
Bohr radius	a_0	5.29167×10^{-11} m
Fine-structure constant	$\alpha = \dfrac{2\pi e^2}{hc}$	7.29720×10^{-3} (= 1/137.0388)
Bohr magneton	μ_β	9.2732×10^{-24} joule/T
Nuclear magneton	μ_N	5.05050×10^{-27} joule/T
Gyromagnetic ratio of proton	γ	2.675192×10^{8} rad/sec-T
Compton wavelength of electron	$h/m_e c$	2.42621×10^{-12} m
Compton wavelength of proton	$h/m_p c$	1.321398×10^{-15} m

INDEX

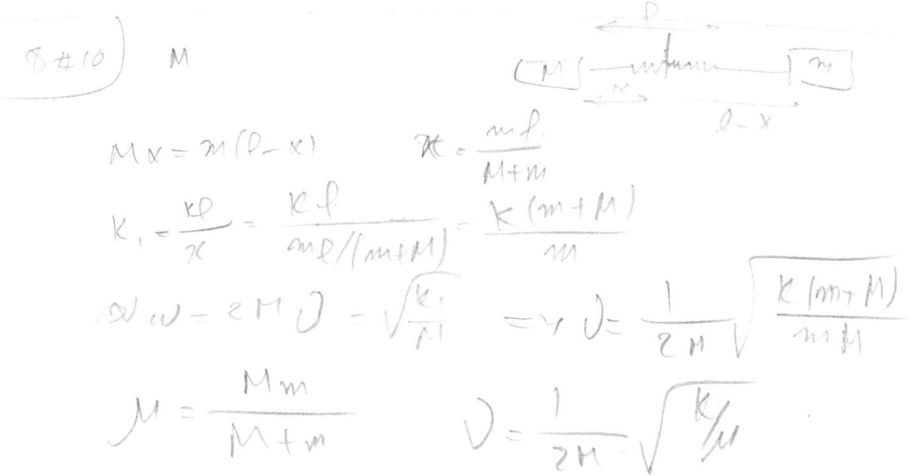